T0180512

Advances in Intelligent Systems and Computing

Volume 1112

The series "Advances in Intelligent Systems and Computing" contains publications on theory, applications, and design methods of Intelligent Systems and Intelligent Computing. Virtually all disciplines such as engineering, natural sciences, computer and information science, ICT, economics, business, e-commerce, environment, healthcare, life science are covered. The list of topics spans all the areas of modern intelligent systems and computing such as: computational intelligence, soft computing including neural networks, fuzzy systems, evolutionary computing and the fusion of these paradigms, social intelligence, ambient intelligence, computational neuroscience, artificial life, virtual worlds and society, cognitive science and systems, Perception and Vision, DNA and immune based systems, self-organizing and adaptive systems, e-Learning and teaching, human-centered and human-centric computing, recommender systems, intelligent control, robotics and mechatronics including human-machine teaming, knowledge-based paradigms, learning paradigms, machine ethics, intelligent data analysis, knowledge management, intelligent agents, intelligent decision making and support, intelligent network security, trust management, interactive entertainment, Web intelligence and multimedia.

The publications within "Advances in Intelligent Systems and Computing" are primarily proceedings of important conferences, symposia and congresses. They cover significant recent developments in the field, both of a foundational and applicable character. An important characteristic feature of the series is the short publication time and world-wide distribution. This permits a rapid and broad dissemination of research results.

** **Indexing: The books of this series are submitted to ISI Proceedings, EI-Compendex, DBLP, SCOPUS, Google Scholar and Springerlink** **

More information about this series at http://www.springer.com/series/11156

Jyotsna Kumar Mandal ·
Somnath Mukhopadhyay
Editors

Proceedings of the Global AI Congress 2019

 Springer

Editors
Jyotsna Kumar Mandal
Department of Computer Science
and Engineering
Kalyani University
Nadia, West Bengal, India

Somnath Mukhopadhyay
Department of Computer Science
and Engineering
Assam University
Silchar, Assam, India

ISSN 2194-5357 ISSN 2194-5365 (electronic)
Advances in Intelligent Systems and Computing
ISBN 978-981-15-2187-4 ISBN 978-981-15-2188-1 (eBook)
https://doi.org/10.1007/978-981-15-2188-1

This Springer imprint is published by the registered company Springer Nature Singapore Pte Ltd.
The registered company address is: 152 Beach Road, #21-01/04 Gateway East, Singapore 189721, Singapore

Preface

Institute of Engineering and Management Kolkata organized the First International Conference on "Global AI Congress" (GAIC 2019), during 12–14 September 2019 at IEM Gurukul Campus. This mega event covered all aspects of artificial intelligence, data science, computer vision, and internet of things, where the scope was not only limited to various engineering disciplines such as computer science, electronics, and biomedical engineering researchers but also included researchers from allied communities like data analytics and management science, etc. The subtracks of the conference were signal processing, intelligent forensics (privacy and security), and intelligent devices and networks.

The volume is a collection of high-quality peer-reviewed research papers received from all over the world. GAIC 2019 could attract a good number of submissions from the different areas spanning over four tracks in various cutting-edge technologies of specialized focus which were organized and chaired by eminent professors. Based on rigorous peer-review process by the technical program committee members along with external experts as reviewers (inland as well as abroad), best quality papers were identified for presentation and publication. The review process was extremely stringent with minimum three reviews for each submission and occasionally up to six reviews. Checking of similarities and overlaps are also done based on the international norms and standards. Out of the submission pool of received papers, only 30% were able to have got berth in this final proceedings.

The organizing committee of GAIC 2019 was constituted with a strong international academic and industrial luminaries and the technical program committee comprising more than two hundred domain experts. The proceedings of the conference were published in Advances in Intelligent Systems and Computing (AISC), Springer. We, in the capacity of the volume editors, convey our sincere gratitude to Springer Nature for providing the opportunity to publish the proceedings of GAIC 2019 in AISC series.

The conference included distinguished speakers such as Dr. Mike Hinchey, President IFIP, Professor, University Of Limerick; Dr. L M Patnaik, Professor, Indian Institute of Science Bangalore; Prof. Goutam Chakraborty, Head, Intelligent Informatics Lab; Dr. Basabi Chakraborty, Iwate Prefectural University, Japan; Dr. Niloy Ganguly, Professor, IIT Kharagpur; Dr. K.K. Shukla, Professor, Department Of Computer Science And Engineering, IIT BHU, India; Dr. Amlan Chakraborty, Dean Of Engg., University of Calcutta, (ACM Distinguished Speaker); Dr. Debasish De, Professor and Director, Department Of Computer Science And Engineering, MAKAUT (WBUT); Dr. Arpan Pal, Principal Scientist and Head of Embedded Systems And Robotics, TCS; Mr. Joy Mustafi, Director and Principal Researcher, Salesforce; and Dr. Koyel Mukherjee, Researcher, IBM Research Lab.

Sincerest thanks are due to Dr. Aninda Bose, Senior Publishing Editor with Springer Nature and Prof. P.K. Roy, APIIT, India, for their valuable suggestions regarding enhancing editorial review process. The editors also thank the other members of the award committee of GAIC 2019 for taking all the trouble to make a very contentious assignment of selecting the best papers from a pool of so many formidable acceptances of the conference.

Special mention of words of appreciation is due to Prof. Debika Bhattacharyya and Prof. Nilanjan Dutta Roy of IEM, Kolkata, for coming forward to host to the conference which incidentally was the second in the series. It was indeed heartening to note the enthusiasm of all faculty, staff, and students of IEM to organize the conference in a professional manner. The involvement of faculty coordinators and student volunteers are particularly praiseworthy in this regard. The editors leave no stone unturned to thank the technical partners and sponsors for providing all the support and financial assistance.

It is needless to mention the role of the contributors. But for their active support and participation, the question of organizing a conference is bound fall through. The editors take this privilege to thank authors of all the papers submitted as a result of their hard work, more so because all of them considered the conference as a viable platform to ventilate some of their latest findings, not to speak of their adherence to the deadlines and patience with the tedious review process. The quality of a refereed volume primarily depends on the expertise and dedication of their viewers who volunteer the associated trouble with smiling face, thankless albeit. The editors are further indebted to the technical program committee members and external reviewers who not only produced excellent reviews but also did these in short timeframes, in spite of their very busy schedule. Because of their quality work, it has been possible to maintain the high academic standard of the proceedings.

The conference may meet its completeness if it is able to attract elevated participation in its fold. A conference with good papers accepted and devoid of any participant is perhaps the worst form of curse that may be imagined of. The editors would like to appreciate the participants of the conference, who have considered the conference a befitting one in spite of all the hardship they had to undergo.

Last but not least, the editors would offer cognizance to all the volunteers for their tireless efforts in meeting the deadlines and arranging every minute detail meticulously to ensure that the conference achieves its goal, academic, or otherwise and that too unhindered.

Happy Reading!!!

Nadia, India   Jyotsna Kumar Mandal
Silchar, India   Somnath Mukhopadhyay

Contents

About the Editors

Jyotsna Kumar Mandal is Former Dean of the Faculty of Engineering, Technology and Management, and a Senior Professor at the Department of Computer Science & Engineering, University of Kalyani, India. Holding a Ph.D. (Eng.) from Jadavpur University, Professor Mandal has co-authored six books: Algorithmic Design of Compression Schemes and Correction Techniques—A Practical Approach; Symmetric Encryption—Algorithms, Analysis and Applications: Low Cost-based Security; Steganographic Techniques and Application in Document Authentication—An Algorithmic Approach; Optimization-based Filtering of Random Valued Impulses—An Algorithmic Approach; and Artificial Neural Network Guided Secured Communication Techniques: A Practical Approach; all published by Lambert Academic Publishing, Germany. He has also authored more than 350 papers on a wide range of topics in international journals and proceedings. His areas of research include coding theory, data and network security, remote sensing and GIS-based applications, data compression, error correction, visual cryptography and steganography, distributed and shared memory, and parallel programming. He is a Fellow of the Institution of Electronics and Telecommunication Engineers, and a member of the IEEE, ACM, and Computer Society of India.

Somnath Mukhopadhyay is currently an Assistant Professor at the Department of Computer Science & Engineering, Assam University, Silchar, India. He completed his M.Tech. and Ph.D. in Computer Science & Engineering at the University of Kalyani, India, in 2011 and 2015, respectively. He has co-authored one book and edited five others. He has published over twenty papers in various international journals and conference proceedings, as well as three book chapters in edited volumes. His research interests include digital image processing, computational intelligence, and pattern recognition. He is a member of the IEEE and IEEE Young Professionals – Kolkata Section, life member of the Computer Society of India, and currently the Computer Society of India's regional student coordinator (RSC) for Region II.

Detection of Leukemia in Blood Samples Applying Image Processing Using a Novel Edge Detection Method

Megha Dutta, Srija Karmakar, Prithaj Banerjee and Rit Ghatak

Abstract In current times, pathologists visually inspect blood cell images under the microscope for the purpose of identifying blood disorders. Identified blood disorders are classified into several blood diseases. Our work aims at studying and designing a model (framework) for the detection of leukemia (blood cancer) and its types using microscopic blood sample images and analyzing them for diagnosis at an earlier stage. Earlier, hematologists used to visually inspect the microscopic images, thereby making the diagnosis process error-prone and time-consuming. However, the use of newly developed automatic image processing systems manages to successfully overcome most of the visual inspection related drawbacks. The early diagnosis of blood cancer will greatly aid in better treatment. In this process, the acquired dataset images are taken as inputs and the images are sent through different image processing techniques such as image enhancement (preprocessing), segmentation, feature extraction and classification. The method proposed is applied to a large number of images of varying quality and is found to provide satisfactory results.

Keywords Leukemia · Image processing · Segmentation · Feature extraction

1 Introduction

Leukemia is a hematologic cancer which develops in the blood tissue and causes rapid generation of abnormal (large) shaped and immature WBC's. The primary cause behind leukemia is the production of a huge number of abnormal lymphocytes (WBCs) by the bone marrow. There is always a healthy balance between red blood cells, white blood cells and platelets in human blood. The onset of leukemia, however, disrupts this balance. Damaged or old blood cells always die, only to be replaced by new ones [1]. The bone marrow of a blood cancer patient produces much more large-sized WBCs compared to that of a normal person's bone marrow. Unlike normal blood cells, they do not get replaced in due time, resulting in a huge abnormal WBC count. Normal functions of the different components of blood are affected due to this

M. Dutta (✉) · S. Karmakar · P. Banerjee · R. Ghatak
Department of CSE, Institute of Engineering & Management, Kolkata, India

© Springer Nature Singapore Pte Ltd. 2020
J. K. Mandal and S. Mukhopadhyay (eds.), *Proceedings of the Global
AI Congress 2019*, Advances in Intelligent Systems and Computing 1112,
https://doi.org/10.1007/978-981-15-2188-1_1

Fig. 1 Example showing different components of a blood stem cell (Image Courtesy [26])

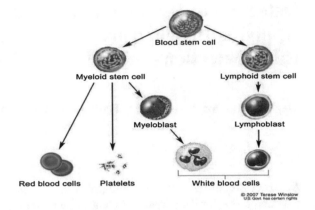

imbalance. Based on the speed of infection development and the severity, leukemia can be classified as either acute (rapid growth and fatal in a short time) or chronic (at nascent treatable stage, not that severe yet). Further classifications can also be made [2]. Acute lymphocytic leukemia [3] will be our main focus of discussion in this paper, as we have succeeded in acquiring dataset sample images for the same. From a biological standpoint, the most widely practiced blood cancer classification by structural analysis is the FAB procedure [4]. An improved variant of the same called immunologic classification [5] is more common presently. Morphologically, healthy WBCs (lymphocytes) have a regular shape, a headphone-shaped nucleus and abundant cytoplasm while infected ones (lymphoblasts) are characterized by scanty cytoplasm and irregular-shaped nucleus. FAB analyzes ALL lymphoblasts into three types, namely L1, L2 and L3. L1 has a regular round nucleus with negligible clefting, scanty and vacuole-less cytoplasm. Clefted, irregular, multiple nuclei with variable cytoplasm are present in L2. Prominent distinguishable vacuoles are present in L3 (Fig. 1).

Statistical data shows that in the last few years, leukemia has been one of the primary causes of death worldwide, irrespective of gender. A delay in detection makes it highly unlikely for ALL patients to survive. This makes early and correct diagnosis all the more important. A traditional setup places heavy importance on the pathologist for accuracy in diagnosis. Bone marrow aspiration, medical history checkup, CBC and cytogenic analysis are some of the commonly used methods in this regard. However, manual detection methods often yield inconsistent and inaccurate results. To remedy this, there has been a lot of research in terms of applying image processing to this field. In addition, image processing enhanced techniques can also help in detecting how far the disease has progressed (the stage it is currently in).

The contributions of the paper are as follows: Firstly, we develop a novel segmentation or rather edge detection approach which provides more accurate results compared to existing approaches. Then we follow the traditional steps of an image processing algorithm and apply our own method in the algorithm implementation.

The result is the detection of the input sample cell to be either healthy or leukemia infected (further stage classification into chronic and acute is done).

The organization of the rest of the paper is as follows:

In Sect. 1.1, we discuss some recently developed related works for leukemia detecting using image processing. We also survey the conventional steps in the algorithm related to our proposed framework namely segmentation, thresholding, etc. In Sect. 2, we developed our own edge detection method. In Sect. 3, we discuss the proposed framework and a unique algorithm to perform the task of detection. Section 4 deals with the results obtained in terms of image outputs and the result accuracy in terms of other methods. Finally, Sect. 5 provides a conclusion to our work, briefly touching upon its future scope.

1.1 Related Work

The introduction already focused on the biological aspects of leukemia and conventional nonimage processing methods related to leukemia detection in blood cell samples. In our literature survey, the focus will be more on the image processing-oriented leukemia detection methods. The steps of the general workflow and most of the important works done to explain and improve these methods are also included. The survey will also encompass most of the novel methods or algorithms tried to improve the steps of the traditional processes or in some cases, the process as a whole.

References [1–3] will elaborate on the biology and morphology of blood cancer. Bennett et al. [4] and Biondi et al. [5] are basically nonimage processing related techniques used for leukemia detection.

Coming to the application of image processing enhanced methods in leukemia detection, thresholding is a very vital step. Many unique works have been designed with a view to modify or improve this step to suit our needs. In [6], Minal D. Joshi, et al. have suggested automatic Otsu's thresholding, along with image enhancement, for segmentation of white blood cells. Lymphocytes and lymphoblasts are distinguished using k-NN classifier. In [7], Ghosh et al. applied the concept of fuzzy divergence to find out a threshold for leukocyte segmentation. Various functions like Cauchy, Gaussian and Gamma have been compared for their benefits and drawbacks and the most suitable ones have been used further. In [8], Dorini et al. have proposed a nucleus extraction scheme using watershed transform, which is further based on the image forest transform. The method, however, gives good results only for round cytoplasm.

Thresholding is closely related to segmentation. The countless number of literary works have been done regarding image segmentation.

In [9], G. EvelinSuji, Y. V. S. Lakshmi, G. Wiselin Jiji et al. have provided a comparative study of different image segmentation algorithms.

In [10], Lim Huey Nee et al. have proposed methods like morphological operations, watershed transformations and thresholding for the purpose of cell segmentation. The method provided satisfactory segmentation results for a sample of 50 images.

In [11], Monica Madhukar, Sos Agaian and Anthony T. Chronopoulos et al. try to segment AML blood smears by applying the popular KMeans algorithm.

In [12], N. H. Abdul Halim et al. have introduced segmentation and contrast stretching by taking into consideration the HIS color space. The same threshold value is used for enhancement purposes.

Feature extraction is also a very important facet of the entire detection process. The accuracy of this step often determines the accuracy of the actual result.

In [13], Fauziah Kasmin et al. feature changes act as classifier inputs. Usually, features such as color, texture and geometrical shapes are considered.

A feasible method to detect ALL has been proposed in [14] by Lalit Mohan Saini and Romel Bhattacharjee et al., which uses features such as circularity, perimeter, a cell's form factor and area for classification purposes.

Speaking of classification, it is the final and most vital step in an image processing related detection setup. There are numerous works on both existing as well as novel classification concepts.

Reta et al. [15] proposed a system for the sole purpose of white blood cell classification. A complete classification system for ALL and AML detection using grayscale level images only has been proposed in [16].

Combining the steps and by performing slight modifications to them or by improving them greatly, entirely new algorithms and methods have been suggested as well.

In [17], Hossein Ghayoumi Zadeh et al. have worked on an image analysis approach for enhancement, smoothing, segmentation, automatic detection and classification of infected cells from normal cells.

In [18], Jaya Sharma, Mashiat Fatma et al. propose application of convoluted neural networks (CNN) to classify healthy and infected cells. It is quite a complex method uniting the fields of computer vision and image processing, and though it is very difficult to execute, the results are great.

In [19], Ruggero Donida Labati et al. have used designed a new blood sample public dataset for the purpose of comparing various existing classification and segmentation algorithms and selecting the best choice out of them. Each dataset image provides figures of merit to compare various algorithm performances.

Our entire work is based on the ALL-IDB image database, which is freely downloadable from [20].

2   Main Contribution

While performing our algorithm for infected WBC cell detection, the main challenge we faced was separating individual WBC cells from clusters. The existing edge detection methods like Roberts [21], Prewitt [22], Sobel [23, 24] or Canny [25] are proposed for transition detection in images. These methods do not solve this problem which led us to device a new modified edge detection algorithm that solved our problem (Fig. 4f).

2.1 Novel Edge Detection Method

Previous edge detection methods [21–25] when applied on our image could not detect the edge separating two WBCs. In [23, 24], a 3×3 matrix was used as the estimate vector for directional derivative whose elements are the difference of density to neighbor. In contrast to previous methods, we used a 5×5 matrix which satisfied our need. Moreover, our edge detector is specially designed to meet our problem, so it works better specific to our image than any existing edge detection methods. In our proposed methods, we use two 5×5 masks along with both X and Y direction for both horizontal and vertical edges.

This mask when kernel convoluted to our image matrix marks the sharp changes in our image and generates an edge matrix.

Since our image of concern is a black and white image converted to its grayscale, the image pixel values are either 0 (black) or 255 (white). Hence, our proposed method does not require any complex directional derivative calculation for the mask cell values. So, our mask cell consists of values $-1, 0, 1$.

As in Fig. 2,

$$I_{ix} = I_x * M_x \tag{1}$$

Similarly,

$$I_{iy} = I_y * M_y \tag{2}$$

Here, * is the kernel convolution function. It multiplies each cell of the mask with its corresponding cell value of the image pixel and the summation of all such product is assigned to the central cell of the mask (I_i).

The mask in both the directions can give absolute mask magnitude when combined together as per the equation.

$$|M| = \sqrt{M_x^2 + M_y^2} \tag{3}$$

I_i detects the edge in the image if it encounters a sharp change of grayscale value as in Fig. 4f. Thus, the mask (M) slides gradually throughout the image-producing edges. The main problem with other edge detection method is that it cannot separate out

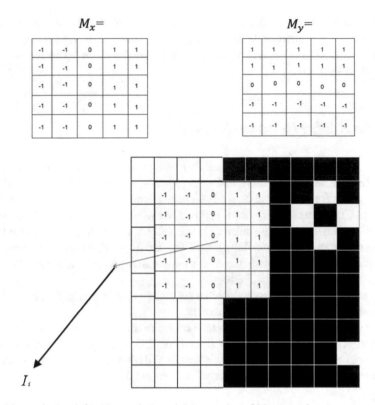

Fig. 2 M_x matrix (top left), M_y matrix (top right), snapshot of M_x on our image

the blobs, so we have adopted filledgegaps[1] algorithm in our edge detection method which in turn fills the gaps in between two adjacent WBCs of our experiment resulting an edge. Algorithm for our proposed edge detection method is as given Sect. 2.1.

2.2 Algorithm for Our Novel Edge Detection Method

Input: Image generated from our image processing step.
Output: Edges detected using our novel method.

Step 1: Masking with M is done to the input image
Step 2: The 5×5 mask is run through the whole image
Step 3: Generated image is then passed to filledgegaps algorithm to fill all the clefts
Step 4: The image from Step 3 is our output edge image.

---

[1] https://www.peterkovesi.com/matlabfns/LineSegments/filledgegaps.m.

3 Proposed Framework

A block diagram of the proposed method is presented in Fig. 3 and an algorithm for the same is given below. Before describing the algorithm in a stepwise manner, let us give a brief overview about the steps involved:

The first step in this process is image acquisition. In this step, blood image from slides will be obtained either from online sources or from a nearby hospital with effective magnification. The obtained images are called datasets. It is necessary that the process must yield satisfactory results over an entire dataset of images in order to certify it as a feasible method. It is recommended that the acquired images be of high pixel quality so that they can yield better results.

The next step is image enhancement, also referred to as image preprocessing. While acquiring dataset images, due to errors in staining, the images are likely to be disturbed by noise. The noise might blur the region of interest, leading to improper segmentation. To remove this noise and to improve quality of image, image preprocessing (enhancement) is required.

In the noise removal process, to ensure a more compact and better study, all WBCs at the image edges and all other non-leukocyte components (RBCs, platelets, etc.) are removed.

Image enhancement is followed by image segmentation. Segmentation refers to the process of distinguishing colors, objects, patterns or textures within an image itself. The main purpose of segmentation in image processing is to obtain a particular part or objects from the entire image so that it can be utilized for subsequent steps. The correctness of the subsequent steps namely feature extraction and image classification depends on the quality of the segmented output.

Segmentation focuses on the nucleus of WBC only as cytoplasm is scanty in infected cell images. Only lymphocytes and myelocytes are examined to determine whether they are blast cells or not.

Our approach has considered different existing segmentation techniques [9] and applied them to both original images and enhanced (noise removed) images to arrive at the most suitable choice. This is arguably the most important process of the entire algorithm and needs to be performed as neatly as possible for accurate end results.

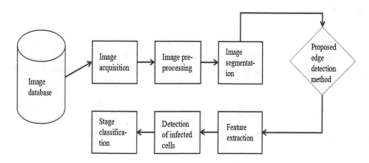

Fig. 3 Flowchart of our framework

In fact, our paper's main contribution lies in proposing a novel segmentation (edge detection) approach which provides greater accuracy when compared to the existing segmentation algorithms or methods. Then we move to feature extraction based on geometrical as well as texture features and postextraction we classify the image to be either that of an infected or a healthy cell. Lastly, based on study of morphological features especially cytoplasm, we classify the cancer to be either chronic or acute.

The algorithm is given in two parts. Part 1 basically deals with acquiring the image sample, breaking it down into its three component color channels, performing thresholding separately in each of the channels and expressing it as the sum of the color planes. Part 2 deals with complementing the image, processing it and then applying edge detection (the improved novel method) on the image. Then feature extraction is carried out and the image is classified into healthy or infected (further stage classified into acute or chronic).

A detailed step by step overview is given below:

Part 1:

Input: A colored blood sample image from dataset
Output: Image expressed as the sum of component planes

- The colored blood slide image is taken as system input. The cell might be healthy or infected; we do not know it yet. Our purpose is to identify whether there are blast cells in the original image and thereby determine if it is healthy or infected.
- Conventional algorithms usually convert the color image into grayscale image in the next step. However, we will not be doing the same.
- Instead, we break down the color image in three color planes namely the red plane, the green plane and the blue plane.
- Then we perform thresholding separately in each of the three channels, and only after it is done, the image is converted to its black and white forms.
- We then express the image as the sum of all color planes by using the AND operation.

Part 2:

Input: Output image from Part 1
Output: Image which has been classified as healthy or infected and further stage classification is done.

- The entire image is complemented (white parts turn black and vice versa). Also, the holes in the image are filled with unwanted small pixels.
- We perform edge detection (a segmentation approach) on the unprocessed image. Generally, state of the art methods are applied but our own research has found our own designed edge detection method to be more accurate than those. So, we use it for edge detection. The underlying purpose of using edge detection on an unprocessed image is to elucidate that applying processing is indeed advantageous in this context.
- Small unwanted pixel groups are removed using morphological operations such as morphological erosion.

- Again, edge detection is used, this time on the processed image.
- The geometrical features such as area and perimeter of segmented cells and texture features are computed. Based on the features extracted in the last step, the cell can be classified as a healthy cell or infected cell.
- For a healthy cell, the image background will be entirely black depicting that there are no blast cells in the image. However, for an infected cell, there will be the presence of white WBC outlines on the black backdrop, depicting the presence of blast cells and thereby, presence of cancer. In fact, the number of white outlines is nothing but the number of blast cells in the sample image. Flowchart of the algorithm is shown in Fig. 3.
- Further stage classification of the detected cancer cell is performed to know the acuteness of the cancer cell.

The entire algorithm will be much easier to understand when visual representation of the obtained results (for both cases) is provided in the result and analysis section.

4 Experiment and Result Analysis

4.1 Datasets

A Canon G5 Powershot camera, aided by an optical microscope, has been used to capture the images for our working dataset. The microscope is magnified within a 300–500 range to capture the images in different zoom angles. Each image has a 2592 * 1944 resolution. The image format is JPG with a color depth of 24 bits. A total of 108 images clicked during 2005 are present in the dataset. A collection of 39,000 blood elements is available in ALL-IDB1 where expert cancer doctors have marked the lymphocytes. A rough estimate of around 510 candidates infected WBC's are present. The dataset is freely downloadable from [20]. The notation is XXX_Y.jpg where a triple-digit integer counter depicts XXX and a Boolean variable (either 0 or 1) depicts Y. For healthy lymphocytes, Y takes a value of 0, and for infected lymphocytes (lymphoblasts), Y takes a value of 1. Thus, it is easy to classify the images as healthy or infected just by looking at the Boolean Y value. The centroids of the lymphoblasts (if any) are documented in a text file IMXXX_Y.xyc, which is linked to the main jpg file.

4.2 Results Related to Own Edge Detection Method

There are many well-known states of art edge detection methods which yield accurate results for any image segmentation process. However, when we tried to apply them, for some images in our dataset, we encountered a problem Fig. 4b–f. In the original processed image shown in Fig. 4a, there is a dumbbell-shaped structure in the middle

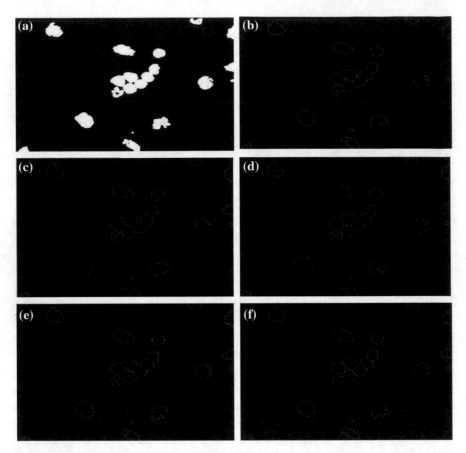

Fig. 4 **a** Original processed image. **b** Roberts on original processed image. **c** Prewitt on original processed image. **d** Sobel on original processed image. **e** Canny on original processed image. **f** Proposed method on original processed image

portion of the image sample. To the naked eye, it appears there is a single large entity (WBC) but in reality, there are seven separate lymphocytes. The only reason it appears to be a single entity is that the separate WBCs are clustered together with no well visible partition (cleft) or blobs separating their edges. When we apply the existing well-known methods on this particular image, they fail to detect the edge outlines of the separate cells and consider the entire collection of seven leukocytes to be a single one. As a result, they fail to trace the separate cell outlines and the edge detection process becomes inaccurate and to a large extent, incorrect. This leads to the necessity of a new edge detection method that can overcome this particular drawback. The main objective of our novel method should be to detect the separate cells and perform separate outline detection on each of them.

Figure 4a–e all display the results of applying various well-known edge detection algorithms on a processed image. However, it clearly shows in the results that the methods have failed in treating the collection of seven WBCs as separate entities and have considered only the outer layer of the collection for outline detection. Figure 4f displays the result of our own approach. It is clearly visible that our algorithm has been successful in considering each leukocyte as a separate entity and edge outlines have been computed for each of them, unlike other methods.

This observation leads to the conclusion that our novel edge detection method provides better accuracy than existing methods in this regard.

4.3 Results Related to Cancer Cell Detection for an Infected Cell

In Fig. 5b, the acquired image is broken up or segregated into its three constituent color planes. In Fig. 5c, thresholding is separately performed for each plane and the image is converted to grayscale form.

A resulting image in Fig. 6A with white outlines on a dark background looks better than the other way round. So, for clarity purposes, the image is complemented and image holes are filled with unwanted small pixels. In Fig. 6B, morphological

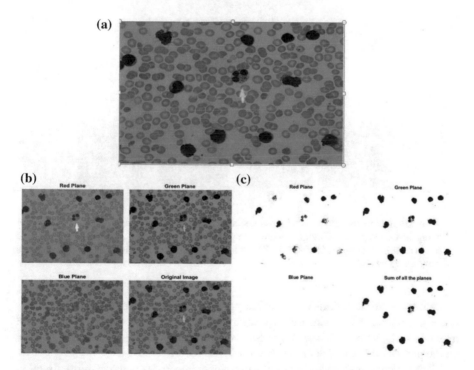

Fig. 5 **a** Original infected cell. **b** Image in three color planes. **c** Thresholding in three planes and summation

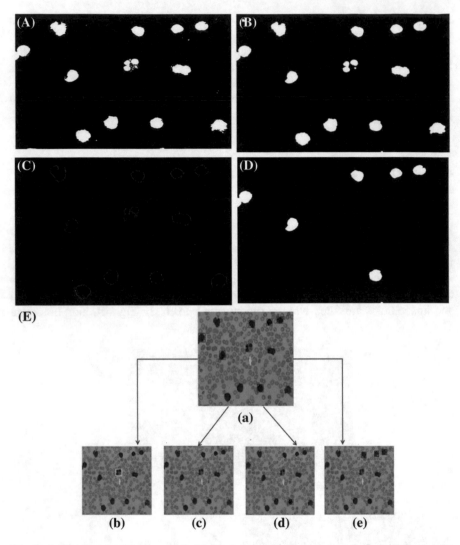

Fig. 6 **A** Complemented image with holes filled along with unwanted small pixels. **B** Image after removing unwanted pixels. **C** Edge detection on processed image using own method. **D** Feature extraction and eliminating healthy WBCs. **E** Identification and count of blast cells in original image and stage classification

operations are used to remove unwanted pixels. Our earlier experimental result has already established that using our own edge detection method instead of conventional methods will give us better accuracy. So, we apply our own edge detection approach on the processed image (as in Fig. 6C, D).

Since the input image was that of a blood cancer infected cell, the feature extraction result picture successfully identified the presence of blast cells in the original image. It also detected the number of blast cells in the image. In fact, the number of white

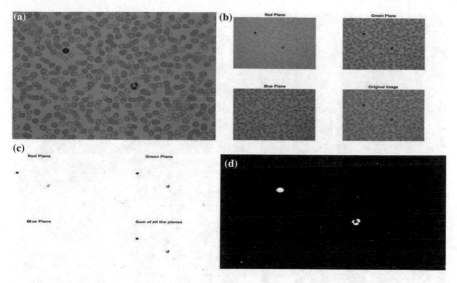

Fig. 7 **a** Original healthy image cell. **b** Image on three color planes. **c** Thresholding in three planes and summation. **d** Complemented image with holes filled with unwanted small pixels

outlines is nothing but the number of blast cells in the sample image (which is five in this case). Figure 6E(b) depicts only the presence of lymphoblasts and Fig. 6E(c–e) represents Stage 1 blood cancer, Stage 2 blood cancer and Stage 3 blood cancer, respectively.

4.4 Results Related to Cancer Detection in Healthy Cells

The same sequence of steps that we applied to the infected cell is applied to the healthy image sample as well (Fig. 7).

Since the input was a healthy cell, after feature extraction and WBC elimination, there were zero blast cells found in Fig. 8d. This goes on to show that the algorithm is capable of producing accurate results.

4.5 Accuracy Calculation in Tabular Form

In the context of our work, accuracy calculation is done by processing some technical parameters. We can either employ a cell test benchmark approach (test returns positive if the cell in question is infected) or an image-level benchmark (test returns positive if image in question has even a single infected cell). Some term elements that need to be discussed in this regard.

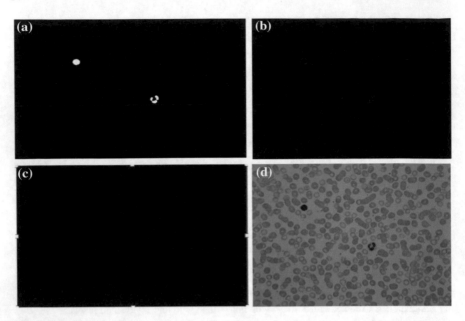

Fig. 8 **a** Image after removing unwanted pixel. **b** Edge detection on processed image using own method. **c** Feature extraction and eliminating healthy WBCs. **d** Identification of blast cells in original image

The number of elements which our test correctly classifies as positive is said to be true positive (t_p) and the elements which the test incorrectly classifies as positive are called false positive (f_p).

Similarly, the number of elements which our test correctly classifies as negative is said to be true negative (t_n) and the elements which the test incorrectly classifies as negative are called false negative (f_n).

The probability that every ALL element has been classified correctly is referred to as sensitivity. In contrast, specificity is defined by the probability of correct classification of non-ALL elements. Mathematically, these terms can be represented as:

$$\text{Sensitivity} = t_p/(t_p + f_n) \tag{4}$$

$$\text{Specificity} = t_n/(t_n + f_p) \tag{5}$$

All these terms are collectively referred to as figures of merit and can be used for accurate calculation for both cell and image tests.

Another mode of calculation using the same parameters is by using the concepts of precision and recall. Precision refers to the percentage of our relevant results. Recall, on the other hand, refers to the percentage of total relevant results that the used algorithm successfully classifies.

Table 1 Accuracy comparison

Method	Accuracy (in %)
K-means clustering	73.1
Watershed transform	71.6
Edge detection using histogram equalizing and linear contrast stretching	74.5
Proposed novel method	86.4

$$\text{Precision} = t_p/(t_p + f_p) \tag{6}$$

$$\text{Recall} = t_p/(t_p + f_n) \tag{7}$$

$$\text{Accuracy} = (t_p + t_n)/(t_p + t_n + f_p + f_n) \tag{8}$$

Taking help from these formulae, we compare the accuracies of results obtained from different methods as well as our own proposed method.

The results from Table 1 clearly show that our own proposed method is more accurate than other methods in this regard.

5 Conclusion

It has been stated several times, in this paper, that the possibility of successful treatment of leukemia is highly dependent on its early detection and diagnosis.

The manual inspection process has been found to be tiring and error-prone and thereby image processing algorithms have been employed to speed up the process and improve accuracy at the same time. We have performed a thorough study of the existing algorithms; the processes involved and even designed our own unique edge detection algorithm to counter a problem that we faced during processing. The algorithm has been found to obtain highly accurate results, thereby assuring that our proposed method is indeed feasible. However, several improvements can be made in future to this work (such as improving dataset quality and extending the process to cancer types other than ALL), in order to allow the work to make a greater impact.

References

1. National Cancer Institute: http://www.cancer.gov/cancertopics/wyntk/leukemia. Last accessed 3 Oct 2013
2. Through the Microscope: Blood Cells—Life's Blood. http://www.wadsworth.org/chemheme/heme. Last accessed 3 Oct 2011
3. https://cancer.org. Last accessed

4. Bennett, J.M., Catovsky, D., Daniel, Marie-Theregse, Flandrin, G., Galton, D.A.G., Gralnick, H.R., Sultan, C.: Proposals for the classification of the acute leukemias French-American-British (FAB) co-operative group. Br. J. Haematol. **33**(4), 451–458 (1976)
5. Biondi, A., Cimino, G., Pieters, R., Pui, C.-H.: Biological and therapeutic aspects of infant leukemia. Blood **96**(1), 24–33 (2000)
6. Joshi, M.D., Karode, A.H., Suralkar, S.R.: White blood cells segmentation and classification to detect acute leukemia. Int. J. Emerg. Trends Technol. Comput. Sci. (2013)
7. Ghosh, M., Das, D., Chakraborty, C., Ray, A.K.: Automated leukocyte recognition using fuzzy divergence. Micron **41**(7), 84 (2010)
8. Dorini, L.B., Minetto, R., Leite, N.J.: White blood cell segmentation using morphological operators and scale space analysis. In: Proceedings of the Brazilian Symposium on Computer Graphics and Image Processing, pp. 294–304, Oct 2007
9. Evelin Suji, G., Lakshmi, Y.V.S., Wiselin Jiji, G.: Comparitive study on image segmentation algorithms. IJACR **3**(3, 11) (2013)
10. Nee, L.H., Mashor, M.Y., Hassan, R.: White blood cell segmentation for acute leukemia bone marrow images. In: International Conference on Biomedical Engineering (ICoBE), Penang, Malaysia, 27–28 Feb 2012
11. Agaian, S., Senior Member, IEEE, Madhukar, M., Chronopoulos, A.T., Senior Member, IEEE: Automated screening system for acute myelogenous leukemia detection in blood microscopic image. IEEE Syst. J. (2014)
12. Abd Halim, N.H., Mashor, M.Y., Abdul Nasir, A.S., Mokhtar, N.R., Rosline, H.: Nucleus segmentation technique for acute leukemia. In: IEEE 7th International Colloquium on Signal Processing and its Applications (2011)
13. Kasmin, F., Prabuwonw, A.S., Abdullah, A.: Detection of leukemia in human blood sample based on microscopic images: a study. J. Theor. Appl. Inf. Technol. (2012)
14. Bhattacharjee, R., Saini, L.M., Robust technique for the detection of acute lymphoblastic leukemia. In: 2015 IEEE Power, Communication and Information Technology Conference (PCITC) Siksha O Anusandhan University, Bhubaneswar, India
15. Reta, C., Robles, L.A., Gonzalez, J.A., Diaz, R., Guichard, J.S.: Segmentation of bone marrow cell images for morphological classification of acute leukemia. In: Proceedings of the Twenty-First International Florida Artificial Intelligence Research Society Conference (2010)
16. Taylor, K.B., Schorr, J.B.: Blood **4** (1978)
17. Zadeh, H.G., Janianpour, S., Haddadnia, J.: Recognition and classification of the cancer cells by using image processing and LabVIEW. Int. J. Theory Eng. **5**(1) (2013)
18. Fatma, M., Sharma, J.: Identification and classification of acute leukemia using neural network. In: 2014 International Conference on Medical Imaging, m-Health and Emerging Communication Systems (MedCom)
19. Labati, R.D., Piuri, V., Scotti, F.: A1I-Idb: the acute lymphoblastic Leukemia image database for image processing (2011)
20. https://homes.di.unimi.it/scotti/all/
21. Aybar, E.: Sobel Edge Detection Method For Matlab. Anadolu University, Porsuk Vocational School, 26410 Eskisehir
22. Vincent, O.R., Folorunso, O.: A descriptive algorithm for sobel image edge detection. In: Proceedings of Informing Science & IT Education Conference (InSITE) (2009)
23. Sobel, I.: An Isotropic 3×3 Gradient Operator, Machine Vision for Three-Dimensional Scenes, pp. 376–379. Freeman, H., Academic Pres, New York (1990)
24. Prewitt, S., Scharr gradient 5×5 convolution matrices Guennadi (Henry) Levkine Email: hlevkin at yahoo.com Vancouver, Canada. First draft, Feb 2011, Second Draft, June 2012
25. Gao, W., Yang, L.: An Improved Sobel Edge Detection. 978-14244-5540-9/10/$26.00 ©2010 IEEE
26. Terese Winslow LLC: Medical and Scientific Illustration. https://www.teresewinslow.com. Last accessed 3 Jan 2020

Biometric-Based Unimodal and Multimodal Person Identification with CNN Using Optimal Filter Set

Goutam Sarker and Swagata Ghosh

Abstract The convolution neural network (CNN) has brought about a drastic change in the field of image processing and pattern recognition. The filters of CNN model correspond to the activation maps that extract features from the input images. Thus, the number of filters and filter size are of significant importance to learning and recognition accuracy of CNN model-based systems such as the biometric-based person authentication system. The present paper proposes to analyze the impact of varying the number of filters of CNN models on the accuracy of the biometric-based single classifiers using human face, fingerprint and iris for person identification and also biometric-based super classification using both bagging and programming-based boosting methods. The present paper gives an insight into the optimal set of filters in CNN model that gives the maximum overall accuracy of the classifier system.

Keywords Person authentication · Biometrics · CNN · Deep learning · Super classification · Bagging · Boosting · Confusion matrix · Precision · Recall · F-score

1 Introduction

Biometric identification [1–4] is a security process that relies on the unique biological characteristics of an individual to verify that he/she really is the individual who he/she claims to be. Biometric identification systems [5, 6] compare the biometric data captured to the learnt authentic data in a database. The most commonly available and used biometrics are face, fingerprint and iris [7–9]. CNN [2, 10–12] is a revolutionary and dramatic change in the field of neural networks, and it has helped computer scientists achieve astounding improvement in lowering the error rates in pattern recognition, image classification and other applications in the field of computer vision. In our works, three different biometric, namely face, fingerprint and iris have been used in the CNN model [2, 11] for person identification, and then, these three

G. Sarker (✉) · S. Ghosh
CSE Department, NIT Durgapur, Durgapur, India
e-mail: g_sarker@ieee.org; goutam.sarker@cse.nitdgp.ac.in

© Springer Nature Singapore Pte Ltd. 2020
J. K. Mandal and S. Mukhopadhyay (eds.), *Proceedings of the Global AI Congress 2019*, Advances in Intelligent Systems and Computing 1112,
https://doi.org/10.1007/978-981-15-2188-1_2

single biometric-based CNN classifiers have been integrated to build the model for biometric-based super classification. In the present paper, we analyze the effect of varying the number of CNN filters for all these CNN-based biometric classifier models and thereafter determine the optimal number of filters that will give the maximum overall accuracy of the system. The proposed system is efficient, effective and useful for future research work.

2 Related Works

A CNN for handwritten character classification is described in [13]. A CNN for synthetic vision systems is described in [14]. Large-scale video classification with CNN system is done in [15]. Evaluation of CNN for visual recognition is described in [16, 17]. Pedestrian detection with CNN is done in [18]. The work [19] is a deep neural network for object detection. A new class of CNN and their applications in face detection is described in [20].

A super classifier in face and facial expression identification with learning-based boosting is described in [21]. The papers [5, 6] are a super classifier for biometric-based person authentication implementing programming-based boosting method. The work [3, 4] a multi-level integrator with programming-based boosting for person authentication with different biometrics. A person authentication system using a biometric-based efficient multi-level integrator using RBFN is developed in [8]. The paper [9] is a super classifier for biometric-based person identification with template matching technique.

A survey paper on the working principles of convolution neural networks is done in [11]. A set of CNN models for single classifiers based on three different biometrics namely human face, fingerprint and iris for person identification is described in [2].

3 Theory of Operation

The present section presents the details of the neural network architectures that are used to implement the application of the CNN-based super classifier biometric authentication system for person recognition.

3.1 Convolution Neural Network (CNN)

The convolution neural network (CNN) is a revolutionary improvement in the field of artificial neural network (ANN) as it not only overcomes the limitations of ANN but also achieves unprecedented accuracy in machine learning and recognition. The CNN comprises mostly of three different types of layers, namely the convolution layer, the pooling layer and the fully connected layers that are stacked together to form the complete CNN architecture [2, 11]. The advantage of using CNN over ordinary ANN is that CNN result is not affected by image transformations like scaling, translation and rotation of images, and furthermore, CNN-based models give much higher overall accuracy in learning and recognition of classifier systems.

3.2 Overview of the System and Algorithms

Preprocessing of Images

The input images for each of the CNN-based single biometric classifier have to be preprocessed before CNN learning as well as recognition. The different steps in preprocessing are shown as below:

(i)   Convert training or test biometric image set into gray scale image set.
(ii) Remove noises from the biometric image set using dilation–erosion method.
(iii) Resize biometric image set to a standard 128×128 pixel size. This is to ensure uniformity of input into CNN model.
(iv) Convert 2D matrix image files into 1D matrix files. This set is the input to the convolution neural network for learning and recognition.

Operation of the System

The network has input image of size 128 * 128 neurons. These are the pixel intensities for the biometric data set images. We create models with varying number of filters and try to compare the training and test accuracies for each case. We take the number of filters of the CNN model as a variable N, and then, we vary the value of N between 5 and 100. Then, a convolution layer with N kernels each of size 3 * 3 (this is also the size of the local receptive field) follows this. This results in N * 126 * 126 hidden feature neurons.

The next step is to perform max pooling with a max-pooling kernel size 2 * 2 with a stride of 2 (no overlapping) across all the N feature or activation maps. The result is a layer of N * 63 * 63 hidden feature neurons. The process of convolution and max pooling is once again repeated, with N kernels of size as before, i.e., 3 * 3.

The final layer of connection is a fully connected layer. So, this layer connects every neuron from the flattened max pooled layer to every one of the five output neurons (as there are five categories for classification or human subjects).

3.3   Algorithm for Training Super Classifier Model

Algorithm for Learning Biometric Images

Input: Set of preprocessed images of face, fingerprint and iris each of size 128 × 128 pixels.
Output: Three CNN models that are trained separately for the identification of face, iris and fingerprint.
Steps for training each CNN model:

1.   Input image of size 128 * 128 into the network.
2.   Convolve with N kernels each of size 3 * 3 (this is also the size of the local receptive field) follows, which results in N * 126 * 126 hidden feature neurons. For first iteration that a model is created N is initially chosen as 5, and at each iteration, the value of N is incremented by five, and thus, N is varied from 5 to 100.
3.   The next step is to perform max pooling with a max-pooling kernel size 2 * 2 with a stride of 2 (no overlapping) across all the N feature or activation maps, and the result is a layer of N * 63 * 63 hidden feature neurons.
4.   The process of convolution and max pooling is once again repeated with N kernels of size 3 * 3. With this, convolution layer yields a layer of N * 61 * 61 hidden feature neurons, while the max-pooling layer yields a layer of N * 30 * 30 hidden feature neurons.
5.   This is now flattened to a size of (900 * N) pixels at the next layer.
6.   The final layer is a fully connected layer. So, this layer connects every neuron from the flattened max pooled layer to every one of the five output neurons for the learning and recognition for each CNN model (as there are five classes for each model in the first case).
7.   As N is varied from 5 to 100 at increments of five, there are total 20 models that are created with number of filters as 5, 10, 15, 20, 25, ..., 90, 95, 100. For each model, the corresponding training and testing accuracies are noted, and graph is plotted.

Bagging Algorithm for Super Classifier Model

Input: Set of preprocessed images of face, fingerprint and iris each of size 128 × 128 pixels.
Output: Trained super classifier model that can correctly recognize and identify persons based on their biometrics.
Steps for training the super classifier model:

1.   Input the preprocessed biometric images to the respective single CNN-based classifiers for face, fingerprint and iris.

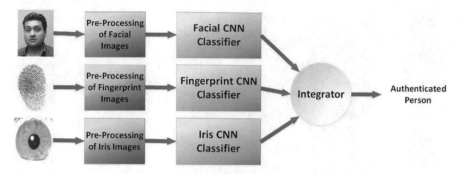

Fig. 1 Super classifier CNN model implementing bagging method

2. The prediction results of the three single CNN classifiers are noted and saved for future uses.
3. Based on these predictions, the final prediction of the classifier is denoted as the predicted result that is concluded by majority of the single classifiers, i.e., by at least two of the three single biometric classifiers.
4. The final conclusion or the combined result of the conclusions of the three single classifiers is declared as the prediction of the super classifier.
5. The number of filters is varied from 5 to 100 (at intervals of five) for the CNN off each biometric in order to get the optimal number of filters that will result in the CNN-based super classifier model with maximum accuracy.
6. Finally, the confusion matrix of the super classifier is obtained implementing the model that uses the optimal number of filters that give maximum overall accuracy (Fig. 1).

Programming-Based Boosting Algorithm for Super Classifier Model

Input: Set of preprocessed images of face, fingerprint and iris each of size 128×128 pixels.
Output: Trained super classifier model that can correctly recognize and identify persons based on their biometrics.
Steps for training the super classifier model:

1. Input the preprocessed biometric images to the respective single CNN-based classifiers for face, fingerprint and iris.
2. The evaluation results of the three single CNN classifiers are saved for future use.
3. Based on these results, some weight is associated with each single biometric-based classifier, and this weight is proportional to the accuracy of that CNN classifier model.

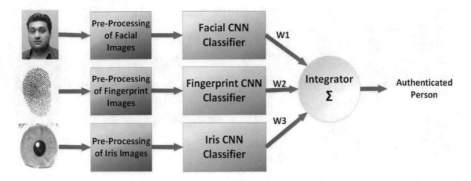

Fig. 2 Super classifier CNN model implementing programming-based boosting method

4. The final conclusion or the prediction of the super classifier is obtained by taking a weighted sum of the weights attached to the predictions of the single classifier models and thereby selecting the predicted class that is associated with the maximum sum of weights of the single classifiers.
5. The number of filters is varied from 5 to 100 (at intervals of five) for the CNN off each biometric in order to get the optimal number of filters that will result in the CNN-based super classifier model with maximum accuracy.
6. Finally, the confusion matrix of the super classifier is obtained implementing the model that uses the optimal number of filters that give maximum overall accuracy (Fig. 2).

Algorithm for Determining the Optimal Filter Set for each CNN model

Input: Plot showing number of CNN filters versus system accuracy.
Output: Number of common or essential filters that give the maximum overall accuracy.
Steps for finding the optimal set of CNN filters:

1. From the plot, determine the number of filters that give the maximum accuracy. More than one such number of filters that give the same maximum system accuracy that is close to the maximum accuracy.
2. Each number of filters corresponds to a visible filter set.
3. Every filter in one filter set is compared with every filter in all the other filter sets to extract the common filter set.

4 Experimental Results

4.1 Platform

The experiments are performed in a system having a dual-core processor of type Intel® Core™ i3-7020U CPU with a clock rate of 2.3 GHz and equipped with 4 GB RAM and 1 TB of hard disk space that runs on Microsoft Windows 10 operating system.

4.2 Input Dataset

Dataset for Facial CNN Classifier
A part of the famous benchmarked databases called "Labeled Faces in the Wild" (LFW) is used. There are 5749 subjects with total number of 13,244 images. Out of these, five subjects are chosen for the present work.

Dataset for Fingerprint CNN Classifier
Fingerprint images from the FVC2002 database are used. Here, the fingerprints of five persons/subjects are chosen each having a total variation of eight images.

Dataset for Iris CNN Classifier
The CUHK Iris image dataset obtained from the archives of the Chinese University of Hong Kong has been used here. Out of all 36 subjects, five classes are chosen, and each subject has a total variation of seven images.

Dataset for Super Classifier
For the dataset of the super classifier, as the correlated dataset for the different biometrics of the same person is currently unavailable, we had to make the assumption that each set of the three biometric namely face, fingerprint and iris of different persons is regarded as of the same one. We chose the collections of biometric images for five different subjects and collected them together to form the dataset of five subjects that are referred to as P1, P2, P3, P4 and P5.

4.3 Performance Evaluation Results

1. **Performance of the CNN-based Face Classifier Model for Person Identification**

The Face Classifier model has been implemented using CNN with the number of filters varying from 5 to 100 at intervals of five as 5, 10, 15, 20, 25, 30, 35, 40, 45,

50, 55, 60, 65, 70, 75, 80, 85, 90, 95, 100. The following plot visualizes the train and test accuracies for this range of number of filters.

From the plot below, we can easily observe that the maximum overall system accuracy of 95% is achieved using number of filters as 100; however, the accuracy of 92%, when the number of filters is 70 or 95, is also quite close to the optimal system accuracy.

From the results of analysis of the impact of varying number of filters on system accuracy, we observe that the common set of filters that give the near-optimal accuracy has altogether 64 common filters. These 64 filters are the most essential filters, and these are present in the filter sets of size 70 and 95. The remaining six filters in case of filter set of size 70 and 31 filters in case of the filter set of size 31 are redundant filters because these filters have no significant impact on the overall system accuracy. Thus, we have determined the number of optimal filters for the face CNN classifier (Fig. 3).

2. Performance of the Fingerprint CNN Classifier Model for Person Identification

The fingerprint classifier model has been implemented using CNN with the number of filters varying from 5 to 100 at intervals of five as 5, 10, 15, 20, 25, 30, 35, 40,

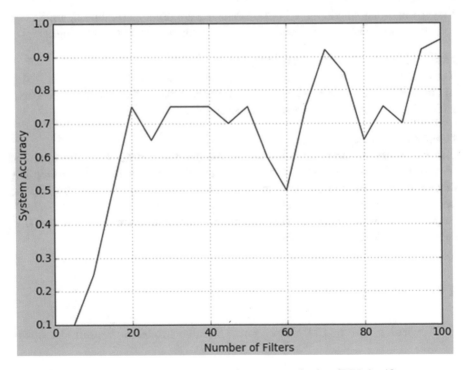

Fig. 3 Plot showing number of filters versus system accuracy for face CNN classifier

45, 50, 55, 60, 65, 70, 75, 80, 85, 90, 95, 100. The following plot visualizes the train and test accuracies for this range of number of filters.

From the plot below, we can easily observe that the maximum overall system accuracy of 85% is achieved using number of filters as 85; however, the accuracies of 80% when the number of filters is 100 and 77% when the number of filters is 95 are also quite close to the maximum overall system accuracy. Furthermore, it can be observed from the following plot that the system accuracy stands more or less at around 75% for the range in which the number of filters varies from 30 to 65.

From the results of analysis of the impact of varying number of filters on system accuracy, we observe that the common set of filters that give more or less the same overall accuracy has altogether 27 common filters. These 27 filters are the most essential filters, and these are present in the filter sets of sizes between 30 and 65. Therefore, the remaining three filters in case of filter set of size 30 or 38 filters in case of the filter set of size 65 are all redundant filters because these filters have no significant impact on the overall system accuracy. Thus, we have determined the number of optimal filters for the fingerprint CNN classifier (Fig. 4).

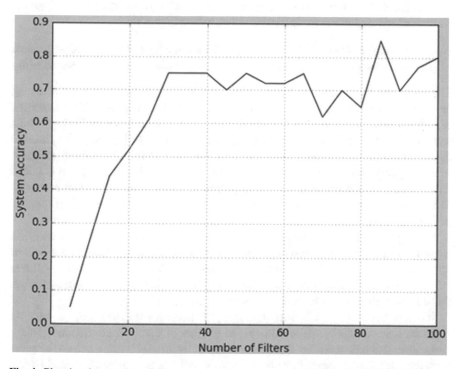

Fig. 4 Plot showing number of filters versus system accuracy for fingerprint CNN classifier

3. **Performance of the CNN-based Iris Classifier Model for Person Identification**

The iris classifier model has been implemented using CNN with the number of filters varying from 5 to 100 at intervals of five as 5, 10, 15, 20, 25, 30, 35, 40, 45, 50, 55, 60, 65, 70, 75, 80, 85, 90, 95, 100. The following plot visualizes the train and test accuracies for this range of number of filters.

From the plot below, we can easily observe that the maximum overall system accuracy of 87% is achieved using number of filters as 65; however, the accuracies of 85% when the number of filters is 85 and 82% when the number of filters is 70 are also quite close to the maximum overall system accuracy. Furthermore, it can be observed from the following plot that the system accuracy stands more or less at around 70% for the range in which the number of filters varies from 40 to 60.

From the results of analysis of the impact of varying number of filters on system accuracy, we observe that the common set of filters that give more or less the same overall accuracy has altogether 31 common filters. These 31 filters are the most essential filters, and these are present in the filter sets of sizes between 40 and 65. Therefore, the remaining nine filters in case of filter set of size 40 or 34 filters in case of the filter set of size 65 are all redundant filters because these filters have no significant impact on the overall system accuracy. Thus, we have determined the number of optimal filters for the iris CNN classifier (Fig. 5).

4. **Performance of the CNN-based Super Classifier Model for Biometric-based Person Identification Implementing Bagging Method of Super Classification**

The bagging-based super classifier model has been implemented using CNN with the number of filters varying from 5 to 100 at intervals of five as 5, 10, 15, 20, 25, 30, 35, 40, 45, 50, 55, 60, 65, 70, 75, 80, 85, 90, 95, 100. For simplifying results and reducing running time, the number of epochs taken in this stage is five, i.e., the model is trained for continuous five iterations. Accuracy of the CNN model increases with increasing number of epochs, so we later set the number of epochs as 60. This latter stage of CNN learning takes a long time but results in higher test accuracy. The following plot visualizes the train and test accuracies for this range of number of filters (Fig. 6).

Since the maximum overall accuracy is recorded when the number of filters is 90, we will consider the confusion matrix and other parameters for when the number of filters for each of the single classifier CNN models is 90. With the number of filters set as 90, we have trained the model for 60 epochs or iterations in the latter stage. For this case, the overall accuracy of the super classifier is 84.25%.

It can be observed that in both the cases, the accuracy remains more or less the same for a certain range of the value of the number of filters N. For instance, in the plot above, it can be seen that the test accuracy of the model is more or less the same even though the plot varies from $N = 15$ to $N = 70$. This can be explained by the fact that a certain number of filters are common for the CNN-based super classifier models that are created with the value of N varying in that range.

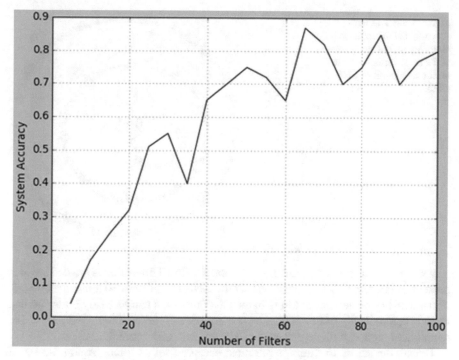

Fig. 5 Plot showing number of filters versus system accuracy for iris CNN classifier

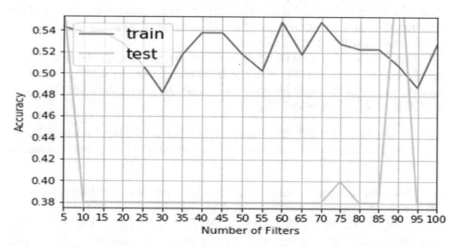

Fig. 6 Number of filters versus accuracy for bagging-based super classification

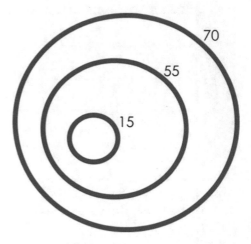

Thus, the 15 filters in the case of CNN model with 15 filters that is created are also present among the 70 filters in the case of the CNN model with 70 filters are created. Thus, with higher number of filters in the CNN model, it can be easily observed that there is commonality of filters. The lower number of filter is nothing but a subset of the higher number of filters present in CNN models with more filters.

Hence, the set of 15 filters is the most essential set of filters. Again the set of 55 filters is more important subset of the 70 filters compared to the remaining ones because the accuracy in all these cases is more or less the same. We can say that the remaining 15 filters that are outside the subset of essential 55 filters in the total set of 70 filters are redundant or not essential.

From this observation, we can safely conclude that the common set of filters are the basic and most essential filters of the CNN model that can help find the most important feature of the input image. This observation proves the commonality of essential filters in a CNN model (Fig. 7).

1. **Performance of the CNN-based Super Classifier Model for Biometric-based Person Identification Implementing Boosting Method of Super Classification**

The boosting-based super classifier model has been implemented using CNN with varying number of filters. The number of filters range from 5 to 100 at intervals of five as 5, 10, 15, 20, 25, 30, 35, 40, 45, 50, 55, 60, 65, 70, 75, 80, 85, 90, 95, 100. To simplify results and reducing running time, the number of epochs taken in this stage is 5, i.e., the model is trained for continuous five iterations. Accuracy of the CNN model increases with increasing number of epochs, so we later set the number of epochs as 60. This latter stage of CNN learning takes a long time but results in higher test accuracy. The following plot visualizes the train and test accuracies for this range of number of filters of this CNN-based model (Fig. 8).

Since the maximum overall accuracy is recorded when the number of filters is any of 45, 55 or 75, we will consider the confusion matrix and other parameters for

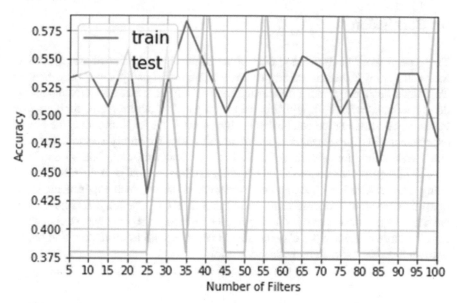

Fig. 8 Number of filters versus accuracy for boosting-based super classification

when the number of filters for each of the single classifier CNN models is 55. With the number of filters set as 55, we have trained the model for 60 epochs or iterations in the latter stage. It is observed that higher accuracy is achieved with higher number of epochs. For this case, the overall accuracy of the super classifier is 92.83%.

In this case, the number of common or essential filters is 43 as determined from the experimental results. The remaining filters in the filter sets are redundant.

Novelty of the Work: In the present paper, we have analyzed the effect of varying number of CNN filters on the overall accuracy of the CNN-based classifier systems. This establishes the concept that the set of CNN filters that give almost same system accuracy is actually more or less a superset of the most essential or common filter set. This common set of filters usually corresponds to the most basic features in any recognized pattern such as straight lines—vertical, horizontal or even diagonal, colored dot patterns, etc. These common features are extracted using the common set of filters in a CNN model. By analyzing the effect of varying number of filters on the accuracy of each of our CNN-based single biometric classifier and super classifier models, we have established that any set of filters that give more or less the maximum overall system accuracy is nothing but more or less a superset of the optimal filter set. Determining optimal filter set dramatically reduces the CNN learning and recognition time. The redundant filters need not be considered for further experimental analysis.

5 Conclusion

The present work analyzes the effect of variation of number of filters of CNN on the accuracy of biometric-based person authentication using super classification. This finds out the optimal number of filters and filter set for a CNN model and detects redundant filter set. With this, the learning and recognition time for biometric-based super classification can be drastically reduced. As for limitations, the present experiment was time-consuming and required a lot of memory space. The results could have been more significant if we had used more training data, or employed several other biometric-based CNN classifier models, or analyzed the CNN models using a large variation of the number of filters. However, such experiments would take immense amount of computation time as well as CPU memory. As future scope, we may use more number of other biometrics. We may also fuse the conclusions of three or more of such super classifiers and integrate them to ultimately find out the optimal filter set for each individual classifier.

References

1. Sarker, G.: A Treatise on Artificial Intelligence. ISBN: 978-93-5321-793-8; 1, 325, 2018
2. Sarker, G., Ghosh, S.: A set of convolution neural networks for person identification with different biometrics. Int. J. Adv. Comput. Eng. Netw. **7**(6) (2019). ISSN(p): 2320-2106, ISSN(e): 2321-2063
3. Kundu, S., Sarker, G.: A multi-level integrator with programming based boosting for person authentication using different biometrics. J. Inf. Process. Syst. **14**(5), 1114–1135 (2016) (SCOPUS & ESCI indexed)
4. Kundu, S., Sarker, G.: A person authentication system using a biometric based efficient multi-level integrator. Int. J. Control Theory Appl. (IJCTA) **9**(40), 471–479 (2017). International Science Press (SCOPUS indexed)
5. Kundu, S., Sarker, G.: A programming based boosting in super classifier for fingerprint recognition. In: International Conference ICCI—2015, held at BIT Mesra, Ranchi, 10–11 Dec 2015, © Springer Science + Business Media Singapore, Advances in Intelligent Systems and Computing, vol. 509, pp. 319–329 (2015). https://doi.org/10.1007/978-981-10-2525-9_31
6. Kundu, S., Sarker, G.: A super classifier with programming based boosting using biometrics for person authentication. In: International Conference ICCI—2015, held at BIT Mesra Ranchi, 10–11 Dec 2015, © Springer Science + Business Media Singapore, Advances in Intelligent Systems and Computing, vol. 509, pp. 331–341 (2015). https://doi.org/10.1007/978-981-10-2525-9_32
7. Kundu S., Sarker, G.: A person identification system with biometrics using modified RBFN based multiple classifier. In: 2016 International Conference on Control, Instrumentation, Energy & Communication (CIEC), held at Kolkata, 20–30 Jan 2016, pp. 125–129. 978-1-5090-0035-7/16/$31.00 © 2016 IEEE
8. Kundu, S., Sarker, G.: An efficient multiple classifier based on fast RBFN for biometric identification. In: Advanced Computing, Networking and Informatics Smart Innovations, Systems and Technologies, vols. 27, 1, pp. 473–482. Springer International Publishing Switzerland, Switzerland (2014). https://doi.org/10.1007/978-3-319-07353-8_55
9. Kundu, S., Sarker, G.: An Efficient Integrator Based on Template Matching Technique for Person Authentication using Different Biometrics. Indian J. Sci. Technol. **9**(42) (2016). https://doi.org/10.17485/ijst/2016/v9i42/93805, Nov 2016. ISSN (Print): 0974-6846 (SCOPUS indexed)

10. Goodfellow, I., Bengio, Y., Courville, A.: Deep Learning, vol. 1. MIT Press, Cambridge
11. Sarker, G.: Some studies on convolution neural network. Int. J. Comput. Appl. (IJCA) **182**(21), 13–22 (2018). Foundations of Computer Science, New York. https://doi.org/10.5120/ijca201891
12. Sarker, G., Ghosh, S.: A convolution neural network for optical character recognition and subsequent machine translation. Int. J. Comput. Appl. (IJCA) **182**(30), 23–27 (2018). Foundations of Computer Science, New York. https://doi.org/10.5120/ijca2018918203
13. Ciresan, D.C., Meier, U., Gambardella, L.M., Schmidhuber, H.: Convolution neural network committees for handwritten character classification. In: 2011 International Conference on Document Analysis and Recognition (ICDAR), pp 1135–1139. IEEE (2011)
14. Farabet, C., Martini, B., Akselrod, P., Talay, S., LeCun, Y., Culurciello, E.: Hardware accelerated convolution neural networks for synthetic vision systems. In: Proceedings of 2010 IEEE International Symposium on Circuits and Systems (ISCAS), pp 257–260. IEEE (2010)
15. Karpathy, A. Toderici, G., Shetty, S., Leung, T., Sukthankar, R., Fei-Fei, L.: Large scale video classification with convolution neural networks. In: 2014 IEEE Conference on Computer Vision and Pattern Recognition (CVPR), pp. 1725–1732. IEEE (2014)
16. Nebaaer, C.: Evaluation of convolution neural networks for visual recognition. IEEE Trans. Neural Netw. **9**(4), 685–696 (1998)
17. Simard, P.Y., Steinkraus, D., Platt, J,C.: Best practices for convolution neural networks applied to visual document analysis, p. 958. IEEE (2003)
18. Szarvas, M., Yoshizawa, A., Yamamoto, M., Ogata, J.: Pedestrain detection with convolution neural networks. In: Intelligent Vehicles Symposium, 2005. Proceedings, pp. 224–229. IEEE (2005)
19. Szegedy, C., Toshev, A., Erhan, D.: Deep neural networks for object detection. In: Advances in Neural Information Processing Systems, pp. 2553–2561 (2013)
20. Tivive, F.H.C., Bouzerdoum, A.: A new class of convolution neural networks (SICoNNets) and their applications of face detection. In: Proceedings of the International Joint Conference on Neural Networks, vol. 3, pp. 2157–2162. IEEE (2003)
21. Bhakta, D., Sarker, G.: A new learning based boosting in multiple classifier for color facial expression identification. In: International Conference ICCI—2015, held at BIT Mesra, Ranchi, 10–11 Dec 2015, © Springer Science + Business Media Singapore, Advances in Intelligent Systems and Computing, vol. 509, pp. 267–277 (2015). https://doi.org/10.1007/978-981-10-2525-9_26
22. Pujari, A.K.: Data Mining Techniques. Universities Press

Bed Expansion in Two-Phase Liquid–Solid Fluidized Beds with Non-Newtonian Fluids and ANN Modelling

Samit Bikas Maiti, Nirjhar Bar and Sudip Kumar Das

Abstract Bed expansion characteristics in non-Newtonian fluid–solid two-phase fluidized bed have been examined. Bed expansion measurements are carried with irregular shaped sized particles and non-Newtonian liquids (aqueous solutions of carboxymethyl cellulose). The bed voidage is found to be increasing with the increasing superficial liquid velocity. It has also been noticed that the value of this increase has been smaller as the particle size increases. Also, increase in the viscous non-Newtonian type flow behaviour has increased in bed voidage. A multilayer perceptron neural network trained with the backpropagation as well as the Levenberg-Marquardt algorithm has been attempted for the proposed analysis. Four types of common transfer functions are used with the use of a single hidden layer. The best performing ANN model has been the Levenberg-Marquardt algorithm applied with transfer function 4 having 11 processing elements within the hidden layer for the predictability related to the bed height ratio.

Keywords Fluidized bed · Bed voidage · Non-Newtonian fluid · ANN · Backpropagation · Levenberg-Marquardt

Nomenclature

AARE   Average Absolute Relative Error, $AARE = \frac{1}{N}\sum_{i=1}^{N}|\frac{(y_i - x_i)}{x_i}|$, dimensionless

S. B. Maiti (✉) · N. Bar · S. K. Das
Department of Chemical Engineering, University of Calcutta,
92, A. P. C. Road, Kolkata, West Bengal 700009, India
e-mail: samit.maiti@gmail.com

N. Bar
e-mail: nirjhar@hotmail.com

S. K. Das
e-mail: skdchemengg@caluniv.ac.in; drsudipkdas@gmail.com

N. Bar
St. James' School, 165, A. J. C. Bose Road, Kolkata, West Bengal 700014, India

© Springer Nature Singapore Pte Ltd. 2020
J. K. Mandal and S. Mukhopadhyay (eds.), *Proceedings of the Global AI Congress 2019*, Advances in Intelligent Systems and Computing 1112,
https://doi.org/10.1007/978-981-15-2188-1_3

H_f Final bed height, m

H_o Static bed height, m

MSE   Mean Squared Error, $\text{MSE} = \frac{1}{N} \sum_{i=1}^{N} (x_i - y_i)^2$, dimensionless

R   Correlation coefficient, $R = \frac{\sum_{i=1}^{N} (x_i - \bar{x})(y_i - \bar{y})}{\sqrt{\sum_{i=1}^{N} (x_i - \bar{x})^2 \sum_{i=1}^{N} (y_i - \bar{y})^2}}$, dimensionless

SD   Standard deviation, $\text{SD} = \sqrt{\sum_{i=1}^{N} \frac{1}{N-1}[|\frac{(y_i - x_i)}{x_i}| - \text{AARE}]^2}$, dimensionless

TF$_1$ Transfer function 1, $f_{1h}(x) = \tanh \beta x = \frac{e^{\beta x} - e^{-\beta x}}{e^{\beta x} + e^{-\beta x}}$, dimensionless

TF$_2$ Transfer function 2, $f_{2h}(x) = \beta x$, where $\begin{cases} \beta x = 1 & \text{for } \beta x > 1 \\ \beta x = -1 & \text{for } \beta x < -1 \end{cases}$, dimensionless

TF$_3$ Transfer function 3, $f_{3h}(x) = \beta x$, where $\begin{cases} \beta x = 0 & \text{for } \beta x < 0 \\ \beta x = 1 & \text{for } \beta x > 1 \end{cases}$, dimensionless

TF$_4$ Transfer function 4, $f_{4h}(x) = \frac{1}{1-e^{-\beta x}}$, dimensionless

χ^2 Chi Squared Test, $\chi^2 = \sum_{i=1}^{N} \frac{(x_i - y_i)^2}{y_i}$, dimensionless

1 Introduction

Fluidization has always been a process where a granular material can be converted into a dynamic fluid-like state static out of a solid-like state. This process occurs at a time in which a fluid (liquid or gas) passes through granular material. The fluidized beds are applicable in the industry owing to their large amount of contact area in between phases. This enhances chemical reactions as well as mass and heat transfer. The hydrodynamic properties like minimum fluidization velocity as well as the bed expansion related to solid particles are very important parameters concerning process simulations and also for simulations concerning the flow behaviour in a fluidized bed particularly in the fluidized bed reactor.

Two-phase fluidized bed concerning liquid–solid has been encountered frequently in number of biochemical, chemical and petrochemical processes. In the recent past, a good amount of progress can be observed while exploring the hydrodynamics related to two-phase fluidized bed. Therefore, the phenomenon of bed expansion has been a significant factor when the design of fluidized beds is to be considered. It has now become the subject of a large number of experimental as well as theoretical investigations. However, several types of fluidized beds presently operate with the non-Newtonian liquids related to polymer and food processing as well as for biotechnology. The bed expansion related to two-phase liquid–solid non-Newtonian systems has not been studied as compared to that of the two-phase liquid–solid Newtonian systems. The hydrodynamic characteristics of Newtonian fluids, which flows through fixed as well as fluidized bed, are being studied and some researchers reviewed both theoretical and empirical type of approaches and a concerning liquid fluidized bed and a large number of literature is available [6, 8, 9, 11, 13]. Different researchers [4, 7, 10] have described many equation concerning the expansion related to solid-liquid fluidized bed.

ANN models are being found to be extensively used in many different fields of engineering [1–3, 5, 12]. Artificial neural networks are extremely powerful tools for the identification of underlying extremely complex relationships in relation to the input and output of a given data. The ANN principals have been derived out of the concepts related to their biological counterparts, i.e. biological neural network. These principles have been formulated on the concept in which a very highly interconnected system related to simple processing elements (neurons or nodes) is capable of learning extremely complex interrelationships that are existing between some inputs as well as output variables for a given data set. This research paper reports the characteristics related to bed expansion for normal fluidization by the use of non-Newtonian liquid along with the test of predictability of neural network in relation to the phenomenon of bed expansion.

2 Experimental

Three different Perspex columns are used for experimental studies. The set-up consisted of a liquid storage tank, flow rate and pressure measuring devices, centrifugal pump along with other accessories (Fig. 1). Table 1 shows the dimensions of the columns. For the prevention of vibration, the entire equipment set-up has been mounted vertically and very tightly. A liquid distribution section is introduced in the lower part of the column, and 1–3 cm-sized glass beads have been used. A 16-mesh stainless steel grid is used to support and enclose the marbles. Trappings are provided in the column to measure pressure drop using either U-tube or inclined manometer, and mercury (Hg) has been used as the manometric fluid. The scale fitted in the column measures bed height. Initially, water is used to test the fluidized column. A $0.45\,\text{m}^3$ rectangular tank is used to store the liquid. The non-Newtonian liquid used is the solutions of sodium salt of carboxylmethyl cellulose (SCMC). Tap water has been circulated in the tank through a copper coil to maintain a constant temperature of the content in the tank. The test liquid has been circulated through a centrifugal pump, and a bypass valve is applied for controlling the flow rate. From the test section, the liquid has been made to return to the tank. A rotameter R (Manufactured by Transducer and Controls Private Limited, Hyderabad, India, with accuracy $\pm2\%$) is used for measuring flow rate and also for the collection of the liquid at a regular time interval at the discharge point of the column. Six different sizes of sand are used as a solid particle (Table 1). The SCMC solution is a time independent pseodoplastic fluid and follows the Power law model. The pipeline viscometer is used to measure the rheological properties of the liquids, and it is also presented in Table 1. Perme-

Table 1 Range of data sets for ANN analysis

Measurement type		Range
Column properties		
Column diameter (m)		0.047–0.0923
Column height (m)		1.3762
Physical properties of liquid		
SCMC solution concentration (kg/m^3)		0.4–0.8
Density (kg/m^3)		$1002.13 \leq \rho \leq 1003.83$
Consistency index (Ns$^{n'}$/m^2)		$0.1222 \leq K' \leq 0.7112$
Flow behaviour index (dimensionless)		$0.6015 \leq n' \leq 0.7443$
Physical properties of sand particle		
Density (kg/m^3)		2650
Particle size mesh	Sphericity (ϕ)	Particle diameter (average) $D_p \times 10^3$ (m)
$-30 + 36$	0.7965	0.538
$-25 + 30$	0.7285	0.651
$-20 + 25$	0.6900	0.774
$-16 + 20$	0.5900	1.0155
$-12 + 16$	0.5610	1.435
$-8 + 12$	0.5015	2.03
Flow rate		
Fluid velocity, U_l (m/s)		0.0019–0.062
Measuring parameter		
Height ratio (dimensionless)		$1.0735 \leq \frac{H_f}{H_o} \leq 3.125$
Total number of data for ANN analysis		436

ability test is used to measure the sphericity of sand particles. The bed expansion is measured from the scale reading fixed in the column.

3 Neural Network Modelling

The analysis is performed using backpropagation (BP) along with Levenberg-Marquardt (LM) algorithm separately. The range of the data used can be observed in Table 1. The training parameters set for BP training algorithm, i.e. learning rate (α) along with momentum coefficient (μ) is 0.7 and 1.00, respectively. Flow behaviour index, (n'), density of the liquid (ρ_l), consistency index (k'), sphericity (ϕ), particle diameter (D_p), tube diameter (D_t) and fluid velocity (U_l) are applied as variables concerning the input parameters. The height ratio $\left(\frac{H_f}{H_o} \right)$ is the output variable. A

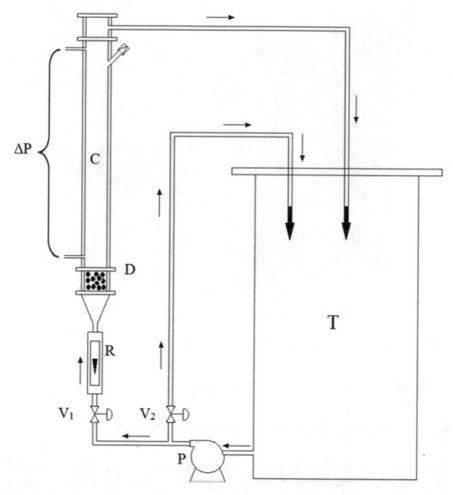

Fig. 1 Set-up of experiment concerned with the present study of fluidization [12]. **C**: main experimental column; **D**: liquid distribution section; **P**: pump pumping liquid to the column; **R**: rotameter; **M**: manometer; **T**: main liquid collecting tank; V_1: valve controlling liquid flow to column;
V_2: valve in the bypass line

total of 436 data points are generated from experiments and have subsequently been
used for the ANN. Figure 2 is the illustrative diagram representing the ANN used.

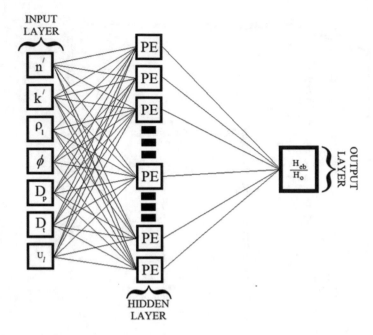

Fig. 2 Schematic diagram of the neural network

4 Results and Its Related Discussions

4.1 The Effect of Different Operating Parameters

At an extremely low fluid velocity, the buoyancy force is less, but gravitational force is more for particles, and these forces are in the opposite direction. The beds behave like a static bed. When the velocity increases, the buoyancy force overcomes the gravitational force, and then bed expands freely. The liquid velocity for which the particles begin expanding is the minimum fluidization velocity. Figure 3 has shown that at a constant column diameter and constant SCMC concentration with the increase in particle size, the minimum fluidization velocity also increases at constant bed weight. Figure 4 has shown that at a constant column diameter and constant particle size with the increase in SCMC concentration, the minimum fluidization velocity decreases at constant bed weight. Figure 5 shows that at constant SCMC concentration and particle size, there is no effect on column diameter on minimum fluidization velocity at constant bed weight. As column diameter increases, then the static bed height is found to be low, and then the expanded bed height has also been found to be low. Figures 3, 4 and 5 shows that the bed height increases with increase of liquid velocity in the fully fluidized condition.

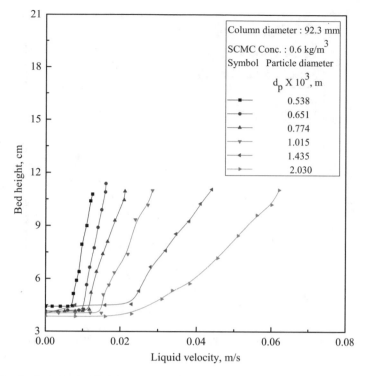

Fig. 3 Graphs depicting bed height versus liquid velocity at constant diameter and concentration

4.2 Modelling Performance

The parameter in relation to optimization is the magnitude related to the least magnitude of cross-validation (CV) MSE reached at the time of training, and it has been presented in Table 2. This optimum number related to the hidden layer nodes associated with the concerned transfer function (TF) is also presented in Table 2. This optimum number of nodes related to the hidden layer has been used for testing predictability. Figure 6 shows the change in the minimum value related to cross-validation MSE to that of the value of epochs concerning LM algorithm. For training with BP algorithm, we followed the same process.

Figure 7a represents the cross-validation curve in which the training is performed with BP algorithm. By observing the curves, the minimum MSEs are observed at a maximum allotted 32,000 epochs in case of TFs 1, 3 and 4. However, the limitation concerning the ANN application software (i.e. Neurosolution 5.07) does not permit us with the training to continued beyond 32,000 epochs. For TF 2, it can be stated that the magnitude of cross-validation MSE attains its minimum before 32,000 epochs. Figure 7b represents the cross-validation curves in which the training is performed with LM algorithm. By observing the curves, the minimum MSEs can be found

S. B. Maiti et al.

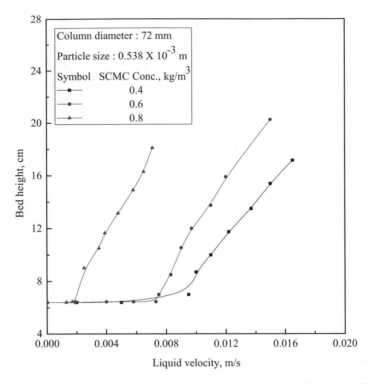

Fig. 4 Graphs depicting bed height versus liquid velocity at constant particle size and diameter

out. From Fig. 7b, it is possible to observe that for TF 1, TF 2 and TF 4, the non-improvement related to the magnitude of MSE value for the 100 consecutive epochs caused the training to be stopped abruptly. The abrupt ceasing of the training for TF 3 can easily be observed in Fig. 7b ahead of reaching the allotted 10^3 epoch as scheduled. It is also observed that the training comes to an immediate halt due to the least cross-validation MSE reaches the threshold value. For this analysis, the threshold value is 10^{-5}. Otherwise, the stopping criterion is based on the improvement capacity compared to the training up to 100 epochs. Table 3 presents the magnitudes related to the error parameters in relation to the analysis concerning the final prediction. Table 3 confirms the extremely good result. This observation also yields the closeness of the data. This closeness of the observed data demands χ^2 test to draw any concluding remarks. The χ^2 test confirms the fact that when transfer function 4 having 11 optimized number of processing elements gives the best possible prediction results when LM algorithm has been used. The near-perfect correlation can be evident by observing Fig. 8. Figure 8 is the comparative representation of experimental to that of the predicted values of height ratio.

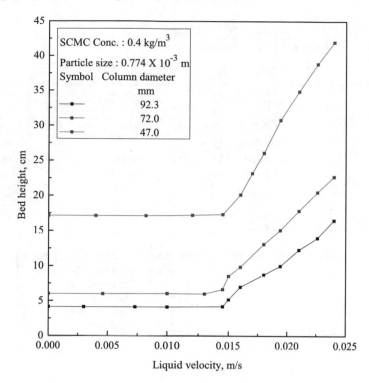

Fig. 5 Graphs depicting bed height versus liquid velocity at constant particle size and concentration

Table 2 Processing elements number within the hidden layer in relation to the 4 transfer functions

TF in hidden layer	Type of algorithm	Optimum numbers of processing elements	Minimum cross-validation error
TF 1	BP	14	2.069×10^{-3}
	LM	7	1.871×10^{-3}
TF 2	BP	10	1.862×10^{-3}
	LM	25	3.039×10^{-3}
TF 3	BP	13	7.279×10^{-3}
	LM	21	4.340×10^{-3}
TF 4	BP	14	8.807×10^{-3}
	LM	11	5.199×10^{-3}

Table 3 Final performance to predict the height ratio

Transfer function	Algorithm	Measurement type				
		AARE	SD	MSE	R	χ^2
TF 1	BP	0.035017	0.023163	0.006670	0.991074	0.145506
	LM	0.038082	0.024902	0.006916	0.990737	0.157970
TF 2	BP	0.041736	0.032973	0.010493	0.987631	0.226093
	LM	0.062142	0.048675	0.016575	0.980207	0.449669
TF 3	BP	0.070043	0.059720	0.026732	0.971665	0.680582
	LM	0.097239	0.089618	0.050733	0.933467	1.339044
TF 4	BP	0.045228	0.029035	0.008900	0.988196	0.206813
	LM	*0.032009*	*0.023292*	*0.006168*	*0.992057*	*0.132703*

Italics indicates best predicted model

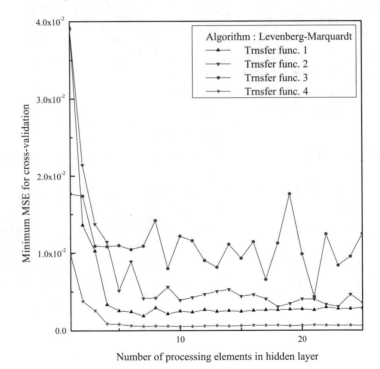

Number of processing elements in hidden layer

Fig. 6 Graph depicting change in the least CV MSE with processing element number in hidden layer

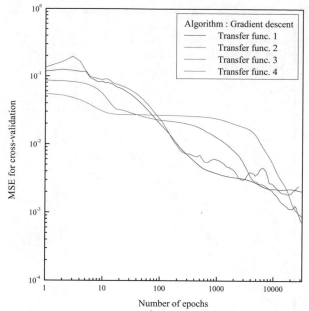

(a) BP algorithm inside hidden as well as output layer

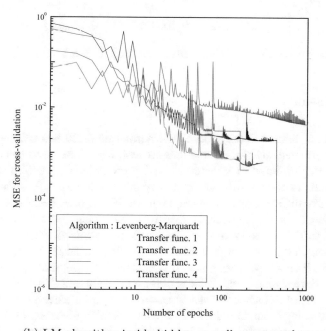

(b) LM algorithm inside hidden as well as output layer

Fig. 7 Cross-validation curves

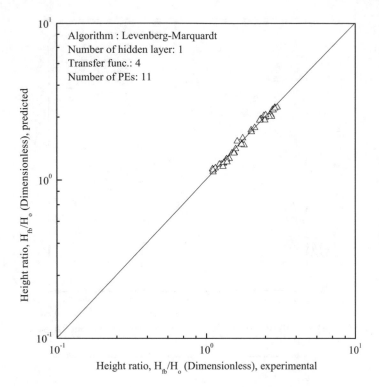

Fig. 8 Final comparison for experimental versus the network predicted

5   Conclusions

Bed expansions related to the solid–liquid normal fluidized bed are being experimentally determined, and the observations are recorded. The experiments have been conducted using different irregular sized sand particles having a different diameter, density, sphericity and three distinct column diameters of 0.092, 0.072 and 0.047 m are used. Three different diluted solutions of the SCMC have been used in the form of non-Newtonian liquids. These experiments have been performed by using the single-sized particles. Bed expansion data are then determined visually by using the scale, fitted upon the outer surface area of the column used in the experiment. An optimized MLP neural network model is developed by using the experimental data in relation to the test of the predictability of the bed height ratio. The ANN modelling proves to be successful when transfer function 4 having 11 processing elements has been applied using a single hidden layer with Levenberg-Marquardt algorithm to predict the bed height ratio. The analysis using statistical parameters indicates that the ANN prediction gives an acceptable result.

References

1. Bar, N., Das, S.K.: Modeling of gas holdup and pressure drop using ANN for gas-non-Newtonian liquid flow in vertical pipe. Adv. Mater. Res. **917**, 244–256 (2014). https://doi.org/10.4028/www.scientific.net/AMR.917.244

2. Bar, N., Biswas, A.B., Biswas, M.N., Das, S.K.: Holdup analysis for gas-non-Newtonian liquid flow through horizontal helical coils—empirical correlation versus ANN prediction. In: Paruya, S., Kar, S., Roy, S. (eds.) International Conference on Modeling, Optimization and Computing (ICMOC 2010), vol. 1298, pp. 104–109 (2010). http://aip.scitation.org/doi/abs/10.1063/1.3516284

3. Bar, N., Biswas, A.B., Das, S.K., Biswas, M.N.: Frictional pressure drop prediction using ANN for gas-non-Newtonian liquid flow through 45° bend. Artif. Intell. Syst. Mach. Learn. **3**(9), 608–613 (2011). http://www.ciitresearch.org/dl/index.php/aiml/article/view/AIML082011021

4. Couderc, J.P.: Incipient fluidization and particulate systems. In: Davidson, J.f., Clift, R., Harrison, D. (eds.) Fluidization, 2nd edn. Academic Press, New York (1985)

5. Das, B., Ganguly, U.P., Bar, N., Das, S.K.: Holdup prediction in inverse fluidization using non-Newtonian pseudoplastic liquids: empirical correlation and ANN modeling. Powder Technol. **273**, 83–90 (2015). https://doi.org/10.1016/j.powtec.2014.12.034

6. Di Felice, R.: Hydrodynamics of liquid fluidisation. Chem. Eng. Sci. **50**(8), 1213–1245 (1995). https://doi.org/10.1016/0009-2509(95)98838-6

7. Jamialahmadi, M., Müller-Steinhagen, H.: Bed voidage in annular solid–liquid fluidized beds. Chem. Eng. Process.: Process Intensif. **31**(4), 221–227 (1992). https://doi.org/10.1016/0255-2701(92)87014-8

8. Jamialahmadi, M., Müller-Steinhagen, H.: Hydrodynamics and heat transfer of liquid fluidized bed systems. Chem. Eng. Commun. **179**(1), 35–79 (2000). https://doi.org/10.1080/00986440008912188

9. Joshi, J.B.: Solid–liquid fluidized beds: some design aspects. Chem. Eng. Res. Des. **61**, 143–161 (1983)

10. Khan, A., Richardson, J.: Fluid–particle interactions and flow characteristics of fluidized beds and settling suspensions of spherical particles. Chem. Eng. Commun. **78**(1), 111–130 (1989). https://doi.org/10.1080/00986448908940189

11. Kunii, D., Levenspiel, O.: Fluidization Engineering, 2nd edn. Elsevier (1991). https://doi.org/10.1016/C2009-0-24190-0

12. Maiti, S.B., Let, S., Bar, N., Das, S.K.: Non-spherical solid-non-Newtonian liquid fluidization and ANN modelling: minimum fluidization velocity. Chem. Eng. Sci. **176**, 233–241 (2018). http://doi.org/10.1016/j.ces.2017.10.050

13. Richardson, J.F.: Incipient fluidization and particulate systems. In: Davidson, J.F., Harrison, D. (eds.) Fluidization, chap. 2, pp. 25–64. Academic Press, New York (1971). https://ci.nii.ac.jp/naid/10012827867/

M-UNet: Modified U-Net Segmentation Framework with Satellite Imagery

Ashish Soni, Radhakanta Koner and Vasanta Govind Kumar Villuri

Abstract In recent years, convolution neural network (CNN) has emerged as a dominant paradigm in machine learning for image processing application. In remote sensing, image segmentation is a very challenging task, and CNN has shown its worth over traditional image segmentation methods. U-Net structure is one of the simplified architectures used for image segmentation. However, it cannot extract promising spatial information from satellite data due to insufficient numbers of layers. In this study, a modified U-Net architecture has designed and proposed. This new architecture is deep enough to extract contextual information from satellite imagery. The study has formulated the downsampling part by introducing the DenseNet architecture, encourages the use of repetitive feature map, and reinforces the information propagation throughout the network. The long-range skip connections implemented between downsampling and upsampling. The quantitative comparison has performed using intersection over union (IOU) and overall accuracy (OA). The proposed architecture has shown 73.02% and 96.02% IOU and OA, respectively.

Keywords Skip connection · Building detection · Aerial imagery · Hybrid architecture · DenseNet architecture

1 Introduction

The remote sensing (RS) analysis has provided vast practical applications like urban planning, agriculture, and disaster management. The recent technological advancement in the field of remote sensing has significantly increased the volume of the

A. Soni (✉)
Remote Sensing and GIS Laboratory, Department of Mining Engineering, Indian Institute of Technology (ISM), Dhanbad, Dhanbad 826004, India
e-mail: ashishsoni@me.ism.ac.in

R. Koner · V. G. K. Villuri
Department of Mining Engineering, Indian Institute of Technology (ISM), Dhanbad, Dhanbad 826004, India
e-mail: rkoner@iitism.ac.in

© Springer Nature Singapore Pte Ltd. 2020
J. K. Mandal and S. Mukhopadhyay (eds.), *Proceedings of the Global AI Congress 2019*, Advances in Intelligent Systems and Computing 1112,
https://doi.org/10.1007/978-981-15-2188-1_4

dataset. Historically, RS is applied to coarse resolution multispectral imagery (e.g., LANDSAT has 30 * 30 m spatial resolution), and now moving toward the high-resolution image analysis (e.g., Quick bird with 2.2 m spatial resolution), to extract objects such as cars and buildings. For RS imagery, semantic labeling is highly applicable. However, it becomes challenging for large dataset.

The real-word object recognized by their spectral signatures like vegetation, soil, water, and many classification approaches designed based on spectral signature [1], whereas spectral signature is not sufficient to distinguish the object. Other classifiers combine the information from the neighboring pixel to increase the performance of the classifier, which is also known as spectral-spatial classification [2]. This classifier is mostly depended upon the datasets, which means that the separability of the classes relies on the spectrum of single as well as neighboring pixels.

There has been a significant improvement in the field of classification and segmentation for RS imagery. In spectral signature analysis-based classification, decision tree, artificial neural network (ANN) [3], and support vector machine (SVM) [4] approaches are applied on multispectral as well as hyperspectral imagery. Other methods like AdaBoost method combine with K-means clustering algorithms to classify the building and widely used in the various pattern recognition algorithms [5, 6].

Yue et al. [7] used spatial-spectral information to classify the hyperspectral imagery; Li and Li [8] designed the CNN architecture to classify the building. Lee et al. [9] used CNN to recognize the attribute of the large-scale dataset, and Sermanet et al. [10] used CNN for detecting house number. For high-resolution image processing, CNN has been involved in scene classification and achieved the OA of 98% [11]. Other than deep learning approach, SVM, random forest (RF) was also performed well in classifying RS imagery [12].

Recently, CNNs [13, 14] have gained attention in the field of visual recognition and achieved the state-of-the-art result in the various number of problems in the computer vision such as pixel-based classification [15, 16]. Pixel-based classification is the active topic in the field of geoscience and RS community. CNN-based pixel classifier derives the relationship between every pixel of imagery with its corresponding ground-truthing. These network architectures are designed to extract more robust and abstract feature from input data, and thus to improve the classification accuracy [17]. Notably, CNN has also applied to various remote sensing applications such as road tracking [18] and land use classification [19].

In segmentation, CNN is evolved and successfully implemented, which requires many thousand training samples. Recently, fully convolution network (FCN) achieved great success in the segmentation task [20–22]. U-Net architecture [21] is one of the effective architectures in segmentation. It consists of upsampling and downsampling path for capturing the context and enables the precise localization, respectively. This end-to-end framework is outperformed well with few training samples and uses a sliding window convolution network. The U-Net works well in extracting the features, however, to increase the performance of intra-slices feature extraction, modification like increased network depth, and hybrid connection in the architecture needed.

Firstly, network depth is increased to extract the high level of feature from the training sample data, known as DenseNet [23]. This architecture is implemented in downsampling part U-Net so that it could inherit the property of both DenseNet and U-Net connection. Integration of densely connected module improved the performance of information flow and parameter efficiency to make the dense architecture converge faster.

Secondly, hybrid connection such as long-range skip connection between upsampling and downsampling part is used. Hence, it preserves the low-level spatial feature for better intra-slice information extraction [24]. In upsampling, we used the conventional U-Net frame instead of DenseNet. It works well with relatively less number of learning parameters. And decreasing the depth in the downsampling end, the architecture will eliminate the requirement of high computational cost and GPU memory consumption.

In summary, this work is used to achieve the following things:

1. Firstly, the pixels of the dataset were adjusted to approximately normal distribution. This preprocessing avoids abnormal gradient behavior and reduces the chances of getting stuck in local optima.
2. Secondly, we designed the DenseNet architecture for downsampling part. In DenseNet, the repetitive use of features between layers results in the effective extraction of features from the satellite imagery.
3. Modified-U-Net (M-UNet) is the end-to-end system, integrated with the characteristics of DenseNet and long-range skip connection by U-Net. This architecture serves the new paradigm for exploiting the 2D context from satellite imagery, i.e., building structure.

2 CNN for Segmentation

CNN architecture is designed for visual computing and pattern recognition task. Image classification is one of the most active research topics in CNN [25]. In RS, the same problem arises where a real object like agriculture, water, road, and building should be categorized in the entire image patch. On the other hand, the pixel-level classifier categorizes every pixel of imagery to some class. In this work, the problem of segmentation has been addressed, where each pixel in the imagery is binary classified, i.e., building or not building.

2.1 U-Net Framework

This architecture works based on an elegant architecture called as FCN. It works with few training patches and yields precise segmentation result. The upsampling operation replaces the pooling operation; thus, the higher resolution of image could be achieved. In the U-net architecture, the upsampling part has a large number of feature channel that allows the network to propagate the information to higher resolution layer. As a consequence, the expansion part is more or less symmetric than the contraction part, which results in U-symmetric. The detailed structure is explained in Table 1.

U-Net is designed with two parts, namely contracting/downsampling and expansion/upsampling. The contracting part follows the conventional convolution network, which consists of the 3×3 kernel (unpadded convolution) for convolution operation. For adding nonlinearly, activation function is implemented, i.e., ReLU (Rectified Linear Unit) and max-pooling operation with kernel of 2×2 for downsampling of input data after each convolution operation. By applying downsampling at each layer, it doubles the size of feature channel (depth). In the expansion part, the feature map gets half at each stage while applying the 2×2 upsampling operation and concatenates with feature map from concerning path. In the last layer, the 1×1 kernel is used to assign feature vector to class. This network does not contain fully connected

Table 1 Architecture of U-Net and modified U-Net (M-UNet) is explained below

Input	Feature size	U-Net-Architecture	Input	Feature size	M-UNet
Downsampling Block_1	128 × 128	Conv_2D (3 × 3) / Batch_norm / Act_Funct] × 2	Downsampling_1 (Denseblock_1)	128 × 128	Batch _norm / Act_Funct / Conv_2D (3 × 3)] × 6
Pooling operation	64 × 64	Max pool (1 × 1, stride-2)	Transition block	128 × 128	Conv_2D (3 × 3)
			Pooling operation	64 × 64	Average pooling (1 × 1, stride 2)
Downsampling Block_2	64 × 64	Conv_2D (3 × 3) / Batch_norm / Act_Funct] × 2	Downsampling_2 (Denseblock_2)	64 × 64	Batch _norm / Act_Funct / Conv_2D (3 × 3)] × 12
Pooling operation	32 × 32	Max pool (1 × 1, stride-2)	Transition block	64 × 64	Conv_2D (3 × 3)
			Pooling operation	32 × 32	Average pooling (1 × 1, stride 2)
Downsampling Block_3	32 × 32	Conv_2D (3 × 3) / Batch_norm / Act_Funct] × 2	Downsampling_3 (Denseblock_3)	32 × 32	Batch _norm / Act_Funct / Conv_2D (3 × 3)] × 24
Pooling operation	16 × 16	Max pool (1 × 1, stride-2)	Transition block	32 × 32	Conv_2D (3 × 3)
			Pooling operation	16 × 16	Average pooling (1 × 1, stride 2)
Downsampling Block_4	16 × 16	Conv_2D (3 × 3) / Batch_norm / Act_Funct] × 2	Downsampling_4 (Denseblock_4)	16 × 16	Batch _norm / Act_Funct / Conv_2D (3 × 3)] × 36
Pooling operation	8 × 8	Max pool (1 × 1, stride-2)	Transition block	16 × 16	Conv_2D (3 × 3)
			Pooling operation	8 × 8	Average pooling (1 × 1, stride 2)
Downsampling Block_5	8 × 8	Conv_2D (3 × 3) / Batch_norm / Act_Funct] × 2	Downsampling_5 (Denseblock_5)	8 × 8	Batch _norm / Act_Funct / Conv_2D (3 × 3)] × 24
Pooling operation	4 × 4	Max pool (1 × 1, stride-2)	Transition block	8 × 8	Conv_2D (3 × 3)
			Pooling operation	4 × 4	Average pooling (1 × 1, stride 2)
Upsampling layer 1	8 × 8	Conv (3 × 3)	Upsampling layer 1	8 × 8	Conv (3 × 3)
Upsampling layer 2	16 × 16	Conv (3 × 3)	Upsampling layer 2	16 × 16	Conv (3 × 3)
Upsampling layer 3	32 × 32	Conv (3 × 3)	Upsampling layer 3	32 × 32	Conv (3 × 3)
Upsampling layer 4	64 × 64	Conv (3 × 3)	Upsampling layer 4	64 × 64	Conv (3 × 3)
Upsampling layer 5	128 × 128	Conv (3 × 3)	Upsampling layer 5	128 × 128	Conv (3 × 3)

The symbol "}" denotes the layer-by-layer feature transformation in the U-Net; however "]" defines the dense block architecture (shown in Fig. 3b) from DenseNet

Note "Act_func" is activation function, "Batch_norm" is batch normalization

layer, it only contains convolution layer. The network predicts the border value of the patches with the help of extrapolation by mirroring the input image. This framework takes less than a second on the latest GPUs. U-Net has some limitations; it uses layer by layer to transfer the feature data. Hence, it cannot preserve previous feature information. In this study, a modified framework is structured to resolve the drawback of U-Net framework.

2.2 Modified U-Net

The modified U-Net (shown in Fig. 1a) is designed based on U-Net structure. In the downsampling part, we used DenseNet architecture instead of layer-by-layer feature extraction. The DenseNet [26] is designed by repetitive densely connected block with different output dimension. Each dense block is directly connected to the subsequent block shown in Fig. 3b. Each layer produced k number of feature map where k is defined as the number of filter in each layer, which is known as the growth rate. The DenseNet is effective than traditional network because it produces fewer output by avoiding redundant feature. The densely connected path transfers maximum information between layers to improve the gradient flow, but the optimal solution is very time-consuming because of deep architecture.

Originally, the DenseNet architecture was designed for the image classification task, whereas DenseNet architecture is used here for segmentation process. Traditional segmentation architecture (i.e., FCN, SegNet) was designed by using the various layers of pooling operation which leads to information loss. We develop the architecture which inherits the property of DenseNet architecture as well as long-range skip connection from. U-Net architecture is known as modified U-Net (M-UNet). Skip connection was implemented between encoding and decoding part which ensure to preserve the low-level information.

Structure is explained in Fig. 1a and Table 1, respectively. The M-UNet contains the convolution layer, pooling layer, dense block, transition layer, and upsampling layer. The dense block is a connection between layers of several micro-blocks, which are directly connected to all the layers (Fig. 3b). The transition layer was used to change the size of the feature map, which contains batch normalization, convolution layer, and pooling operation (Fig. 1d). The upsampling part is same as the basic U-Net structure containing the bilinear interpolation with convolution operation.

2.3 Evolution Metric

Overall Accuracy (OA)

OA is the most intuitive performance measure, and it is simply a ratio of correctly predicted observation to the total observations. Accuracy (Eq. 1) is a great measure, but only when you have symmetric datasets; hence, Jaccard index was introduced

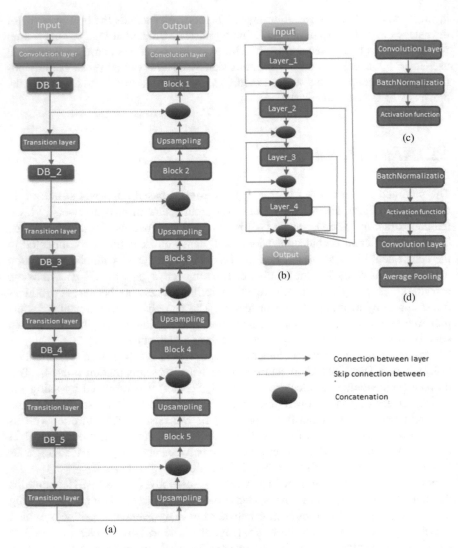

Fig. 1 Modified U-Net framework with downsampling part consists of **a** DenseNet framework and upsampling part consists U-Net structure, **b** dense block with K = 4, **c** convolution layer operation explained in sequence, **d** transition layer

$$\text{Accuracy} = \text{TP} + \text{TN}/(\text{TP} + \text{FP} + \text{TN} + \text{FN}) \qquad (1)$$

Jaccard Index

To assess the performance of the framework, we used the Jaccard index, which is also known as IOU (intersection over union). It is an object-based quantitative measure typically used for imbalanced datasets. Herein, building is defined as the object, and Jaccard index defines the number of corrected pixels predicted by network, i.e., building with respect to total number pixels classified in either of them (Eq. 2)

$$\text{Jaccard Index}/\text{IOU} = TP/(TP + FP + FN) \qquad (2)$$

where

TP Number of pixels correctly predicted as building and ground truth is also building.
TN Number of pixels not predicted as building and ground truth is also not building.
FP Number of pixels correctly predicted as the building but ground truth is not building.
FN Number of pixels not predicted as the building also ground truth is not building.

The Jaccard indices are divided into two categories: individual and overall. The individual Jaccard index represents the score of particular tile and overall index refers average score of all tiles.

2.4 Implementation Details

The whole model is implemented on the Keras package [27] with the tensor flow [28] as backend. The stochastic gradient descent optimizer was used with an initial learning rate of 0.001. The U-Net model took around 25-h using FUJITSU Workstation CELSIUS M740 which has NVIDIA Quadro K2200 GPU with the dedicated memory of 4096 MB, whereas the modified U-Net took 31-h training time under the same setting. In test phase, the total processing time was around 6 tiles/s (i.e., 128 × 128 × 3).

3 END-to-END Framework

In remote sensing image segmentation, most of the supervised training algorithms depend upon the small number of isolated pixels [29]. Training of the CNN relies on the reliability of reference dataset and on the fact of inter-separability between classes by observing the spectral signature of multichannel datasets.

3.1 Fine-Tuning

Before the commencement of training, the weight value is randomly initialized with a normal distribution where mean is zero and the standard deviation is minimum. CNN consists of a large number of weights and comparatively less amount of labeled data, so initializing the random weight may lead to local minima. So, to resolve this problem, the architecture trained with massive labeled data then uses the same architecture for the training on different datasets using pre-trained weight; this whole operation is known as *fine-tuning*.

In literature, the fine-tuning is a common procedure performed in the neural network and applied in variety of applications [11, 30, 31]. The idea behind this technique is to use the pre-trained model and train it for different datasets except for the last fully connected layer, because the final layer consists of neuron, which directly depends upon the number of classes. The idea behind fine-tuning that low-level feature (e.g., edge detection in different direction) can be re-used for the different application instead of running the framework from scratch. In the condition of small training datasets, the additional measure is considered to avoid the overfitting, commonly taken as early stopping (stop operation after few iterations), and readjust the weight at lower layer by decreasing the learning rate sometimes knows as step size.

3.2   Dataset Used

CNN is a supervised classification technique where quality and amount of training data play a sensitive role to learn the network parameters properly. A large source of free data is available like OpenStreetMap and Google map, but the availability of data varies from region to region. Some areas have limited coverage or misregistration, which can directly affect the classification. In this study, Inria dataset [32] was used for the detection of building in various regions like Chicago and Vienna. Ortho-rectified color images have been used with a spatial resolution of 0.3 m, and referenced data are semantically labeled to building and non-building class. The images contain a dissimilar urban landscaper which ranges from very densely populated areas (i.e., Vienna) to low densely populated area (i.e., Tyrol) [32]. The image tiles were used for training and testing of framework. The images used in the network are shown in Fig. 2.

3.3   Preprocessing and Experiment on the Framework

For image preprocessing, we truncated the image size as per the requirement of the network, i.e., 128 × 128 (RGB with the spatial resolution of 0.3 m) and for removing the irrelevant details. Generally, the digital images have pixel value ranged from 0 to 255 (consider 8-bit), whereas neural network weight parameter is initialized to [0, 1], which results abnormal gradient behaviors during the training, and to avoid this problem, the process of normalization of pixels is necessary [33]. The pixel value is adjusted over the curve of the normal distribution, which facilitates the progress of gradient descent [34]. As the note suggested by Krizhevsky et al. [35], the data augmentation, i.e., flipping and rotation, was applied, but it did not affect the OA.

Fig. 2 Fragment of the Inria dataset for training the network

During experiment phase, the U-Net framework is trained from scratch and further fine-tuned the weight by training on a different set of tile. The training hyperparameter is kept under the similar configuration in fine-tuning, but training process is stopped after 100 epochs. This same process has been done on the M-UNet, trained until the validation score doesn't change and loss not decreasing. For the assessment of accuracy, Jaccard index was used, which accurately defines the percentage of overlap of the predicted pixel with respect to reference pixel. The IOU of M-UNet is improved after several iterations and fine-tuning. In Table 2, first rows indicate the performance of the Conventional U-Net framework, which is trained from scratch. The significance of the pre-trained U-Net is shown where the IOU and accuracy are approximately 11% and 1%, respectively.

After replacing the conventional U-Net framework with DenseNet structure, the depth of the network is increased significantly high. The increase in the depth of network helps to learn the parameter efficiently, and the performance of the modified

Table 2 Segmentation result by proposed method and comparison with well-defined framework and description explain the preprocessing of datasets

Model	Overall IOU	IOU	OA	Description
U-Net	63.43	66.34	94.92	Scratch/without normalization
U-Net	71.40	72.10	95.75	Fine-tuned/without normalization
U-Net	71.84	72.64	95.83	Fine-tuned/with normalization
M-UNet	**73.03**	**73.52**	**96.01**	Fine-tuned/normalization

structure is significantly improved to 1.19% in IOU (shown in Table 2, bold font). As the experiment validates, the issue is not in having a large amount of high-resolution data but with proper designing of the network and carefully fine-tuning increases the accuracy. Our multilayer deep architecture works effectively in predicting the building while keeping a reasonable number of parameter. This end-to-end framework performs pixel-level classification on the images directly. This architecture can effectively scale the multiple number of classes and multiple bands.

Figure 3c–f illustrates the result of the visual fragment of test datasets. The M-UNet yields better classification map with respect to conventional framework. The result generated by M-UNet exhibits shaper angle and straighter lines. The entire result of M-UNet with normalization is depicted in Fig. 3f. The entire prediction of test dataset with the hardware configuration is taken around 6 tile/s.

4 Conclusion

CNN is standard classier in the field of image processing for its potentiality of learning complex patterns. This network is revised and redesigned to deal with the problems in pixel level.

The study proposed end-to-end architecture by analyzing the various state-of-the-art models. Despite their learning capability, the applicability of CNN is more realistic in the field of remote sensing. The DenseNet architecture works effectively as feature detector for building and by careful integration with long-skip connection with upsampling shows effective improvement in the pixel-level classification. This investigation showed that proposed model is performed better than the previous U-Net model in terms of accuracy. This is achieved by analyzing the role played by each layer and repetitive use of previous feature map. Detail analysis shows the effective applicability of fine-tuning the framework.

The proposed framework shows that CNN can be used as end-to-end architecture for large RS datasets and will produce accurate pixel-level classification. Additionally, in the future study, this network architecture can be extended in large-scale detector for other objects too.

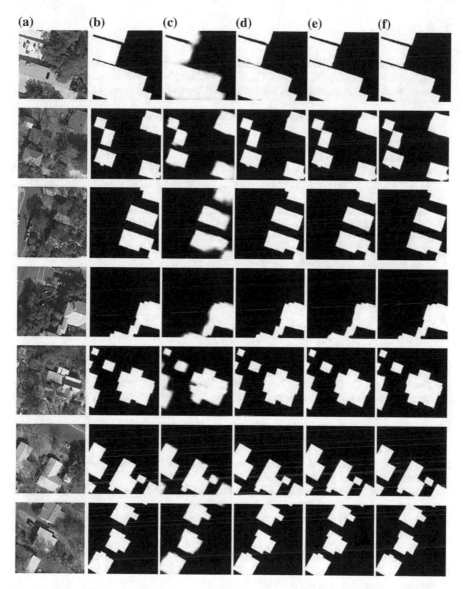

Fig. 3 **a** RGB color tile of Inria datasets, **b** ground truth data, **c** U-Net training from scratch, **d** fine-tuned U-Net framework, **e** normalized the input dataset and feed to fine-tuned U-Net framework for further training, **f** M-UNet with normalized input data

References

1. Lillesand, T.M., Kiefer, R.W.: Remote Sensing and Image Interpretation (1999)
2. Maggiori, E., Tarabalka, Y., Charpiat, G., Alliez, P.: Convolutional Neural Networks for Large-Scale Remote-Sensing Image Classification. IEEE Trans. Geosci. Remote Sens. **55**, 645–657 (2017)
3. Civco, D.L.: Artificial neural networks for land-cover classification and mapping. Int. J. Geogr. Inf. Syst. **7**, 173–186 (1993)
4. Pal, M., Mather, P.M.: Support vector machines for classification in remote sensing. Int. J. Remote Sens. **26**, 1007–1011 (2005)
5. Zheng, J., Cui, Z., Liu, A., Jia, Y.: A K-means remote sensing image classification method based on AdaBoost. In: 2008 Fourth International Conference on Natural Computation, pp. 27–32 (2008)
6. Viola, P., Jones, M.: Rapid object detection using a boosted cascade of simple features. In: Proceedings of 2001 IEEE Computer Society Conference *on* Computer *Vision and* Pattern Recognition, CVPR 2001, vol. 1, pp. I-511–I-518 (2001)
7. Yue, J., Zhao, W., Mao, S., Liu, H.: Spectral-spatial classification of hyperspectral images using deep convolutional neural networks. Remote Sens. Lett. **6**, 468–477 (2015)
8. Li, B.-Q., Li, B.: Building pattern classifiers using convolutional neural networks. In: Proceedings of International Joint Conference on Neural Networks, IJCNN'99, vol. 5 (Cat. No. 99CH36339) (1999)
9. Lee, S., Zhang, H., Crandall, D.J.: Predicting geo-informative attributes in large-scale image collections using convolutional neural networks. In: 2015 IEEE Winter Conference on Applications of Computer Vision, pp. 550–557 (2015)
10. Sermanet, P., Chintala, S., LeCun, Y.: Convolutional neural networks applied to house numbers digit classification, pp. 3288–3291 (2012)
11. Wang, J., Luo, C., Huang, H., Zhao, H., Wang, S.: Transferring pre-trained deep CNNs for remote scene classification with general features learned from linear PCA network. Remote Sens. **9** (2017)
12. Attarchi, S., Gloaguen, R.: Classifying complex mountainous forests with L-Band SAR and landsat data integration: a comparison among different machine learning methods in the Hyrcanian forest. Remote Sens. **6**, 3624–3647 (2014)
13. Fukushima, K., Miyake, S.: Neocognitron: a self-organizing neural network model for a mechanism of visual pattern recognition (1982)
14. LeCun, Y., Bottou, L., Bengio, Y., Haffner, P.: Gradient-based learning applied to document recognition. Intell. Signal Process. **86**, 306–351 (2001)
15. Lin, G., Shen, C., Hengel, A., van dan Reid, I.: Efficient piecewise training of deep structured models for semantic segmentation, pp. 3194–3203 (2015)
16. Long, J., Shelhamer, E., Darrell, T., Long, J., Darrell, T.: Fully convolutional networks for semantic segmentation, vol. 39, pp. 640–651 (2017)
17. Zhu, X.X., Tuia, D., Mou, L., Xia, G.-S., Zhang, L., Xu, F., Fraundorfer, F.: Deep learning in remote sensing: a review, pp. 1–60 (2017)
18. Wang, J., Song, J., Chen, M., Yang, Z.: Road network extraction: a neural-dynamic framework based on deep learning and a finite state machine. Int. J. Remote Sens. **36**, 3144–3169 (2015)
19. Sevo, I., Avramovi, A.: Convolutional neural network based automatic object detection on aerial images. IEEE Geosci. Remote Sens. Lett. **13**, 740–744 (2016)
20. Havaei, M., Davy, A., Warde-Farley, D., Biard, A., Courville, A., Bengio, Y., Pal, C., Jodoin, P.M., Larochelle, H.: Brain tumor segmentation with deep neural networks. Med. Image Anal. **35**, 18–31 (2017)
21. Ronneberger, O., Fischer, P., Brox, T.: U-Net: Convolutional networks for biomedical image segmentation, pp. 1–8 (2015)

22. Prasoon, A., et al.: Deep feature learning for knee cartilage segmentation using a triplanar convolution neural network. Lecture Notes Computer Science (including its subseries Lecture Notes in Artificial Intelligence (LNAI) and Lecture Notes in Bioinformatics (LNBI)), vol. 8150, pp. 599–606 (2013)
23. Mocsari, E., Stone, S.S.: Colostral IgA, IgG, and IgM-IgA fractions as fluorescent antibody for the detection of the coronavirus of transmissible gastroenteritis. Am. J. Vet. Res. **39**, 1442–1446 (1978)
24. Li, X., Chen, H., Qi, X., Dou, Q., Fu, C.-W., Heng, P.A.: H-DenseU-Net: hybrid densely connected U-Net for liver and liver tumor segmentation from CT volumes, pp. 1–10 (2017)
25. Goodfellow, I.J., Bulatov, Y., Ibarz, J., Arnoud, S., Shet, V.: Multi-digit Number Recognition from Street View Imagery using Deep Convolutional Neural Networks. CoRR. abs/1312.6, 1–13 (2013)
26. Huang, G., Liu, Z., Van Der Maaten, L., Weinberger, K.Q.: Densely connected convolutional networks. In: IEEE Conference on Computer Vision and Pattern Recognition (2017)
27. Chollet, F., et al.: Keras (2015)
28. Bonnin, R.: Building Machine Learning Projects with TensorFlow (2016)
29. Tarabalka, Y., Fauvel, M., Chanussot, J., Benediktsson, J.A.: SVM- and MRF-based method for accurate classification of hyperspectral images. IEEE Geosci. Remote Sens. Lett. **7**, 736–740 (2010)
30. Scott, G.J., England, M.R., Starms, W.A., Marcum, R.A., Davis, C.H.: Training deep convolutional neural networks for land-cover classification of high-resolution imagery. IEEE Geosci. Remote Sens. Lett. **14**, 549–553 (2017)
31. Tajbakhsh, N., Shin, J.Y., Gurudu, S.R., Hurst, R.T., Kendall, C.B., Gotway, M.B., Liang, J.: Convolutional neural networks for medical image analysis: full training or fine tuning? IEEE Trans. Med. Imaging **35**, 1299–1312 (2016)
32. Maggiori, E., Tarabalka, Y., Charpiat, G., Alliez, P.: Can semantic labeling methods generalize to any city? The Inria aerial image labeling benchmark to cite this version, pp. 3226–3229 (2017)
33. Zhong, Y., Fei, F., Liu, Y., Zhao, B., Jiao, H., Zhang, L.: SatCNN: satellite image dataset classification using agile convolutional neural networks. Remote Sens. Lett. **8**, 136–145 (2017)
34. Yann, L.: Efficient backprop. In: Neural Networks: Tricks of the Trade, vol. 53, pp. 1689–1699 (1998)
35. Krizhevsky, A., Sutskever, I., Hinton, G.E.: ImageNet classification with deep convolutional neural networks. In: Proceedings of 25th International Conference on Neural Information Processing Systems, vol. 1, pp. 1097–1105 (2012)

Study and Analysis of Various Heuristic Algorithms for Solving Travelling Salesman Problem—A Survey

Roneeta Purkayastha, Tanmay Chakraborty, Anirban Saha
and Debarka Mukhopadhyay

Abstract Combinatorial optimization problems operate on a set of problems, which have a range of feasible solutions either discrete in nature or can be minimized to discrete. The primary objective comprises finding the best solution. One of the prevalent problems of combinatorial optimization is the travelling salesman problem (TSP). In this article, we have studied four heuristic algorithms for solving TSP: (a) genetic algorithm combined with ant colony optimization, local-opt and remove sharp heuristics, (b) nearest neighbour, (c) simulated annealing and (d) tabu search. We have compared the best-known optimal value for specific TSP instances with these four algorithms.

Keywords Combinatorial optimization · Genetic algorithm · Ant colony optimization · Nearest neighbour · Meta-heuristic · Simulated annealing · Travelling salesman problem · Tabu search

1 Introduction

Among the routing problems, the travelling salesman problem (TSP) is one of the most basic problems. This problem is NP-hard in nature [9]. A well-defined weighted interconnection among cities is stated as an input for TSP. The shortest path, which visits all cities exactly one time and comes back to the starting node, has to be computed. The problem can be divided into two different types based on the features

R. Purkayastha (✉) · T. Chakraborty · A. Saha · D. Mukhopadhyay (✉)
Adamas University, Kolkata, West Bengal 700126, India
e-mail: roneeta.nitd@gmail.com

D. Mukhopadhyay
e-mail: debarka.mukhopadhyay@gmail.com

T. Chakraborty
e-mail: tanmay.chakraborty@ieee.org

A. Saha
e-mail: anirban.saha@ieee.org

© Springer Nature Singapore Pte Ltd. 2020
J. K. Mandal and S. Mukhopadhyay (eds.), *Proceedings of the Global AI Congress 2019*, Advances in Intelligent Systems and Computing 1112,
https://doi.org/10.1007/978-981-15-2188-1_5

satisfied by the distances. If the distances in the problem fulfil the condition that the distance from city i to city j is equal to that from city j to city i for all cities i and j, then it is known as symmetric. If the same condition is not satisfied, it is known as asymmetric.

The presentation of the problem can be done in the form of a mathematical model. $G = (V, B)$ is considered to be a graph, which is complete and directed, where $V = \{1, \ldots, n\}$ denotes the set of nodes and B denotes the set of arcs. A distance or cost c_{ij} is assigned to each arc $(i, j) \in B$. x_{ij} is a decision variable which is binary and is fixed as one if and only if arc (i, j) is utilized in the solution. The formulation of the problem can be done as:

$$\min \ z = \Sigma_{(i,j) \in B} c_{ij} x_{ij} \tag{1}$$

subject to

$$\Sigma_{i=1}^{N} x_{ij} = 1 \quad \forall \quad j \in \{2, \ldots, N\} \tag{2}$$

$$\Sigma_{j=1}^{N} x_{ij} = 1 \quad \forall \quad i \in \{2, \ldots, N\} \tag{3}$$

$$\Sigma_{i \in S} \Sigma_{j \in S} x \leq S - 1 \quad \forall \quad S \subset V \tag{4}$$

$$x_{ij} \in \{0, 1\} \quad \forall \quad i, j \in \{2, \ldots, N\} \tag{5}$$

The aim of the problem is to minimize the overall distance travelled in (1). As observed from (2) and (3), each node must be traversed exactly once. Equation 4 represents the elimination of sub-tours. The $|S|$ indicates the count of elements in the subset S. An overview of the TSP algorithms can be found in [7].

Two types of algorithms exist for solving this type of problem [8]:

(a) Exact algorithm and
(b) Heuristics and approximation algorithm.

Small size of problems can be solved by the first type of algorithm, i.e. exact algorithm. For exact algorithms, the execution time varies within a polynomial factor of $O(n!)$, where n denotes the number of cities. This solution is suitable for only up to 20 cities. Held–Karp algorithm solves the problem in time $O(n^2 2^n)$, which requires exponential space. The time complexity can be reduced within a polynomial factor of 2^n i.e $O(2^n)$ and polynomial space. The second type of algorithm is fit for larger problem instances. This type of algorithm strives to search an approximation of the solution.

The paper is structured as follows: first, we outline the basic travelling salesman problem (TSP); then, we outline the literature survey of various heuristic algorithms to solve TSP, and finally, we do a comparative study of the optimal value found by these algorithms for specific TSP instances.

2   Literature Survey on Methods to Solve TSP

Optimization algorithms can be divided into two groups: deterministic algorithms and stochastic algorithms [10]. Deterministic algorithms follow an identical path towards a solution and result in the identical output for the given input. This behaviour is observed in hill-climbing algorithm. The stochastic algorithms display irregularity and execute disparate path with respect to the best solution for a given input.

Stochastic algorithms are additionally classified into two types: heuristic and meta-heuristic algorithms. Heuristic algorithms compute solution by 'trial and error', that is not guaranteed to find the global optimum, but these algorithms result in good promising result. Advancement of heuristic algorithms with an integration of local search intelligence along with the concept of randomization has led to the evolution of meta-heuristic algorithms.

Innumerable efforts were accomplished in solving the travelling salesman problem ever since 1930s [5], utilizing either deterministic or non-deterministic approach. The deterministic algorithms have drawbacks over stochastic algorithms since the calculation time gets larger exponentially with increase in count of city. Hence, heuristic approach is more dependable than deterministic approach because of the exploration processes and lower computing time. It is not guaranteed that the heuristic search will identify the precise solution. It strives to search a near-optimal solution. For all these reasons, we have focused only on stochastic algorithms in our study. Some of the heuristic algorithms considered for our study are enlisted below:

(a) **Genetic algorithm combined with ant colony optimization, local-opt and remove sharp heuristics**:

Genetic algorithm is a randomized algorithm, which utilizes a crossover function, a fitness function and a mutation function. An ideal fitness is the actual length of the solution.

The ant colony optimization (ACO) algorithm is one of the most prominent optimization algorithms, influenced from the real-life ant colonies [2]. The individual operation of the ant is well-known and restricted, but when the behaviour of the ant colony is considered as a whole, then it portrays a much more organized and structured pattern of operation. This algorithm is apt for solving combinatorial optimization problems efficiently. In this system, an individual ant does not communicate with other ants directly. Instead, they communicate with the aid of a chemical substance, called pheromone, which they lay along the path they traverse from their nest to the food source location. Other ants may choose this path depending on two primary factors: pheromone intensity and heuristic weights. Ants tend to choose the path with stronger pheromone intensity values discarding other paths. The optimal solution is computed within a reasonable amount of time with similar several iterations. We utilize the ACO algorithm for computing the best path in the TSP in a reasonable amount of time. We briefly introduce the probability equations for the pheromone intensity and other parameters.

An ant in city i has to pick the next city j it traverses, from those cities which it has not already traversed at each time t. The probability of choosing a particular city j depends on the length of path from i to j and the quantity of pheromone presents on the edge between city i and j. Let τ_{ij} denote the amount of pheromone on the edge from city i to j. Let η_{ij} indicate the visibility of j from i defined as:

$$\eta_{ij} = \frac{1}{distance(i, j)} \tag{6}$$

It is observed from the following equation that as the product of τ_{ij} and η_{ij} becomes bigger, it becomes more probable that j will be picked up as the city which is the successor. The visibility and trail are now weighted by parameters β and α, and we stand at the following formula. Here, $p_{ij}(t)$ denotes the probability of picking city j from city i at time t and alw_k denotes the set of cites that are still untraversed for ant k:

$$p_{ij}(t) = \begin{cases} \dfrac{\left[\tau_{ij}(t)\right]^{\alpha} * \left[\eta_{ij}\right]^{\beta}}{\Sigma_{k \epsilon alw_k} \left[\tau_{ik}(t)\right]^{\alpha} * \left[\eta_{ik}\right]^{\beta}} & \text{if } j \epsilon \ alw_k \\ \qquad\qquad 0 & \text{otherwise} \end{cases} \tag{7}$$

It has been observed that the optimal solution may not be found in all cases from the probabilistic choice. In the initial stages of computation, a little worse than optimal solution may be constructed in some cases. Some sub-optimal arcs may be augmented by pheromone intensity. The augmentation of deposited pheromones may cause inactive behaviour which may lead to failure of the algorithm.

The level of pheromone has to be updated for each of the edges of the route. The evaporation factor ρ ascertains that pheromone is not assembled in a limitless manner. It denotes the amount of old pheromone which is moved forward to the algorithm's next iteration. Thus, we arrive at the following pheromone-level-update equation in which the amount of pheromones accumulated by each ant which utilized this current edge are summed to the old pheromone left after evaporation:

$$\tau_{ij}(\text{new}) = \rho * \tau_{ij}(\text{old}) + \Sigma_{k=1}^{m} \triangle \tau_{ij}^{k} \tag{8}$$

where m is the number of ants and $\triangle \tau_{ij}^{k}$ is the amount of pheromone deposited by ant k onto the edge from i to j at the current iteration. $\triangle \tau_{ij}^{k}$ depends on a constant Q divided by the path length traversed by ant k indicated by L^k.

$$\triangle \tau_{ij}^{k} = \begin{cases} \dfrac{Q}{L} & \text{if } k\text{th ant uses} \\ \text{edge}(i, j) & \text{in its tour (between time } t \text{ and } t + n) \\ \qquad 0 & \text{otherwise} \end{cases} \tag{9}$$

Ant colony system has a drawback on the speed for convergence for the path of shortest length. In this paper [8], the proposed nested hybrid ant colony system has been discussed in the following three main parameters:

(a) decreased count of calculations for the process of convergence,
(b) reduction in count of repetitions and
(c) remove the count of unnecessary states.

The proposed algorithm utilizes both local-opt and remove sharp heuristics. The local-opt considers all possible arrangements of cities and arranges them based on the minimum distance between the cities. The remove sharp algorithm scales down the tour cost if any city has been inserted in a bad position using a NEARLIST. It reinserts the city in the correct position to keep the cost minimum. All the ants are initialized at city r (starting city) and travel all the cities once in various routes and finally the tour ends at city r. This action causes local update of pheromone on all explored paths and a global update of pheromones on the shortest path. The initial parameter values for ACS are chosen as $\alpha = 0.2$, $\beta = 0.8$ and $\rho = 0.5$. The time complexity for the proposed algorithm is $O(n^2)$, and the space complexity results in $O(n)$. As depicted in Table 1 [8], the performance of nested hybrid ACS is compared with ACS, genetic algorithm (GA), hybrid GA (HGA), evolutionary programming (EP) and by simulated annealing (SA).

(b) **Nearest Neighbor**: This is a basic heuristic algorithm built on greedy approach. The procedure begins by picking up an arbitrary city and appends the nearest unvisited city to the last city in the traversal, and this process continues till all the cities become visited [6]. The steps of the algorithm can be enlisted as:

1. Let an arbitrary city, n, be selected and city n_0 is taken as start city.
2. The nearest unvisited city is selected.
3. The city, which is currently visited, is marked as a visited city.
4. If there exists any city which is unvisited, then move directly to step 2.
5. The execution is terminated if the entire list of cities become visited.

The advantage of the algorithm is that it computes a short tour in less time, but it may not be the optimal one. This algorithm may fail to compute shorter routes which may be present owing to its greedy nature. At certain instances, it may even fail to compute a feasible tour although there is an availability of tour traversal.

(c) **Simulated Annealing**: This is an optimization algorithm which is based on the annealing process of metal that cools and freezes into crystalline state. The algorithm was first developed by Kirkpatrick, Gelatt and Vecchi in 1983. If we want to search an optimum solution in TSP, then the solution x is represented as permutation p. The different operators such as reversion, insertion and swap are used to search the candidate solution y [11]. The SA algorithm as described in [4, 11] is based on selected nodes i and j, then the multiple operators are applied on the routes for randomization. The smallest distance extracted from this procedure is taken as the best candidate solution.

After calculating the fitness value which is based on the distance of generated tour, a comparison is done with the best solution. If the existing solution satisfies the condition of having lower fitness, then the best solution is updated with the existing solution. But if there is no improvement in the solution, then we have to follow the transition probability ρ to update the current solution.

$$\rho = e^{\frac{\Delta F}{T}} > rand \tag{10}$$

where ΔF denotes the difference between the existing and the best solution. T denotes the value of temperature which is being modified with each execution as depicted in (11):

$$T = T_0 \alpha \tag{11}$$

The temperature value is modified utilizing geometrical cooling schedule with T_0 as the previous temperature and initial temperature, α denotes the cooling constant which may be fixed in the range $\alpha = 0.7$–0.99. According to this study, this value is fixed at 0.99.

Simulated annealing attempts to divert from the local optimums by accepting worsening moves with a given probability at each iteration if no better moves are available. This characteristic of simulated annealing prevents it from getting stuck in sub-optimal regions of search space.

(d) **Tabu Search**: The search history is indicated as tabu list in this search technique with the primary aim of finding the optimal value. It is an algorithm capable of performing local search exhaustively, and these local search solutions are prevented from being recycled by the algorithm. This procedure results in an improvement of efficiency of optimum search. If there exists a rule violation by the prospective solution, then it is marked as taboo. The solution will not hold any probability of being viewed by the algorithm [1]. According to this study, a variation of tabu search with three operators: insertion, swap and reversion is executed for improvement of the travelling salesman problem solution. At first, the three operators carry out the rules and a comparison is done with nodes of other series. Subsequently, the taboo list is modified.

As stated earlier, the travelling salesman problem (TSP) can be categorized into two groups: symmetric TSP and asymmetric TSP. The tabu search for TSP does not address large problem sizes in most of the related works. The achievement of tabu search in the field of asymmetric TSP is restricted when compared to that of symmetric TSP. The dearth of specific problem instances for specific heuristics in asymmetric TSP is one of the probable reasons for the limited accomplishment of tabu search in this context.

Table 1 Comparison table of proposed nested hybrid ant colony system for standard instances of TSP

Name of problem	Ant colony system	Genetic algorithm	Hybrid genetic algorithm	Evolutionary programming	Simulated annealing	Nested hybrid ant colony system	Best-known solution from library of TSP
Oliver 30	420 (423.74)	421 (N/A)	(NA)	420 (423.74)	424 (N/A)	420 (423.74)	420 (423.74)
Att48	(NA)	(NA)	10,628	(NA)	(NA)	10,628 (10,630.76)	10,628
Berlin 52	(NA)	(NA)	7542 (7544.37)	(NA)	(NA)	7542 (7544.37)	7542 (7544.37)
Eil50	425 (427.96)	428 (N/A)	(NA)	426 (427.86)	443 (N/A)	425 (427.96)	425 (N/A)
Eil51	426	496	426 (428.87)	(NA)	(NA)	426 (429.98)	426 (429.98)
Eil75	535 (542.31)	545 (N/A)	(NA)	542 (549.18)	580 (N/A)	535 (542.31)	535 (N/A)
Eil76	560 (562.6)	(N/A)	538 (544.37)	(NA)	(NA)	538 (545.39)	538 (545.39)
Eil101	672.1	(NA)	629 (640.975)	(NA)	(NA)	629 (642.31)	629 (642.31)

Table 2 Comparison chart of optimum values of different sized travelling salesman problem instances

Number	Instance of TSP	Optimal value	Size
1	eil51	426	Small
2	berlin52	7542	Small
3	st70	675	Small
4	eil76	538	Small
5	pr76	108,159	Small
6	rat99	1211	Small
7	kroA100	21,282	Moderate
8	eil101	629	Moderate
9	ch130	6110	Moderate
10	ch150	6528	Moderate
11	rat195	2323	Big
12	d198	15,780	Big
13	a280	2579	Big
14	rd400	15,281	Big
15	pcb442	50,778	Big

3   Comparison of Optimum Values of Various TSP Instances

In this section, we have compared the optimum values of different sized TSP problem instances. Three ranges of size have been chosen for the benchmark problems: smaller (n is less than 100), moderate (n is between 100 and 190) and bigger range: (n is greater than 190). Specific TSP instances with different ranges of size and true optimum value are indicated in Table 2 [3]. The crucial parameters used in each of the discussed heuristic algorithms for solving TSP are indicated in Table 3 [3].

The graph showing the comparison of optimal values with different TSP instances is depicted in Fig. 1. The TSP instances of eil51, berlin52, st70, eil76, pr76, rat99, kroA100, eil101, ch130, ch150, rat195, d198, a280, rd400, pcb442 are depicted along the horizontal axis and their corresponding optimal values are plotted along the vertical axis. It is observed from the graph that the TSP instance of pr76, a small-sized problem, has the maximum optimal value of 108,159.

4   Conclusion

Throughout this paper, we have introduced the travelling salesman problem (TSP) and discussed its problem formulation. Two types of algorithms are available for solving this type of problem, and we outlined their merits and drawbacks. In this

Fig. 1 Graph for TSP instances versus optimal value

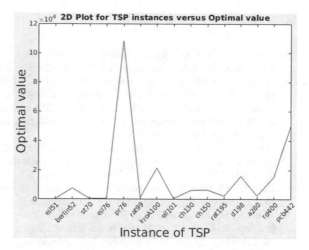

Table 3 Initialization of parameters for each algorithm

Algorithm	Parameter initialization
NN	Number of repetition = 10,000
	Traversed route = problem size
GA	Number of repetition = 10,000
	Population = 300
	Mutation = swap, slide, flip
	Selection = Roulette-Wheel
SA	Number of repetition = 10,000
	Initial temperature = 0.025
	alpha = 0.99
TS	Number of repetition = 10,000
	Action list = 30
ACO	Number of repetition = 10,000
	Number of ants = 200
	α(pheromone exponent) = 1
	β(heuristic exponent) = 1
	ρ(evaporation rate) = 0.95

paper, we focus on only heuristic approaches for solving TSP and compare their optimum values with respect to different TSP instances. For the future work, we plan to implement the ant colony optimization, one of the most prominent optimization algorithms, for solving vehicle routing problem (VRP), one of the important path estimation and optimization problems and evaluate the performance. In real-world applications, there are several additional constraints pertaining to VRP. We are aiming to focus on one of these constraints and optimize the problem solution.

References

1. Basu, S.: Tabu search implementation on traveling salesman problem and its variations: a literature survey. Am. J. Oper. Res. **2**(2), 163 (2012). https://doi.org/10.4236/ajor.2012.22019
2. Gaertner, D., Clark, K.L.: On optimal parameters for ant colony optimization algorithms. In: IC-AI (2005). http://citeseerx.ist.psu.edu/viewdoc/summary?doi=10.1.1.96.6751
3. Halim, A.H., Ismail, I.: Combinatorial optimization: comparison of heuristic algorithms in travelling salesman problem. Arch. Comput. Methods Eng. **26**(2), 367–380 (2019). https://doi.org/10.1007/s11831-017-9247-y
4. Hasegawa, M.: Verification and rectification of the physical analogy of simulated annealing for the solution of the traveling salesman problem. Phys. Rev. E **83**(3), 036708 (2011). https://doi.org/10.1103/PhysRevE.83.036708
5. Held, M., Hoffman, A.J., Johnson, E.L., Wolfe, P.: Aspects of the traveling salesman problem. IBM J. Res. Dev. **28**(4), 476–486 (1984). https://doi.org/10.1147/rd.284.0476
6. Kizilateş, G., Nuriyeva, F.: On the nearest neighbor algorithms for the traveling salesman problem. In: Advances in Computational Science, Engineering and Information Technology, pp. 111–118. Springer, Heidelberg (2013). https://doi.org/10.1007/978-3-319-00951-3_11
7. Laporte, G.: The traveling salesman problem: an overview of exact and approximate algorithms. Eur. J. Oper. Res. **59**(2), 231–247 (1992). https://doi.org/10.1016/0377-2217(92)90138-Y
8. Sahana, S.K.: Hybrid optimizer for the travelling salesman problem. Evol. Intell. **12**, 179–188 (2019). https://doi.org/10.1007/s12065-019-00208-7
9. Shweta, K., Singh, A.: An effect and analysis of parameter on ant colony optimization for solving travelling salesman problem. Int. J. Comput. Mob. Comput. **2**, 222–229 (2013)
10. Yang, X.S.: Nature-Inspired Metaheuristic Algorithms. Luniver Press, Bristol, UK (2008)
11. Zhan, S., Lin, J., Zhang, Zj, Zhong, Yw: List-based simulated annealing algorithm for traveling salesman problem. Comput. Intell. Neurosci. **2016**, 8 (2016). https://doi.org/10.1155/2016/1712630

Question Answering System-Based Chatbot for Health care

Sharob Sinha, Suraj Mandal and Anupam Mondal

Abstract Chatbot helps to provide automated as well as instant output at the absence of human intervention. It is more essential in an emerging domain like health care to manage the emergency condition without the presence of medical experts. In this research, we are motivated to develop a health-care chatbot system to recognize diseases from user-provided health conditions or symptoms. This research helps to overcome the above-mentioned challenges in partially. Primarily, these challenges are introduced due to the rapid development of information and communication technology. On the other hand, the chatbot industry is rapidly growing while promising to cut the costs. Also, less involvement of domain experts and lack of automated information extraction system introduced more difficulties in this task. Hence, we have employed an unsupervised machine learning technique to build this chatbot. Additionally, we have prepared an experimental dataset that assists in validating the output of the proposed system. Primarily, this system recognized a set of diseases from the user given a set of symptoms and vice versa.

Keywords Chatbot · Health care · Machine learning · Question answering

1 Introduction

The development in the fields of communication and information technology has made artificial intelligence (AI) systems more complicated. Nowadays, artificial

S. Sinha (✉) · A. Mondal
Department of Computer Science & Engineering, Institute of Engineering & Management, Kolkata, India
e-mail: sharobsinha@gmail.com

A. Mondal
e-mail: anupam.mondal@iemcal.com

S. Mandal
Department of Electronics and Telecommunication Engineering, Jadavpur University, Kolkata, India
e-mail: suraj.mandal738@gmail.com

© Springer Nature Singapore Pte Ltd. 2020
J. K. Mandal and S. Mukhopadhyay (eds.), *Proceedings of the Global AI Congress 2019*, Advances in Intelligent Systems and Computing 1112,
https://doi.org/10.1007/978-981-15-2188-1_6

intelligent systems are trying to imitate human actions such as taking decisions and performing daily chores. In the field of artificial intelligence, there are some hybrid and some adaptive methods available which are making the systems complex.

Years ago, the landscape of health care was dramatically different from what it looks like today. The chatbot industry is rapidly growing while promising to cost cuts, but since the AI field is not completely explored, there are some fear that this approach is filled with risks and it is safer to let bots handle only non-vital tasks.[1]

Yet, for patients living in remote areas with limited access to health-care facilities and control the emergency situations like natural disaster, these tech tools could make a notable difference in the quality of their lives. We have observed that the immediate assistance for each of the emergency situations is difficult for the medical practitioners due to the increase in population day by day. Hence, we were motivated to develop an automated chatbot in the domain of health care that provides a virtual assistant. This system is able to recognize a set of diseases for a set of user-provided symptoms and vice versa.

In order to design this system, we have addressed the following objectives:

- Preparation of a dataset that refers to a set of diseases and their corresponding symptoms.
- A disease–symptom matrix representation for building the system.
- An algorithm has been designed for recognizing similar diseases from a set of symptoms and vice versa.
- Finally, we have validated the proposed system using an agreement analysis.

The rest of the paper presents a detailed background study related to chatbot application in Sect. 2. Sections 3 and 4 describe the proposed system and the obtained result from this system in details. Finally, Sect. 5 illustrates the concluding remarks and future scopes in this research.

2 Related Work

In 1950, Alan Turing's famous article "Computing Machinery and Intelligence" was published, which proposed what is now called the Turing test as a criterion of intelligence [1].

The Turing test, developed by Alan Turing in 1950, is a test of a machine's ability to exhibit intelligent behaviour, equivalent to, or indistinguishable from, that of a human. Turing proposed that a human evaluator would judge natural language conversations between a human and a machine designed to generate human-like responses. The evaluator would be aware that one of the two partners in conversation is a machine, and all participants would be separated from one another. The conversation would be limited to a text-only channel such as a computer keyboard and screen so the result would not depend on the machine's ability to render words as speech.

---

[1]https://chatbotsmagazine.com/how-chatbots-will-shape-the-future-of-healthcare-fa8e30cebb1c.

If the evaluator cannot reliably tell the machine from the human, the machine is said to have passed the test. The test results do not depend on the ability to give correct answers to questions, only how closely one's answers resemble those a human would give.

The classic historic early chatbots are ELIZA [2] (1964–1966, Joseph Weizenbaum) and PARRY (1972, Kenneth Colby). Though the term "ChatterBot" was originally coined by Michael Mauldin (creator of the first Verbot (Verbal-Robot), Julia) in 1994 to describe these conversational programs [3], the tradition of ELIZA, the first bot that mimicked a Rogerian therapist, is proudly continued by modern-day conversational interfaces. Some help you book appointments; others remind us to take our pills or assist us in refilling our prescriptions. Here are a few examples of what medical chatbots can do:

Florence: This chatbot nurse tells us to take our medicine, gives us instructions if we forgot to take a pill, monitors our health and can help us find specialists and book appointments for us [4].
Your. MD: It replaces the assistant of a general practitioner, asks about symptoms and puts enough questions approved by health professionals to identify a condition probabilistically then sets up appointments, referring to a physician [5].
Safedrugbot: This is a messaging app which helps doctors to take notice of possible side effects of drugs during breastfeeding and helps to keep mothers safe [6].
Babylon Health: Another conversational health-care assistant with the feature of booking a doctor [7].
SimSensei: Though it is still in its experimental phase, it uses voice and face recognition to mimic a therapist [8], also interacting with the patient at deeper levels.
Infermedica: Infermedica claims that it leverages AI and machine learning to power the symptom-checker chatbot, Symptomate. It interprets patient symptoms, recommends a diagnosis and provides health information related to identified condition [9].

In a recent study, Mrs. Rashmi Dharwadkar considered the opinion for a medical chatbot which helps users to submit the problem about their health and be recommended a solution without the hassle of going to a hospital [10].

Similarly, in another paper, a medical chatbot was considered by Kyo-Joong Oh which adapted to a conversational service to understand the emotional dialogue of the person and suggests help [11].

In a paper [12] published in 2017, an idea was proposed that using artificial intelligence that when symptoms were inputted, the chatbot would suggest the nearest diseases. Based on deep-learning techniques, a chatbot aimed to investigate the effectiveness of human machines interaction for the paradigms for eHealth application.

This study was suggested by Amato [13]. A Paediatric Generic Medicine Consultant chatbot was proposed by Benilda Eleoner having a basic idea to have a conversational chatbot which is designed to prescribe generic medicines for children [14].

In an old study, Fabian Merges proposed a study for a knowledge-based medical system which can foster knowledge for transfer and network building [15].

3 Methodology

In order to design the chatbot, we have employed two well-known machine learning approaches over a simple keyword or similar word pattern matching technique. Both of the approaches have been applied to our prepared dataset. In the following subsections, we have discussed the dataset preparation and model building in details.

3.1 Dataset Preparation

A proper dataset is essential in building an automated chatbot application. The objective of the research was to develop a chatbot which can recognize diseases for a set of symptoms [16–18]. Hence, the required dataset was all existing diseases along with their symptoms. Initially, we have crawled a dataset from a website.[2] Thereafter, we have applied a pre-processing step to clean and prepare our experimental dataset that has been employed to build the proposed chatbot.

Pre-processing: While web-scrapping the disease and symptom links, it is observed that there were many duplicate links which opened the same website. Hence, using Python script we have removed all the duplicates links and considered only the unique links. Dataset also contained diseases which had less than five symptoms. We have excluded those diseases for better accuracy results. Similarly, we have excluded the diseases which did not have proper symptoms present in it. The final dataset was containing 439 diseases and 1909 symptoms.

We have observed that on an average, each disease contains around 12 symptoms in our experimental dataset. Thereafter, we converted the dataset into a sparse matrix. It stores data that contains a large number of zero-valued elements (in our case symptoms, i.e. 12 out of 1909 on an average). Additionally, it helps to save a significant amount of memory and speed up the processing of that data. Sparse matrices also have significant advantages in terms of computational efficiency. Unlike operations with full matrices, operations with sparse matrices do not perform unnecessary low-level arithmetic, such as zero-adds ($x + 0$ is always x). The resulting efficiencies can lead to dramatic improvements in execution time for programs working with large amounts of sparse data [1x]. The dimension of the sparse matrix was 439 (disease) * 1909 (symptoms), where the corresponding symptoms of the diseases were represented with "1" and the rest were filled with "0".

Visualization: In order to visualize the complete dataset in 2D graphical form, initially the row of the sparse matrix was divided into two (X, Y) equal parts. Since X

---

[2]https://www.medicinenet.com/.

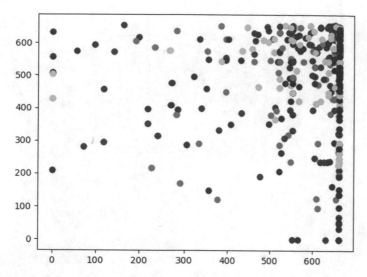

Fig. 1 A plot diagram to visualize dataset

and Y were represented in binary form, we have converted them into their equivalent decimal form. Thereafter, it has been observed that the decimal values were very large; hence, we have taken the log base 10 values. Finally, each point $(X1, Y1)$, $(X2, Y2)$ ... (Xn, Yn) was plotted as shown in Fig. 1.

3.2 Module Building

In order to build the model, we have employed an unsupervised machine learning approach to identify similarity between diseases and symptoms. So, we have used two different clustering techniques. The cluster analysis is a branch of machine learning that groups the data that has not been labelled, classified or categorized. Instead of responding to feedback, cluster analysis identifies commonalities in the data and reacts based on the presence or absence of such commonalities in each new piece of data [19].

We have extracted the number of closely relatable diseases according to the symptoms that are provided by the user. Primarily, we have used symptoms as a feature for a set of diseases and diseases as a feature for a set of symptoms without initial labelling. Both of the feature sets have been applied on K-means and Mini Batch K-means clustering techniques [20–23]. These clusters assist in identifying a group of similar diseases with their corresponding symptoms and vice versa. The overall steps of this module building are discussed in the following algorithm.

Step 1: Initially, we have generated the sparse matrix.

Step 2: Thereafter, we have employed K-means and Mini Batch K-means approach on the processed sparse matrix to assign the proper groups of diseases.

Step 3: After a detailed observation, we have set a threshold K value as 7 which was incremented up to K value 12 to extract the correct group from our dataset (D).

$$D = \langle S_1, S_2, \ldots, S_n : Di \rangle$$

Step 4: A user has provided a set of symptoms (Si) to identify common diseases. In that occasion, we have employed Euclidean distance technique to compute the similarity between Step 3 generated group of diseases along with the input symptoms.

$$Si = \langle S'_1, S'_2, \ldots, S'_n \rangle$$

$$\text{Similarity} = \sqrt{\left(S_1 - S'_1\right)^2 + \left(S_2 - S'_2\right)^2 + \cdots + \left(S_n - S'_n\right)^2}$$

Step 5: In our case, we have set a threshold value > 0.6 for similarity measurement.

Step 6: Finally, we have considered similar diseases which score more than 0.6. On the other hand, it is also applicable for predicting similar symptoms from a set of users given diseases under the proposed chatbot.

Additionally, we have provided the flowchart of the proposed module in detail (Fig. 2).

4   Result Analysis

In order to validate both of the clusters of the proposed system, we have performed a comparison-based agreement between the annotated and predicted output. The output has been produced in the form of accuracy. We have identified similar diseases as well as symptoms from user-given symptoms and diseases, individually. In this process, we have varied K values from 2 to 12 for both K-means and Mini Batch K-means. We have noticed that the range of K values 2–6 not provided an adequate output for the proposed system. So, we have presented rest of K values in the range of 7–12 for both of the techniques with their accuracy that is shown in Table 1.

We have observed that among these K values, K value 9 provides a better output over rest of the K values. It is also noticed that Mini Batch K-means clustering outperforms over K-means. Figure 3 shows the overall performance of the Mini Batch K-means clustering when K value is 9.

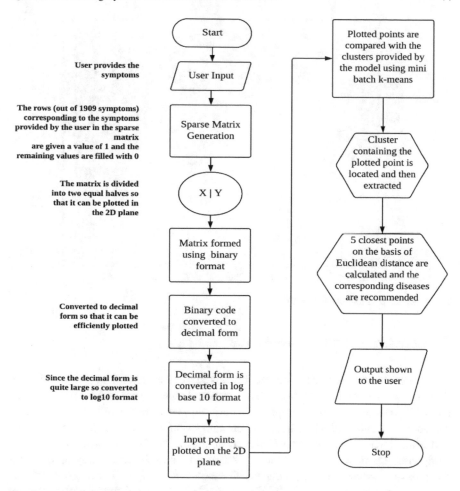

Fig. 2 Flowchart illustrating the steps of the working module

	K-values	Mini Batch K-means	K-means
Table 1 A comparative result analysis for various K-values under Mini Batch K-means and K-means	7	80	74
	8	83	78
	9	87	84
	10	81	76
	11	80	72
	12	82	75

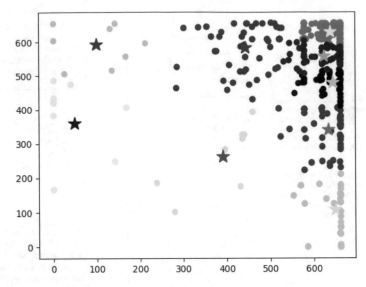

Fig. 3 Mini batch K-means ($K = 9$)

5 Conclusion and Future Scope

The research was primarily focused on developing a chatbot in the health-care domain. To prepare the chatbot, we have prepared an experimental dataset that contains 439 diseases and 1909 symptoms. Thereafter, we have converted in the form of sparse matrix and applied on two different clustering approaches, namely K-means and Mini Batch K-means. Both of the clusters provide similar diseases as well as symptoms. Thereafter, we have taken a set of inputs in the form of symptoms and diseases and processed with Euclidean distance to identify a similar group of diseases and symptoms, respectively. Finally, we have validated the proposed chatbot which depends on K value selection where we have observed K value 9 provides 87% accuracy using Mini Batch K-means clustering.

Additionally, we have observed that this chatbot is unable to predict a set of diseases with unknown symptoms due to dependency on the sparse matrix. Hence, we have planned to manage this difficulty in our future research. The proposed chatbot may assist in designing various automated applications like lexicon design, concept identification, relationship extraction, recommendation system, etc., in health care [24–26].

References

1. Turing, A.M.: Computing machinery and intelligence. In: Epstein, R., Roberts, G., Beber, G. (eds.) Parsing the Turing Test. Springer, Dordrecht (2009)
2. Weizenbaum, J.: ELIZA—a computer program for the study of natural language communication between man and machine. Commun. ACM **9**(1), 36–45 (1966)
3. Mauldin, M.L.: Chatterbots, tinymuds, and the turing test: entering the loebner prize competition. In: AAAI, vol. 94, pp. 16–21 (1994)
4. Ni, L., Lu, C., Liu, N., Liu, J.: Mandy: towards a smart primary care chatbot application. In: International Symposium on Knowledge and Systems Sciences, pp. 38–52. Springer, Singapore (2017)
5. Brindha, G.: Emerging trends of telemedicine in India. Indian J. Sci. Technol. **6**(sup 5) (2013)
6. Meskó, Bertalan, Hetényi, Gergely, Győrffy, Zsuzsanna: Will artificial intelligence solve the human resource crisis in healthcare? BMC Health Serv. Res. **18**(1), 545 (2018)
7. Al–Juboury, A.W., AL-Assadi, M.K., Ali, A.M.: Seroprevalence of Hepatitis B and C among blood donors in Babylon Governorate-Iraq. Med. J. Babylon **7**(1–2), 121–129 (2010)
8. DeVault, D., Artstein, R., Benn, G., Dey, T., Fast, E., Gainer, A., Georgila, K., Gratch, J., Hartholt, A., Lhommet, M., Lucas, G.: SimSensei Kiosk: a virtual human interviewer for healthcare decision support. In: Proceedings of the 2014 International Conference on Autonomous Agents and Multi-agent Systems, pp. 1061–1068. International Foundation for Autonomous Agents and Multiagent Systems (2014)
9. Okokpujie, K.O., Orimogunje, A., Noma-Osaghae, E., Alashiri, O.: An Intelligent Online Diagnostic System With Epidemic Alert. Int. J. Innov. Sci. Res. Technol. **2**(9) (2017)
10. Dharwadkar, R., Deshpande, N.A.: A Medical ChatBot. Int. J. Comput. Trends Technol. (IJCTT) **60**(1), 41–45 (2018). ISSN: 2231-2803. www.ijcttjournal.org. Published by Seventh Sense Research Group
11. Hsu, H.-H., Huang, N.-F.: Xiao-Shih: the educational intelligent question answering bot on Chinese-based MOOCs. In: 2018 17th IEEE International Conference on Machine Learning and Applications (ICMLA), pp. 1316–1321 (2018)
12. Wen, M.-H.: A conversational user interface for supporting individual and group decision-making in stock investment activities. In: 2018 IEEE International Conference on Applied System Invention (ICASI), pp. 216–219 (2018)
13. Amato, F., Marrone, S., Moscato, V., Piantadosi, G., Picariello, A., Sansone, C.: Chatbots Meet eHealth: automatizing healthcare. In: WAIAH@ AI* IA, pp. 40–49 (2017)
14. Comendador, B.E.V., Francisco, B.M.B., Medenilla, J.S., Nacion, S.M.T., Serac, T.B.E.: Pharmabot: a pediatric generic medicine consultant chatbot. J. Autom. Control. Eng. **3**(2), 137–140 (2015). https://doi.org/10.12720/joace.3.2.137-140
15. Merges, F., Holland, A., Schneider, S., Fathi, M.: Knowledge-based medical system integration to foster knowledge transfer and network building. In: 2011 IEEE International Conference on Systems, Man, and Cybernetics, pp. 2951–2957. IEEE (2011)
16. Mondal, A., Das, D., Cambria, E., Bandyopadhyay, S.: Wme 3.0: an enhanced and validated lexicon of medical concepts. In: Proceedings of the Ninth Global WordNet Conference (2018)
17. Mondal, A., Das, D., Cambria, E., Bandyopadhyay, S.: Wme: sense, polarity and affinity based concept resource for medical events. In: Proceedings of the Eighth Global WordNet Conference, pp. 242–246 (2016)
18. Mondal, A., Cambria, E., Feraco, A., Das, D., Bandyopadhyay, S.: Auto-categorization of medical concepts and contexts. In: 2017 IEEE Symposium Series on Computational Intelligence (SSCI), pp. 1–7. IEEE (2017)
19. Pedregosa, F., Varoquaux, G., Gramfort, A., Michel, V., Thirion, B., Grisel, O., Blondel, M., Prettenhofer, P., Weiss, R., Dubourg, V., Vanderplas, J.: Scikit-learn: machine learning in python. J. Mach. Learn. Res. **12**(Oct), 2825–2830 (2011)
20. Hartigan, J.A., Wong, M.A.: Algorithm AS 136: a k-means clustering algorithm. J. R. Stat. Society. Ser. C (Appl. Stat.) **28**(1), 100–108 (1979)

21. Sculley, D.: Web-scale k-means clustering. In: Proceedings of the 19th International Conference on World Wide Web, pp. 1177–1178. ACM (2010)
22. Jain, A.K.: Data clustering: 50 years beyond K-means. Pattern Recogn. Lett. **31**(8), 651–666 (2010)
23. Feizollah, A., Anuar, N.B., Salleh, R., Amalina, F.: Comparative study of k-means and mini batch k-means clustering algorithms in android malware detection using network traffic analysis. In: 2014 International Symposium on Biometrics and Security Technologies (ISBAST), pp. 193–197. IEEE (2014)
24. Mondal, A., Cambria, E., Das, D., Hussain, A., Bandyopadhyay, S.: Relation extraction of medical concepts using categorization and sentiment analysis. Cogn. Comput. 1–16 (2018)
25. Mondal, A., Das, D., Bandyopadhyay, S.: Relationship extraction based on category of medical concepts from lexical contexts. In: Proceedings of 14th International Conference on Natural Language Processing (ICON), pp. 212–219 (2017)
26. Mondal, A., Chaturvedi, I., Das, D., Bajpai, R., Bandyopadhyay, S.: Lexical resource for medical events: a polarity based approach. In: 2015 IEEE International Conference on Data Mining Workshop (ICDMW), pp. 1302–1309. IEEE (2015)

Improvement of Packet Delivery Fraction Due to Wormhole Attack by Modified DSR and AODV Algorithm

Sayan Majumder

Abstract Mobile ad hoc network (MANET) is the assembly of communication endpoints but without any basic facilities. These mobile nodes must be connected through a particular routing protocol. Dynamic Source Routing and ad hoc on-demand distance vector routing algorithms are mostly used Reactive routing algorithms which can connect the mobile nodes efficiently. MANET is exposed to so many attacks among which Wormhole attack is very much complicated, where an identical tunnel is created by attackers. In this paper, we have used Mean Absolute Deviation (MAD) Correlation technique on both routing protocols AODV and DSR and modified them. Then, we have computed Packet Delivery Fraction (PDF), caused by Wormhole attack, with our modified routing algorithms and make a comparative study between them. Modified DSR is proved to give better results with our algorithm. The relation between packets dropped and packets sent is also estimated using regression analysis and accuracy is calculated. MATLAB simulator is used here to do the simulation.

Keywords DSR · AODV · PDF · Wormhole · MAD Correlation

1 Introduction

Mobile ad hoc network (MANET) is decentralized wireless network which has no particular foundation. To move in or out from network, it has mobile nodes with mobility features. Communication between starting node to destination node can be occurred if multi-hop fashion is maintained by the nodes of MANET [1]. There are so many nodes act as neighbours or intermediate nodes between source node and destination node, when they are not in same range of transmission. The maintaining of connection between nodes is not so easy, as the data or packets routing protocols among nodes of wireless network structure is very much complex. We face so many problems regarding routing protocols as the topology of MANET is dynamic in

S. Majumder (✉)
Department of Computer Science & Engineering, Dream Institute of Technology, Kolkata, India
e-mail: SayanMajumder90@gmail.com

© Springer Nature Singapore Pte Ltd. 2020
J. K. Mandal and S. Mukhopadhyay (eds.), *Proceedings of the Global AI Congress 2019*, Advances in Intelligent Systems and Computing 1112,
https://doi.org/10.1007/978-981-15-2188-1_7

nature. Vulnerability of MANET, due to different attacks like Sybil, Wormhole, Black hole, Jellyfish, etc., is also established, among which Wormhole attack [2] is the dangerous one. A tunnel is created between source and destination nodes, where one or more harmful mobile nodes may be present, which actually snatch the packets and change the data in it or replicates it. Attackers can show that the time taken to pass the packets through Wormhole or fake tunnel is less than the original route but there are so many algorithms to detect and mitigate this type of attacks.

We can classify these routing algorithms into three parts, viz. Proactive, Reactive and Hybrid protocols. We have chosen DSR [3] and AODV [4] algorithms, which are Reactive routing protocols. Features of Reactive routing algorithms,

1. Routing structure is mostly flat.
2. Routing overhead is low compared to other two.
3. Route acquisition is on demand.
4. Mobility is route maintenance.
5. Storage, bandwidth and power requirement is also low.
6. Periodic updates are not needed.
7. Latency is high due to flooding.

A routing table is maintained by AODV protocol which keeps track of one entry per route, and it also omits loops by sequence numbers. DSR is an unblended on-demand routing protocol, where the path is computed whenever it is necessary. Both algorithms support multi-hop functionality.

Different machine learning algorithms and statistical analysis are already used to detect and mitigate Wormhole attack in MANET. The routing technique which uses multipath function was used by SAM [5]. Cross Correlation [6] method is another statistical approach in this field. Correlated mobility was also described well by Khin Sandar Win to detect Wormhole Attack [7]. We have already proved Absolute Deviation Covariance and Correlation in AODV routing protocol to detect and mitigate this attack.

Now, in this paper we have applied this Absolute Deviation about Mean [8] technique on DSR routing protocol also to improve the Packet Delivery Fraction (total packets dropped/total packets sent) caused by Wormhole attack and made a comparative study between modified AODV and DSR algorithms.

Wormhole attack is detected by computing Correlation Coefficient between packets sent and packets dropped using our AD Correlation method. If Correlation Coefficient is high, then we can consider it as a malicious node.

2 Methodology

AD Correlation about expectation of random variable u and v,

$$D_{\text{CORR}}(u, v) = 1/4\left[(D(u_a + v_a))^2 - (D(u_a - v_a))^2\right]$$

where

$$u_a = [u - E(u)]/D(u), \quad v_a = [v - E(v)]/D(v)$$

AD Correlation algorithm will be used in DSR and AODV routing protocol,

1. Generation of node z, set $z = 0$ to z_{max}.
2. Set $z_k = d_k$ (node number).
3. Generation of course from d_k to destination.
4. Send RREQ message that is route request message.
5. Because of analysis of link frequency, receive RREP message that is route reply.
6. If the connection is suspected, jump into step 8.
7. Else, continue the packets forwarding process.
8. Using trust data from RREP, compute $D_{CORR}(u, v)$ within packets dropped to send.
9. If the value of Correlation Coefficient is much greater than normal, jump into step 11.
10. Else, continue the packet forwarding.
11. Report intruder.

For neighbouring nodes, each node calculates the MAD Correlation Coefficient value for packets sent by the node to packets dropped. For all routes, this information is gathered. Here, X and Y are two different variables, which define packets dropped and packets sent, respectively. We have done MATLAB simulation to compute the value of MAD Correlation Coefficient.

Lastly, if we can put packets sent to packets dropped on graph paper as points, then the collection of points must follow certain type of curve. The points are linear, which is already proved from correlation, so the best fitting straight line is presented by MLS technique (method of least square). We have computed this regression [9] curve for both routing protocols [10], DSR and AODV and also found the accuracy percentage.

3 Simulation and Result

Figure 1 depicts the distribution technique.

In Fig. 2, we have established attacking situation by Wormhole, where red and blue are two malicious nodes between total of nine mobile nodes. Intruder attacks the packets passed through blue and red coloured nodes or route through a fake tunnel to reach destination.

In this figure, Wormhole attack is depicted using AODV routing protocol.

DSR is Source Routing protocol, where AODV follows multi-hop routing algorithm. The meaning of Source Routing is nothing but source node knows the total path scenario to reach the destination and this route is captured in a small memory, named as route cache. Among two phases of DSR algorithm, route discovery phase

Fig. 1 Node distribution
scenario

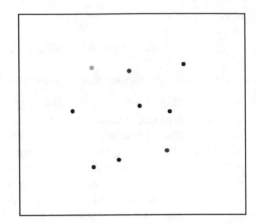

Fig. 2 Wormhole attack in
AODV

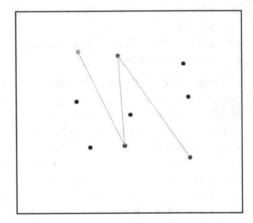

is shown by Figs. 3 and 4, where shortest distance from node 2 to node 5 is calculated
as 55.2187 and shortest distance from node 5 to node 1 is computed as 51.302. Route
maintenance is established using Fig. 5.

Fig. 3 Route discovery
(2–5) by DSR for Wormhole
attack

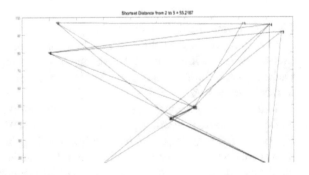

Fig. 4 Route discovery
(5–1) by DSR for Wormhole
attack

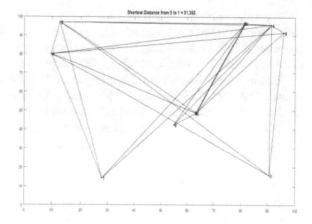

Fig. 5 Route maintenance
by DSR for Wormhole attack

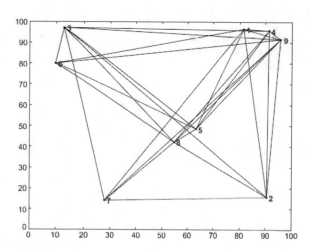

Parameters	Descriptions
Used protocols	AODV, DSR
Time taken	260, 290 s
Area to simulate	270 × 250 pixel
Nodes taken	9
Suspected nodes	2
Wormhole	1

Figure 6 depicts the Packet Delivery Fraction in normal condition on Mean Absolute Deviation DSR and Mean Absolute Deviation AODV, i.e., no attack was present between mobile nodes. Along X axis, number of nodes are shown, and along Y axis, PDF is calculated for both modified AODV and DSR.

Figure 7 establishes the packet dropped by Wormhole attack. It is obvious from picture that our modified algorithm with Mean Absolute Deviation gives better result in case of DSR than AODV, when PDF is concerned.

Here are the comparisons on the basis of PDF and Wormhole attack between two routing protocols shown below.

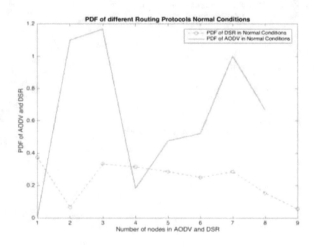

Fig. 6 PDF of modified AODV and DSR in normal condition

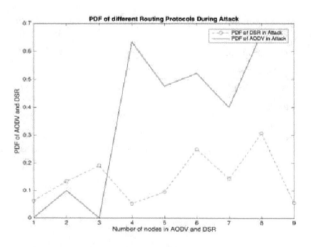

Fig. 7 PDF of modified AODV and DSR during Wormhole attack

S. No.	Mean Absolute Deviation—DSR algorithm	Mean Absolute Deviation—AODV algorithm
1.	During attack, first 3 nodes have dropped more packets in DSR	During attack, first 3 nodes have dropped less packets in AODV
2.	After third node, packet dropping is reduced in DSR	After third node, packet dropping goes very high in AODV
3.	Performs well in Wormhole attack if compared to without attack condition	Performs well also in attack, but not as much efficient to stop packet dropping as DSR
4.	Time taken is 290 s	Time taken is 270 s
5.	Value of MAD Correlation Coefficient is 0.96 or higher	Value of MAD Correlation Coefficient is 0.95 or higher

In our next Fig. 8, we have calculated the AD Correlation Coefficient for DSR algorithm, as we have already knew that our modified DSR algorithm gives better result for PDF in case of Wormhole attack. The value of coefficient is 0.96 or higher. Along X axis, number of nodes in DSR are plotted, whereas along Y axis, AD Correlation is calculated.

Now, we have proved that our modified DSR algorithm follows a mostly linear curve as the Correlation value is near about 1. This relation between packets dropped to packets sent during Wormhole attack in our modified DSR algorithm is estimated by regression analysis. We have used method of least square to find the regression curve. Along X axis, False positive rate is considered, and along Y axis, True positive rate is established (Fig. 9).

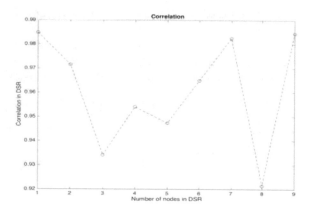

Fig. 8 AD Correlation in modified DSR

Fig. 9 Regression curve for
modified DSR

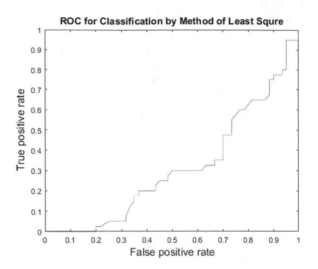

4 Conclusion

We can say that, we have done a comparative study between AODV and DSR routing algorithm, modified with Mean Absolute Deviation statistical approach. AD
Correlation really changed the classic routing protocols into a robust one.

Our main aim was to reduce the number of dropped packets due to Wormhole
attack in those two algorithms. We have successfully reduced that with our Absolute
Deviation approach, and made a comparative study. According to this study, Dynamic
Source Routing protocol works better than ad hoc on-demand distance vector routing
protocol with our Absolute Deviation modification over Karl Pearson's Correlation
Coefficient. Later, we have also proved the curve by logistic regression using MLS
technique, where True positivity is increased.

Three main advantages of modified DSR over modified AODV are:

1. Number of packets dropped due to Wormhole attack is minimal with modified
   DSR algorithm than modified AODV.
2. AD Correlation Coefficient of modified DSR is 0.96, which is higher than that
   of modified AODV.
3. In further, this modified DSR algorithm with AD Correlation can be used for
   other attacks also.

References

1. Xing, F., Wang, W.: On the survivability of wireless ad hoc networks with node misbehaviors and failures. In: IEEE Trans. Dependable Secur. Comput. **7**(3), 284–299 (2010)
2. Ji, S., Chen, T., Zhong, S.: Wormhole attack detection algorithms in wireless network coding systems. IEEE Trans. Mob. Comput. **14**(3), 660–674 (2015)
3. Beaubrun, R., Molo, B.: Performance evaluation of DSR in multi-services ad hoc networks. In: Park, J.H., Chen, H.H., Atiquzzaman, M., Lee, C., Kim, T., Yeo, S.S. (eds.) Advances in Information Security and Assurance, ISA 2009. Lecture Notes in Computer Science, vol. 5576. Springer, Berlin (2009)
4. Moudni, H., Er-rouidi, M., Mouncif, H., El Hadadi, B.: Performance analysis of AODV routing protocol in MANET under the influence of routing attacks. In: 2016 International Conference on Electrical and Information Technologies (ICEIT), Tangiers, pp. 536–542 (2016)
5. Qiana, L., Songa, N., Li, X.: Detection of wormhole attacks in multi-path routed wireless ad hoc networks: a statistical analysis approach. J. Netw. Comput. Appl. (2005)
6. Patnaik, G.K., Gore, M.M.: Trustworthy path discovery in MANET—a message oriented cross-correlation approach. In: Workshops of International Conference on Advanced Information Networking and Applications, pp. 170–177 (2011)
7. Jia, R., Yang, F., Yao, S., Tian, X., Wang, X., Zhang, W., Xu, J.: Optimal capacity–delay tradeoff in MANETs with correlation of node mobility. IEEE Trans. Veh. Technol. **66**(2), 1772–1785 (2017)
8. Majumder, S., Bhattacharyya, D.: Mitigating wormhole attack in MANET using absolute deviation statistical approach. In: 2018 IEEE 8th Annual Computing and Communication Workshop and Conference (CCWC), Las Vegas, NV, pp. 317–320 (2018)
9. Venkataraman, R., Pushpalatha, M., Rama Rao, T.: Regression-based trust model for mobile ad hoc networks. IET Inf. Secur. **6**(3), 131–140 (2012)
10. Alslaim, M.N., Alaqel, H.A., Zaghloul, S.S.: A comparative study of MANET routing protocols. In: 2014 Third International Conference on e-Technologies and Networks for Development (2014)

Air Pollution Forecasting Using Multiple Time Series Approach

K. N. Tejasvini, G. R. Amith, Akhtharunnisa and H. Shilpa

Abstract Air pollution forecasting helps to take precautionary measures in order to maintain public health. Time series algorithms and software such as Prophet package are used to forecast the air quality level in environment. A univariate data makes it easy to apply time series algorithms such as auto regressive integrated moving average (ARIMA) and Naive Bayes. The results of Naive Bayes, Prophet package and ARIMA are represented in 2-D plane just to better understand and analyze the output. Comparison of all the above said algorithms and software is done to come up with the best approach.

Keywords ARIMA · Naive Bayes · Prophet package · Means absolute percentage error (MAPE)

1 Introduction

Industries, vehicle exhausts and crop burning are some of the main cause for severe air pollution. The focus of this paper is to find the best-suited technique among ARIMA, Naive Bayes and Prophet package software in order to forecast the air pollution. Increase of population is also a major concern as far as increase in parallel pollution. Particulate matter 10 (PM10), particulate matter 2.5 (PM2.5), carbon monoxide (CO), carbon dioxide (CO_2), nitrogen oxide (NO_x) and ozone (O_3) are some of the primary pollutants.

K. N. Tejasvini (✉) · G. R. Amith · Akhtharunnisa · H. Shilpa
Department of CSE, Ramaiah Institute of Technology, Bengaluru, India
e-mail: tejasvinigowda10@gmail.com

G. R. Amith
e-mail: amithgr01@gmail.com

Akhtharunnisa
e-mail: akhtharazisa@gmail.com

H. Shilpa
e-mail: shilpahariraj@msrit.edu

© Springer Nature Singapore Pte Ltd. 2020
J. K. Mandal and S. Mukhopadhyay (eds.), *Proceedings of the Global AI Congress 2019*, Advances in Intelligent Systems and Computing 1112,
https://doi.org/10.1007/978-981-15-2188-1_8

Increase in these pollutants leads to health issues especially in children and elders. In this alarming situation, it is essential to record the pollutants concentration in atmosphere using censors. Recorded data is later used to create public awareness and better preparation in addition to thorough analysis. However, there are very little efforts as far as forecasting air pollution using historical data or past data is concerned.

Data analysis techniques where the data is preprocessed, transformed and modeled with the goal of retrieving useful knowledge out of the data are fallowed for analysis. The air pollution level measured as per the presence of CO in the atmosphere. PM2.5 is a very small dust particle of size 2.5 μm; whereas, PM10 is a dust particle of size 10 μm.

Indian standards of air quality are little less as compared to the standards set by the World Health Organization. The existing systems record the air quality of particular area then categorize air quality as different in rages; for example, if air quality index (AQI) is between 51 and 100, then air quality is moderate.

2   Literature Survey

W. Geoffrey Cobourn suggests designs of a model to forecast the PM2.5 level in atmosphere. The model he developed is based on the nonlinear regression and back trajectory concentrations techniques [1]. Corani used feed-forward neural networks, pruned neural networks and lazy learning to predict air quality in Milan city [2]. Giorgio uses other two models, (1) neural networks and (2) lazy learning (LL) approaches to compare with his primary model. Ana Russo et al. focus on forecast air quality using optimal neural network and stochastic variables [3].

Dan Wei et al. predict the air quality in Beijing city using classification and support vector machine (SVM) algorithm [4]. L. Wald observes the air quality using thermal infrared data quite uniquely. Most of the research papers prediction is based on size of particulate matters mainly PM2, PM2.5 and PM 10 level. Air quality in the urban air is computed as per prediction, interpolation and reasonable analysis. By using the prediction, valuable information is forecasted to the people [5].

S. A. Rizwan et al. magnitude and effect of air pollution in Delhi provides insight about the effect of air pollution on health in Delhi and some of the measures that government took to control the effect of pollution [6]. Forecasting and time series analysis of air pollutants in several areas of Malaysia by Ibrahim et al. In order to forecast air quality, **ARIMA** approach was employed to model the time for 1 h CO and NO$_x$ in atmosphere [7].

In tentative identification, past data is used to identify model. In time series analysis to forecast air quality indices in Thiruvananthapuram by Naveen and Anu [8], (1) air quality index (AQI) of Thiruvananthapuram city has been calculated, (2) ARIMA and SARIMA models are used to calculate AQI and (3) optimization of forecasted data. C. R., Aditya et al. work with logistic regression which is used to find whether data is polluted or not. Auto regression is used to forecast dust particle level. In their proposed system, based on atmospheric value, particulate matter 2.5 level is detected [9].

In, "Deep Air Learning: Interpolation, Prediction, and Feature Analysis of Fine-grained Air Quality" by Qi et al. [10]. A four-step procedure of ARIMA is iteratively applied on the model. Comparative analysis is carried based on the statistics. Krzysztof Siwek et al. attempt to predict the particulate matter 2.5 levels for the data sets having traffic parameter with respect to weather condition [11].

Integrated ARIMA provides uniform weights to all growths. For last observed weight, maximum weights provided and minimum weights provided to the growths at the start of the sample [11]. Data mining methods for prediction of air pollution generation and prognostic data feature are selected for the next day forecast with respect to trends; main task is to select of features [11].

3 System Design

Figure 1 represents diagram for forecasting air pollution using time series analysis. In the diagram, cylinder represents data set, which is divided into two separate set of data, i.e., training data for training and test data is for testing the model. After separating data into training and testing data, training data passed to another block where Firstly, data pre-processing takes place. In the preprocess stage, data is prepared, i.e., null values are removed; however, outlier not be deleted as it may have negative

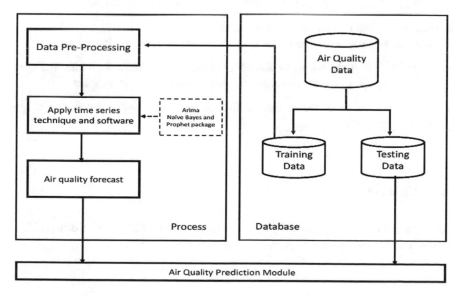

Fig. 1 System design

impact on the final result. In the next step, time series techniques are applied results of which is forecasted, and in addition to time series approaches, a software Prophet package is applied on the data set. Dotted box shows the algorithms and software which are used to train the model. Finally, model will be tested using test data.

4 Implementation Details

4.1 Problem Statement

In real time, data is recorded in the areas which are mostly organized, and these are made public by the pollution control agencies. Based on the human knowledge and experiences come to conclusion whether the level of the pollution has a severe impact on the people. We are using Naive Bayes, ARIMA and Prophet package to predict the level of the air pollution.

4.2 Proposed Method

Our proposed model is based on time series techniques (1) Naive Bayes, (2) ARIMA and (3) Prophet package software using mean absolute percent error (MAPE) and RMSE to find the absolute error in terms of percentage in prediction.
Reason for choosing time series rather than regression technique is.

- Time series data contains data with respect to time.
- There are two types of time series approach based on kind of data it uses.

  - Multivariate method used for the data which contains more than one attribute.
  - Univariate data used for the data set which contains single attribute.

- Univariate time series techniques in this paper as our data set contains single attribute that is CO.

4.3 Data Set

Data set is downloaded from Open Government Data (OGD) platform [12]. It contains single parameter CO [12].

4.4 Implementation

Data Preprocessing. Raw data has to be preprocessed before applying algorithms. In this approach,

- Identifying the null values, outliers and handling them properly. Interquartile range (IQR) method is used to identify the outliers.

$$IQR = Q_3 - Q_1$$

Output:

$$Q_1: \text{quantile}(\text{pair\$CO}, 0.25)$$
$$254.690416$$

$$Q_3: \text{quantile}(\text{pair\$CO}, 0.75)$$
$$75\%$$
$$14.49138$$

$$IQR(\text{pair\$CO})$$
$$[1]\,9.800961$$

Below arrange values

$$- 10 \text{ to } 23 \text{ are the outlier values.}$$

- Log applied to increase value of attributes. This leads to good trends.
- Naive Bayes assigns the last value. Generally, next point and the last observed point are equal.

$$dd = tt\$NO_x$$
$$tv\$naive = dd\big[\text{length}(dd) - 1\big]$$

- Finding the RMSE value to find error in prediction,

$$RMSE = \text{sqrt}\sum i = 1N1N(p - a)2$$

- Prophet is a software for forecasting time series data. It is based on an additive model where nonlinear trends are fit with daily, weekly and year wise seasonality.
- The ARIMA forecasts for a stationary time series in linear order.

5   Result

Time series data is collected with respect to time. CO depends on time. This is used for air quality forecasting. For example, in market business time series is needed for analyzing the next year sale, stack prediction, etc.

Figure 2 shows the May–October month having very high level of air pollution. Due to burning of crackers, in addition to vehicular traffic pollution and industries pollution, generally pollution increases during this time period.

Figure 3 represents the future air pollution forecasting. Blue line shows the prediction of future year and previous year also. This plot represents daily, weekly, monthly and yearly in graphical format. It shows the forecast errors with respect to density (Fig. 4).

Figures 5 and 6 represent the ARIMA result. Errors in forecasting data are shown in normalized form. Figure 6 represents error rate in year wise graph from year 2016 to 2019. In the graph, we can increase it.

From Fig. 6, we assume that the next expected point is equal to the last observed point. Yellow color shows the Naive of all the predictions are equal to the last observed point. We can calculate how accurate our predictions are using RMSE. RMSE is the standard deviation of the residuals. Residuals are a measure of how far from the regression line data points are.

Figure 7 predicts the air pollution future level using Prophet package. According to MAPE, this package gives the low error rate compared to other used algorithms.

Comparison
See Table 1.

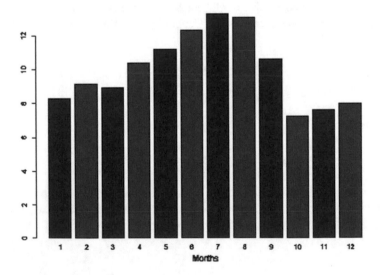

Fig. 2 Air pollution CO trend every month

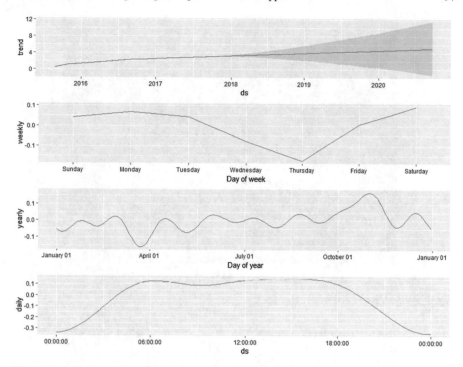

Fig. 3 Prediction weekly and day wise prediction

Fig. 4 Histogram representing normalized form

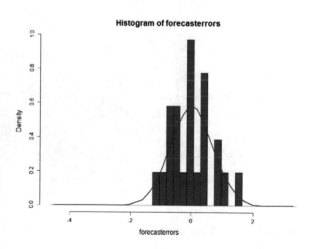

6 Conclusion

Among the time series approaches such as ARIMA, Naive Bayes and software Prophet package, Prophet package with strong seasonal effects and several seasons of historical data works better than ARIMA and Naive Bayes. Prophet is robust. It

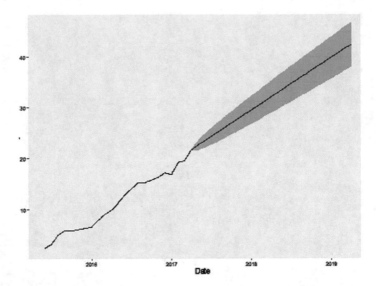

Fig. 5 ARIMA prediction trend every year

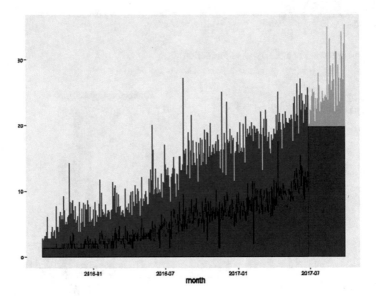

Fig. 6 Naive result and error rate in data set

also handles missing data, shifts in the trend and handles outliers well compared to ARIMA and Naive Bayes. Future Enhancement—If any sudden changes in weather happens the existing model fails forecast air quality accurately. As a future enhancement, we hope to design a module that can even on varying weather condition gives

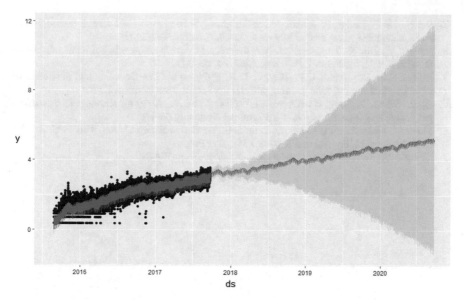

Fig. 7 Prediction using prophet graphical representation between the years 2016 and 2020

Table 1 Comparison table

Characteristics	Prophet package	ARIMA	Naive Bayes
Efficiency	More	Medium	Less than ARIMA
Performance	High	Medium	Less
Time taken for execution (min)	2.782859	3.633825	17.90942
MAPE	0.7272662	2.215956	2.604805

accurate result. We hope to create an API, and it helps us to apply the all the algorithms at a time on the input data.

References

1. Cobourn, G.: An enhanced PM2.5 air quality forecast model based on nonlinear regression and back-trajectory concentrations. Atmos. Environ. (2010)
2. Corani, G.: Air quality prediction in Milan: feed-forward neural networks, pruned neural networks and lazy learning. Ecol. Model. (2005)
3. Russo, A., Raischel, F., Lind, P.G.: Air quality prediction using optimal neural networks with stochastic variables. Atmos. Environ. (2013)
4. Wei, D.: Predicting air pollution level in a specific city (2014)
5. Wald, L.: Observing air quality over the city of Nantes by means of Landsat thermal infrared data. Int. J. Remote Sens. **20**(5), 947–959 (1999)
6. Rizwan, S.A., Nongkynrih, B., Gupta, S.K.: Air pollution in Delhi: its magnitude and effects on health. IJCM (2013)

7. Ibrahim, M.Z., Zailan, R., Ismail, M., Safiih Lola, M.: Forecasting and time series analysis of air pollutants in several area of Malaysia. Am. J. Environ. Sci. (2009)

8. Naveen, V., Anu, N.: Time series analysis to forecast air quality indices in Thiruvananthapuram District, Kerala, India. Int. J. Eng. Res. Appl. **07**, 66–84 (2017)

9. Aditya, C.R., Deshmukh, C.R., Nayana, D.K, Vidyavastu, G.P.: Detection and prediction of air pollution using machine learning models. Int. J. Eng. Trends Technol. (2018)

10. Qi, Z., Wang, T., Song, G., Hu, W., Xi, (Mark) Zhang, L.: Deep air learning: interpolation, prediction, and feature analysis of fine-grained air quality (2017)

11. Siwek, K., Osowski, S.: Data mining methods for prediction of air pollution. Warsaw University of Technology, pl. Politechniki 1, 00-661 Warsaw, Poland

12. https://data.gov.in/catalog/historical-daily-ambient-air-quality-data

Deep Learning-Based Smart Attendance Monitoring System

Rohit Halder, Rajdeep Chatterjee, Debarshi Kumar Sanyal
and Pradeep Kumar Mallick

Abstract In this paper, we have addressed an approach for accurate smart attendance monitoring system designed on the basis of deep learning algorithm. This approach will spectate the entry and exit of people into an institute or university. When a person approaches a surveillance camera near the entrance, automatically his/her face will be recognized and the entry time will be stored. Similarly while exiting, their faces will be recognized in another deep learning model embedded surveillance camera and the exit time will be stored. Our approach will help the institute to provide attendance even for attending a lecture for a percentage of time. The smart attendance monitoring system provides an advantage over the traditional approach of signing or giving bio-metric and is more reliable. This is based on real-time approach and consumes no additional time of the institute, university, or its students and faculties.

Keywords Attendance · Deep learning · Convolution neural network (CNN) · Image processing · Smart classroom

1 Introduction

In today's life, our world is mostly digitized, our genre shifted from human laboring background to machine-operated systems. We have adopted the technique of smart approach in many fields like banking, health, and tourism. From the late 1990s, data scientists have worked on and developed algorithms of computer vision [1]. However, we have felt an absence of a smart academic system. For keeping a track of attendance, even a bio-metric system [2] permits the authority to record the entry and exit time only. But sometimes it does not measure the actual time they are present, if intermittent entry and exits are not allowed. In this paper, we have come

R. Halder (✉) · R. Chatterjee · D. K. Sanyal · P. K. Mallick
School of Computer Engineering, KIIT Deemed to be University, Bhubaneswar 751024, India
e-mail: rhaldar9@gmail.com

P. K. Mallick
e-mail: pradeepmallick@gmail.com

© Springer Nature Singapore Pte Ltd. 2020
J. K. Mandal and S. Mukhopadhyay (eds.), *Proceedings of the Global AI Congress 2019*, Advances in Intelligent Systems and Computing 1112,
https://doi.org/10.1007/978-981-15-2188-1_9

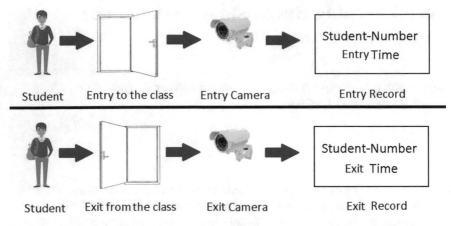

Fig. 1 Pictorial representation of our proposed model

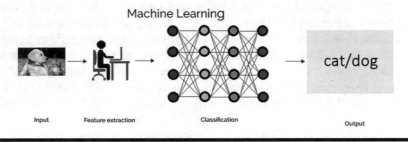

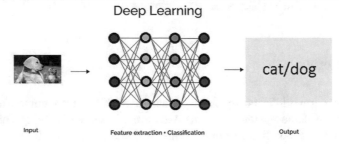

Fig. 2 Machine learning versus deep learning

forward with a prototype for smart attendance system. In this prototype, the student gets his/her attendance by automated recognition of his/her face in the surveillance cameras as shown in Fig. 1.

Our prototype is a deep learning (DL) based approach, because it is really difficult for traditional machine learning (ML) algorithms to work on multi-dimensional tensors. Moreover in case of ML, we need to extract the features and choose the right classifier based on the handcrafted features, shown in Fig. 2. But in DL, we simply train the model end-to-end with the images and their labels. The suitable features are automatically learned by the model. DL forms a bridge step between ML algorithms and artificial intelligence (AI) [3].

A deep neural network [4] includes input layers, hidden layers, and output layers. Two or more layers can be merged, added, subtracted, multiplied, or divided. An artificial neural network [5] mimics a neural network (neurons) [6] in human brain. Common examples of neural networks include convolutional neural network (CNN), recurrent neural network [7], and simple dense neural networks. CNNs are the most widely used ones while dealing with image objects. In our implementation, we have used CNN to obtain the desired result.

This section is followed by theoretical background section, where the theoretical background from the basics of image processing to training deep learning models has been explained. The section is further followed by Experimental Setup which describes the configuration of our machine, datasets, its preprocessing, and the model itself. Then we have a section where we have tried to visually represent what the paper is all about, the results of our research, to make it easily understandable to all who are new to this field. Finally, we have the conclusion section to provide the gist of our paper and related research works also help us develop the smarter academic system.

2 Background on Deep Learning-Based Image Recognition

An image is nothing but a two-dimensional array. It has many formats—gray scale, Blue-Red-Green (BRG), Hue-Saturation-Value (HSV) [8], etc. For a gray-scale image, it is pretty simple, and it is only a matrix with each cell having values from 0 to 255, with 0 as black and 255 as white rest all values depict a particular shade of gray. For images in BRG or HSV formats, we have a little complex structure. It is the same matrix with cells, but here, each cell contains three values. For BRG, it is the values for Blue component, Red Component, and Green component each ranging from 0 to

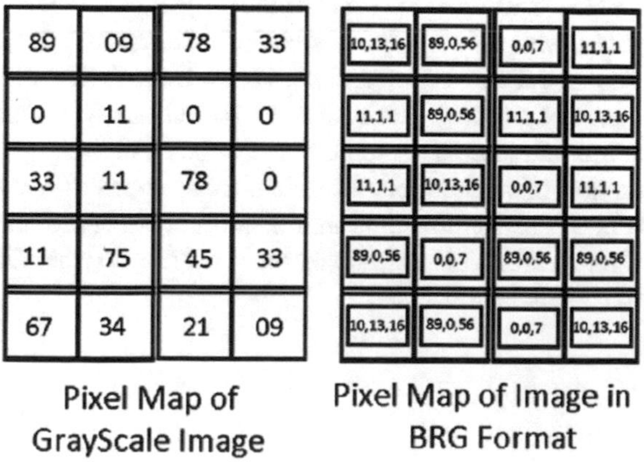

Fig. 3 Gray-scale versus BRG formats

Pixel Map of GrayScale Image

Pixel Map of Image in BRG Format

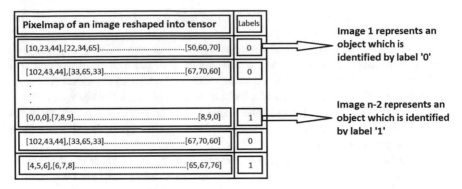

The images must be pre-processed to such a matrix before feeding it to any artificial neural network

Fig. 4 Glimpse of how the preprocessed data looks like

255 that is a 3 dimensional matrix as referred in Fig. 3. Similarly, HSV is the hue, saturation, and value. In our model, we have used the BRG format of images. So with respect to python programming language, BRG format is like having a list of three values in each of the cells of a 2D array.

Before training a model especially with images, we need to process the dataset. By processing, we mean, converting each image into a numpy array. The values of each pixel for a row are stored in a list. So for n number of rows, we get n numbers of separate lists, now combining the n number of rows in a single list. So by now, we have got a list of lists. Thus for each image, we stored the data in a row. For m number of images, we get a matrix, whose each row represents an image. Now we need to append each row with a label value to represent the class of object as shown if Fig. 4. It is pretty easy, its like a list containing the image and its label. This is not that complex, as numpy package has provided us with inbuilt functions to do so. The label values are used to separate two classes. The label values may either be integer numbers or we can even apply one-hot encoding. By now, the dataset is completely ready to be trained by the model.

Now coming to the brain of the project, it is the model. We have used a sequential model which is already available in the Keras modules, using tensor-flow background. The model contains neural networks. Neural networks are like layers which may be dense or shallow. Each layer of neural network contains nodes which calculates some value based on some characteristics or weights. The values then pass through an activation function. There are many types of activation functions, but the most relied ones for convolutional neural networks are *relu* for hidden layers and either *sigmoid* (for binary classification) or *softmax* (when there are more than two classes) for output layer [9, 10].

2.1 Convolutional Layer

In this layer, we pass a kernel, i.e., an $n \times n$ matrix over the entire image pixels. The kernels have a value in each of its cell which when processed with the original image helps to produce certain characteristics, which help us to identify the images of the same object while predicting [11], referred in Fig. 7. The basic convolution operation is shown in Fig. 5.

2.2 Max Pooling-2D Layer

This layer is used to extract the features highlighted by the convolutional layer [12] as referred in Fig. 6. There are min pooling and average pooling also. But max pool often gives the best result.

Convolution Operation in a CNN

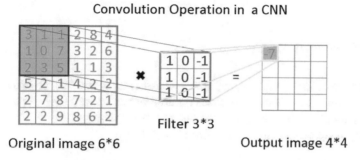

Fig. 5 Basic convolution operation

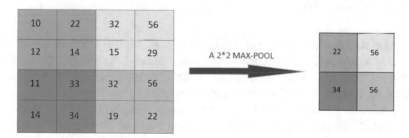

Fig. 6 Basic max-pool operation

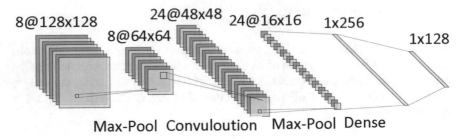

Fig. 7 A general CNN model for image processing

2.3 Dense Layers

Fully connected layers are hidden layers, which provide detailed analysis of the characteristic values. Dense layers are mainly used for linearity-based operation on a layer's input vector [13].

2.4 Number of Nodes

One of the most important hyperparameters that define the architecture of the artificial neural network is the number of nodes (neuron or perceptron) [14]. A perceptron is basically used for computing the weights of various features fed to a layer. These perceptrons are batched together to form layers [15].

2.5 Activation Function

Activation function is itself a vast field of study. Activation function allows us to determine a threshold and to activate a particular node(s). And hence, the final prediction will be better for an advanced activation function. Activation functions may be sigmoid, rectified linear [16], softmax depending on the dataset [17].

There are no hard and fast rules for the setup of these layers. In our project, we have obtained a stable setup of layers by hit and trial and then varied the essential hyperparameters like number of convolutional layers [10], number of dense layers, and number of nodes over a range. Then we studied the plotted graphs for each network in the tensor-board to get the best accuracy and minimum loss. Again both the factors must be taken into consideration for the optimized results. A CNN layer adds functionality over other dense layers because of its mechanism. A CNN does the handcrafted feature extraction automatically. In a convolutional neural network, a kernel (filter) is passed over the pixels. In each layer, we get a dot product of the input pixel and the kernel and accordingly pass the result into the other layer. On passing through more than one convolutional layers, we get a clear feature map, in our case a distinct eye, nose, eyebrow, lips, etc. Hence, we are able to extract features from the image, which will later used to predict the classes. However, if there are too many layers, there might be a gradient explosion or a vanishing gradient problem [18]. The efficiency of feature extraction increases with the number of nodes and then saturates or might even decrease at the cost of a greater training time.

3 Experimental Setup

3.1 System Configuration

Our system configuration includes Intel Core i5 7th generation processor, 8 GB RAM, dedicated graphics, Windows 10, Anaconda environment preinstalled. Both Jupyter notebook and sublime text editors were used in the process. A python version of 3.6 and 1.8 version of tensor-flow [19] was installed, including other modules like numpy, OpenCV, and Keras.

3.2 Our Dataset

It includes 500 images for each of 50 person, that is, a total of $500 \times 50 = 25,000$ images. The images for the dataset were clicked in real time—left faced, right faced, as well as center faced with their concern, so as to train our model with facial expressions as well. The images were collected in a time span of 1–2 months to record even the slightest changes in their beards, mustaches, and hair styles. The images were originally of 2304×4096 pixels; however, they were later reshaped to 50×50 and fed to the model. The images were saved in the format of "person-id_label.jpg".

3.3 Preprocessing of the Dataset

The images were reshaped into 50×50 pixels, and the labels were extracted accordingly from the file name and fed to the CNN model in the form of the four-dimensional tensor, that is, a matrix with two columns: The first column had the information of the image in BRG format and the second column had the values of the respective labels associated with the different classes of object Fig. 4. A model generally looks like as shown in Fig. 7.

3.4 Our Model

Our initial deep learning model consisted of two convolutional layers of 64 nodes with a window of 3×3, followed by a max pooling of window 2×2. The output from the convolutional layer is flattened. Two fully connected dense layers of 64 nodes have been added to gain the additional features before the output layer. A dropout of 0.25 has been added before the final layer to avoid vanishing gradient problem. The activation function softmax [20] has been used for final classification. The loss function used is sparse categorical crossentropy. The optimizer used in this approach is Adam. Experiments have been done with variation in these hyperparameter values (as shown in Fig. 1) to get the model with the best performance.

Model-11 has been represented in Fig. 8.

3.5 Motion Triggered System

Only when the motion is detected by the surveillance cameras, our smart attendance monitoring system is activated and takes real-time attendance. Video is nothing but a sequence of frames. By motion triggered system, we mean whenever there is an appreciable amount of change in pixel intensity between the ith frame and $(i + 1)$th frame, the model activates itself [21]. The change in pixel intensity is of two types.

- local change: surveillance cameras are fixed.
- global change: surveillance cameras are not fixed. And the change in the pixel intensities is due to two major reasons.

    - Vibrating tree leaves, movement of water (Static variation)
    - Due to actual movement of considerable objects (Dynamic variation).

There are many ways for motion detection, and one of the most common technique is background subtraction [22]. In this technique, we generally record the background at different times of a day at different intensity conditions. Then compare the real time frame with the pre-recorded frame corresponding to the similar time and intensity conditions as shown in Fig. 9.

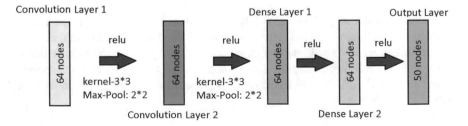

Fig. 8 A schematic representation of Model-11

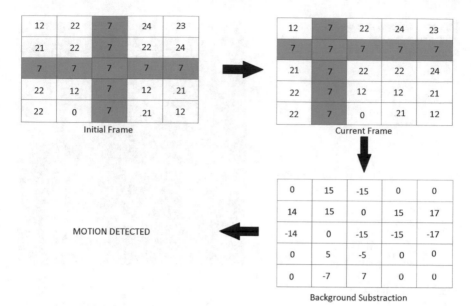

Fig. 9 Background subtraction

4 Result and Analysis

The research was carried out on different models distinguished by the choice of the hyperparameters which include the number of convolutional layers, the number of dense layers, and the number of nodes shown in Figs. 11, 12, 13, 14, 15, and 16, respectively. Based on validation accuracy for detecting faces, we identified the best model. The face detection is shown in Fig. 10.

Our basic approach is to create a smart attendance monitoring system. We have considered a small session of 15 min out of which the first and the last 2.5 min are kept as buffer time (for any delay due to emergency) in favor of students. The attendance is on the basis of a patch of 10 min starting from 2.5th minute to 12.5th minute as shown in Fig. 18. We do not report an exhaustive demonstration here. Instead, we illustrate three prominent cases. All cases are tested with three sample students (Fig. 17).

Case 1: Students enter at different time of the lecture, attend the lecture until its completion, and then leave.

Case 2: Students enter at different time of the lecture, attend the lecture, two of the students leave during the session, and rejoin the session after some interval of time.

Fig. 10 Detected face with student number and confidence

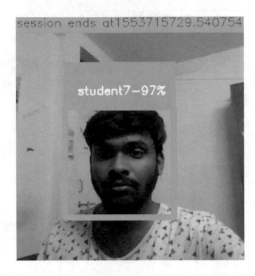

Fig. 11 Validation accuracy due to variation in convolutional layers

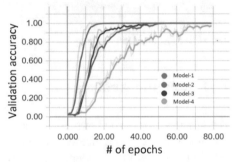

Fig. 12 Validation loss due to variation in convolutional layers

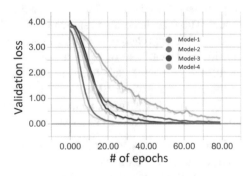

Fig. 13 Validation accuracy due to variation in dense layers

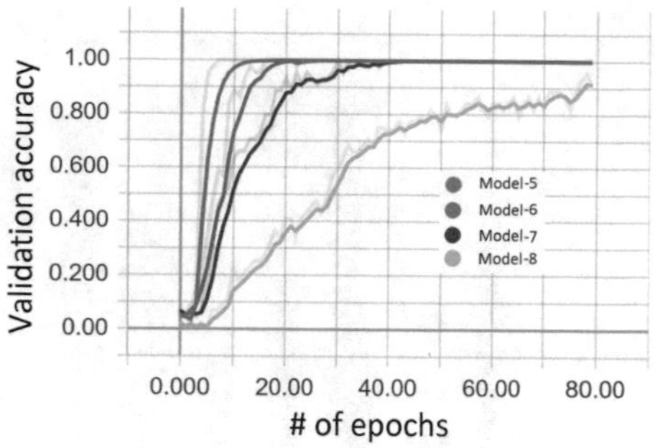

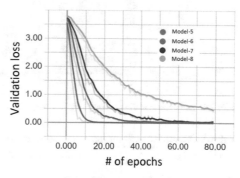

Fig. 14 Validation loss due to variation in dense layers

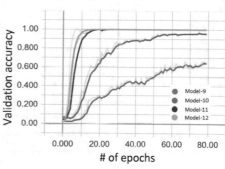

Fig. 15 Validation accuracy due to variation in number of nodes

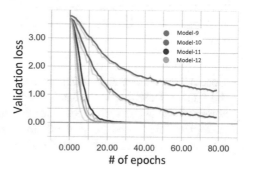

Fig. 16 Validation loss due to variation in number of nodes

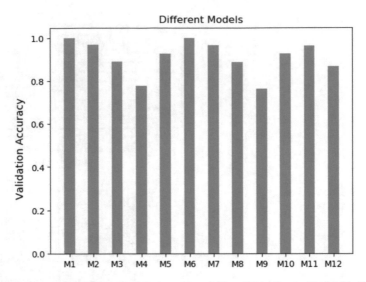

Fig. 17 Accuracy of the different models that are used. Here, M-*i* refers to Model-*i* in Table 1

Table 1 Various models for our CNN-based smart attendance system

Model name	No. of convolutional layers	No. of dense layers	No. of nodes
Model-1	2	2	64
Model-2	3	2	64
Model-3	4	2	64
Model-4	5	2	64
Model-5	2	1	64
Model-6	2	2	64
Model-7	2	3	64
Model-8	2	4	64
Model-9	2	2	16
Model-10	2	2	32
Model-11	2	2	64
Model-12	2	2	128

Case 3: Students enter at different time of the lecture, attend the lecture, two of the students leave during the session, and only one student rejoins the session after some interval of time, while the other students bunk the rest of the lecture. In all the cases, the time line of the students' movements is captured as shown in Figs. 19, 20, and 21 and recorded in a database for further use.

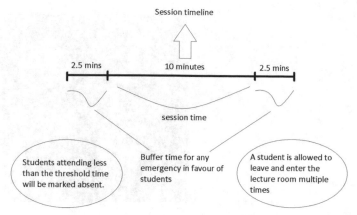

Session timeline

| 2.5 mins | 10 minutes | 2.5 mins |

session time

Students attending less than the threshold time will be marked absent.

Buffer time for any emergency in favour of students

A student is allowed to leave and enter the lecture room multiple times

Here we have taken 75% of the session time as threshold

Fig. 18 Time line of the session

Fig. 19 Case 1

```
SESSION: 15:00 hrs- 15:10 hrs
ENTER STATUS: STUDENT 1
ENTER TIME: 14:58 hrs
ENTER STATUS: STUDENT 2
ENTER TIME: 14:59 hrs
ENTER STATUS: STUDENT 3
ENTER TIME: 15:00 hrs
EXIT STATUS: STUDENT 1
EXIT TIME: 15:10 hrs
EXIT STATUS: STUDENT 2
EXIT TIME: 15:11 hrs
EXIT STATUS: STUDENT 3
EXIT TIME: 15:12 hrs
ATTENDANCE STATUS:
STUDENTS 1 : TIME PRESENT 10.00 mins - PRESENT
STUDENTS 2 : TIME PRESENT 10.00 mins - PRESENT
STUDENTS 3 : TIME PRESENT 10.00 mins - PRESENT
```

Fig. 20 Case 2

```
SESSION: 15:00 hrs- 15:10 hrs
ENTER STATUS: STUDENT 1
ENTER TIME: 14:58 hrs
ENTER STATUS: STUDENT 2
ENTER TIME: 14:59 hrs
ENTER STATUS: STUDENT 3
ENTER TIME: 15:00 hrs
EXIT STATUS: STUDENT 1
EXIT TIME: 15:02 hrs
EXIT STATUS: STUDENT 2
EXIT TIME: 15:03 hrs
ENTER STATUS: STUDENT 1
ENTER TIME: 15:04 hrs
ENTER STATUS: STUDENT 2
ENTER TIME: 15:05 hrs
EXIT STATUS: STUDENT 1
EXIT TIME: 15:10 hrs
EXIT STATUS: STUDENT 2
EXIT TIME: 15:11 hrs
EXIT STATUS: STUDENT 3
EXIT TIME: 15:12 hrs
ATTENDANCE STATUS:
STUDENTS 1 : TIME PRESENT 8.00 mins - PRESENT
STUDENTS 2 : TIME PRESENT 8.00 mins - PRESENT
STUDENTS 3 : TIME PRESENT 10.00 mins - PRESENT
```

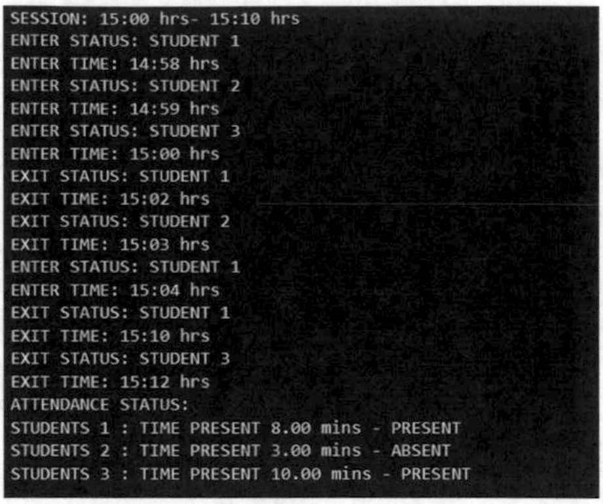

Fig. 21 Case 3

```
SESSION: 15:00 hrs- 15:10 hrs
ENTER STATUS: STUDENT 1
ENTER TIME: 14:58 hrs
ENTER STATUS: STUDENT 2
ENTER TIME: 14:59 hrs
ENTER STATUS: STUDENT 3
ENTER TIME: 15:00 hrs
EXIT STATUS: STUDENT 1
EXIT TIME: 15:02 hrs
EXIT STATUS: STUDENT 2
EXIT TIME: 15:03 hrs
ENTER STATUS: STUDENT 1
ENTER TIME: 15:04 hrs
EXIT STATUS: STUDENT 1
EXIT TIME: 15:10 hrs
EXIT STATUS: STUDENT 3
EXIT TIME: 15:12 hrs
ATTENDANCE STATUS:
STUDENTS 1 : TIME PRESENT 8.00 mins - PRESENT
STUDENTS 2 : TIME PRESENT 3.00 mins - ABSENT
STUDENTS 3 : TIME PRESENT 10.00 mins - PRESENT
```

5 Conclusion

In this paper, smart attendance monitoring system has been proposed. The approach takes real-time pictures from surveillance camera to detect the presence. The features from the images are extracted and compared with the features as per the dataset. This approach can be used in colleges for attendance of student in a university or in any corporate offices or even in the security departments for secretly monitoring people and observing their behavior. The concept will further improvised in future though the current approach provides a high amount of accuracy, of detecting the faces as well as recording the time line of the students, ranging from 96 to 98%.

References

1. Hartley, R., Zisserman, A.: Multiple View Geometry in Computer Vision. Cambridge University Press, New York, NY (2003)
2. DiMaria, P.C., Madsen, J.: Biometric time and attendance system with epidermal topographical updating capability. US Patent 5,959,541, 28 Sept 1999
3. Russell, S.J., Norvig, P.: Artificial Intelligence: A Modern Approach. Pearson Education Limited, Malaysia (2016)
4. Larochelle, H., Bengio, Y., Louradour, J., Lamblin, P.: Exploring strategies for training deep neural networks. J. Mach. Learn. Res. **10**, 1–40 (2009)
5. Haykin, S.: Neural Networks, vol. 2. Prentice Hall, New York (1994)
6. McEwen, B.S., Gould, E., Orchinik, M., Weiland, N.G., Woolley, C.S.: Oestrogens and the structural and functional plasticity of neurons: implications for memory, ageing and neurodegenerative processes. In: Ciba Foundation Symposium 191—Non-reproductive Actions of Sex Steroids, pp. 52–73. Wiley, Chichester (2007)
7. Graves, A., Mohamed, A.R., Hinton, G.: Speech recognition with deep recurrent neural networks. In: IEEE International Conference on Acoustics, Speech and Signal Processing, pp. 6645–6649. IEEE (2013)
8. Acharya, T., Ray, A.K.: Image Processing: Principles and Applications. Wiley, New York, NY (2005)

9. Krizhevsky, A., Sutskever, I., Hinton, G.E.: ImageNet classification with deep convolutional neural networks. In: Advances in Neural Information Processing Systems, pp. 1097–1105 (2012)

10. Oquab, M., Bottou, L., Laptev, I., Sivic, J.: Learning and transferring mid-level image representations using convolutional neural networks. In: Proceedings of the IEEE Conference on Computer Vision and Pattern Recognition, pp. 1717–1724 (2014)

11. LeCun, Y., Bengio, Y., Hinton, G.: Deep learning. Nature 521(7553), 436 (2015)

12. Boureau, Y.L., Ponce, J., LeCun, Y.: A theoretical analysis of feature pooling in visual recognition. In: Proceedings of the 27th International Conference on Machine Learning (ICML-10), pp. 111–118 (2010)

13. Huang, G., Liu, Z., Van Der Maaten, L., Weinberger, K.Q.: Densely connected convolutional networks. In: Proceedings of the IEEE Conference on Computer Vision and Pattern Recognition, pp. 4700–4708 (2017)

14. Ruck, D.W., Rogers, S.K., Kabrisky, M., Oxley, M.E., Suter, B.W.: The multilayer perceptron as an approximation to a Bayes optimal discriminant function. IEEE Trans. Neural Netw. 1(4), 296–298 (1990)

15. Szegedy, C., Liu, W., Jia, Y., Sermanet, P., Reed, S., Anguelov, D., Erhan, D., Vanhoucke, V., Rabinovich, A.: Going deeper with convolutions. In: Proceedings of the IEEE Conference on Computer Vision and Pattern Recognition, pp. 1–9 (2015)

16. Xu, B., Wang, N., Chen, T., Li, M.: Empirical evaluation of rectified activations in convolutional network. arXiv preprint arXiv:1505.00853 (2015)

17. Glorot, X., Bengio, Y.: Understanding the difficulty of training deep feedforward neural networks. In: Proceedings of the Thirteenth International Conference on Artificial Intelligence and Statistics, pp. 249–256 (2010)

18. Huang, G., Liu, Z., van der Maaten, L., Weinberger, K.Q.: Densely connected convolutional networks. In: The IEEE Conference on Computer Vision and Pattern Recognition (CVPR) (2017)

19. Abadi, M., Barham, P., Chen, J., Chen, Z., Davis, A., Dean, J., Devin, M., Ghemawat, S., Irving, G., Isard, M., et al.: Tensorflow: a system for large-scale machine learning. In: 12th USENIX Symposium on Operating Systems Design and Implementation (OSDI 16), pp. 265–283 (2016)

20. Jang, E., Gu, S., Poole, B.: Categorical reparameterization with Gumbel-Softmax. arXiv preprint arXiv:1611.01144 (2016)

21. Courtney, J.D.: Motion based event detection system and method. US Patent 5,969,755, 19 Oct 1999

22. Elgammal, A., Harwood, D., Davis, L.: Non-parametric model for background subtraction. In: European Conference on Computer Vision, pp. 751–767. Springer, Berlin, Heidelberg (2000)

Mutation-Based Chaotic Gravitational Search Algorithm

Moujinjir Mukherjee, Suman Mitra and Sriyankar Acharyya

Abstract Gravitational search algorithm, a stochastic, nature-inspired algorithm, is designed on the basis of gravitational kinematics of physics. To promote an appropriate stability between exploration and exploitation and to diminish the weakness that was affecting the search to reach global optima, a new variant of GSA named as mutation-based chaotic gravitational search algorithm (MCGSA) is introduced in this work. This new variant of GSA consists of five mutation strategies. These five mutations are applied to the best solution in a successive manner. To justify the overall functioning of the proposed method, it has been evaluated on CEC2005 benchmark suit and has been compared with four other modified versions derived from original GSA, namely standard GSA (SGSA), chaotic-based GSA (CGSA), crossover-based GSA (CROGSA), and self-adaptive GSA (GGSA). The comprehensive result of MCGSA has outperformed other GSA variants.

Keywords Gravitational search algorithm · Gravitational law · Meta-heuristics · Mutation · Chaotic map · Exploitation and exploration

1 Introduction

The real-life optimization problems [1] are hard to solve in polynomial time by deterministic algorithms. Hence, nature-inspired heuristic algorithms, called meta-heuristics [2], are used to solve such problem. There are many meta-heuristic algorithms, but 'no free lunch theorem' [3] coined by D. H. Wolper states that there is no particular algorithm that can show the superiority in all optimization problems.

M. Mukherjee · S. Mitra (✉) · S. Acharyya
Maulana Abul Kalam Azad University of Technology, Haringhata, Nadia, West Bengal, India
e-mail: smsumanmitra@gmail.com

M. Mukherjee
e-mail: moujinjir.mukherjee@gmail.com

S. Acharyya
e-mail: srikalpa8@gmail.com

© Springer Nature Singapore Pte Ltd. 2020
J. K. Mandal and S. Mukhopadhyay (eds.), *Proceedings of the Global AI Congress 2019*, Advances in Intelligent Systems and Computing 1112, https://doi.org/10.1007/978-981-15-2188-1_10

Rashedi et al. [4] have introduced GSA in 2009. It is population-based meta-heuristic algorithm which is influenced by Newton's laws of gravity and motion [5]. GSA considers its agents as celestial bodies, moving towards the agent having heavier mass. These bodies pull each other towards itself with a gravitational force which directly varies with the product of their masses and varies inversely with the square of the length of separation between them. The performance of any meta-heuristic algorithm depends on the appropriate mixture of exploration and exploitation characteristics of the algorithm. After its inception, GSA has gone through many modifications to improve its performance. After standard GSA having gained popularity, Rashedi again introduced binary GSA (BGSA) [6] in 2010. Later on, Sarafrazi improved the diversification and intensification abilities of the GSA by introducing an operator called 'disruption' [7]. Again, GSA was modified by Khajehzadeh et al. [8] and the modified version had the potential to expedite convergence speed, while improving the quality of the solution. To enhance the convergence characteristics of GSA, self-adaptive GSA [9] was proposed by Ji et al. Another variant called chaotic GSA (CGSA) [10] was proposed by S. Mirjalili and A. H. Gandomi 2017 which improves the performance of GSA by introducing randomness in chaotic variables. In 2017, B. Yin et al. upgraded the performance of GSA by integrating the crossover mechanism into standard GSA to produce CROGSA [11].

In this research works, a modified version of chaotic gravitational search algorithm, namely mutation-based chaotic gravitational search algorithm (MCGSA), has been proposed to enhance the exploration capability of the search process in CGSA. The best solution obtained from CGSA is successively passed through five different mutation strategies to enhance the diversification in a controlled way and thus enabling MCGSA to avoid premature convergence. The newly introduced MCGSA has been evaluated on CEC2005 benchmark functions, and its performance has been compared with that of its variants, namely SGSA, CROGSA, CGSA, and GGSA. It has been observed that MCGSA has outperformed its variants in most of the benchmark functions of CEC2005.

The paper is structured as follows: Sects. 2, 2.1, 2.2, and 2.3 discuss the basic GSA, chaotic GSA, and proposed MCGSA, respectively. Section 3 is about experimental result and discussion, and Sect. 4 marks the end of this paper with a brief conclusion.

2   Basic Gravitational Search Algorithm (GSA)

Gravitation law is one of the basic laws of nature. It deals with the force between any two particles having masses M_1 and M_2. The force (F) acting between them linearly varies with the product of their masses and varies inversely with the square of the separation length R between them. G is known as gravitational constant.

$$F = G\frac{M_1 M_2}{R^2} \tag{1}$$

Newton's law states that the acceleration is dependent only on the force F applied to it as well as on the mass of the particle.

$$a = \frac{F}{M} \qquad (2)$$

Every particle pulls each other by a force of gravitation as a result it creates the global movement of the particles in the search space. Particles having heavy mass attracts all the lighter particles towards itself causing greater acceleration upon them. Therefore, the heavy particle corresponds to the good solution. GSA steps are explained below:

We have a set of N candidate solution. The ith candidate solution has been initialized as follows:

$$X_i = \left(x_i^1, \ldots x_i^d, \ldots x_i^n\right) \qquad (3)$$

Here, $1 \le i \le N$, n represents the number of decision variables, and x_i^d is the dth decision variable of the ith candidate solution. The mass of the particle is treated as solution fitness. The mass of the particle i is defined as given below:

$$m_i(t) = \frac{\text{fit}_i(t) - \text{worst}(t)}{\text{best}(t) - \text{worst}(t)} \qquad (4)$$

Here, $\text{fit}_i(t)$ stands for fitness of ith candidate solution in the generation t. For minimization problem, worst(t) is the maximum of all fitness values in tth generation; best(t) is the minimum of all fitness values in the generation t. Similarly, for maximization problem, worst(t) represents minimum fitness value, and best(t) stands for maximum fitness in tth generation.

$$\text{best}(t) = \max_{i \in \{1,2,\ldots N\}} \text{fit}_i(t) \qquad (5)$$

$$\text{worst}(t) = \min_{i \in \{1,2,\ldots N\}} \text{fit}_i(t) \qquad (6)$$

After this step, the masses are normalized between 0 and 1. To normalize it, we have the following equation:

$$M_i(t) \frac{m_i(t)}{\sum_{i=1}^{N} m_j(t)} \qquad (7)$$

The next step is the calculation of force, that is exerted on particle i by particle j. We define the force at tth generation by the following equation:

$$F_{ij}^d(t) = G(t) \frac{M_{pi}(t) * M_{aj}(t)}{R_{ij}(t) + \epsilon} \left(x_j^d(t) - x_i^d(t)\right) \qquad (8)$$

where $M_{aj}(t)$ and $M_{pi}(t)$ stand for active and passive gravitational masses related to particle j and i, respectively. $G(t)$ stands for gravitational constant, and T is the maximum number of generations. $\in$ is usually a very small number. Between particle i and j, $R_{ij}(t)$ is the Euclidian distance.

$$G(t) = G_0 \exp(-\alpha * t/T) \tag{9}$$

Let K_{best} be a set of K candidate solutions with best fitness starting from the beginning of N candidate solutions. We can say that K_{best} particles will attract the other particles. K_{best} is linearly decreasing with time. So the total force acting on a particle i by K_{best} solutions can be defined as

$$F_i^d(t) = \sum_{j \in K\text{best}, j \neq i} \text{rand}_j F_{ij}^d(t) \tag{10}$$

Here, $\text{rand}_j \in (0, 1)$ is a random number drawn from uniform distribution. From the law of motion, the acceleration of particle i, during generation t in the direction d can be defined as

$$a_i^d(t) = \frac{F_i^d(t)}{M_{ii}(t)} \tag{11}$$

$M_{ii}(t)$ stands as inertial mass of the particle i. The next step is to calculate the velocity v. Velocity of each candidate solution is updated by the following equation:

$$v_i^d(t+1) = \text{rand}_i \times v_i^d(t) + a_i^d(t) \tag{12}$$

$\text{rand}_i \in (0, 1)$ is a random number generated from uniform distribution. This process integrates randomness to the search process. The position of solution is updated by the following equation:

$$x_i^d(t+1) = x_i^d(t) + v_i^d(t+1) \tag{13}$$

2.1 Chaotic Gravitational Search Algorithm

Basic GSA suffers from stagnation to local optima and slow convergence rate. To overcome this problem, S. Mirjalili and A. H. Gandomi have improved the performance of standard GSA by embedding chaotic map into basic GSA [10]. Experimental results showed that sinusoidal map [10] performs better than other chaotic maps. Therefore, for this experiment, sinusoidal map is taken into consideration.

$$x_{i+1} = ax_i^2 \sin(\pi x_i), \quad a = 2.3, \text{range}(0, 1) \tag{14}$$

The chaotic map will be applied to G, the gravitational constant which is the most important parameter to control the stability of exploration and exploitation. In this approach, the range of normalization is decreased proportionately with respect to the generations as shown below:

$$V(t) = \text{MAX} - \frac{t}{T}(\text{MAX} - \text{MIN}) \tag{15}$$

Now, the random chaotic variable is normalized from the range (a, b) to the range $(0, V(t))$

$$\text{crn}^{\text{norm}}(t) = \left(\frac{(\text{Crn}(t) - a) * V(t)}{(b - a)} \right) \tag{16}$$

where t stands for the present generation, $\text{crn}(t)$ is the chaotic random number, T represents maximum generations, [Max, Min] denotes the adaptive interval which is 20 and 1e$-$10. $[a, b]$ is the range of sinusoidal map, normalized to $[0, V(t)]$ in each generation, while $V(t)$ is decreased along with generations. G, the gravitational constant, can be updated as follows:

$$G(t) = \text{crn}^{\text{norm}}(t) + G_0 * e^{\left(-\alpha \frac{t}{T}\right)} \tag{17}$$

α is the constant having initial value as 20. G_0 is a parameter which states the initialized value of the constant of gravitation, t represents the current generation, and T stands for the maximum generations.

2.2 Mutation-Based Chaotic GSA (MCGSA)

Here, to explore the search space more, we propose another version of GSA, namely mutation-based chaotic GSA (MCGSA). In this case, five successive mutations, namely Gaussian mutation, Cauchy mutation, opposition-based mutation, DE/Best/1-based mutation [12], Weibull-based mutation [13], have been applied on the best solution obtained from chaotic GSA. The best solution produced in each generation is passed through successive mutations. At first, we pass the best solution through Gaussian mutation. If the mutant solution 'f_{mutant}' is better than the best solution 'f_{b_sol}', then the best solution gets updated to f_{mutant}; otherwise, the old best solution is retained. The new solution (updated or retained) is passed to the next (Weibull-based) mutation strategy. The same process is followed afterwards. Equipped with the mutation strategies, MCGSA enhances the exploration capability of the search process. It saves the problem from getting trapped into local optima.

Algorithm: MCGSA

```
initialize G₀ = 100; α = 20;
Randomly choose Np particles in initial population;
    //each particle xᵢ is a vector containing d components
for each particle i = 1 to Np
    evaluate fitness(i); // fitness of each particle i
end
Calculate initial best fitness and worst fitness;
    //using equation (5) and (6)
while termination criteria is not satisfied
    for t = 1 to maxgeneration
      for each particle i = 1 to Np
          compute the mass of each particle using equation (4);
      end
      generate a chaotic random number crn and normalise it to Crnnorm;
      Update gravitation constant G(t); // accordingly to equation (17)
```

$$K_{best} = N_p - \left(N_p * \frac{t}{max\_iteration} \right);$$

```
      for every particle iₚ = 1 to Np
        for every particle jₚ = 1 to Np
          if (iₚ ≠ jₚ)
            compute the force F between particle i and j;
                    // using equation (8) and (10);
          end
        end
      end
      For every particle, we compute the acceleration, velocity and posi-
          tion; //based on equation(11),(12)and (13)respectively
      Order the candidate solutions and get the best candidate solution;
      Now apply five successive mutations on the best solution;
      Return the best solution.
    end
```

---

3 Experimental Study

The performance of the algorithms, namely SGSA [4], CGSA [10], GGSA [14], CROGSA [11], and the proposed MCGSA, is evaluated using CEC2005 benchmark functions [12]. Each algorithm is executed over 30 independent runs. The experiment is performed in a single computer having Intel Core i7 processor and 12 GB RAM 64 bit operating system and MATLAB, version 16.

3.1 Initialization of Parameters

In this experiment, there are 50 particles/candidate solutions. The initial value for gravitational constant (G_0) is taken as 100. Another important parameter, α, used in calculating the force is initially set to 20. In CROGSA, there is a crossover rate which is assigned to 0.1. The total number of generations is assigned to 6000 for each independent run.

3.2 Experimental Results

In this experiment, the results of MCGSA have been computed and its results are measured with that of four other peer versions of GSA, namely SGSA, CROGSA, CGSA, and GGSA. The performance of each algorithm is measured by taking 30 independent runs over benchmark suit CEC2005 comprising 25 functions for dimension 30. The computational results are presented in Table 1. It is noticed that MCGSA outperforms CGSA on 13 benchmark functions out of 25, and it is equally competitive

Table 1 Computational results of functions with dimension $D = 30$ in CEC 2005 benchmarks

Functions	Algorithm	Minima	Maxima	Mean	Median	Std. dev.
$F1$	MCGSA	4.24E−07	5.56E−06	1.93E−06	1.53E−06	1.30E−06
	GGSA	5.72E+04	1.79E+05	1.11E+05	1.18E+05	3.39E+04
	CGSA	6.46E−07	1.07E−05	3.11E−06	2.70E−06	2.46E−06
	CROGSA	7.21E−19	1.77E−18	1.19E−18	1.12E−18	2.3171E−19
	SGSA	5.32E−18	1.81E−17	1.06E−17	1.10E−17	2.92E−18
$F2$	**MCGSA**	**1.07E−05**	**2.19E−04**	**4.85E−05**	**4.01E−05**	**3.92E−05**
	GGSA	7.84E+06	1.07E+06	2.19E−05	1.65E+05	1.89E+05
	CGSA	3.56E−05	3.67E−04	1.27E−04	1.20E−04	7.23E−05
	CROGSA	6.82E−18	2.30E+04	2.19E+03	2.08E+02	4.93E+03
	SGSA	4.30E+01	2.46E+02	1.16E+02	1.06E+02	4.50E+01
$F3$	**MCGSA**	**8.67E+04**	**2.41E+05**	**1.47E+05**	**1.40E+05**	**3.58E+04**
	GGSA	1.32E+09	6.52B+09	3.63E+09	3.81E+09	1.69E+09
	CGSA	1.16E+05	3.59E+05	2.60E+05	2.70E+05	5.65E+04
	CROGSA	5.18E+05	1.06E+07	2.98E+06	2.04E+06	2.36E+06
	SGSA	5.16E+05	1.63E+06	1.02E+06	1.01E+06	2.72E+05
$F4$	MCGSA	4.28E+03	1.94E+04	1.14E+04	1.18E+04	3141.845
	GGSA	7.51E+04	1.55E+06	3.60E+05	2.51E+05	3.61E+05
	CGSA	6.85E+03	1.89E+04	1.14E+04	1.07E+04	2.79E+03
	CROGSA	5.34E+03	5.93E+04	2.59E+04	2.25E+04	1.32E+04
	SGSA	1.69E+04	3.20E+04	2.32E+04	2.30E+04	3.17E+03
$F5$	**MCGSA**	**1.95E−02**	**3.60E+02**	**3.73E+01**	**5.30E−02**	**9.42E+01**
	GGSA	4.01E+34	7.05E+04	5.41E+04	5.41E+04	7.62E+03
	CGSA	1.45E−02	1.84E+02	6.80E+00	7.28−02	3.36E+01
	CROGSA	1.31E−03	1.29E+04	4.98E+03	4.81E+03	2.80E+03
	SGSA	2.61E+03	6.99E+03	4.57E+03	4.26E+03	1.21E+03
$F6$	**MCGSA**	**9.11E−01**	**2.86E+02**	**4.36E+01**	**1.95E+01**	**6.11E+01**
	GGSA	1.86E+10	1.62E+11	8.19E+10	7.98E+10	4.32E+10
	CGSA	1.03E+01	6.86E+02	5.02E+01	2.12E+01	1.21E+02

(continued)

Table 1 (continued)

Functions	Algorithm	Minima	Maxima	Mean	Median	Std. dev.
	CROGSA	1.98E+01	1.56E+07	5.94E+05	2.32E+01	2.86E+06
	SGSA	2.12E+01	9.11E+02	2.98E+02	1.93E+02	2.68E+02
F7	MCGSA	4.73E+03	6.07E+03	5.19E+03	5.11E+03	3.82E+02
	GGSA	1.25E+04	5.79E+04	1.66E+04	1.50E+04	8.03E+03
	CGSA	1.16E+04	1.35E+04	1.27E+04	1.28E+04	5.90E+02
	CROGSA	4.70E+03	5.60E+03	4.80E+03	4.70E+03	2.03E+02
	SGSA	1.15E+04	1.34E+04	1.25E+04	1.27E+04	5.94E−02
F8	MCGSA	2.01E+01	2.06E+01	2.02E+01	2.02E+01	1.05E−01
	GGSA	2.00E+01	2.01E+01	2.01E+01	2.00E+01	2.74E−02
	CGSA	2.01E+01	2.03E+01	2.02E+01	2.02E+01	6.47E−02
	CROGSA	2.06E+01	2.09E+01	2.08E+01	2.08E+01	7.39E−02
	SGSA	2.00E+01	2.11E−01	2.01E+01	2.00E+01	2.01E−01
F9	MCGSA	2.10E+01	7.24E+01	5.14E+01	5.24E+01	1.06E+01
	CGSA	1.76E+02	3.33E+02	2.37E+02	2.26E+02	4.55E+01
	CGSA	3.37E+01	8.47E+01	5.20E+01	5.16E+01	1.24E+01
	CROGSA	1.46E+02	6.65E+02	2.17E+02	1.94E+02	1.15E+02
	SGSA	8.32E+02	1.17E+03	1.01E+03	1.03E+03	8.65E+01
F10	MCGSA	2.39E+01	8.27E+01	4.40E+01	4.25E+01	1.20E+01
	CGSA	3.95E+02	8.61E+02	5.72E+02	5.75E+02	9.85E+01
	CGSA	2.73E+01	7.63E+01	4.28E+01	4.19E+01	1.15E+01
	CROGSA	1.48E+02	2.89E+02	2.27E+02	2.35E+02	3.74E+01
	SGSA	1.48E+03	2.31E+03	1.92E+03	1.92E+03	1.92E+02
F11	MCGSA	2.67E+01	3.73E+01	3.40E+01	3.47E+01	2.53E+00
	GGSA	7.81E−03	8.74E+00	1.88E+00	1.39E+00	2.39E+00
	CGSA	2.99E+00	9.83E+00	6.34E+00	6.13E+00	1.42E+00
	CROGSA	4.66E+01	5.42E+01	4.97E+01	4.98E+01	1.81E+00
	SGSA	5.27E−04	3.45E+00	8.99E−01	1.173E−03	1.21E+00
F12	**MCGSA**	**5.46E−01**	**2.43E+03**	**2.92E+02**	**1.28E+02**	**5.22E+02**
	GGSA	9.70E+00	4.75E+04	9.65E+03	6.51E+03	1.10E+04
	CGSA	9.35E−01	3.12E+03	4.60E+02	2.05E+02	6.50E+02
	CROGSA	8.51E+03	6.59E+05	1.44E+05	2.51E+02	1.27E+05
	SGSA	9.60E−01	5.46E+03	1.39E+03	3.91E+02	1.64E+03
F13	**MCGSA**	**1.91E+00**	**1.23E+01**	**4.34E+00**	**3.56E+00**	**2.82E+00**
	GGSA	1.65E+00	6.42E+00	3.68E+00	3.63E+00	1.03E+00
	CGSA	1.79E+00	1.12E+03	1.12E+02	3.32E+01	2.27E+02

(continued)

Table 1 (continued)

Functions	Algorithm	Minima	Maxima	Mean	Median	Std. dev.
	CROGSA	3.11E+00	1.37E+01	4.99E+00	4.38E+00	2.31E+00
	SGSA	2.32E+00	6.60E+01	1.28E+01	4.36E+00	1.70E+01
$F14$	MCGSA	1.32E+01	1.39E+01	1.36E+01	1.36E+01	1.91E+01
	GGSA	1.45E+01	1.48E+01	1.46E+01	1.46E+01	5.52E−02
	CGSA	1.33+01	1.45E+01	1.39E+01	1.39E+01	3.30E+−01
	CROGSA	1.24E+01	1.40E+01	1.32E+01	1.32E+01	3.77E−01
	SGSA	1.37E+01	1.44E+01	1.40E+01	1.40E+01	1.92E−01
$F15$	MCGSA	3.03E+01	4.66E+01	3.18E+02	3.06E+02	3.36E+01
	GGSA	4.66E+02	1.38E+02	9.54E+02	9.50E+02	2.38E+02
	CGSA	3.03+02	4.00E+02	3.17E+02	3.06E+02	2.11E+01
	CROGSA	4.24E+02	9.27E+02	5.65E+02	5.76E+02	9.77E+01
	SGSA	3.00E+02	5.00E+02	3.15E+02	3.00E+02	4.03E+01
$F16$	**MCGSA**	**3.41E+01**	**5.01E+02**	**1.78E+02**	**7.28E+01**	**1.74E+02**
	GGSA	4.00E+02	9.00E+02	6.14E+02	6.08E+02	1.34E+02
	CGSA	4.97E+01	5.46E+02	2.35E+02	8.60E+02	2.01E+02
	CROGSA	1.89E+02	5.81E+02	3.93E+02	3.58E+02	1.37E+02
	SGSA	2.96E+01	5.44E+02	3.04E+02	4.00E+02	2.09E+02
$F17$	**MCGSA**	**4.29E+01**	**5.69E+02**	**2.62E+02**	**7.59E+01**	**2.32E+02**
	GGSA	1.07E+03	2.08E+03	1.50E+03	1.51E+03	2.45E+02
	CGSA	6.23E+01	5.96E+02	2.89E+02	1.39E+02	2.32E+02
	CROGSA	1.89E+02	6.55E+02	4.22E+02	3.60E+02	1.58E+02
	SGSA	3.42E+01	5.65E+02	2.80E+02	2.88E+02	2.35E+02
$F18$	MCGSA	9.05E+02	9.07E+02	9.05E+02	9.05+E02	4.44E−01
	GGSA	9.00E+02	1.39E+03	1.12E+03	1.22E+03	2.02E+02
	CGSA	9.05E+02	9.07E+02	9.05E+02	9.05E+02	3.72E−01
	CROGSA	9.05E+02	9.09E+02	9.07E+02	9.06E+02	1.03E+00
	SGSA	9.03E+02	9.06E+02	9.03E+02	9.03E+02	5.18E−01
$F19$	MCGSA	9.05E+02	9.09E+02	9.06E+02	9.05E+02	8.30E−01
	GGSA	9.00E+02	1.42E+03	1.13E+03	1.19E+03	1.93E+02
	CGSA	9.05E+02	9.07E+02	9.05E+02	9.05E+02	5.80E−01
	CROGSA	9.05E+02	9.09E+02	9.06E+02	9.05E+02	1.17E+00
	SGSA	9.03E+02	9.04E+02	9.04E+02	9.03E+02	2.09E−01
$F20$	MCGSA	9.05E+02	9.07E+02	9.05E+02	9.05E+02	4.02E+−01
	GGSA	9.00E+02	1.39E+03	1.11E+03	1.13E+03	2.07E+02
	CGSA	9.05E+02	9.07E+02	9.06E+02	9.05E+02	5.67E−01

(continued)

Table 1 (continued)

Functions	Algorithm	Minima	Maxima	Mean	Median	Std. dev.
	CROGSA	9.05E+02	9.10E+02	9.07E+02	9.07E+02	1.20E+00
	SGSA	9.03E+02	9.07E+02	9.04E+02	9.03E+02	9.92E−01
$F21$	**MCGSA**	**5.00E+02**	**9.93E+02**	**5.98E+02**	**5.22E+02**	**1.52E+02**
	GGSA	8.013E+02	1.38+03	1.27E+03	1.28E+03	9.47E+01
	CGSA	5.00E+02	9.32E+02	6.15E+02	5.22E+02	1.49E+02
	CROGSA	5.43E+02	1.04E+03	7.67E+02	7.79E+02	1.17E+02
	SGSA	5.00E+02	1.77E+03	6.54E+02	5.41E+02	2.59E+02
$F22$	**MCGSA**	**8.02E+02**	**8.37E+02**	**8.19E+02**	**8.19E+02**	**8.79E+00**
	GGSA	1.01E+03	1.38E+03	1.12E+03	1.10E+03	9.79E+01
	CGSA	8.03E+02	8.64E+02	8.23E+02	8.21E+02	1.37E+01
	CROGSA	7.99E+02	8.85E+02	8.30E+02	8.25E+02	2.39E+01
	SGSA	768E+02	8.40E+02	8.24E+02	826E+02	1.55E+01
$F23$	MCGSA	5.34E+02	1.05E+02	7.78E+02	8.64E+02	1.47E+02
	GGSA	1.30E+03	1.99E+03	1.46E+03	1.38E+03	1.75E+02
	CGSA	5.34E+02	1.77E+03	7.75E+02	6.84E+02	2.92E+02
	CROGSA	7.87E+02	1.79E+03	9.29E+02	8.67E+02	2.29E+02
	SGSA	5.34E+02	1.08E+03	6.58E+02	5.62E+02	1.52E+02
$F24$	MCGSA	2.00E+02	9.40E+02	4.26E+02	3.85E+02	2.17E+02
	GGSA	1.31E+03	1.49E+03	1.36E+03	1.34E+03	4.23E+01
	CGSA	2.00E+02	9.40E+02	5.86E+02	6.19E+02	2.43E+02
	CROGSA	5.21E+02	9.57E+02	9.24E+02	9.40E+02	7.78E+01
	SGSA	2.00E+02	9.33E+02	4.56E+02	3.19E+02	2.74E+02
$F25$	**MCGSA**	**2.09E+02**	**2.09E+02**	**2.09E+02**	**2.09E+02**	**1.34E−01**
	GGSA	1.38E+03	1.71E+03	1.46E+03	1.45E+03	7.21E+01
	CGSA	2.09E+02	2.09E+02	2.09E+02	2.09E+02	1.21E−01
	CROGSA	2.10E+02	2.13E+02	2.11E+02	2.11E+02	9.48E−01
	SGSA	2.10E+02	2.12E+02	2.11E+02	2.11E+02	3.70E−01

7 functions. It has also been observed that MCGSA outperforms GGSA, CROGSA, SGSA on 23, 21, and 16 functions, respectively.

The results indicate that MCGSA performs significantly better across other peer variants of GSA.

3.3 Convergence Analysis

It is observed from the graphical presentation that MCGSA has a better convergence rate compared to SGSA, GGSA, CGSA, and CROGSA avoiding premature convergence and preventing from being trapped to local minima (Fig. 1).

4 Conclusion

This paper presents mutation-based chaotic gravitational search algorithm (MCGSA) which aims at maintaining proper adjustment between exploration and exploitation phases of the process of search, avoiding premature convergence. One of the chaotic maps, named sinusoidal map, was embedded into GSA to boost the performance of GSA resulting in the variant CGSA. In further modification of CGSA, five mutation strategies, namely Gaussian mutation, Cauchy mutation, Weibull-based mutation, opposition-based mutation, and DE/Best/1 mutation, have been embedded in CGSA resulting in MCGSA. The performance of MCGSA is comparatively better than that of SGSA, GGSA, CROGSA, and CGSA. To conduct the comparative study, CEC2005 benchmark suit consisting of 25 benchmark functions has been considered. The chaotic map promotes good exploration ability of GSA because it changes the value of G abruptly, preventing from getting trapped into local optima. The successive mutation strategies help enhancing the exploration capability of the search process without disturbing its exploitation ability. The method will have to be tested on real-life problems in future like feature selection [15].

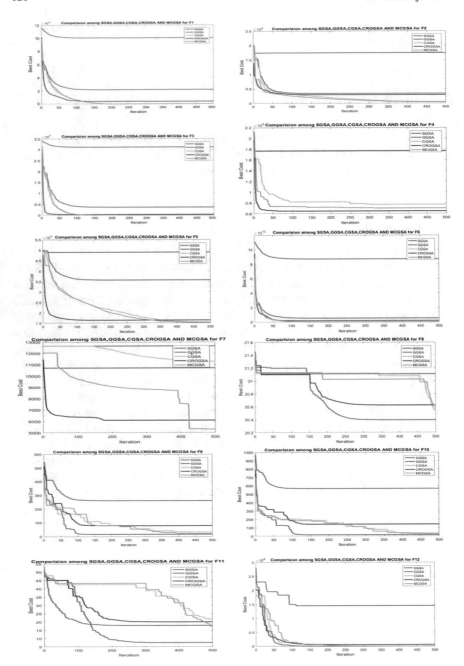

Fig. 1 Convergence graph

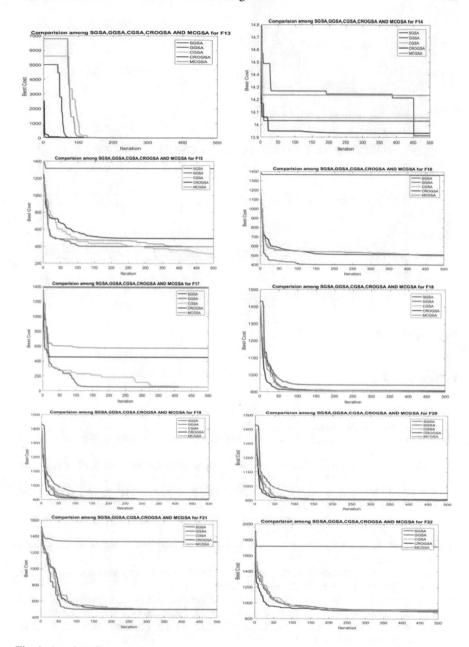

Fig. 1 (continued)

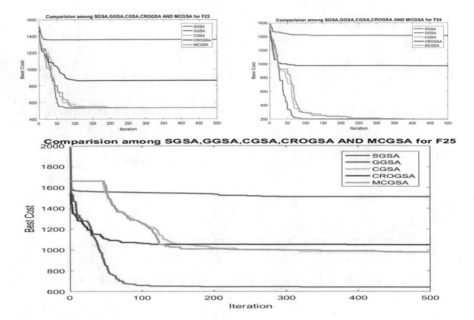

Fig. 1 (continued)

References

1. Rakshit, P., Konar, A., Das, S.: Noisy evolutionary optimization algorithms—a comprehensive survey. Swarm Evol. Comput. **33**, 18–45 (2017)
2. Boussaïd, I., Lepagnot, J., Siarry, P.: A survey on optimization metaheuristics. Inf. Sci. **237**, 82–117 (2013)
3. Wolpert, D.H., Macready, W.G.: No free lunch theorems for optimization. IEEE Trans. Evol. Comput. **1**(1), 67–82 (1997)
4. Rashedi, E., Nezamabadi-Pour, H., Saryazdi, S.: GSA: a gravitational search algorithm. Inf. Sci. **179**(13), 2232–2248 (2009)
5. Kirillov, A.A., Turaev, D.: Modification of Newton's law of gravity at very large distances. Phys. Lett. B **532**(3–4), 185–192 (2002)
6. Rashedi, E., Nezamabadi-Pour, H., Saryazdi, S.: BGSA: binary gravitational search algorithm. Nat. Comput. **9**(3), 727–745 (2010)
7. Sarafrazi, S., Nezamabadi-Pour, H., Saryazdi, S.: Disruption: a new operator in gravitational search algorithm. Sci. Iran. **18**(3), 539–548 (2011)
8. Khajehzadeh, M., Taha, M.R., El-Shafie, A., Eslami, M.: A modified gravitational search algorithm for slope stability analysis. Eng. Appl. Artif. Intell. **25**(8), 1589–1597 (2012)
9. Ji, J., Gao, S., Wang, S., Tang, Y., Yu, H., Todo, Y.: Self-adaptive gravitational search algorithm with a modified chaotic local search. IEEE Access **5**, 17881–17895 (2017)
10. Mirjalili, S., Gandomi, A.H.: Chaotic gravitational constants for the gravitational search algorithm. Appl. Soft Comput. **53**, 407–419 (2017)
11. Yin, B., Guo, Z., Liang, Z., Yue, X.: Improved gravitational search algorithm with crossover. Comput. Electr. Eng. **66**, 505–516 (2018)
12. Suganthan, P.N., Hansen, N., Liang, J.J., Deb, K., Chen, Y.P., Auger, A., Tiwari, S.: Problem definitions and evaluation criteria for the CEC 2005 special session on real-parameter optimization. KanGAL Rep. **2005005**, 2005 (2005)

13. Jordehi, A.R.: Enhanced leader PSO (ELPSO): a new PSO variant for solving global optimisation problems. Appl. Soft Comput. **26**, 401–417 (2015)
14. Mirjalili, S., Lewis, A.: Adaptive gbest-guided gravitational search algorithm. Neural Comput. Appl. **25**(7–8), 1569–1584 (2014)
15. Taradeh, M., Mafarja, M., Heidari, A.A., Faris, H., Aljarah, I., Mirjalili, S., Fujita, H.: An evolutionary gravitational search-based feature selection. Inf. Sci. **497**, 219–239 (2019)
16. Cárdenas-Montes, M.: Weibull-based scaled-differences schema for differential evolution. Swarm Evol. Comput. **38**, 79–93 (2018)

SCap Net: A Capsule Network Based Approach for Person Re-identification

Nirbhay Kumar Tagore and Arnab Mondal

Abstract Automatic individual re-identification proof in a multi-camera observation arrangement is significant for powerful following and checking swarm development. As of late, a couple of deep learning-based re-ID methodologies have been created which are precise be that as it may, time-escalated, and in this way unacceptable for reasonable applications. Capsule networks have demonstrated empowering results on de facto benchmark computer vision datasets, for example, MNIST, CIFAR, and smallNORB. This paper proposes a SCap Net, a deep network architecture for person re-identification. With this SCap Net, we present Siamese Capsule networks, another variation that can be utilized for pairwise learning errands. The model is prepared to utilize contrastive loss with l2-normalized capsules encoded pose features. The proposed technique has been assessed on three publicly available datasets: VIPeR, CUHK01, and CUHK03. Results obtained are presented in the CMC curves show that our methodology resolves the orientation and pose based dissimilar image problem and also improves the Rank 1 accuracy from various existing approaches.

Keywords Person re-identification · Siamese network · Capsule network

1 Introduction

The extensive use of camera networks in public spaces such as such as railway stations, airports, shopping malls, hospitals, and office buildings has led to the generation of large amounts of video data that need to be processed either manually or through automated means. Automated analysis of video data generated in large

N. K. Tagore
Department of Computer Science and Engineering, IIT (BHU) Varanasi, Varanasi, India
e-mail: nirbhaykrtag.rs.cse17@itbhu.ac.in

A. Mondal (✉)
Department of Computer Science and Engineering, NIT Durgapur, Durgapur, India
e-mail: arnabmondal39478@gmail.com

© Springer Nature Singapore Pte Ltd. 2020
J. K. Mandal and S. Mukhopadhyay (eds.), *Proceedings of the Global AI Congress 2019*, Advances in Intelligent Systems and Computing 1112,
https://doi.org/10.1007/978-981-15-2188-1_11

133

amounts not only reduces the cost and time taken for surveillance but also leads to significant improvement in the quality of surveillance [1]. The surveillance mechanism requires tracking and monitoring people across multiple cameras at different locations and different time instants over different durations as one of its fundamental tasks. This problem of matching person images taken across different cameras with non-overlapping fields of view to establish correspondence between disconnected tracks is referred to as person re-identification in computer vision literature.

Person re-identification is a challenging task as the appearance of a person which may vary significantly across camera views owing to variations in view angle, illumination, occlusions, pose and devices in acquisition. Person re-identification involves learning robust and discriminative visual descriptors to describe the appearance of a person based on visual features such as color, texture, and spatial orientation. The quality of the visual data may be affected by factors such as frame rate, resolution, imaging conditions, and angle of view which makes the automatic extraction of reliable detectors a very difficult task. Person re-identification may be treated as a closed set problem as in identity retrieval problems or an open set problem as in multi-camera tracking problems.

In this paper, we introduce a novel framework for short-period re-ID where the appearance of a person does not change for a given time duration. The proposed approach uses a typical deep network architecture based on Capsule networks to address the person re-identification problem. The use of capsule layers allows the network to perform feature extraction from the available images of a person irrespective of their shape and orientation. The remainder of the paper is organized as follows. Section 2 provides a briefing of the existing work in this field while Sect. 3 describes the proposed approach and the underlying network architecture in detail. Section 4 describes the datasets used to conduct the experiment, details of implementation and results obtained. Section 5 concludes the paper with insights into some of the open problems and directions for future research.

2 Related Work

Person re-identification has been a field of intense research in computer vision over the past decade or so. Earlier research techniques made use of manually monitored images and combined the relationship between cameras with the process of matching person images taken across multiple cameras for re-identification. Automated person re-identification on the other hand involves two principal steps—extraction of discriminative and reliable descriptors for the person based on appearance and personality traits and use of suitable distance metrics to establish a match between the person images from different cameras. In the following subsections, we introduce the state-of-the-art methods in re-identification starting from passive approaches not requiring machine learning to modern deep learning approaches using deep learning models.

2.1 Initial Re-identification Methods

Erstwhile methods can be broadly subdivided into passive and active methods. Passive methods do not make use of machine learning techniques for descriptor extraction and matching. These methods use single or multiple images to extract features such as color, texture, position, and gradients to form the appearance model. Re-identification is carried out with by matching these features with the help of some suitable distance metric. A feature-based appearance model based on color and shape is proposed in [2] where the descriptor is formed by segmenting the blob into the color Gaussian model, multiple polar bins and edge pixel counts. Spatial covariance descriptors are extracted from different parts of the human body detected using histogram of oriented gradients (HOG) in [3] are matched by computing covariance matrix distance. SIFT features are used to generate the re-ID model in [4] while HS color histograms determined from each human body part are used to generate the appearance model in [5]. Other passive approaches used to generate the descriptor for person re-identification involve Gabor and LBP features as in [6], probabilistic histograms extracted from fuzzy color features in [7] and color position histograms in [8].

Active methods carry out descriptor extraction and matching with the help of either supervised or unsupervised learning techniques. A brightness transfer function (BTF) which establishes correspondence between images of an object appearing in two different cameras with disjoint fields of view is used in [9] to model the appearance of the object. Shape and appearance context models in which the human silhouette is divided into parts on the context of shape are used to form person descriptors in [10]. Codewords characterizng appearance based on HOG features are used to construct the appearance descriptors based on their spatial occurrence in [11] while appearance words based on features like SIFT and average RGB color are used to describe a persons appearance in [12]. Adaboost classifiers are used in [13] to determine features with the most discriminative characteristics to find the most fitting appearance model while partial least squares (PLS) technique is employed to learn discriminative color, texture, and edge features in [14].

Active approaches involving learning suitable distance metrics may be used to achieve increased efficiency of correspondence. A large margin nearest neighbor (LMNN-R) distance metric used in [15] maximizes the distance between true matches while minimizing the distance between false matches. A probabilistic relative distance learning approach is employed in [16] to maximize the probability of a matching pair lying at a smaller distance as compared to a non-matched pair. SVM-based rankers learnt from features like color and texture are used to determine the rank of a match in [13]. A set-based discriminative ranking (SBDR) model proposed in [17] uses an iterative metric learning approach to determine the distance between a pair of images in such a way that a true match pair lies at a smaller distance than a false matching pair. Lab colors, HSV, and LBP features learnt from a person image are used to generate a metric in [18] and a similar pairwise metric discussed in [19] uses the large margin nearest neighbor network.

2.2   Deep Learning-Based Methods

The use of deep learning methodologies to confront the person re-identification problem has gained overwhelming popularity among researchers across the globe. One
of the first approaches in this field was [20] which employed a filter pairing neural
network (FPNN) to tackle discrepancy in images resulting from misalignment, occlusions, and background clutter. A dataset known as CUHK03 was introduced to
evaluate the network performance while a convolutional neural network (CNN) was
used to perform feature extraction. A deep learning framework based on relative distance comparison was used in [21] while a non-negative low-rank and sparse graph
were used to represent appearances of the human silhouette in [22]. The latter approach is based on the assumption that all images of a person within range of a single
camera will lie in the same low-rank sub-space. Sub-space clustering is performed
using normalized cut (NCut), which is used to evaluate unique discriminative features for each subject while correspondence between subjects from different camera
views is established using a cross view quadratic determinant. Another multi-shot
re-identification technique described in [23] makes use of a similarity metric based
on reference points to perform the re-identification task.

Similarity between two given signature samples in [24] is evaluated using a
Siamese neural network which provides a similarity score at its output neurons representing the similarity between the input signatures. The deep learning approach
in [25] uses Siamese network to compare color and texture features for person re-
identification while an improvement to this architecture proposed in [26] uses four
convolutional layers to extract higher level features. Learning approaches based on
partition-based appearance models described in [27] use Siamese Convolutional Neural Networks (SCNN) to compare features from different parts of the human body
while cosine similarity is used to calculate the matching score. An unsupervised
approach in [28] involves learning spatio-temporal features of pedestrians with the
help of a visual classifier which is fused with spatio-temporal features learnt from
test data using Bayesian fusion model to perform re-identification. Another approach
in [29] focuses on trying to maximize the number of correct matches in a camera
network. A previously trained CNN model has been used to represent features in
this approach and a similarity matrix for every two persons is calculated using cosine similarity metric. The approach maximizes global similarity while minimizing
certain constraints arising due to inter-camera inconsistencies.

3   Proposed Approach

In this section, we introduce a new part-based deep network architecture named SCap
Net for person re-identification. Our goal is to produce a framework for short-term
re-identification scenario where the appearance of a person does not change for a
given period of time. We introduce a typical Siamese network architecture explained

in [26] using the capsule layers explained in [30]. This combination of capsule layers with the network allows the network to learn the features from the images those are orientation and shape independent.

3.1 Capsule Networks

Capsule networks were first introduced by Hinton et al. in [30] as an improvement to the existing neural network architecture. A capsule is a group of neurons in which the values of the instantiation parameters of a particular object or entity are represented in the form of an activity vector. The magnitude of the vector represents the probability of the entity's presence within the image while the pose of the vector defined by the spatial orientation of the entity in the image with respect to the other entities represents its instantiation parameters. This was opposed to the previously used scalar outputs in conventional neural networks. Pooling operation used by CNNs for routing between layers results in viewpoint invariance of neurons which leads to loss of information regarding the spatial relationship between features. To overcome this limitation, pooling was replaced by a dynamic routing mechanism to learn the spatial relationships between entities lying at a lower level and send the output capsule to the appropriate parent capsule in the next higher level.

As far as the architecture of the network is concerned, there are two convolutional layers that are used as initial input representations for the first capsule layer that are then routed to a final class capsule layer. The initial capsule layers allow reuse and replication of the knowledge learnt from local feature representations in other parts of the receptive field. An Iterative Dynamic Scheme is used to determine the capsule inputs. A transformation W_{ij} is made to the output vector u_i of the capsule layer C_i^L. The length of the output vector u_i gives the probability of detection of the entity by the lower level capsule while the direction of the vector represents the instantiation parameters of the entity. The output capsule u_i is transformed to a prediction vector $\hat{u}_{j|i}$ by multiplying the it with a weight matrix W_{ij} such that $\hat{u}_{j|i} = W_{ij}u_i$. This prediction vector is weighed by a coupling coefficient c_{ij} to obtain $s_j = \sum_i c_{ij}\hat{u}_{j|i}$ where the coupling coefficient of each capsule $\sum_j c_{ij} = 1$. The coupling coefficients are obtained from a sigmoid function using log prior probabilities b_j followed by a softmax function $c_{ij} = e^{b_{ij}} / \sum_k e^{b_{ik}}$. If the prediction vector $\hat{u}_{j|i}^L$ gives a high scalar product on multiplying with $\hat{u}_{j|i}^{L+1}$, the coupling coefficient for the possible parent capsule c_{ij} is increased and that of the remaining potential parent capsules is decreased. The output vector length is represented as the probability of an entity being present in an image using non-linear normalization as shown in the equation which is given as follows:

$$v_j = \frac{\|s_j\|^2}{1 + \|s_j\|^2} \frac{s_j}{\|s_j\|} \tag{1}$$

where v_j is the vector output of the capsule j and s_j is its total input. The agreement between the output v_j of the capsule j and the prediction $\hat{u}_{j|i}$ made by capsule is a scalar product $a_{ij} = v_j\hat{u}_{j|i}$ that is added to the log prior b_{ij}.

As mentioned in [30], the capsule layers work well for the different shape and orientation of the images. This approach is motivated from the human visual cortex for image recognition, as human eye and brain coordination for the image recognition and identification is not dependent on the particular size and orientation of the images. It just analyzes the complete information of features available in the image irrespective of the orientation and size. The detailed architecture of the proposed SCap Net can be seen in Fig. 2. Herein the network we used four layers of convolution, two with the pooling before the feature comparison and two without pooling just after the feature comparison from the previous two layers. The capsule layers named primary capsules are introduced in conv2_d layer to better learn the relative information between the different features extracted after the first layer.

3.2   Siamese Capsule Network (SCap Net) for Person Re-identification

Dataset Preprocessing: Datasets that we used for performance evaluation of our proposed approach are preprocessed by dividing the images into three parts to obtain the correlated image parts of the same image. This splitting is done on the basis of change in color information of the person image, which is evaluated by first converting the RGB image to HSV and plotting the hue value of each image and whenever there is a rapid change in hue value, we split the image from that hue point. In this way, we find the three correlated image segments for each image pair of the same person. These three correlated image segments of each image pair are used to train a Siamese network of three layers, one layer is for each image pair segment. The architecture of the network is explained in the next section.

SCap Net training: Siamese networks are used very frequently in the field of person re-identification. Siamese networks are good in learning the similarity and dissimilarity between a pair of images. We also used a traditional architecture of a Siamese network, we called it Simple Siamese Capsule Network (Scap Net) that can be seen in Fig. 1. The network is of four layers of convolution in which two layers are tied convolved and rest two are normal convolution layers. After the first two convolution layers, we used a feature difference layer which learns the first two-layer differences and similarity of the given image pair. This feature difference is further distilled with two other convolution layers, and these two layers are without pooling which means there is no more feature shrinking after the first two layers.

The input image pairs are processed with this SCap Net architecture to learn the significant features. Each of the four layers of SCap Net architecture is followed by two fully connected layers, first for correlated image pair and other for concatenating these feature tensors of correlated pairs of the same image. The learned model is used

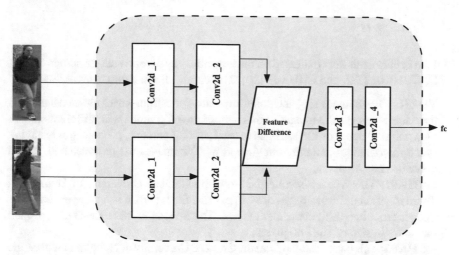

Fig. 1 Siamese Capsule Net (SCap Net) architecture with four convolution boxes and one feature difference layer

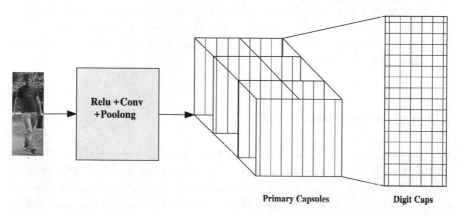

Fig. 2 Capsule architecture at Conv2d_2 layer for retaining the relative information of the images

to obtain the results. The overall architecture of the proposed network can be seen in Fig. 1. The results are like whether the two input given pair of images belongs to the same person or not and also the similarity score of the testing identity with all the top gallery identities (Fig. 2).

4 Experiments and Results

In this section, we explain the details of the experiments, datasets used, and settings. We have used three publicly available datasets to evaluate the performance of our proposed work and compared the results to some of the existing state-of-the-art methods.

4.1 Datasets

All the experiments are conducted on three publicly available datasets named VIPeR [31], CUHK01 [32], and CUHK03 [20]. The details of the datasets are below:

VIPeR: VIPeR stands for Viewpoint Invariant Pedestrian Recognition dataset. As the name suggests, the images in this dataset are of persons with different camera viewpoints. There are 1264 images in total, of 632 persons, each person is having exactly two images of a different viewpoint. The images in this dataset are scaled down to 128 * 48 pixels.

CUHK01: This dataset is collected from the Chinese University of Hongkong. There is a total of 3884 images of 971 persons. Each person is having exactly four images from two different camera views. The images in this dataset are cropped manually in size of 168 * 60 pixels.

CUHK03: This dataset is also from the Chinese University of Hongkong with some large number of persons' identities and viewpoint variations. It is much larger and well-prepared than the earlier one and stored in a .MAT file. It has 13,164 images of 1360 individuals captured from six different surveillance cameras. Each individual is observed from two disjoint camera views. On an average, there are five to eight images per person in each view. The dataset contains three cells in .MAT file: detected, labeled, and testsets.

4.2 Implementational Details

We used Tensorflow for implementation and conducted all experiments on NVIDIA TITAN Xp GPU (Graphics Processing Unit) driver version 390.67 with capacity 12,195 MiB. We used l2 regularizer to optimize the networking rate. To take care of the overfitting problem we adopt 0.0005 weight decay factor at each convolution layer. Batch size is set to 150 for CUHK03 dataset. Training is done till the network stops converging, for our best we found that training till 80,000 epochs is optimal for our SCap Net.

We trained our network on two big datasets, namely CUHK03 and CUHK01 and reported the Rank 1 accuracy with CMC curve. Although VIPeR is a very small dataset in comparison with the other datasets, we evaluated our network on this dataset also to check whether our network is capable of cross-dataset experiment or not and found that it surpasses all the existing state of the art on this dataset.

We compared our proposed approach with existing works implementing Siamese networks [26, 33, 34] as well as another non-Siamese network-based approach [20]. We performed a rank-based performance-based analysis of our proposed method along with some of the state-of-the-art approaches. Figure 3a–c displays the Cumulative Matching Characteristic (CMC) curves for the different re-identification approaches [20, 26, 33, 34] on the three datasets. These figures depict the variation of matching rates on a scale of 0–1 with increment in rank between 1 and 10. Results

Fig. 3 CMC on **a** CUHK01;
b CUHK03; **c** VIPeR
datasets

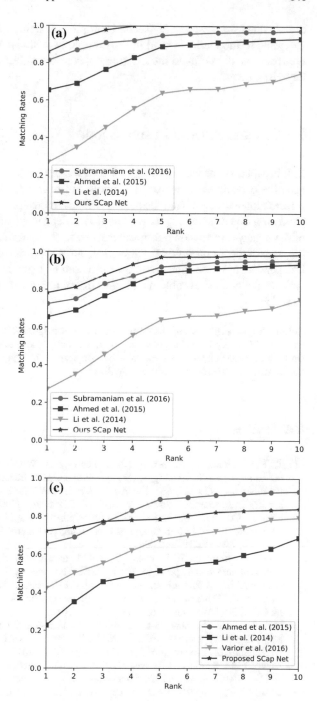

are obtained for a randomly selected sample consisting of 100 test identities corresponding to the methods used in the comparative study. We find that the proposed approach mostly eclipses the existing approaches in terms of Rank 1 accuracy for any given dataset.

5 Conclusions and Future Scope

In this paper, we propose a SCap Net architecture for person re-identification which uses the capsule architecture of convolution neural network. This capsule vector information extracted from the person images improves the performance of our network in terms of accuracy as well as it is independent of the scale and orientation of the images. Results on public datasets show that our method outperforms most of the state of the art proposed till now in terms of rank-based accuracy and orientation independence. In the future, our approach can be extended to some real-time applications like open set problem and high illumination changing conditions and other situations.

Acknowledgements The authors would like to acknowledge NVIDIA for supporting their research with the TITAN Xp Graphics processing unit. The authors also express their sincere gratitude to Dr. Pratik Chattopadhyay, Assistant Professor, Department of Computer Science and Engineering, IIT (BHU), Varanasi for his valuable support.

References

1. Tu, P.H., Doretto, G., Krahnstoever, N.O., Perera, A.A., Wheeler, F.W., Liu, X., Rittscher, J., B. Sebastian, T., Yu, T., Harding, K.G.: An intelligent video framework for homeland protection. In: Unattended Ground, Sea, and Air Sensor Technologies and Applications IX, vol. 6562, p. 65620C. International Society for Optics and Photonics (2007)
2. Kang, J., Cohen, I., Medioni, G.: Object reacquisition using invariant appearance model. In: Proceedings of the 17th International Conference on Pattern Recognition, 2004. ICPR 2004, vol. 4, pp. 759–762. IEEE (2004)
3. Corvee, E., Bremond, F., Thonnat, M., et al.: Person re-identification using spatial covariance regions of human body parts. In: 2010 7th IEEE International Conference on Advanced Video and Signal Based Surveillance, pp. 435–440. IEEE, Boston, MA, USA (2010)
4. Jüngling, K., Bodensteiner, C., Arens, M.: Person re-identification in multi-camera networks. In: CVPR 2011 Workshops, pp. 55–61. IEEE (2011)
5. Bedagkar-Gala, A., Shah, S.K.: Multiple person re-identification using part based spatio-temporal color appearance model. In: 2011 IEEE International Conference on Computer Vision Workshops (ICCV Workshops), pp. 1721–1728. IEEE (2011)
6. Zhang, Y., Li, S.: Gabor-LBP based region covariance descriptor for person re-identification. In: 2011 Sixth International Conference on Image and Graphics, pp. 368–371. IEEE (2011)
7. D'Angelo, A., Dugelay, J.-L.: People re-identification in camera networks based on probabilistic color histograms. In: Visual Information Processing and Communication II, vol. 7882, p. 78820K. International Society for Optics and Photonics (2011)

8. Cong, D.-N.T., Khoudour, L., Achard, C., Meurie, C., Lezoray, O.: People re-identification by spectral classification of silhouettes. Sig. Process. **90**(8), 2362–2374 (2010)
9. Javed, O., Shafique, K., Shah, M.: Appearance modeling for tracking in multiple non-overlapping cameras. In: 2005 IEEE Computer Society Conference on Computer Vision and Pattern Recognition (CVPR'05), vol. 2, pp. 26–33. IEEE, San Diego, CA, USA, USA (2005)
10. Wang, X., Doretto, G., Sebastian, T., Rittscher, J., Tu, P.: Shape and appearance context modeling. In: 2007 IEEE 11th International Conference on Computer Vision Workshops (ICCV Workshops), pp. 1721–1728. IEEE, Rio de Janeiro, Brazil (2011)
11. Dalal, N., Triggs, B.: Histograms of Oriented Gradients for Human Detection (2005)
12. Lowe, D.G.: Distinctive image features from scale-invariant keypoints. Int. J. Comput. Vis. **60**(2), 91–110 (2004)
13. Gray, D., Tao, H.: Viewpoint invariant pedestrian recognition with an ensemble of localized features. In: European Conference on Computer Vision, pp. 262–275. Springer, Berlin, Heidelberg (2008)
14. Schwartz, W.R., Davis, L.S.: Learning discriminative appearance-based models using partial least squares. In: 2009 XXII Brazilian Symposium on Computer Graphics and Image Processing, pp. 322–329. IEEE, Rio de Janeiro, Brazil (2009)
15. Dikmen, M., Akbas, E., Huang, T.S., Ahuja, N.: Pedestrian recognition with a learned metric. In: Asian Conference on Computer Vision, pp. 501–512. Springer, Berlin, Heidelberg (2010)
16. Zheng, W.-S., Gong, S., Xiang, T.: Reidentification by relative distance comparison. IEEE Trans. Pattern Anal. Mach. Intell. **35**(3), 653–668 (2012)
17. Wu, Y., Minoh, M., Mukunoki, M., Lao, S.: Set based discriminative ranking for recognition. In: European Conference on Computer Vision, pp. 497–510. Springer, Berlin, Heidelberg (2012)
18. Ojala, T., Pietikäinen, M., Mäenpää, T.: Multiresolution gray-scale and rotation invariant texture classification with local binary patterns. IEEE Trans. Pattern Anal. Machi. Intell. **7**, 971–987 (2002)
19. Hirzer, M., Roth, P.M., Bischof, H.: Person re-identification by efficient impostor-based metric learning In: 2012 IEEE Ninth International Conference on Advanced Video and Signal-Based Surveillance, pp. 203–208. IEEE, Beijing, China (2012)
20. Li, W., Zhao, R., Xiao, T., Wang, X.: Deepreid: deep filter pairing neural network for person re-identification. In: Proceedings of the IEEE Conference on Computer Vision and Pattern Recognition, pp. 152–159 (2014)
21. Ding, S., Lin, L., Wang, G., Chao, H.: Deep feature learning with relative distance comparison for person re-identification. Pattern Recogn. **48**(10), 2993–3003 (2015)
22. Zheng, A., Zhang, X., Jiang, B., Luo, B., Li, C.: A subspace learning approach to multishot person reidentification. IEEE Trans. Syst. Man Cybern.: Syst. **99**, 1–10 (2018)
23. Zhou, J., Su, B., Wu, Y.: Easy identification from better constraints: multi-shot person re-identification from reference constraints. In: Proceedings of the IEEE Conference on Computer Vision and Pattern Recognition, pp. 5373–5381 (2018)
24. Bromley, J., Guyon, I., LeCun, Y., Säckinger, E., Shah, R.: Signature verification using a "Siamese" time delay neural network. In: Advances in Neural Information Processing Systems, pp. 737–744 (1994)
25. Ge, Y., Gu, X., Chen, M., Wang, H., Yang, D.: Deep multi-metric learning for person re-identification. In: 2018 IEEE International Conference on Multimedia and Expo (ICME), pp. 1–6. IEEE, San Diego, CA, USA (2018)
26. Ahmed, E., Jones, M., Marks, T.K.: An improved deep learning architecture for person re-identification. In: Proceedings of the IEEE Conference on Computer Vision and Pattern Recognition, pp. 3908–3916 (2015)
27. Li, D., Chen, X., Zhang, Z., Huang, K.: Learning deep context-aware features over body and latent parts for person re-identification. In: Proceedings of the IEEE Conference on Computer Vision and Pattern Recognition, pp. 384–393 (2017)
28. Lv, J., Chen, W., Li, Q., Yang, C.: Unsupervised cross-dataset person re-identification by transfer learning of spatial-temporal patterns. In: Proceedings of the IEEE Conference on Computer Vision and Pattern Recognition, pp. 7948–7956 (2018)

29. Lin, J., Ren, L., Lu, J., Feng, J., Zhou, J.: Consistent-aware deep learning for person re-identification in a camera network. In: Proceedings of the IEEE Conference on Computer Vision and Pattern Recognition, pp. 5771–5780 (2017)
30. Sabour, S., Frosst, N., Hinton, G.E.: Dynamic routing between capsules. In: Advances in Neural Information Processing Systems, pp. 3856–3866 (2017)
31. Gray, D., Brennan, S., Tao, H.: Evaluating appearance models for recognition, reacquisition, and tracking. In: Proceedings of the IEEE International Workshop on Performance Evaluation for Tracking and Surveillance (PETS), vol. 3, pp. 1–7. Citeseer (2007)
32. Li, W., Zhao, R., Wang, X.: Human reidentification with transferred metric learning. In: Asian Conference on Computer Vision, pp. 31–44. Springer, Berlin, Heidelberg (2012)
33. Varior, R.R., Shuai, B., Lu, J., Xu, D., Wang, G.: A Siamese long short-term memory architecture for human re-identification. In: European Conference on Computer Vision, pp. 135–153. Springer, Cham (2016)
34. Subramaniam, A., Chatterjee, M., Mittal, A.: Deep neural networks with inexact matching for person re-identification. In: Advances in Neural Information Processing Systems, pp. 2667–2675 (2016)

An Embedded System for Gray Matter Segmentation of PET-Image

Khakon Das, Dipankar Khorat and Samarendra Kumar Sharma

Abstract This paper presents a complete embedded system for segmentation of gray matter of the human brain from Positron emission tomography (PET) image using 8-bit embedded Atmel microcontroller operating at low power and a frequency of 16 MHz. Hardware implementation of the embedded system for gray matter segmentation (ESGMS) of PET image is chosen as hardware design is more efficient by adopting the principles of software development. The ideology of parallelism achieved by the sixth level of software pipelining and two levels of dedicated hardware pipelining. Using lifting technique, the wavelet function is generated and denoised the same simultaneously, generally known as second-generation wavelet. The proposed architecture reduced number of memories through in-place approach as the resultant of the operands is overwritten on the operand location. For realization, Haar wavelet is used as mother wavelet in the design which is compiled lifting technique to perform split and update. Haar mother wavelet with the lifting scheme procures the denoised image for future operations. The wavelets confer a better output image, which is established with altering different statistical measures. With our proposed system, we can obtain peak signal to noise ratio (PSNR) about 107 dB. For segmentation, unsupervised clustering algorithm (K-mean) is used and each segment is superimposed with different colors. The gray matter of the human brain in the PET image is leveled using pink color.

Keywords Positron emission tomography · Denoising · Segmentation · K-mean · Wavelet transform · Embedded system

K. Das (✉) · D. Khorat
RCC Institute of Information Technology, Kolkata 700015, India

S. K. Sharma
Department of Health and Family Welfare, Kolkata, West Bengal, India

© Springer Nature Singapore Pte Ltd. 2020
J. K. Mandal and S. Mukhopadhyay (eds.), *Proceedings of the Global AI Congress 2019*, Advances in Intelligent Systems and Computing 1112,
https://doi.org/10.1007/978-981-15-2188-1_12

1 Introduction

Conventionally, diagnostic imaging in morphological constructions was limited to the imagining of irregularities. Since the contrast acquired to differentiate pathophysiological abnormality defines the efficiency of an imaging modality, it was necessary to develop extremely delicate techniques for imaging changes in brain functioning. In general, any radiation that is passing through the skeletons, skin, muscle, and the tissue of the human body can be used as a signal (information) source for imaging and diagnostics. In PET imaging [1] systems, the high energy photons are emitted due to the annihilation of positrons and electrons, which provides useful information about the object. During the signal collection period, noise is also generated and added to the corresponding images [2]. Noise is a fundamental characteristic, and it is found in all types of images. Noise decreases the visibility level of some structures and entities, especially those that have quite low contrast. Operations on images are done level by level to enhance the image information without changing the original image information content [3]. The main focus is to reduce the noise to a clinically acceptable level of medical imaging. Denoising is a vital step for boosting the quality of medical images (PET, CT, and MRI). Therefore, the type of the noise of the signal should be recognized, and its extent and statistical properties should be studied to design optimal denoising techniques for the recognized type of noise [4–6]. However, details of image information like edge, curve, etc. are lost during denoising, which may be retrieved by using image edge preservation [7], image contrast enhancement [8, 9]. Mallat gave the concept of discrete wavelet transform (DWT) for the first time in 1989. In the modern era, the DWT is used broadly for medical imaging to denoise the image and reproduce a clinically acceptable image. Several modifications of discrete wavelet transform have been made to amend its performance. Sweldens gave an incipient concept of lifting scheme [10] in 1996, which is planarity predicated on in-place calculation [11] and thereby eliminates the requirements for auxiliary memory. Das et al. already designed a hybrid algorithm to denoise medical images using modified Haar wavelet and fusion of two CT and PET images [12] and an embedded implementation of early started hybrid denoising was designed by Das et al. [13]. The k-means, first used in 1967 by James MacQueen [14], is a simple unsupervised learning technique that capable of solving the well-known problems of clustering. Hardware implementation of discrete wavelet transform was made to reduce the processing cost, time, and power as it involves sizably voluminous computations [15, 16] The k-means is fast, robust, and easier to understand and efficient enough to be implemented for faster processing with discrete wavelet transform in a hardware system. This article is divided into six subsections where Sect. 2 contains a related study regarding lifting scheme, wavelet, k-means algorithm, and a brief insight of the embedded system. In Sect. 3, design methodology describes the embedded system design and Sect. 4 deals with the result, including design benefit, statistical result analysis, and visual output analysis. Section 5 contains a detailed discussion, and Sect. 6 concludes the article.

2 Related Study

In this section, we perform related study which is most relevant to our proposed work.

2.1 Lifting Scheme

Using the lifting scheme, we can design wavelets as well as perform the discrete wavelet transform at the same time. In an implementation, time of lifting scheme by merging these steps, the wavelet filters are designed while performing the wavelet transform that is why it is called the second-generation wavelet transform. Wim Sweldens introduce this lifting scheme technique [10] In this scheme, factorization of any DWT is done to reduce the number of mathematical operations (by approximating the factor of two). In Fig. 2, there is a most straightforward version of a wavelet transform, shown with the lifting scheme. Wavelet construction of wavelet transform is performed in the predict step of lifting scheme, and this act as a high-pass filter. Scaling function is done in the update step by which we get a smoother version of the data [11]. It is also noteworthy that the wavelet transform can now be calculated by way of in-place calculation. It means that for a given finite-length signal with n samples, we need exactly n memory cells, each of them capable to store one sample, to compute the transform (Fig. 1).

2.2 Wavelet

A wavelet is a waveform whose amplitude begins at zero, then increases after that decreases to zero. Wavelets are purposefully made to have particular characteristics that make them useful for image and signal processing. Wavelets can be coalesced with kenned portions of a damaged signal to extract information from the unknown portions of a damaged signal. Wavelet is a great mathematical tool by using it; we can able to extract information from different types of data like images and signals [17]. To achieve quick denoising of a noisy signal, the discrete wavelet transform is easy to implement [18, 19]. Haar wavelet is defined by a simple mathematical function which

Fig. 1 Lifting scheme

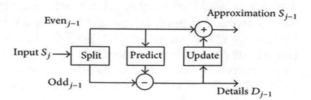

Fig. 2 Wavelet transform
(Haar)

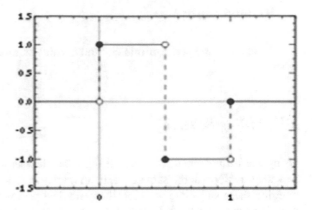

generates the series of rescaled square formed. Wavelet analysis sanctions a target function over an interval to be represented in terms of an orthogonal substructure, as similar to Fourier analysis [20]. The Haar wavelet's mother wavelet function $\Psi(t)$ can be described as

$$\Psi(t) = \begin{cases} 1, & \text{if } 0 \leq t < \frac{1}{2} \\ -1, & \text{if } \frac{1}{2} \leq t < 1 \\ 0, & \text{otherwise} \end{cases} \quad \text{here } t \text{ denotes time}$$

Its scaling function $\varphi(t)$ can be described as

$$\varphi(t) = \begin{cases} 1, & \text{if } 0 \leq t < 1 \\ 0, & \text{otherwise} \end{cases}$$

2.3 K-Mean Clustering Algorithm

An unsupervised learning method (K-means) is used to solve the problem of clustering where k is the number of clusters known as a priori. To define the k centers is the central concept, for each of every cluster. Randomly select "p" cluster center point after calculating the distance between the cluster centers and each data point. Assign the cluster level to a data point, which cluster center has the minimum distance from that data point. This process is repeated for all data points. Recalculate the new cluster center using $W_i = (1/p_i) \sum_{j=1}^{p_i} X_i$ where, p_i is the number of data points in ith cluster. Here, the set of data points is $V = v1, v2, v3, \ldots, vn$ and the set of centers is $W = w1, w2, \ldots, wp$. We calculate the distance between each data point and another new obtained cluster centers again. If no reassigned data point was found, then stop, otherwise, the process continues from assigning the data point.

Finally, this algorithm aspires at minimizing an objective function, which is known as the squared error function given by [14] $J(w) = \sum_{i=1}^{P} \sum_{j=1}^{p_i} (\|v_i - w_j\|)^2$ where, $\|v_i - w_j\|$ is the Euclidean distance between v_i and w_j, p_i is the number of data points in ith cluster. p is the number of cluster centers.

2.4 Embedded System

It is convenient to program an embedded system specifically for the function it is designed to do. It can, therefore, have a lot of distinct CPU architectures. In these processors, the most common word length is in the 8- to16-bit range. Based on their clock frequency, storage size, and voltages, embedded processors are differentiated. We use an embedded processor having a capacity of 4 KB storages [21]. There are hundreds of embedded processors available. Embedded processors can be distinguished using the architectures despite the vast amount of available designs. The most important is the performance of the system; it measures inherently by a combination of throughput, number of cycles, clock period.

3 Design and Methodology

K. Das et al. already developed a technique for denoising medical images using CT-PET image fusion and modified Haar wavelet and implemented an embedded system of the same. Here, we propose an alternative approach with PET image to implement the embedded system for medical image segmentation as a positron emission tomography image gives us excellent information about the organ and soft tissues. We have used modified Haar wavelet as mother wavelet and applied the wavelet transformation using the lifting technique through split and update of wavelet, which can also be called the second-generation wavelet. The memory space is reduced through the in-place calculation in the lifting scheme. Now, we need to cluster the denoised image to segment out the gray matter of the human brain, for which we used the very well established and unsupervised clustering method, k-mean. Hardware implementation of the ESGMS using PET is preferred because hardware presents a proper scope of parallelism and pipelining over software. ESGMS using PET can give good result even a high level of noise.

3.1 Embedded System Design

Figure 3 describes the basic spinal circuitry of the proposed embedded system using Atmel microcontroller. The circuit is switched on 5 V dc supply, which is connected to the Vin pin of 28-pin PDIP Atmel Atmel328/P and the circuit is grounded through

Fig. 3 Circuit diagram for
SD card access module

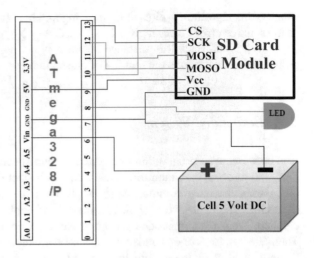

GND pin. We have connected the digital pins to interface the SD card with the microcontroller. To power on the SD card, the 5 V supply is used, and the pin-10, pin-11, pin-12, and pin-13 are used for chip select (CS), Master Out Slave In (MOSI) TxD data, Master In Slave Out (MISO) RxD data, and Serial Clock (SCK), respectively. The full process is based on a synchronous serial data protocol, Serial Peripheral Interface (SPI) which microcontrollers used to communicate over a limited distance with one or more peripheral devices quickly. Through SPI, communication between two microcontrollers can be established. In the case of SPI connection, a master device, i.e., a microcontroller controls the external devices.

In MOSI, the master line is for sending data to the external devices. SCK is the clock pulses to synchronize data transmission, and in the case of MISO, the master is receiving data through the slave line. Figure 4 shows the designed embedded system of our proposed method. We have used one TFT monitor to display the denoised and segmented PET image where pink color represents the gray matter of the human brain, as shown in Fig. 4.

4 Experimental Result

In this section, we perform in-depth analysis of various output result. In Sect. 4.1, we discuss about design benefit, and in Sect. 4.2, statistical result analysis is discussed. On the other hand, Sect. 4.3 is about visual output and analysis.

Fig. 4 Working model of embedded denoising and segmentation of PET-medical image using wavelet and K-mean

4.1 Design Benefit

In our proposed model, a single SD card is used for storing the result after completion of each process, both input and output data. At the very first stage, the input data are read from the SD card to the register after getting the input data processor performs the required operation on that and sends back the output data to the internal register. After that, the output result is written back to the SD card. In this way, a single cycle execution of our algorithm completed and the operation can repeatedly be done for multiple cycles and pixels of the given image. Fig. 5a describes the method for multiple pixels and multiple cycle formats in our proposed algorithm. In this method, we considered an image of n (128 × 128) pixels where the processor needs "n" times card read operations to read every pixel and after performing required operation write back "n" times to SD Card. In our proposed methodology in each level of denoising, the pixels are denoised, after a certain number of denoising level significant data loss may arise.

We have examined and found that in the sixth denoising level, we get a denoised image which is clinically acceptable. To perform the sixth level of denoising system need $6n$ times registers and SD card read and write operation. This process becomes a huge time taking because of the too many operations, i.e., read and write on SD card. The drawback is resolved using loop optimization, as shown in Fig. 5b where all denoised pixels of sixth denoising level stored in the register and written back into the SD card at ones. Through that process, we achieved optimized loop benefits. The flow of the process comparing Fig. 5a with Fig. 5b. We can conclude that in Fig. 5b, the time required to fetch data from the SD card and store the output data to the SD card is much less than that of Fig. 5a.

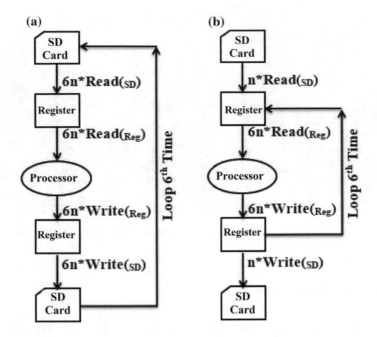

Fig. 5 **a** Multiple cycle format of proposed algorithm. **b** Multi cycle execution of proposed algorithm with optimization loop [13]

4.2 Statistical Result Analysis

We have used 8-bit unsigned integer image of size 128×128. After applying the proposed method, we have calculated the time required for sixth level denoising including SD card read/write and also calculated the number of clocks required for denoising, which we have shown in tabular form through Tables 1 and 2, respectively.

Figure 6 shows the benefit of loop optimization in terms of time used in our proposed work where the vertical axis represents time in seconds for SD card read/write and horizontal axis presents single level, sixth level and optimized sixth level of denoising.

In Fig. 7, we have shown the time benefit for processing using the modified discrete wavelet transform.

Table 1 Time required for sixth level denoising

Run	SD read (s)	DWT-M (s)	SD write (s)	Total time (s)
1	104.527644	0.401132	42.419712	147.348488
2	103.339440	0.400492	42.387420	146.127352
3	103.699636	0.400392	42.409640	146.509668

Table 2 Number of clock required for sixth level denoising

DWT-M (s)	Clocks = (time * frequency) (Atmel 328p, 16 MHz)	Avg. clocks
0.401132	6,418,112	6,410,752
0.400492	6,407,872	
0.400392	6,406,272	

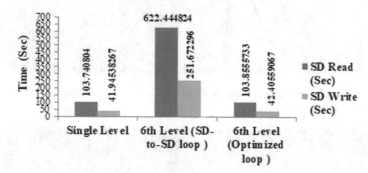

Fig. 6 Chart for benefit of loop optimization

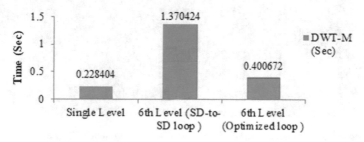

Fig. 7 Chart for benefit of modified DWT processing

Here also, vertical axis represents time in seconds for SD card read/write for DWT-M operation, and horizontal axis presents single level, sixth level and optimized sixth level of denoising in the chart in Fig. 7.

4.3 Visual Output and Analysis

Here in Fig. 8, we have shown the output observed in the display device of our proposed embedded system where Denoised and Segmented Gray Matter for Patient 1. Figure 8a represents the raw image of size 128×128 and depth information is of 8 bits. Now, we applied the modified DWT and performed the sixth level of denoising of the input image where we used Haar as mother wavelet, shown in Fig. 8b. In

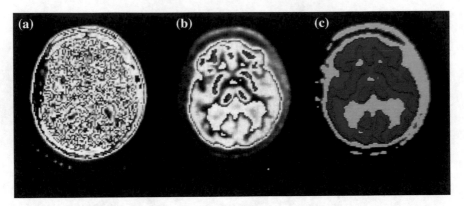

Fig. 8 Embedded system-generated output of denoised and Segmented Gray Matter for Patient 1 using proposed method

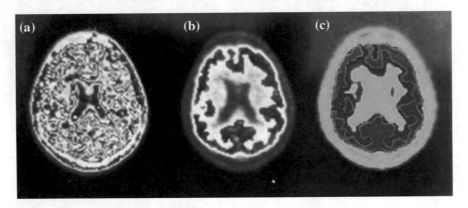

Fig. 9 Embedded system-generated output of denoised and Segmented Gray Matter for Patient 2 using proposed method

Fig. 8c is the output after segmentation, where we used an unsupervised clustering method, i.e., k-means. Here, we superimposed different colors for different clusters where pink color signifies the gray matter of the brain of Patient 1. Moreover, the same proposed method was applied to Patient 2 and shown in Fig. 9, where a, b, and c signify the same terminologies, respectively.

Table 3 gives the details about the time and number of clocks required for the complete execution of our proposed work which includes different execution times like SD card read, denoising and segmentation, SD card write, LCD display. This proposed algorithm was applied to original PET image of some patients, and here, we have presented the statistical parameter (i.e., MSE, PSNR, and UIQI) data using the brain PET image of two patients. From the result as presented Table 4, we can say that we have achieved good quality the denoising output images based on the statistical parameters like MSE, PSNR. From universal image quality index (UIQI), we can say that few improvements is required.

Table 3 Time and number of clock calculation for complete execution of proposed algorithm

Features	SD read (input + denoising)	Denoising (M-DWT) and (K-mean)	SD write (denoised + cluster)	LCD display	Total
Clocks	3,336,559,616	8,870,336	1,376,439,296	70,320,064	4,792,189,312 × 10^6
Time (s)	208.534976	0.554396	86.027456	4.395004	299.511832

Table 4 Comparison study with benchmark existing techniques (DWT) and proposed method based on MSE, PSNR, and UIQI

Algorithm		PET-image (Patient 1)			PET-image (Patient 2)		
		MSE	PSNR (dB)	UIQI	MSE	PSNR (dB)	UIQI
Basic DWT	Input image	537,052.74	±18.27074	0.49312	202,646.73	±9.80519	0.49923
	Level 1	50.08105	+62.03613	0.76160	64.67578	+60.11476	0.75050
	Level 6	43.96118	+63.83012	0.95319	43.79785	+63.50054	0.94070
Modified DWT	Input image	537,052.74	±18.27074	0.49312	202,646.73	±9.80519	0.49923
	Level 1	42.33423	+63.79577	0.69068	29.08185	+67.05716	0.67821
	Level 6	0.29688	+106.87813	0.74180	0.24139	+108.67507	0.73512

5 Discussion

In our proposed model, we deal with the gray matter segmentation of human brain from the PET image. In this particular research, we came across that there was a huge time required while using SD to SD format. We overcome the problem with register to register format and performed sixth level denoising; this is how we optimized the loop as well as save the clock, shown in Fig. 5a and b. As we already mentioned in Fig. 8 and Fig. 9 that for denoising purpose, we efficiently develop the modified DWT to input PET image using Haar wavelet as the mother wavelet. Based on the statistical image quality measurement parameters, i.e., MSE, PSNR, UIQI, we have achieved a better-denoised image for further consideration of our algorithm. Using the admired unsupervised clustering algorithm, k-means, we get a segmented image where the different regions of the human brain are represented using a different color as shown in Figs. 8 and 9 for two randomly chosen patients. For our consideration, we have used color pink to represent the gray matter of the brain. From Fig. 6, it is relevant to articulate that only using a few more seconds than the single cycle (SD to SD) format, we can achieve the result using sixth level loop optimization (register to register format). As per the UIQI is a concern, the closer the UIQI to unity, the better the denoised image is [22]. From Table 4, the input image UIQI is 0.49923, which is got enhanced with the level of denoising. For Level-1, we get UIQI 0.69068 and 0.67821 for Patient 1 and Patient 2, respectively, whereas, after Level-6, we get UIQI 0.74180 and 0.73512, respectively. From the UIQI, result is indicative that we get a result, good enough for clinical consideration. The MSE, we gain from our result is small enough to replicate that the denoised image has significant assessment quality [23]. The PSNR [24] is such a remarkable feature for any design. We achieved PSNR about +106.87813 dB and +108.67507 dB for Patient 1 and Patient 2, respectively, which is significantly increased through Level-1 to Level-6 from the primary input image PSNR of ± 18.27074 dB and ± 9.80519 dB respectively. From the PSNR viewpoint, the result we get after applying our proposed method is adequate in the field of gray matter segmentation of human brain.

Pros and Cons of proposed method: K-means can characterize the regions adequately. FCM identifies only three tissue classes, whereas K-means identifies all the six classes. Segmentation of brain with histogram guided initialization of cluster, K-means can cluster the regions comparatively better than FCM [25, 26]. Even wavelet guided initialization of cluster can also able to reduce the clustering time, we use wavelet guided initialization.

6 Conclusions

In our proposed model, we aim to design an embedded system for gray matter segmentation of human brain. For this particular purpose, we have worked on several PET images of the human brain of patients from RG Kar Medical College and Hospital, Kolkata. On those PET images, we applied our proposed method where we use Loop optimization for timing advancement, modified DWT for denoising, and k-means algorithm for segmentation. We able to achieve a better result in terms of MSE, PSNR, and UIQI mentioned in Sect. 5. After the final gray matter segmentation, the result was visually inspected by medical professionals from RG Kar Medical College and Hospital, Kolkata, Rabindranath Tagore International Institute of Cardiac Sciences, Kolkata and Apollo Gleneagles Hospitals, Kolkata and clarified that the output images are good enough and clinically acceptable.

Acknowledgements We would like to acknowledge Dr. Dinesh Jhaluka, Consultant Neurosurgeon, RG Kar Medical College and Hospital, Kolkata, Dr. Haseeb Hassan, Consultant Neurologist, Rabindranath Tagore International Institute of Cardiac Sciences, Kolkata and Dr. Punit Sharma, Professor, Apollo Hospitals Educational and Research, Kolkata for their valuable comments throughout the research and verification and validation of the output results.

References

1. Bailey, D.L., Townsend, D.W., Valk, P.E., Maisey, M.N.: Positron-Emission Tomography. Basic Sciences. Springer, London (2005)
2. Sonka, M., Hlavac, V., Boyle, R.: Image Processing, Analysis and Machine Vision. Springer, USA (1993)
3. Yu, S., Muhammed, H.H.: Noise Type Evaluation in Positron Emission Tomography Images (2016)

4. Unser, M., Aldroubi, A.: A review of wavelets in biomedical applications. Proc. IEEE **84**(4), 626–638 (1996)
5. Nowak, R.D.: Wavelet-based rician noise removal for magnetic resonance imaging. IEEE Trans. Image Process. **8**(10), 1408–1419 (1999)
6. Portilla, J., Strela, V., Wainwright, M.J., Simoncelli, E.P.: Image denoising using scale mixtures of Gaussians in the wavelet domain. IEEE Trans. Image Process. **12**(11), 1338–1351 (2003)
7. Verma, K., Singh, B.K., Thoke, A.S.: An enhancement in adaptive median filter for edge preservation. Procedia Comput. Sci. **48**, 29–36 (2015). International Conference on Computer, Communication and Convergence (ICCC 2015)
8. Singh, S., Pal, P.: Contrast enhancement of medical images: a review. IJRDO: J. Health Sci. Nurs. 1(4), 32–35 (2016). ISSN: 2456-298X
9. Nancy, S.K.: Contrast enhancement of medical images: a review. IOSR: J. Comput. Eng. (IOSR-JCE) **9**(6), 84–88 (2013)
10. Daubechies, I., Sweldens, W.: Factoring wavelet transforms into lifting steps. J. Fourier Anal. Appl. **4**(3), 247–269 (1998)
11. Kuzume, K., Niijima, K., Takano, S.: FPGA-based lifting wavelet processor for real-time signal detection. Sig. Process. **84**(10), 1931–1940 (2004)
12. Das, K., Maitra, M., Sharma, P., Banerjee, M.: Early started hybrid denoising technique for medical images. In: Bhattacharyya, S., Mukherjee, A., Bhaumik, H., Das, S., Yoshida, K. (eds.) Recent Trends in Signal and Image Processing, pp. 131–140. Springer, Singapore (2018)
13. Das, K., Maitra, M., Banerjee, M., Sharma, P.: Embedded implementation of early started hybrid denoising technique for medical images with optimized loop. In: Mandal, J.K., Bhattacharya, D. (eds.) Emerging Technology in Modelling and Graphics, pp. 295–308. Springer, Singapore (2019)
14. Macqueen, J.: Some methods for classification and analysis of multivariate observations. In: 5th Berkeley Symposium on Mathematical Statistics and Probability, pp. 281–297 (1967)
15. Chilo, J., Lindblad, T.: Hardware implementation of 1d wavelet transform on an FPGA for infrasound signal classification. IEEE Trans. Nucl. Sci. **55**(1), 9–13 (2008)
16. Motra, A.S., Bora, P.K., Chakrabarti, I.: An efficient hardware implementation of DWT and IDWT. In: TENCON 2003. Conference on Convergent Technologies for Asia-Pacific Region, vol. 1, pp. 95–99 (2003)
17. Haar, Alfred: Zur theorie der orthogonalen funktionensysteme. Math. Ann. **69**(3), 331–371 (1910)
18. Sweldens, W.: The lifting scheme: a construction of second generation wavelets. SIAM J. Math. Anal. **29**(2), 511–546 (1998)
19. Das, K., Daschaklader, D., Roy, P.P., Chatterjee, A., Saha, S.P.: Epileptic seizure prediction by the detection of seizure waveform from the pre-ictal phase of EEG signal. Biomed. Signal Process. Control **57**, 101720 (2020)
20. Sweldens, Wim: Wavelets and the lifting scheme: a 5 minute tour. Z. Angew. Math. Mech. **76**, 41–44 (1996)
21. Alldatasheet.com: Atmega328p datasheet (16/20 pages). atmel: Atmel® avr® 32-bit microcontrollers (2019)
22. Wang, Z., Bovik, A.C.: A universal image quality index. IEEE Signal Process. Lett. **9**(3), 81–84 (2002)
23. Wang, Z., Bovik, A.C.: Mean squared error: love it or leave it? A new look at signal fidelity measures. IEEE Signal Process. Mag. **26**(1), 98–117 (2009)
24. Huynh-Thu, Q., Ghanbari, M.: Scope of validity of PSNR in image/video quality assessment. Electron. Lett. **44**(13), 800–801 (2008)
25. Nimeesha, Rajaram: Automated brain tumour segmentation techniques—a review. Int. J. Comput. Sci. Inf. Technol. Res. Excell. **3**(2), 60–65 (2013)
26. Angulakshmi, M., Lakshmi Priya, G.G.: Automated brain tumour segmentation techniques—a review. Int. J. Imaging Syst. Technol. **27**(1), 66–77 (2017)

Estimation of Resemblance and Risk Level of a Breast Cancer Patient by Prognostic Variables Using Microarray Gene Expression Data

Madhurima Das, Biswajit Jana, Suman Mitra and Sriyankar Acharyya

Abstract Breast cancer is a common type of cancer affecting women worldwide. Continuous efforts are being made for the identification of significant genes for prognosis of breast cancer. A microarray gene expression dataset contains tens of thousands of genes. Identification of smaller subset of disease-causing genes from a large gene expression dataset is a challenging task for the researchers. Here, a variant of Cox proportional hazard regression model, namely 1d-DDg method, has been applied to select the predictive genes for breast cancer patients. Here, in this paper, instead of using the Euclidian distance, the Manhattan distance has been used to estimate the risk level of a query (newly admitted) patient by matching similarity with the reference (existing) patients. The time complexity of Manhattan distance is better compared to Euclidian distance. The post-surgery disease recurrence and the level of risk of a breast cancer patient can be reliably predicted using personified prognosis.

Keywords Breast cancer · Cox proportional hazard regression model · 1d-DDg · Manhattan distance

1 Introduction

Cancer has become a big threat to mankind globally. It is the second leading cause of death worldwide and considered for 9.6 million deaths [1] in 2018. Cancer is

M. Das (✉) · B. Jana · S. Mitra · S. Acharyya
Department of Computer Science and Engineering, Maulana Abul Kalam Azad
University of Technology, Haringhata, West Bengal, India
e-mail: madhurimadas64@gmail.com

B. Jana
e-mail: biswajit.cseng2012@gmail.com

S. Mitra
e-mail: smsumanmitra@gmail.com

S. Acharyya
e-mail: srikalpa8@gmail.com

© Springer Nature Singapore Pte Ltd. 2020
J. K. Mandal and S. Mukhopadhyay (eds.), *Proceedings of the Global
AI Congress 2019*, Advances in Intelligent Systems and Computing 1112,
https://doi.org/10.1007/978-981-15-2188-1_13

159

a disease which is formed in the tissue level [2, 3]. Data [3] suggest that prostate and breast cancers are very common in men and women, respectively. Breast cancer includes 6.28% of all cancers [4] among Indian women in 2018.

The main objective of our work is to diagnose the breast cancer disease of the patients at the preliminary stage so that the proper treatment to the patients would be provided. A particular proportion of breast cancer can be cured by chemotherapy and radiotherapy, surgery or prognosis by detecting the cancer earlier.

Most of the studies of breast cancer are based on microarray data analysis [5]. The gene expression datasets are available in public repositories, i.e. The Cancer Genome Atlas (TCGA) [6], National Centre for Biotechnology Information (NCBI).

Tarca et al. [7] discussed about the molecular tools and fingerprinting which help to characterize the reproductive processes and disorders of a gene sample. Hansebout et al. [8] proposed that the concept of prognosis is not used only for deciding proper treatment but also focuses on the opinion of patients or relatives to determine whether the patient is suitable for surgery or not. Nounou et al. [9] considered that the advancement in molecular biology helps to understand the breast cancer in a better way. Dobbin and Simon [10] developed an algorithm to determine the proportion of splitting the dataset in an optimal way by simulating under different conditions which determine the best splitting strategies. Tang et al. [11] investigated the potential roles, clinic pathological functions and prognostic values of let-7 miRNA family in HG-SOC. Bao and Davidson [12] discussed about the breast cancer diagnosis, prognosis in order to optimize and individualize breast cancer treatment. Chen et al. [13] illustrated their proposed methodology with an example of real cancer dataset using ample conditions for both types of data, i.e. completely observed data and data with missing covariates. Ades et al. [14] discussed the improvement and modification of the treatment of breast cancer from the ancient period to future approaches like molecular profiling, gene signatures.

Here, in this paper, the experiment has been performed with the breast cancer patient dataset. The analysis from the experiment suggests that there are some specific genes, which are selected using Cox proportional hazard regression, are responsible for the breast cancer disease. A model based on 1d-DDg [15] has been implemented where both the query (newly admitted) and reference (existing) patients are categorized into two risk groups, i.e. low risk and high risk, based on those selected genes. The level of risk evaluation and matching homogeneity between a query patient and the reference patients is calculated using Manhattan distance [16]. The treatment of the newly admitted, i.e. query patient will be suggested accordingly.

The rest of the paper has been organized as follows. Section 2 discusses the personified prognosis using gene expression analysis. The algorithms COX_SELECTGENE for selecting the disease responsible genes and SIMILAR_PATIENTS for matching the homogeneity and risk level of patients have been discussed in Sect. 3. Experimental results and discussion are presented in Sect. 4. Section 5 concludes the paper.

2 Personified Prognosis by Gene Expression Analysis

Selection of the predictive genes of a query patient as well as reference patient based on clinical data and gene expression data has been carried out by 1d-DDg [15] method. Then, the Manhattan distance has been applied to calculate the risk level of a query patient by matching the similarity with the reference patients. A patient's clinical information involved with disease risk such as disease relapse, overall survival, body mass index, age is referred to clinical data, whereas the molecular expressions of the variables are obtained by microarray analysis of individual DNA, mRNA and protein. TCGA, NCBI are the independent sources where microarray gene expression data as well as clinical data are available as an individual dataset. Two breast cancer datasets with accession number GSE2990 and GSE45255 obtained from National Centre for Biotechnology Information (NCBI) data portal containing normalized miRNA and mRNA expression profiles (Table 2) have been used in our paper. The description of these datasets is presented in Table 1. Each of these datasets was fabricated through a microarray experiment. In Table 2, each column is referred to individual sample (patient) containing microarray gene expression values of all genes. Each newly admitted patient sample generally consists of 22,268 genes. After removing the affymetrix probes, 22,215 genes prevailed.

The clinical information is available for both the datasets. The GSE2990 is referred to reference patient training set containing 189 existing reference patients, and GSE45255 dataset is referred to 139 newly admitted query sample patients. Clinical information of patients resides under the phenotypic column of the dataset.

This phenotype data consist of detail information about individual patient. Each dataset has its own microarray gene expression value for each patient. Gene expression values of both datasets undergo a preprocessing stage called normalization using

Table 1 Microarray gene expression datasets

S. No.	Dataset accession No.	Collected from (site)	No. of genes (n)	No. of sample patients (m)	Last updated
1	GSE2990	GEO, NCBI	22,283	189	06 September 2013
2	GSE45255	GEO, NCBI	22,268	139	21 March 2013

Table 2 Structure of microarray gene expression dataset

Gene	Reference GSE2990 dataset				Query GSE45255 dataset			
	Sample$_{R1}$	Sample$_{R2}$	…	Sample$_{Rm}$	Sample$_{Q1}$	Sample$_{Q2}$	…	Sample$_{Qt}$
G1	$g_{1,1}^{R1}$	$g_{1,1}^{R2}$	…	$g_{1,1}^{Rm}$	$g_{1,1}^{Q1}$	$g_{1,1}^{Q2}$	…	$g_{1,1}^{Qt}$
G2	$g_{2,1}^{R1}$	$g_{2,1}^{R2}$	…	$g_{2,1}^{Rm}$	$g_{2,1}^{Q1}$	$g_{2,1}^{Q2}$	…	$g_{2,1}^{Qt}$
⋮	⋮	⋮	…	⋮	⋮	⋮	…	⋮
G$_n$	$g_{n,1}^{R1}$	$g_{n,1}^{R2}$	…	$g_{n,1}^{Rm}$	$g_{n,1}^{Q1}$	$g_{n,1}^{Q2}$	…	$g_{n,1}^{Qt}$

Table 3 Combined table of phenotypic data and Z-score value of gene

Sample Id	Age	Survival time	Relapse	Grade	Gene1	Gene2	...	Gene n
GSM001	a_1	t_1	0	1	$g_{1,1}$	$g_{1,2}$	$\cdots$	$g_{1,n}$
GSM002	a_2	t_2	1	1	$g_{2,1}$	$g_{2,2}$	$\cdots$	$g_{2,n}$
:	:	:			:	:		:
GSM00m	A_m	t_m	0	3	$g_{m,1}$	$g_{m,2}$	$\cdots$	$g_{m,n}$

Eq. (1)

$$Z_{\text{score}} = \frac{x_{\text{gene}} - \mu}{\sigma} \tag{1}$$

where

x_{gene} is the value to be standardized, i.e. gene expression value

μ is the mean

σ is standard deviation

Z_{score} is 'Z-score' or standard score.

A mathematical model has been designed where a new affected breast cancer patient would be equivalent to a reference patient in the training cohort based on homogeneity and similarities in their prognostic signature vectors (PSVs). This prognostic signature vector (PSV) is derived from prognostic binary variable vector (PBVV) [16] which is originated from predictive gene selection and risk classification method. This could help us in precision medicine for clinical purpose. To predict the suggestive genes responsible for breast cancer, COX_SELECTGENE algorithm has been used. The normalized value of Z-score is combined with the phenotypic data of each sample, and it is represented in Table 3. After using the Cox proportional hazard regression model, a number of genes are selected based on their hazard ratio along with the P-value of respective genes. To remove the false positive from those selected genes, we have applied Benjamini–Hochberg FDR step-up [17] method, and finally, the most suggestive and significant genes responsible for breast cancer are selected. Each of these genes is survival-significant that assigned each patient with a value of 1 or 2 representing low or high-risk subgroup, respectively (Fig. 1).

3 Methods

Cox proportional hazard regression model has been used here to select the most significant genes responsible for breast cancer, and it has been discussed in Sect. 3.1. Similarity measurement and analysing the risk level using those significant genes have been discussed in Sect. 3.2.

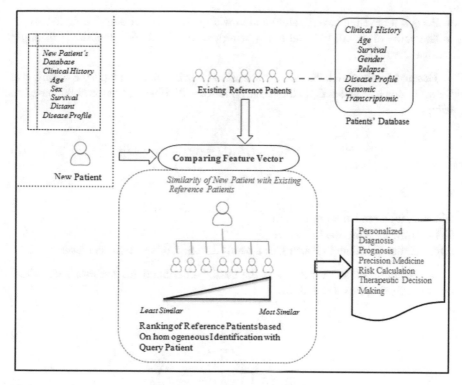

Fig. 1 Schema for estimation of predictive variables of breast cancer [16]

3.1 Cox Proportional Hazard Regression Model

Cox proportional hazard regression model is used to identify survival-significant genes of breast cancer patients globally [18]. Here, overall survival (OS) and recurrence of disease (event) for each patient are fitted to gene expression data. Using this method [11], individual gene is identified, and each identified gene arranged the patients into two individual groups of low risk and high risk. This Cox proportional hazard regression model represents the effect of independent variables as a multiplier of a common baseline hazard function $H_0(t)$.

$$H(X_{m,t}) = H_0(t) \exp\left[\sum_{m=1}^{p} X_{ml}\beta_l\right] \tag{2}$$

where

- $X_m = (x_{m1}, x_{m2}, ..., X_{mp})$ is a predictor variable. Here, X_{ml} is denoted as the gene expression values of p patients under an individual gene l.
- $H(X_m, t)$ is the hazard rate at time t for x_m.

- $H_0(t)$ is a baseline hazard rate function where all the covariates denote to zero.
- The value of β will be determined using partial likelihood function; β is single covariate here.

Hazard Ratio (HR): The Cox proportional hazard model relates the hazard rate for individuals at value X_m to the hazard rate for individuals of items at the baseline value.

$$\text{HR}(X_m) = h(X_m, t)h_0(t) = \exp\left[\sum_{m=1}^{p} X_m \beta_1\right] \tag{3}$$

If,

HR $= 1$: then No Effect in Hazard
HR < 1: then Deduction in Hazard
HR > 1: then Escalate in Hazard, as a result duration of survival decreases.

When there are m events, L_m is the partial likelihood for the mth event time, calculated using Eqs. (4) and (5).

$$L_p(\beta) = \prod_{m=1}^{s} L_m \tag{4}$$

$$L_p(\beta) = \prod_{m=1}^{s} \left(\frac{e^{\beta x_l}}{\sum_{j \in R(t_i)} e^{\beta x_l}}\right) \delta_l \tag{5}$$

where δ is the censoring variable. If $\delta = 1$, then event exists else, $\delta = 0$ then censored, and $R(t_1)$ is the risk set at time t_1.

Algorithm 1 COX_SELECTGENE
Take input for reference patient dataset **GSE2990** and query patient dataset **GSE45255** from National Centre for Biology Information (NCBI). Here, breast cancer dataset is selected which contains microarray gene expression data.
   Set parameters **idx**, discard
   Initialize **gset, x, meta_data, final, cox_data, res, annotlookup, ss**

1. Load platform data **GSE2990** from GEO to **gset** which are containing the gene expression data and phenotypic data.
2. Expression data of **gset** are stored to variable **x**.
3. Extract the relevant phenotypic data, i.e. **pdata** from **gset** and store it in **idx**.
4. A data frame named as **meta_data** is created where the related **pdata** is stored.
5. Remove samples containing NA value from **idx**.
6. Convert the expression data x to z-score and filter it to match the samples in our **pdata** using equation number (1).
7. Merge the **idx** containing **pdata** and z-score object. Store it in **cox_data**.

8. Change the column names of **pdata** in **cox_data** to '**Age**', '**Distant.RFS**', '**ER**', '**Grade**', '**Node**', '**Size**', '**Time.RFS**'.
9. Prepare phenotypic data by taking two columns '**Distant.RFS**' and '**Time.RFS**' which are essential for calculating the hazard of each gene.
10. Calculate coefficient β using equation number (5).
11. Calculate hazard of each gene of m number of patients at time t using equation number (3).
12. Calculate **P-value** using equation number (8).
13. All results of **coxph()** function for **n** number of genes of **m** number of patients are loaded in **res**.
14. Hazard ratio of every gene which is greater than 1 is stored in final along with their respective **P-value**.
15. Selected gene names, gene id and P-values are cumulated in **annotlookup**.
16. Apply equation number (9) on the selected genes to select the most survival suggestive genes. Set cut-off value of modified P-value to 0.01, and the modified P-value is stored in **ss**.
17. Categorize low risk and high risk of each patient based on modified individual gene expression values.

The $L_p(\beta)$ is associated to Chi-square distribution with a degree of freedom which is equal to the number of predictor variables evaluated. The P-value is used to determine the weight of every selected gene, and this P-value from gene expression dataset has been calculated using Eqs. (6)–(8).

$$SE_{gene} = \frac{variance_{gene}}{\sqrt{n}} \tag{6}$$

where n is the number of patients.

$$SE_{gene} = \frac{e^\beta - 1}{e^\beta} \tag{7}$$

$$P = 1 - (p\,chisq(\beta/SE_{gene})^2, 1) \tag{8}$$

The Benjamini–Hochberg method [17, 19] is a dynamic tool to reduce the false discovery rate. Sometimes, small P-values are the reason of incorrect rejection of true null hypotheses, and B–H procedure helps to avoid the false positives. To apply the Benjamini–Hochberg procedure, the individual P-values are kept in ascending order, and a rank is assigned to each P-value, i.e. smallest one is assigned as rank 1, second smallest is assigned as rank 2 and so on. Then, individual P-value is calculated using Eq. (9) which is named as modified P-value.

$$FDR_{B-H} = (r/t)^* Q_{fdr} \tag{9}$$

where

r = individual P-value's rank
t = total number of tests
Q_{fdr} = false discovery rate.

3.2 Similarity Measurement Using the Predictive Genes

Once the responsible genes for breast cancer are selected, P-value of those respective genes is converted to weight using negative logarithmic function using Eq. (10) where m is the number of selected genes.

$$wt_m = -\log_{10} p_m \tag{10}$$

The average weighted risk (AWR) of each reference patient as well as query patient is calculated using 15 genes, and the formulation is calculated using Eq. (11). Where m is the number of genes from 1 to y and l is the number of patients.

$$\text{AWR}_{\text{patient}} = \left[\sum_{m=1}^{y} wt_m r_{m,l} \Big/ \sum_{m=1}^{y} wt_m \right] \tag{11}$$

For each patient, the risk value of the predictive genes is rescaled and represented by 1 for high risk and -1 for low risk.

$$DP_l = dt_{1,l}\ dt_{2,l} \cdots dt_{m,l}$$
$$Dt = \text{Traspose}(DP_l) \tag{12}$$

Then, multiply the weight vector represented in a diagonal matrix and rescaling vector of risk for individual patient as given in Eq. (13).

$$Ad_j = \begin{pmatrix} wt_1 & 0 & 0 & \cdots & 0 \\ 0 & wt_2 & 0 & \cdots & 0 \\ \cdots & \cdots & \cdots & \cdots & \cdots \\ \cdots & \cdots & \cdots & \cdots & \cdots \\ 0 & 0 & 0 & \cdots & wt_y \end{pmatrix} \begin{pmatrix} dt_{,l\,1} \\ dt_{,l\,2} \\ \vdots \\ \vdots \\ dt_{,l\,j} \end{pmatrix} = \begin{pmatrix} ad_{,l\,1} \\ ad_{,l\,2} \\ \vdots \\ \vdots \\ ad_{,l\,j} \end{pmatrix} \tag{13}$$

The prognostic signature vector where each lth patient has mth number of genes is defined in Eq. (14).

$$PSV_l = [vt_{1,l}\ vt_{2,l} \cdots vt_{m,l}]$$
$$Vt_l = \text{transpose}(PSV_l) \tag{14}$$

where $vt_{m,l} = \sum\limits_{x=1}^{m} ad_{x,l}$.

Finally, Manhattan distance [20] is calculated by comparing the predictive vector of each newly admitted query patient hth with the existing reference patient lth using Eq. (15).

$$f(Vt_l, Vt_h) = |Vt_l - Vt_h| = \sum_{m=1}^{y} (vt_{m,l} - vt_{m,h}) \tag{15}$$

Algorithm 2 SIMILAR_PATIENTS

1. Here, the selected genes are considered as predictive variables. Calculate the weight of each of this variable from Equation number (10).
2. Calculate the average weighted risk of each patient in reference cohort from Equation number (11).
3. Rescale the risk vector 2 to 1 for high risk and 1 to −1 for low risk by Equation number (12).
4. Calculate the adjustment vector by Equation number (13).
5. Calculate the *PSV* for each reference patient jth of the training cohort from Equation number (14).
6. Repeat steps 1–6 for newly admitted query patients also.
7. Calculate Manhattan distance between reference patients and query patient using Equation number (15).

4 Experimental Results and Discussion

Hardware and software specification, data of breast cancer have been described in Sect. 4.1, and Sect. 4.2 includes result and discussion.

4.1 Experimental Setup

The entire experiment has been done on the computer having Intel Pentium dual-core CPU with 3.0 GHz, 1 TB hard disk and inbuilt 32 GB RAM. All the programs are written in R (version 3.5.0) and MATLAB R2015a (Windows version 64 bit).

Table 4 Weight of predictive variables

Serial No.	Variable name	External gene name	P-value	Weight
1	X203413_at	GNL3	0.001261497	2.89911395
2	X218499_at	IQCK	0.00137392	2.862038582
3	X220148_at	ALDH8A1	0.001535615	2.813717554
4	X204337_at	RRM2	0.001620391	2.790380165
5	X203217_s_at	HIF1AN	0.001659678	2.779976249
6	X218979_at	ZWILCH	0.002193855	2.658792048
7	X206055_s_at	NEK3	0.002425024	2.615283972
8	X203426_s_at	PAIP1	0.002637717	2.578771764
9	X208821_at	SNRPB	0.002780781	2.55583327
10	X222060_at	ZWINT	0.00284769	2.545507298
11	X204309_at	PSMA4	0.002964404	2.528062606
12	X202503_s_at	KRT8P12	0.003123523	2.505355315
13	X219485_s_at	PCLAF	0.003253353	2.487668841
14	X202534_x_at	CYP11A1	0.003269836	2.485474072
15	X218982_s_at	ABHD11	0.003385724	2.470348476

4.2   Results and Discussion

The Cox proportional hazard regression model has been applied to select the most significant genes statistically for breast cancer. The selected genes along with their P-values and weight are represented in Table 4.

The final outcome of the experiment is based on matching the similarity between a query patient and reference patients.

Here, the graph is suggesting that finding out the most similar reference patient using prognostic binary variable vector (PBVV) depicts no clear relationship between two patients, and the measure of similarity or dissimilarity is neither explanatory nor productive (Fig. 2a). In Fig. 2a, the prognostic binary variable vector of any patient is represented on a graphical format where X-axis refers to the number of predictive variables and Y-axis indicates the binary values of high risk versus low risk.

Again, the prognostic signature vector values are plotted on a graphical format (Fig. 2b) where X-axis indicates the number of predictive variables and Y-axis represents the modified value of PSV. Here, from the graph, we can presume that the newly admitted query patient GSM1100064 is mostly similar to the reference patient GSM65805.

To estimate the similarity between any two lines, we can simply calculate the sum of difference of the coordinate, and this is proportional to Manhattan distance. The ranking of patients based on Manhattan distance is presented in Table 5. Here, the actual value of the calculation is not important because the calculative value is useful for ranking the patients. So, we can use the quantitative value of Manhattan distance

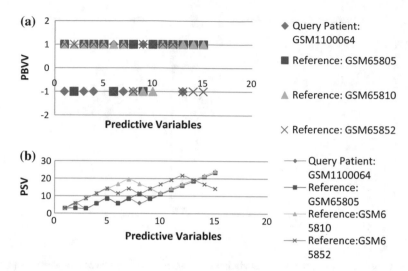

Fig. 2 Graphical representation of **a** PBVV and **b** PSV across 15 variables comparing one query patient and three reference patients

Table 5 Ranking of patients based on Manhattan distance

Reference patient	Manhattan_Distance GSM1100064 (query patient)	Rank
GSM65805	200.63	1
GSM65810	262.13	3
GSM65852	245.48	2

to compare the reference-query patients in the decreasing order in terms of similarity instead of Euclidean distance [20, 21]. Eventually, the most similar reference patient is recognized for risk prediction to the query patient. From Table 5, we can say that the query patient is most similar with the reference patient GSM65805.

Here, the Euclidean distance as well as Manhattan distance between all the reference patients and a single query patient has been plotted where X-axis represents the average weight of all the 125 reference patients and Y-axis represents the Euclidean distance (Fig. 3a) and Manhattan distance (Fig. 3b) between a single query and all the reference patients. Instead of using Euclidean distance, we have used Manhattan distance for its fast calculation of all the parameters as well as generating fast smallest path [21].

5 Conclusion

In this paper, 15 statistically significant genes are selected which are responsible for breast cancer. These genes are further used to measure the similarity between

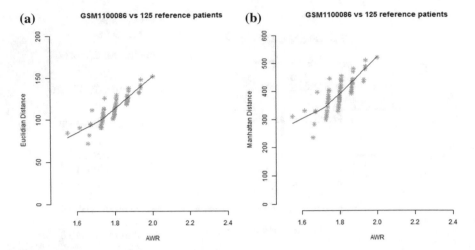

Fig. 3 Graphical representation of **a** Euclidian and **b** Manhattan distance of a single query patient to 125 reference patients using predictive vector

a query and reference patient using Manhattan distance. The Manhattan distance is applied here to find the level of risk of a query patient. The capability of the algorithms COX_SELECTGENE and SIMILAR_PATIENTS for matching the similarity between an unknown query patient and a known reference patient is worthwhile when the information of reference patients is already known. There must be an authentic clinical setting for collecting the tissue samples and processing those samples to create microarray data. Also, all the query and reference data must be measured with the same instrument and processed using uniform parameters and normalized procedures. The COX_SELECTGENE can be easily used in any other disease to select the significant genes.

References

1. International Agency for Research on Cancer: Latest global cancer data: Cancer burden rises to 18.1 million new cases and 9.6 million cancer deaths in 2018 (2018)
2. Ali, I., Wani, W.A., Saleem, K.: Cancer scenario in India with future perspectives. Cancer Ther. **8** (2011)
3. Hassanpour, S.H., Dehghani, M.: Review of cancer from perspective of molecular. J. Cancer Res. Pract. **4**(4), 127–129 (2017)
4. Bray, F., Ferlay, J., Soerjomataram, I., Siegel, R.L., Torre, L.A., Jemal, A.: Global cancer statistics 2018: GLOBOCAN estimates of incidence and mortality worldwide for 36 cancers in 185 countries. CA: Cancer J. Clin. **68**(6), 394–424 (2018)
5. Berrar, D.P., Dubitzky, W., Granzow, M. (eds.): A practical Approach to Microarray Data Analysis, pp. 15–19. Kluwer Academic Publishers, New York (2003)
6. Wang, Z., Jensen, M.A., Zenklusen, J.C.: A practical guide to the cancer genome atlas (TCGA). In: Statistical Genomics, pp. 111–141. Humana Press, New York, NY (2016)

7. Tarca, A.L., Romero, R., Draghici, S.: Analysis of microarray experiments of gene expression profiling. Am. J. Obstet. Gynecol. **195**(2), 373–388 (2006)
8. Hansebout, R.R., Cornacchi, S.D., Haines, T., Goldsmith, C.H.: How to use an article about prognosis. Can. J. Surg. **52**(4), 328 (2009)
9. Nounou, M.I., ElAmrawy, F., Ahmed, N., Abdelraouf, K., Goda, S., Syed-Sha-Qhattal, H.: Breast cancer: conventional diagnosis and treatment modalities and recent patents and technologies. Breast Cancer **9** (2015). BCBCR-S29420
10. Dobbin, K.K., Simon, R.M.: Optimally splitting cases for training and testing high dimensional classifiers. BMC Med. Genomics **4**(1), 31 (2011)
11. Tang, Z., Ow, G.S., Thiery, J.P., Ivshina, A.V., Kuznetsov, V.A.: Meta-analysis of transcriptome reveals let-7b as an unfavorable prognostic biomarker and predicts molecular and clinical subclasses in high-grade serous ovarian carcinoma. Int. J. Cancer **134**(2), 306–318 (2014)
12. Bao, T., Davidson, N.E.: Gene expression profiling of breast cancer. Adv. Surg. **42**, 249–260 (2008)
13. Chen, M.H., Ibrahim, J.G., Shao, Q.M.: Maximum likelihood inference for the Cox regression model with applications to missing covariates. J. Multivar. Anal. **100**(9), 2018–2030 (2009)
14. Ades, F., Tryfonidis, K., Zardavas, D.: The past and future of breast cancer treatment—from the papyrus to individualised treatment approaches. ecancermedicalscience **11** (2017)
15. Motakis, E., Ivshina, A.V., Kuznetsov, V.A.: Data-driven approach to predict survival of cancer patients. IEEE Eng. Med. Biol. Mag. **28**(4), 58–66 (2009)
16. Ow, G.S., Tang, Z., Kuznetsov, V.A.: Big data and computational biology strategy for personalized prognosis. Oncotarget **7**(26), 40200 (2016)
17. Benjamini, Y., Yekutieli, D.: The control of the false discovery rate in multiple testing under dependency. Ann. Stat. **29**(4), 1165–1188 (2001)
18. Fox, J.: Cox proportional-hazards regression for survival data. An R and S-PLUS companion to applied regression (2002)
19. Broët, P., Kuznetsov, V.A., Bergh, J., Liu, E.T., Miller, L.D.: Identifying gene expression changes in breast cancer that distinguish early and late relapse among uncured patients. Bioinformatics **22**(12), 1477–1485 (2006)
20. Kouser, K., Sunita, A.: A comparative study of K means algorithm by different distance measures. Int. J. Innov. Res. Comput. Commun. Eng. **1**(9), 2443–2447 (2013)
21. Sharma, S.K., Kumar, S.: Comparative analysis of Manhattan and Euclidean distance metrics using A* algorithm. J. Res. Eng. Appl. Sci. **1**(4), 196–198 (2016)

A Model for Classification of Skin Disease Using Pretrained Convolutional Neural Network

Riddhi Kumari Bhadoria and Suparna Biswas

Abstract Skin problem is growing fast all over the country. It is one of the most common types of diseases where some can be painful and some can cause fatal to human life. Everyone should pay attention towards this alarming and emerging health problem which is spreading fast due to numerous reasons like pollution, global warming, ultraviolet rays, etc. To avoid delay in treatment, we have developed a model which will classify the disease using image dataset. The model uses the deep learning approach to get trained for classification. It works on convolutional neural network (CNN) with fine-tuned transfer learning using GoogleNet network. Using pretrained network, the model is trained to classify different skin diseases. The implementation result of training using MATLAB 2018 obtains accuracy of 96.63% with dataset of 4000 images into eight different classes.

Keywords Deep learning · Skin disease · CNN · Transfer learning · GoogleNet · Pretrained network

1 Introduction

In last few decades, the need for skincare has shown tremendous growth across the globe. Due to increase in awareness towards skin disease, now people have started paying attention towards this problem which was left neglected in past. Skin disease is a very common problem in India, because of which many ignore it and sometimes it may result in some severe condition [1]. It can be benign or malignant depending upon its type. Skin is such a vital organ of a human being, and since it is easily

R. K. Bhadoria (✉)
Department of Information Technology, Maulana Abul Kalam Azad
University of Technology, Kolkata, West Bengal, India
e-mail: riddhibhadoria@gmail.com

S. Biswas (✉)
Department of Computer Science, Maulana Abul Kalam Azad University of Technology,
Kolkata, West Bengal, India
e-mail: mailtosuparna@gmail.com

© Springer Nature Singapore Pte Ltd. 2020
J. K. Mandal and S. Mukhopadhyay (eds.), *Proceedings of the Global
AI Congress 2019*, Advances in Intelligent Systems and Computing 1112,
https://doi.org/10.1007/978-981-15-2188-1_14

visible, people tend to spend a lot of money; hence, skincare market is expanding at a high rate. People have also started visiting dermatologist for expert advice. Instantly detecting skin disease may become very difficult sometimes, and since people pay less attention, they end up visiting dermatologist when things go out of their hand. Need for early detection of skin disease is very much needed by looking at the current scenario [2].

Skin disease is one of the most neglected conditions around the globe. Due to the fact that all skin disease is not severe like acne, blister, hives, etc., people often tend to ignore it [1]. Skin disease can be painful or painless. In many cases, it may be hereditary. The climatic changes like global warming, pollution, ultraviolet light are also contributing in increase of skin disease cases. Young people are the most visited to doctor as they are more concern about their appearance. This has also hiked the cosmetic market in city side. The early detection of skin disease is much needed in such scenarios where people pay such less attention towards skin-related ailments. They want to spend least amount of their time on the checkups for skin-related ailments. Recently, many works have been done in the field of convolutional neural network (CNN) to detect skin cancer [1, 2]. This technique does not limits itself in the area of medical science but also helps in agricultural field. Automatic fruit harvesting system is developing to reduce the cost in agricultural field and detection of crop disease which remains undetected with human eye [3, 4]. They have worked in the area with different techniques like transfer learning, SVM classifier, and multilayer perceptron [5] to help in increasing the survival rate. The idea of using smartphone as a technique in the field of health care is also growing rapidly. This has given us an idea to develop a smartphone-based architecture which will be easily accessible for clinical purpose [5–7]. Current development in deep learning area has inspired a lot of people to work in this field. Also, now the involvement of GPU along with CPU has attracted many of researchers as it reduces a large amount of time. Since our model works on pretrained network, we did some research before choosing any network. GoogleNet outperformed VGG-16, VGG-19 by winning ILSVRC competition [1].

In this paper, an algorithm is proposed which will be capable of classifying some of the severe skin diseases accurately without the help of an expert, so that an early treatment can be started rather than spending time in unnecessary pathological tests [8–10]. To make it work, a model has to be trained with number of different types of diseases using convolutional neural network (CNN) architecture [1, 5, 11]. Further, to test the model, an image will be fed to the model as input and the model will classify it into appropriate category in which it falls.

The rest of the paper is divided in different sections. Section 2 gives the brief discussion about related works. Section 3 is about the problem definition and Sect. 4 describes proposed framework. Section 5 describes methodology and Sect. 6 presents the detail of experimental results and discussion, and finally, Sect. 7 concludes the work.

2 Literature Survey

In 2016, Liao [1], deep convolutional neural network is used to classify the skin diseases. Twenty-two types of different categories of skin diseases are taken as the dataset. Dataset is also taken from two different sources. The top accuracy of training model to classify the skin disease is 91.0%.

In 2017, Nguyen et al. [2] proposed a framework for the early detection of skin cancer disease. This work is about the most deadly skin cancer, melanoma, which is only curable if detected and treated at early stage. The detection of disease can be done by a specialist only but since the number of specialist is not much, an system which can help in detection of affected disease with help in reducing unnecessary cost of different tests and biopsies. This paper proposes combine methods of hand-coded features, sparse code method, and SVM with different machine-learning technique which further detects the affected area with skin lesions. In the proposed method, using visual reorganization system, the disease is detected in two steps. Segmentation separates the affected area of skin from the normal part. Then, using different machine-learning approaches, the classification of disease is carried out. Features of the segmented area are found out in the order to apply SVM classifier. The proposed model was evaluated on large image dataset of dermoscopic imageset with 900 training images and 379 testing images.

In 2013, Bourouis et al. [5] presented this paper in which artificial neural network is used to identify the skin cancer using images of skin. This paper proposes a mobile-based model to identify normal or abnormal skin type. Neural network is trained and tested with normal and abnormal skin dataset. Multilayer perceptron algorithm is used to train model. Using mobile neural network algorithm, detection of normal skin from abnormal was 96.50%.

In 2017, Alarifi et al. [11] presented a paper in which skin type is classified using convolutional neural network and machine-learning technique. Classification is done into three skin types—normal, spotted, and with wrinkles. Caffe framework was used in this experiment to implement CNN architecture with GoogleNet. Result after training and testing was evaluated using different performance metric like sensitivity, false negative rate, f-measure, recall, precision, and accuracy. Both traditional machine-learning and deep learning approaches are used for this experiment.

In the same year 2017, in work done by Menegola et al. [12], a model is developed and trained using transfer-learning approach from scratch to detect the melanoma with the help of deep neural network. Transfer-learning technique is used in order to train a model from scratch. Dataset was taken from different ISBI challenges, VGG-16 model was taken to carry out the experiment, and the model was trained at retinopathy and also on large dataset on ImageNet. Impact of training on large and small dataset is learned by the experiment. In the year 2018, Codella et al. [13] presented a work that focuses on the automated detection of melanoma disease in human skin. The experiment is carried out in three parts, i.e. lesion segmentation, feature extraction and classification. The classification decision was based on the area under curve (AUC) measurement, whereas segmentation result was based on Jaccard index, dice coefficient, and pixel-wise accuracy (Table 1).

Table 1 Comparative literature review

Author, year	Methodology/algorithm	Achievement	Conclusion	Drawback
Liao, H., 2016	Transfer learning using VGG16, VGG19, GoogleNet	Using deep CNN process, the classification into 22 diseases is done using different pretrained networks	Highest accuracy of 91.0%	Experiment was carried out without any variation because of which the accuracy cannot be trusted
Nguyen, N. C., et al. 2017	Visual reorganization system—segmentation and classification	Using deep leaning with different machine-learning approach, detection of skin cancer by segmentation and classification method in image	Highest accuracy of 76%	As the experiment was carried out on public challenge, it has fixed amount of data which implies insufficient dataset
Bourouis, A., et al. 2013	Multilayer perceptron algorithm	Mobile neural network algorithm is developed to detect skin cancer. It can be used at any location	Using mobile detection of skin disease has become feasible	The system only performs on android OS
Alarifi, J. S., et al. 2017	Convention machine-learning and deep learning approaches are used	Using deep neural network, skin type is classified using normal, spot, and wrinkle	CNN can perform better than traditional machine-learning technique	Experiment was carried out on small dataset, whereas deep learning requires larger dataset to perform accurately
Menegola, A., et al. 2017	Transfer learning using VGG—16 model	Impact of large and small dataset is studied	Accuracy of 79%	Dataset of retinopathy provides worst result
Codella, N., et al. 2018	Lesion segmentation, feature extraction, and disease classification using Jaccard index, dice coefficient, and pixel-wise accuracy	Research for automated detection of melanoma	Result shown in average Jaccard index does not tell how many images fall into segmentation	Dataset is biased. Not all the labels were represented categorically

3 Problem Definition

Skin disease is a very common problem across the globe. It is one of the most neglected problems until it is concerned with exposed body parts, people tend to pay less attention. They often try not to waste their time in its proper examination as they find it not harmful or do not even consider it as a disease. This negligence sometime turns to be the worst-case scenario. In India, rural area is the part where people do not even want to consider it as a medical emergency since there are many families who cannot afford this expensive treatment. Primary step for the treatment of skin disease is consulting with the dermatologist. They examine the affected area which can be any body part like face, hands, legs, thigh, etc., by looking at it with their tools but it becomes very hard to identify sometimes as many disease resembles same symptoms. In order to get the best conclusion, different test needs to be done like biopsy. Biopsy is a process in which a sample is taken from the affected part of the skin and further sent to the laboratory for examination. Test or biopsy can be very costly. To reduce this cost and time, there is a need for an automated system which will be helping in the classification of the skin diseases based on image of affected area, so that the treatment can be initiated early. And it will also save the unnecessary cost of any patient who is not suffering from any harmful skin disease.

4 Proposed Work

In this paper, we have developed a model which is capable of classifying a skin disease into the classes it has been trained for. We have used the deep learning approach to develop this model. The model is based on convolutional neural network (CNN) architecture. Implementation of model is done using transfer-learning process and our model runs on GoogleNet as pretrained network. Since CNN is a supervised-learning technique, the proposed model will be trained using the dataset provided with input and output for its learning.

Selection of pretrained network is a very crucial step. We have taken GoogleNet as our pretrained network. GoogleNet is a pretrained network trained as the subset of ImageNet database with millions of images. It can classify an input into thousands of categories, and it has been trained for like mug, house, car, book, etc. There are different pretrained networks available like ResNet, Inception v3, AlexNet, etc., out of which GoogleNet is one of the best and accurate as it was the winner of ILSVRC 2014 competition [3]. It has 22 hidden layers in its network (Fig. 1).

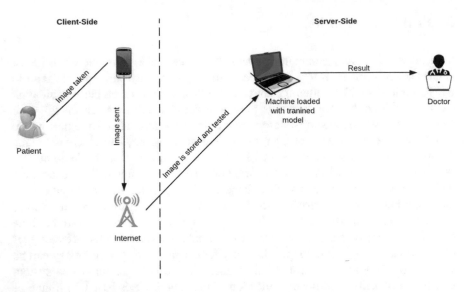

Fig. 1 Architecture of proposed model

5 Methodology

Our proposed model uses deep learning approach to train the model for specific purpose. It works based on convolutional neural network (CNN) algorithm, which is a type of machine-learning technique, using images as dataset for model training and testing. CNN is used to find patterns in images and identify features. It learns from the images itself and do feature selection, feature extraction, and segmentation type of actions automatically. The model works in two phases, first is training and the second is classifying.

Training phase is illustrated with the help of a flowchart given in Fig. 2.

Dataset: The very first step is to select the database and sort it into classes for which the classification will be done. And then, label it accordingly. The dataset for the training purpose is taken from DermNet NZ (https://www.dermnetnz.org). It has more than 25,000 clinical and dermatopathology images. This is one of the largest clinical image databases available for study and research purpose. We have taken images as dataset for eight classes of skin disease. Each class of disease contains 500 image dataset. The details are given in Table 2.

We have taken total of 4000 data as dataset for the training purpose of our model. The dataset is divided in such a way that 70% will be used for training purpose of the model and the rest 30% will be used for the validation of the model simultaneously. Hence, 2800 data will be used to train the model and 1200 data will be used to validate the proposed model. This way, each class will go with 350 data for training and 150 data for testing purpose. Now, the pretrained network is loaded into the machine. For our model, we have taken GoogleNet as the pretrained network. Network is fine-tuned with the transfer learning to get faster and accurate result. First phase is to

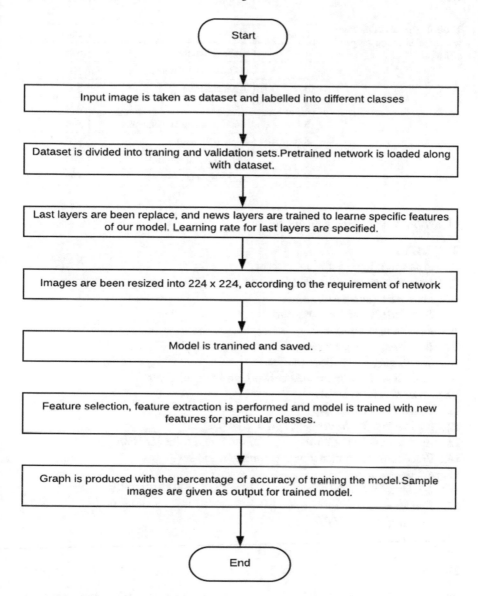

Fig. 2 Flowchart of training phase

train model with desired features. Algorithm 1 is implemented for training the model with specified featured of different skin disease classes. It also yields the result of accuracy with which the model has been trained.

Table 2 Classes of disease with number of images per class

Class	Name	Dataset
1	Actinic keratosis	500
2	Basal cell carcinoma	500
3	Erythroderma	500
4	Lentigo	500
5	Lentigo maligna melanoma	500
6	Lichen Planus	500
7	Palmoplantar pustulosis	500
8	Xanthelasma	500

Algorithm 1 For training the model,

1. BEGIN:
2. Procedure: Read and store dataset
3. *ImageDatastore = labelled images*
4. *Imds = imageDatastore*
5. **If** *includeSubfolder = true* **then**
6. *FileExtension = '.jpg'*
7. *LabelSource = 'foldername'*
8. *Set training and validation data*
9. Procedure: Load pretrained network and replace layer
10. *network = GoogleNet*
11. *inputsize = 224 × 224*
12. *remove loss 3 – 'classifier', 'prob' and 'output'*
13. *replace with 'fullyConnected', 'Softmax' and 'classification'*
14. Procedure: Train model with parameters and save
15. *Set parameters*
16. *MiniBatchSize = 20*
17. *Epochs = 10*
18. *LearningRate = 0.001*
19. *Save model*
20. *show accuracy*
21. END

Second phase is the classification, where an input image is taken and it is passed down the model to find out it belongs to which class of skin disease. Algorithm 2 is implemented to do the classification.

Algorithm 2 For classification of image,

1. BEGIN:
2. Procedure: Load the saved trained model
3. *network = saved model*
4. *imagesize = inputsize*

5. Procedure: Read and resize image
6. *Imread = inputimage*
7. *I = imread*
8. *Show image I*
9. *Resize I*
10. Procedure: Classify image with accuracy
11. *Classify label*
12. *Show image with %*
13. END

6   Experimental Results and Discussion

Our experiment was implemented on MATLAB. After training the model using transfer learning with pretrained GoogleNet network, using certain parameter, it yields graph with the percentage of accuracy. The simulation parameters with value are given in Table 3.

Simulation parameters given in Table 3 are used to simulate our proposed algorithm. Using these parameters, our system is trained to classify the type of skin diseases. After implementing the training procedure, a graph is obtained which shows the accuracy as well as loss. Our system is trained with 96.63% accuracy with a dataset of 4000 images. The whole experiment was carried out on CPU itself so it took 2659 min to train the model.

Training Phase
See Figs. 3 and 4.

Classification Phase
Our model is tested using different simulation parameters as to obtain the best result out of all. The training and validation of data distribution are kept same in all cases.

Table 3 Simulation parameters

Parameter	Value
Mini-batch size	20
Epochs	10
Pixel range	30×30
Learning rate	0.001
Frequency	3
Iteration	3200
Training set	2800
Testing set	1200
Hardware resource	CPU

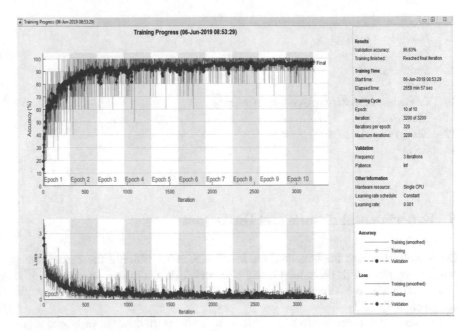

Fig. 3 Graph obtained by training the model

The other parameters like epochs, learning rate, iteration, etc., have been changed to get the best result (Fig. 5; Table 4).

Learning rate of 0.001 is ideal for our proposed model as it gives the accuracy of 96.63%. The model is trained well in this case and provides the best result out of three scenarios. In the first scenario, we can see that the model was trained fair and obtained 92.26% accuracy. Whereas in the second scenario, our model scores the best result with accuracy of 96.63% and was trained well. But the third scenario was way out of the line as it obtained worst accuracy of 12.50%. With the learning rate of 0.01, the model got saturated at this point and did not learnt at all. Due to poor learning, it yields the worst result.

7   Comparison

In paper [4], Alarifi et al. proposed a method to classify the skin type into three classes, i.e. normal, spot and wrinkle skin using deep learning approach. There sole purpose was to differentiate the skin types with an appropriate accuracy. They used dataset of 164 images that consist of only facial part. The whole experiment was carried out on Caffe framework using GoogleNet in python and with the help of sequential minimal optimization, the classification was done. It achieved accuracy of 89%. Also, Han et al. [14] use the same technique in their paper to classify the 12 skin diseases into benign or malignant. With the help of deep learning in Caffe

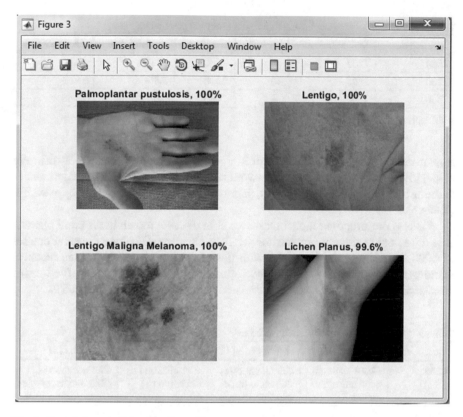

Fig. 4 Sample image of classification with accuracy

(a) **(b)**

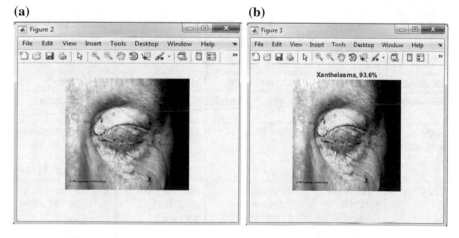

Fig. 5 Classification of image using trained model. **a** Input image after resize and according to requirement of model **b** classification result with accuracy of 93.6%

Table 4 Result (% of accuracy) using different simulation parameters

Training data	Validation data	Epochs	Iteration	Learning rate	Accuracy (%)
2800	1200	6	1680	0.0001	92.26
2800	**1200**	**10**	**3200**	**0.001**	**96.63**
2800	1200	6	1680	0.01	12.50

Bold represents the best one

framework they achieve an accuracy of 91% with ResNet as pretrained network. Wu et al. [15] take different pretrained network to work with their model. Using transfer learning to classify into six categories, their model attains the best accuracy of 92.9% (Table 5).

Whereas our proposed model classifies skin disease into eight different classes. It addresses skin disease problem rather than skincare. Our model uses a dataset of 4000 images to train the model with deep learning approach. It uses transfer-learning framework with CNN architecture. Using fine-tuned network, the images are classified into different skin diseases and achieves an accuracy of 96.63%.

Table 5 Comparative result of related works with our proposed model

Parameter	Proposed model	Alarifi et al. [11]	Han et al. [14]	Wu et al. [15]
Class	Classification is done into eight classes	Classification is done into three classes	Classification is done into two classes	Classification is done into six classes
Problem	Skin disease	Skincare	Skin disease	Skin disease
Dataset	4000 images	164 images	17,125 images	4394 images
Limitation of dataset	No such limitation, image can be of any of human body	Requires only facial image for dataset	Recorded images from review board	Only facial images are used as dataset
Technique	Deep learning	Deep learning	Deep learning	Deep learning
Framework	Transfer learning	Caffe framework	Caffe framework	Transfer learning
CNN architecture	GoogleNet	GoogleNet	ResNet	ResNet50, Inception v3, DenseNet 121, Xception, Inception-ResNet-v2
Classifier	Fined tuned network GoogleNet	Sequential minimal optimization (SMO)	Fined tuned pretrained network	Fined tuned pretrained network
Accuracy (%)	96.63	89	91	92.9

8 Conclusion

Our model is ideally designed for medical diagnosis in health care for skin diseases. Purpose like skin-related problems which are left neglected in all means. Image as input is the key which makes it user-friendly. The model is trained well to classify with best accuracy to save time and cost in all means. Due to its supervised-leaning approach, it learns with the labelled data and more the data more accurate result. It does not involve the manual feature selection or extraction as it learns automatically from the image dataset which makes it faster and easier. This model preserves privacy and helps people to express problem keeping anonymity in compared to traditional questionnaire-based method.

References

1. Liao, H.: A Deep Learning Approach to Universal Skin Disease Classification. University of Rochester, Department of Computer Science, CSC (2016)
2. Codella, N.C., Nguyen, Q.B., Pankanti, S., Gutman, D.A., Helba, B., Halpern, A.C., Smith, J.R.: Deep learning ensembles for melanoma recognition in dermoscopy images. IBM J. Res. Dev. **61**(4/5), 5-1 (2017)
3. Muhammad, N.A., Ab Nasir, A., Ibrahim, Z., Sabri, N.: Evaluation of CNN, Alexnet and GoogleNet for fruit recognition. Indones. J. Electr. Eng. Comput. Sci. **12**(2), 468–475 (2018)
4. Coulibaly, S., Kamsu-Foguem, B., Kamissoko, D., Traore, D.: Deep neural networks with transfer learning in millet crop images. Comput. Ind. **108**, 115–120 (2019)
5. Bourouis, A., Zerdazi, A., Feham, M., Bouchachia, A.: M-health: skin disease analysis system using Smartphone's camera. Procedia Comput. Sci. **19**, 1116–1120 (2013)
6. Ge, Z., Demyanov, S., Chakravorty, R., Bowling, A., Garnavi, R.: Skin disease recognition using deep saliency features and multimodal learning of dermoscopy and clinical images. In: International Conference on Medical Image Computing and Computer-Assisted Intervention, September, pp. 250–258. Springer, Cham (2017)
7. Jaleel, J.A., Salim, S., Aswin, R.B.: Computer aided detection of skin cancer. In: 2013 International Conference on Circuits, Power and Computing Technologies (ICCPCT), March, pp. 1137–1142. IEEE (2013)
8. Pomponiu, V., Nejati, H., Cheung, N.M.: Deepmole: deep neural networks for skin mole lesion classification. In: 2016 IEEE International Conference on Image Processing (ICIP), September, pp. 2623–2627. IEEE (2016)
9. Premaladha, J., Ravichandran, K.S.: Novel approaches for diagnosing melanoma skin lesions through supervised and deep learning algorithms. J. Med. Syst. **40**(4), 96 (2016)
10. Han, D., Liu, Q., Fan, W.: A new image classification method using CNN transfer learning and web data augmentation. Expert Syst. Appl. **95**, 43–56 (2018)
11. Alarifi, J.S., Goyal, M., Davison, A.K., Dancey, D., Khan, R., Yap, M.H.: Facial skin classification using convolutional neural networks. In: International Conference Image Analysis and Recognition, July, pp. 479–485. Springer, Cham (2017)
12. Menegola, A., Fornaciali, M., Pires, R., Bittencourt, F.V., Avila, S., Valle, E.: Knowledge transfer for melanoma screening with deep learning. In: 2017 IEEE 14th International Symposium on Biomedical Imaging (ISBI 2017), April, pp. 297–300. IEEE (2017)

13. Codella, N.C., Gutman, D., Celebi, M.E., Helba, B., Marchetti, M.A., Dusza, S.W., Kalloo, A., Liopyris, K., Mishra, N., Kittler, H., Halpern, A.: Skin lesion analysis toward melanoma detection: a challenge at the 2017 international symposium on biomedical imaging (ISBI), hosted by the international skin imaging collaboration (ISIC). In: 2018 IEEE 15th International Symposium on Biomedical Imaging (ISBI 2018), April, pp. 168–172. IEEE (2018)
14. Han, S.S., Kim, M.S., Lim, W., Park, G.H., Park, I., Chang, S.E.: Classification of the clinical images for benign and malignant cutaneous tumors using a deep learning algorithm. J. Invest. Dermatol. **138**(7), 1529–1538 (2018)
15. Wu, Z., Zhao, S., Peng, Y., He, X., Zhao, X., Huang, K., Wu, X., Fan, W., Chen, M., Li, J., Huang, W., Chen, X., Li, J.: Studies on different CNN algorithms for face skin disease classification based on clinical images. IEEE Access **7** (2019)

Improvement of Packet Delivery Fraction Due to Discrete Attacks in MANET Using MAD Statistical Approach

Sayan Majumder and Debika Bhattacharyya

Abstract Nowadays so many wireless networks are used rapidly, among which mobile ad hoc network (MANET) is placed in the front row. As it has no concept or use of wire, it is unshielded to different kinds of attacks. In this paper, we have analyzed the PDF that is packet delivery fraction of three different kinds of attacks, viz. Sybil, Black Hole, Worm Hole, and improved them. We have to first count the number of packets dropped to number of packets sent. Then we can find the ratio or fraction of the delivery during the attacks and improve the PDF using our algorithm of a newly statistical approach, named as absolute deviation correlation about median (MAD). Then a comparative study is structured between PDF of these three attacks. Mat Lab simulator is used here to get the result.

Keywords MANET · Absolute deviation · Median · PDF

1 Introduction

Wireless network is decentralized in manner, where nodes are present to deliver the data packets from source to destination. The nodes in between starting and ending nodes carry forward the data as a multi-hop technique. Mobile ad hoc network (MANET) follows this transmission of data using some protocols or routing algorithms. As MANET is not a static type in nature, it is unprotected to different kinds of attacks like Black Hole, Worm Hole [1], Jelly Fish, Sybil, DOS, etc.

In case of Worm Hole attack [2] in MANET [3], an artificial passage is created between originating and terminating nodes. In that passage, there is high chance of presence of one or more affected nodes to make a duplicate copy of it or grab it. The fake passage may take less amount of time to route the packets, but the packets

S. Majumder (✉)
Dream Institute of Technology, Kolkata, India
e-mail: Sayanmajumder90@gmail.com

D. Bhattacharyya
Institute of Engineering & Management, Kolkata, India
e-mail: bdebika@iemcal.com

© Springer Nature Singapore Pte Ltd. 2020
J. K. Mandal and S. Mukhopadhyay (eds.), *Proceedings of the Global AI Congress 2019*, Advances in Intelligent Systems and Computing 1112, https://doi.org/10.1007/978-981-15-2188-1_15

become replicated. We can also calculate this number of packets sent to number of packets dropped by this attack, which is termed as packet delivery fraction (PDF) and also can improve this PDF factor using absolute deviation about median approach.

Black Hole attack is mainly occurred due to a fake presentation by the attackers as they present the shortest path to the route but data are snatched without forwarding them. Route request is sent by source node and then corresponding neighboring node sends the route reply message. But in case of this attack, malicious node sends wrong reply message to the source node. So, the starting node continuously sends the data packets to that harmful affected node and the packets get dropped by Black Hole attack.

In mobile ad hoc network, attackers can also attack through the nodes which communicate between each other without the central jurisdiction, and a malicious node acts as multiple identities, which is termed as Sybil attack. Each node is known to others with the help of messages. To transmit messages to particular nodes, Sybil attackers may take various uniformities.

To mitigate discrete attacks in MANET, so many statistical methods [4] have adopted already. We can use these statistical approaches as good machine learning techniques. Cross-correlation methods [5] find the trusted routes in case of Worm Hole attack. Absolute deviation about mean correlation is also a good approach to mitigate attacks in MANET, which is also proved later by LAD and MLS regression analysis. Packet delivery fraction is the quality of service by MANET, which also can be improved by machine learning techniques [6].

In this paper, we have used MAD correlation technique to improve the packet delivery fraction, i.e., number of packets sent to destination nodes to number of packets dropped by different attacks like Worm Hole, Black Hole, and Sybil attacks. MAD stands for median absolute deviation technique which gives better result with correlation technique. A number of packets sent and dropped are calculated first for three attacks. Then we have used our MAD correlation algorithm to improve the PDF. At last, a comparison is also established between these three attacks. Ad hoc on-demand distance vector (AODV) routing protocol [7] is used here.

2 Proposed Methodology

AD correlation about median of random variable a and b,

$$\text{DCORR}(a, b) = 1/4[(D(a_1 + b_1))^2 - (D(a_1 - b_1))^2]$$

where $a_1 = [a - \text{Median}(a)]/D(a)$ and $b_1 = [b - \text{Median}(b)]/D(b)$

AD correlation algorithm will be used in AODV routing protocol:

1. Origination of node s, set $s = 0$ to s_{max}.
2. Set $s_k = d_k$ (node number).
3. Generation of route path from d_k to destination.

4. Send route request message from source node.
5. Receive reply message from route.
6. If the interrelation is malicious, jump into step 8.
7. Else, continue the packets dispatching process.
8. Using trust data from route reply compute DCORR(a, b) within packets dropped to send.
9. If the value of absolute deviation correlation coefficient is much more than normal value, jump into step 11.
10. Else, continue the packet dispatching.
11. Report intruder.

Above algorithm is executed for all the three different attacks, viz. Worm Hole, Black Hole, and Sybil attacks to improve the PDF. Each node determines the median deviation correlation coefficient for adjoining nodes between sources to destination to improve the dropped packets to send packet ratio than in normal condition. For all routes, this information is gathered. Here, a and b are two different variables, which define packets dropped and packets sent, respectively. We have done MATLAB simulation to compute the value of MAD correlation coefficients for all the three types of attacks and then made a comparative study between them.

3 Simulation and Result

We have used MATLAB simulation technique to prove the MAD correlation technique to improve the packet delivery fraction by different attacks. AODV routing algorithm is used here with our modified correlation technique. Then Worm Hole, Black Hole and Sybil attacks are analyzed.

Node distribution scenario is presented by Fig. 1.

Worm Hole attack [8] is established in Fig. 2. Worm Hole attack scenario is created

Fig. 1 Node distribution scenario

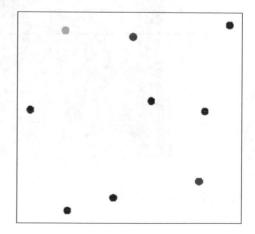

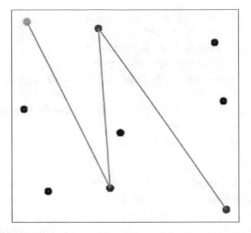

Fig. 2 Worm Hole attack

in between red- and blue-colored nodes. The nodes are hazardous. All the packets between red and blue nodes will be sent directly due to attacks.

Using median absolute deviation correlation technique, at first, we have found the correlation coefficient values. The values are found to be more than 0.95.

Then we have used this method in AODV routing protocol to improve the PDF due to this attack. Figure 3 depicts the MAD correlation coefficient values.

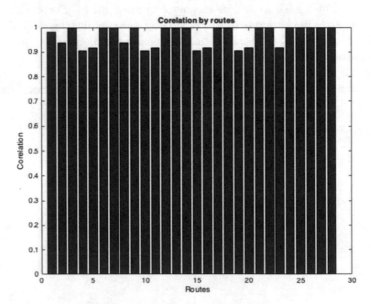

Fig. 3 Correlation coefficient values

The number of packets sent from starting node to ending node is calculated in case of Worm Hole attack, and then our modified algorithm is used to improve this PDF, which is shown by Fig. 4.

Without our algorithm, the PDF was very high as shown by red-colored graph, and using our algorithm, PDF is improved as a number of dropped packets are minimized.

Sybil attack [9] is shown in Fig. 5. In this attack, we have shown base network, distance to zero and multiple identities. Sybil attack is achieved by multiple identities of a harmful node.

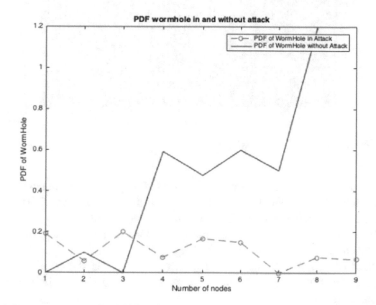

Fig. 4 PDF improvement using MAD AODV

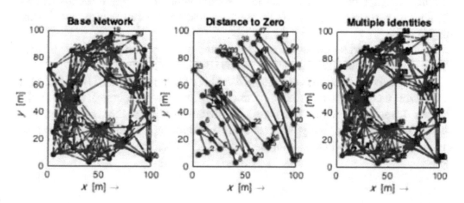

Fig. 5 Sybil attack

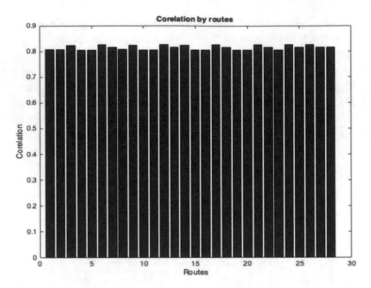

Fig. 6 Correlation coefficients for Sybil attack

Using median absolute deviation correlation technique, at first, we have found the correlation coefficient values. The values are found to be more than 0.80 in case of this attack. Figure 6 depicts the MAD correlation coefficient values.

PDF is calculated in case of Sybil attack and then our modified algorithm is used to improve this PDF, which is shown by Fig. 7.

Without our algorithm, the PDF [10] was very high as shown by red-colored graph, and using our algorithm, PDF is improved as a number of dropped packets are minimized.

Black Hole attack is proved by Fig. 8, where Black Hole is shown by red color. Shortest distance between node 2 and 7 is calculated. This type of attack discards the packets instead of forwarding them to destination. So, this is also called packet drop attack.

In this case, correlation coefficient [11] values are more than 0.75. Figure 6 depicts the MAD correlation [12] coefficient values (Fig. 9).

Packet delivery fraction is calculated for Black Hole attack [13]. Using our MAD correlation technique with AODV routing algorithm, we can improve the fraction. Packets dropped due to this attack are minimized. In Fig. 10, the red-colored graph shows that the PDF [14] is high without using our algorithm, whereas the blue-colored graph depicts that the PDF is improved using our algorithm.

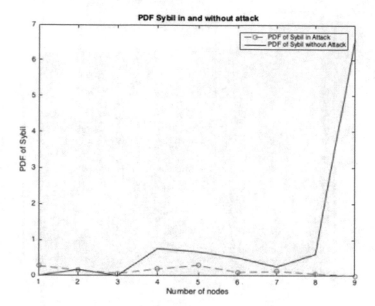

Fig. 7 PDF improvement using MAD AODV

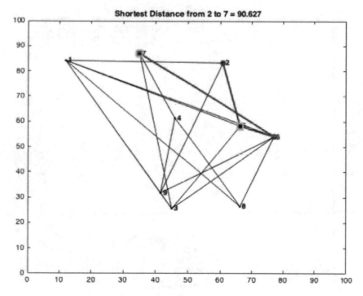

Fig. 8 Black Hole attack

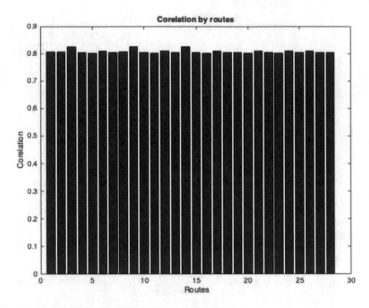

Fig. 9 Correlation coefficient for Black Hole attack

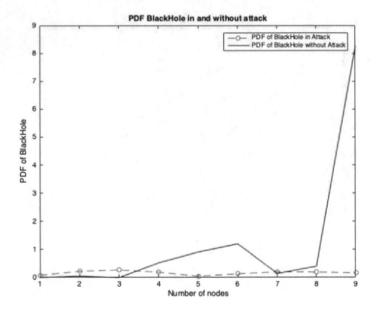

Fig. 10 PDF improvement using MAD AODV

Table 1 Comparison between three attacks

Worm Hole attack	Sybil attack	Black Hole attack
MAD correlation coefficient values are 0.95 and higher	MAD correlation coefficient values are 0.80 and higher	MAD correlation coefficient values are 0.75 and higher
PDF is improved by 82%	PDF is improved by 72%	PDF is improved by 70%
Simulation time 240 s	Simulation time 190 s	Simulation time 180 s
Pattern of packets dropped to packets sent follows mostly linear	Pattern of packets dropped to packets sent follows mostly linear	Pattern of packets dropped to packets sent follows mostly linear

4 Conclusion

We have first analyzed Worm Hole attack and then found the PDF by this attack, which is later optimized by 82% using our MAD correlation algorithm with AODV. The value of MAD correlation coefficient is 0.95 or higher, which is much better than normal Karl Pearson's correlation coefficient (Table 1).

Then, PDF is also calculated by Sybil attack and 72% improvement is done using our algorithm. 0.82 or higher is the coefficient rate for this attack.

Black Hole attack is very crucial one, which is also known as packet drop attack, and packets dropped by this attack are also improved by 70% using MAD technique. Correlation coefficient value for this attack is 0.75 or higher.

The main advantages of our MAD correlation coefficient are:

1. It gives higher correlation coefficient rate than normal Karl Pearson's correlation coefficient for all the three attacks.
2. With AODV routing protocol, our median absolute deviation technique improves the PDF by 82, 72 and 70% for all the three attacks.
3. We can find a mostly linear pattern between packets dropped to packets sent, which can be estimated and proved later.
4. Median gives better correlation coefficient than mean absolute deviation.

References

1. Majumder, S., Bhattacharyya, D.: Mitigating wormhole attack in MANET using absolute deviation statistical approach. In: 2018 IEEE 8th Annual Computing and Communication Workshop and Conference (CCWC), Las Vegas, NV, pp. 317–320 (2018)
2. Roy, D.B., Chaki, R., Chaki, N.: A new cluster-based wormhole intrusion detection algorithm for mobile ad-hoc networks. In: IJNSA, pp. 44–52 (2009)
3. Xing, F.: On the survivability of wireless ad-hoc networks with node misbehaviors and failures. IEEE Trans. Secur. Comput. **7**(3), 284–299 (2010)
4. Qiana, L., Songa, N., Li, X: Detection of wormhole attacks in multi-path routed wireless ad hoc networks: a statistical analysis approach. J. Netw. Comput. Appl. (2005)

5. Patnaik, G.K., GORE, M.M.: Trustworthy path discovery in MANET—a message oriented cross-correlation approach. Workshops of International Conference on Advanced Information Networking and Applications, pp. 170–177 (2011)

6. Jia, R., Yang, F., Yao, S., Tian, X., Wang, X., Zhang, W., Xu, J.: Optimal capacity–delay tradeoff in MANETs with correlation of node mobility. IEEE Trans. Veh. Technol. **66**(2), 1772–1785 (2017 Feb)

7. Alslaim, M.N., Alaqel, H.A., Zaghloul, S.S.: A comparative study of MANET routing protocols. In: 2014 Third International Conference on e-Technologies and Networks for Development (2014)

8. Win, K.S.: Analysis of detecting wormhole attack in wireless networks. Int. J. Electron. Commun. Eng. **2**(12), 2704–2710 (2008)

9. Abbas, S., Merabti, M., Llewellyn-Jones, D., Kifayat, K.: Lightweight Sybil attack detection in MANETs. IEEE Syst. J. **7**(2), 236–248 (2013)

10. Majumder, S., Bhattacharyya, D.: Comparative study between modified DSR and AODV routing algorithms to improve the PDF due to wormhole attack in MANET. Int. J. Sci. Res. Rev. **8**(1), 1095–1110

11. Majumder, S., Bhattacharyya, D.: Adopting machine learning technique to mitigate various attacks in MANET—a survey report. Int. J. Sci. Res. Rev. **8**(6), 288–295

12. Cadger, F., Curran, K., Santos, J., Moffett, S.: MANET location prediction using machine learning algorithms. In: International Conference on Wired/Wireless Internet Communications, pp. 174–185 (2012)

13. Tripathi, A., Mohapatra, A.K.: Mitigation of Blackhole attack in MANET. In: 2016 8th International Conference on Computational Intelligence and Communication Networks (CICN), Tehri, pp. 437–441 (2016)

14. Majumder, S., Bhattacharyya, D.: Relation estimation of packets dropped by wormhole attack to packets sent using regression analysis. In: Mandal, J., Bhattacharya, D. (eds.) Emerging technology in modelling and graphics. Advances in intelligent systems and computing, vol. 937. Springer, Singapore (2020)

IoT Based Smart Posture Detector

Greeshma Karanth, Niharika Pentapati, Shivangi Gupta and Roopa Ravish🆔

Abstract The growing technology in the world is rapidly transforming the way people lead their lives. Industrialization and urbanization have brought an enormous increase in sedentary lifestyle to the modern world. Indulged in technology, people are often found abandoning their good posture and being hunched over for really long hours. Good posture is of utmost importance for leading a healthy lifestyle and it is said that back pain is the third most common reason for people to visit the doctor. Yet, knowingly or unknowingly, people compromise on one of the most essential traits of what makes them human; the ability to walk upright. The aim of this paper is to provide a feasible solution to this problem by presenting a wearable device that recognizes the posture of the person and sends live data on the phone through an app. It records the posture and classifies it as Good, Okay or Bad. It also gives the statistics and overall feedback of how it can be improved.

Keywords Posture detection · Back posture · Arduino · Accelerometer · IoT · Micro-controller · Posture classification · Posture correction

1 Introduction

According to Stanford Health News, 'posture' is defined as the way your muscles and skeleton hold the body erect [1]. In a series of papers in "The Lancet", it was registered that 540 million people across the world suffer from lower back pain [2]. This concludes the severity of back pain. Although the effect is mainly to the lower region of the spinal cord, various issues have started to cause chronic pain in the lower middle and neck regions of the body. Back pains are usually periodic and do not last long. However, they occur very frequently and can have a huge impact on the body in the long run. Negligence toward maintaining a good posture can cause

G. Karanth · N. Pentapati · S. Gupta (✉) · R. Ravish
Department of Computer Science and Engineering, People's Education Society University,
Bengaluru, India
e-mail: rooparavish@pes.edu

© Springer Nature Singapore Pte Ltd. 2020
J. K. Mandal and S. Mukhopadhyay (eds.), *Proceedings of the Global AI Congress 2019*, Advances in Intelligent Systems and Computing 1112,
https://doi.org/10.1007/978-981-15-2188-1_16

several life-threatening conditions such as shoulder pains, back pains, reduced lung function, gastrointestinal pains, scoliosis and postural syndrome.

The consequences of a prolonged bad posture greatly affect the day-to-day activities and have paved its path to the work life as well. It is often said that back pain is a common reason for a day off at work. The estimated number of working days lost in 2013/14 due to back disorders was 2.8 million (UK) [3]. A survey was conducted by Spine Universe in 48 states of America with 606 patients who were suffering from back pain. It was found that 29% of the adults are convinced that the cause of their problem is work-related and has nothing to do with their postural activities [4].

Majority of the people do not seek medication immediately and wait till they feel acute pain. The treatment usually involves some rest and sometimes a doctors prescription. Medical treatment often revolves around physiotherapy and surgery depending on the severity of the issue. Hence, lots of money is annually spent on such treatments which would be unnecessary if proper precautions had been taken. One such way would be to timely monitor the back posture to avoid such difficulties. The Smart Posture Detector (SPD) implemented in this paper helps in achieving this purpose. It detects back posture and warns the user if it is a bad one. Feedback and statistics of the posture are displayed on the mobile device and hence constantly reminding the user to correct their posture. The currently available systems are very expensive and this limits their usage on a wider scale, whereas, the SPD is a very economical and robust solution to the problem.

Further sections of this paper are organized as follows. Section 2 discusses related work in this field. Section 3 specifies the components that are required to build the SPD device and describes its implementation. Section 4 discusses the results and statistics of using this device on three different people. Finally, Sect. 5 concludes the paper.

2   Literature Survey

Posture is the position in which a person holds their body upright against gravity when standing, sitting or walking [5]. There are two approaches to determine posture, image-processing based and sensor-based. A sensor-based approach is implemented in two ways. One is by calculating the pressure distribution over different weight distributing surfaces. The other is by calculating the angular difference between current posture and a predetermined good posture. The author states that the determination of sitting posture is mainly dependent on a chair and four sitting positions. Strength and ultrasonic sensors are installed inside the chair to acquire data. This is then processed using principle component analysis to determine the condition of posture [6]. This approach solves the problems faced by a person when sitting and it is advantageous for users like students who are seated at a desk for a long period. But, it does not help a person correct their posture or practice good posture throughout.

In another sensor-based approach to identify good sitting posture, two accelerometers are placed on different parts of a person's spine. With the help of two other sensors the goniometer and electrogoniometer, the angle is calculated and posture is determined. This design is simple, effective and wearable but is only implemented on sitting positions [7]. An application-specific approach to posture detection, where the device warns computer users when they lean too close to the computer, is the "Postuino". The device is not wearable but is instead placed next to the computer and when the distance between the computer, user and device falls below a specified threshold, the device will alert the user. This is an innovative approach with a popular application and although it helps the user keep a safe distance from their electronic screen (i.e., computer), it does not help the user correct their actual back posture if the user is standing or walking [8, 9].

In another approach, an actuator is used as a bio-mechanical posture detection device. This actuator shows sensory activity through an avatar in the application with which the device communicates. It assesses the user's posture in the state in which the user is, i.e., sitting, standing, lying, etc., to identify the user's movement state or transition between movement states. While this might be a more advanced and applicable posture detector, it is also a more complicated and non-economical solution to the problem [10].

Another technique uses inertial sensors for human posture detection in order to calculate three-dimensional angles of the human arm and hand. This information is recorded and later used to reconstruct it on a computer. This approach does not help rectify the back posture and does not give any feedback on how to correct it [11]. However, an intelligent chair can be adopted in order to classify and correct the posture of the user. Neural Networks are trained to classify the posture based on pre-trained standard postures. Though the approach to the problem is creative, it does not suggest correction for all postures of the user [12].

A novel solution is presented for wireless and wearable posture recognition based on a custom-designed wireless body area sensor network (WBASN), called WiMo-CA. Here, sensors are represented by triaxial integrated micro-electro-mechanical system (MEMS) accelerometers. WiMoCA sensing node is designed to be wearable and low-power. It has a modular architecture to ease fast replacement and update of each component. The proposed method provides the complete implementation of a distributed posture recognition application [13, 14].

The objective of an additional approach is to detect user's postural changes, not to measure the pressure at each point precisely. The author Ricardo Barba et al., state that the current mechanisms to detect postural changes are usually expensive, which greatly limits their use in effective computing. They have ruled out commercial solutions for two basic reasons: the need of adjusting the size of the sensors; the cost. To cope with the aforementioned challenges, their approach consists of combining several simple sensors so that they can be used together to form a posture sensor cushion [15].

Another approach uses three main sensors—Accelerometer, Gyrometer and Bluetooth module. In this approach, the accelerometer measures the tilt of the body, the gyrometer measures the movement in the body and Bluetooth helps in connecting the belt with the phone app designed to display the readings [16]. The Arduino has been programmed for different gestures and positions of the body that a person undergoes in everyday life. Although seeming efficient, the belt measures the tilt of the lumbar region, when in fact, the actual tilt happens at the thoracic (i.e., the upper back) region of the spine. This could result in inaccuracy in posture correction [17].

One of the implementations suggests use of sensors (acceleration sensor) embedded in a smartphone contrary to the ones that require separate hardware components for the same. However, attaching the sensor to the phone and the phone to a belt is not a feasible solution as there is a risk of dropping the phone every time any rigorous activity is performed. Moreover, people use their phones extremely frequently, and for the most part of the day, it will be on their hands and not the belt [18].

3 Implementation

Having studied the abovementioned advances, this paper attempts to overcome the said challenges and implements an optimized approach for posture detection which helps the user maintain a healthy posture. Below is the implementation of the SPD which is positioned on the user's upper back to detect various postures and provide valuable feedback to help rectify it.

3.1 Components and Connections

Implementation of SPD consists of Arduino UNO as shown in Fig. 1 and different sensors as shown in Fig. 2a.
The sensors used and their applications are

- An Inertial Measurement Unit (IMU) sensor MPU9250 whose inbuilt accelerometer is used to detect relative spinal position of the user
- A Bluetooth module HC-05 to transfer signals from the accompanying Posture Detector android app to control the working of the posture detector and provide feedback
- A mini piezo buzzer to indicate a continued presence of bad back posture
- Three light-emitting diode (LED) lights to indicate the different statuses of posture.

Supplementing components as shown in Fig. 2b include jumper wires, resistors to support LED lights and a breadboard on which to place the components.

Various sensors are connected on the breadboard as shown in Fig. 3 and this is placed on the user's back. The mini piezo buzzer is placed in a wristband on the users hand to alert them of prolonged bad posture.

Fig. 1 Arduino UNO

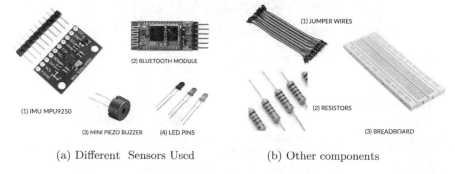

(a) Different Sensors Used   (b) Other components

Fig. 2 Different components

3.2 Working

The SPD is placed on the back of the user. The optimal location for the device to produce accurate results is the position where the spine curves, exactly a few inches below the neck as shown in Fig. 4.

Based on the posture of the user, the IMU sends positional readings, according to z-axis, to the Arduino. These readings are obtained from the accelerometer and are then categorized into three ranges such as good, bad and okay posture. Correspondent to the values received, the Arduino sets the LED lights on the SPD. Red light indicates bad posture, white light indicates okay posture and green light indicates good posture. The Arduino also sends data inferred from the values received to the android app and this transaction happens through the Bluetooth module.

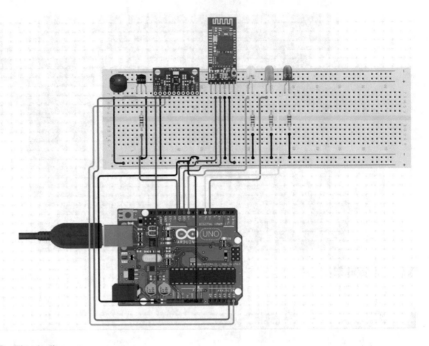

Fig. 3 Circuit diagram

The purpose of the android app is to enable ease of access and readability to the user. The layout of the Android app is shown in Fig. 5. The user can control the working of the SPD by switching it on and off. After the posture monitoring device is switched off, a pie chart is obtained through the values sent to the app by the Arduino. This pie chart will display the fraction of good, bad and okay posture of the user for the period of time when the device was on. According to the pie chart, the app will also prompt some helpful advice regarding the user's back posture.

3.3 Setup and Use

The technological devices needed to run the SPD are the Arduino Integrated Development Environment (IDE) and an Android phone. The program, which runs on Arduino IDE, is uploaded to the Arduino before the device is used. Then, the android app for the SPD is installed on the phone. After placing the device in the appropriate position on the users back, Bluetooth is enabled on the android phone. The phone is then connected to the SPD by establishing a Bluetooth connection between the phone and the Bluetooth module HC-05 situated on the posture detector to send and receive data signals from the Arduino. By clicking on the On and Off button on the app, the user decides when to start taking the readings and when to stop respectively.

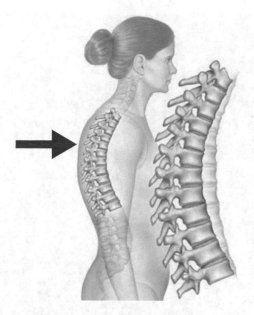

Fig. 4 Position of placement of device

Fig. 5 Android app layout

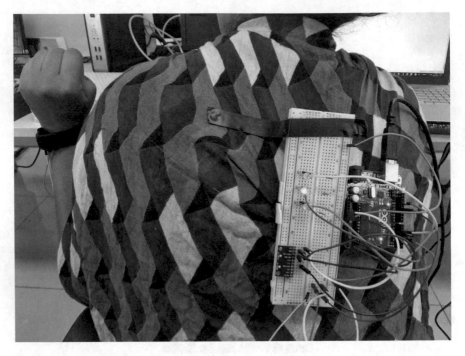

Fig. 6 SPD working model

4 Experiments and Results

Figure 6 shows the working model of the SPD. The SPD was tested and has proved to provide successful results. The outcomes and feedback of testing the SPD on three different test subjects are shown in Fig. 7.

In the pie chart, green, red and blue colors are used for good posture, bad posture and okay posture respectively. In Fig. 7a, the test user 1 had excellent posture because of which the pie chart was mostly green. In reality, a dominating mix of green and blue is also considered good posture. Figure 7b, which shows the results for user 2, has a pie chart where blue is more evidently seen than red or green. As shown in the advice, this depicts an acceptable posture but it can be improved to gain better posture. Figure 7c shows the pie chart of user 3 who has really bad posture. Such users comprise the target audience of this device. In the pie charts of such users, more red can be seen than the other colors. This means that the user has a posture that is dangerous to their health and is hence advised to fix it before it becomes a problem.

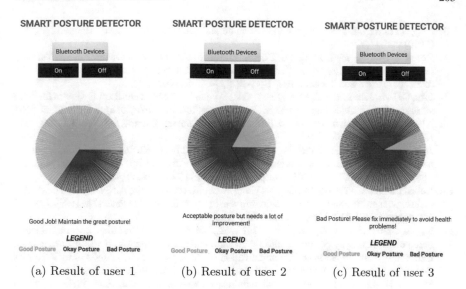

(a) Result of user 1   (b) Result of user 2   (c) Result of user 3

Fig. 7 Results of three different test subjects

5 Conclusion

One of the dangerous effects of urbanization has been the increase in the duration of time a person is seated. This results in them not having a good posture. Bad posture and subsequent back pains have been found to be one of the leading reasons for doctor visits. This paper provides a robust and feasible solution for detecting and correcting unhealthy back posture. This is a wearable device (SPD) implemented using Arduino micro-controller, sensors and an Android mobile phone. It identifies and provides feedback such as good, okay and bad posture for a user sitting, standing or walking. The SPD is advantageous to the user in ways such as helping them stay healthy and take advanced precautions to avoid back problems. Further improvements can be made to the device in terms of compactness and marketability.

References

1. Stanford Health News (2012 January 1) https://news.sanfordhealth.org/orthopedics/what-is-posture/
2. Healio Primary Care https://www.healio.com/family-medicine/pain-management/news/online/%7B86d8d0a9-d55f-4bd1-b7a7-98ee70dc1fd9%7D/more-than-1-in-10-worldwide-suffer-lower-back-pain/
3. Cooper, E.: Health and safety statistics for Great Britain (2015). https://www.posturepeople.co.uk/health-and-safety-statistics-for-great-britain/
4. SpineUniverse, Chronic+ Back and Neck Pain in America 2015 Survey Results, vol. 16, Issue 1. https://www.practicalpainmanagement.com/pain/spine/chronic-back-neck-pain-america-2015-survey-results

5. May, S., Lomas, D.: Posture, the lumbar spine and back pain. Int. Encycl. Rehabil., pp. 1–8 (2010)
6. Rosero-Montalvo, Paul, Jaramillo, Daniel, Flores, Stefany, Peluffo, Diego, Alvear, Vanessa, Lopez, Milton: Human sit down position detection using data classification and dimensionality reduction. J Adv. Sci. Technol. Eng. Syst. **2**(3), 749–754 (2017)
7. Lim, C.C., Basah, S.N., Ali, M.A., Fook, C.Y.: Wearable posture identification system for good sitting position. J. Telecommun. Electr. Comput. Eng. **10**(No. 1–16) (2018). e-ISSN: 2289-8131
8. Alattas, R.: Postuino: bad posture detector using Arduino. Int. J. Innov. Sci. Res. **3**(2), 208–212 (2014 June). ISSN 2351-8014
9. Alattas, R., Elleithy, K.: Detecting and minimising bad posture using Postuino among engineering student. In: 2nd International Conference on Artificial Intelligence, Modelling and Simulation, Madrid, Spain (2014). https://doi.org/10.1109/AIMS.2014.55
10. Robert Chang, A., Perkash, M., Charles Wang, C., Martin Hauenstein, A.: System and method of biochemical posture and feedback including sensor normalization. In: Patent No.: US 9,128,521 B2, Applicant: LUMO Bodytech, Inc., Palo Alto, CA (US) (2015)
11. Ni, W., Gao, Y., Lucev, Z., Pun, S.H., Cifrek, M., Vai, M.I., Du, M.: Human posture detection based on human body communication with muti-carriers modulation. In: 39th International Convention on Information and Communication Technology, Electronics and Microelectronics (MIPRO), Opatija, Croatia (2016). https://doi.org/10.1109/MIPRO.2016.7522151
12. Martins, L., Lucena, R., Belo, J., Almeida, R., Quaresma, C., Jesus, A.P., Vieira, P.: Intelligent chair sensor classification and correction of sitting posture. In: Roa, R.L. (ed.) XIII Mediterranean Conference on Medical and Biological Engineering and Computing, IFMBE Proceedings, vol. 41. Springer, Cham (2013)
13. Farella, E., Benini, L., Acquaviva, A.: A wireless body area sensor network for posture detection. In: International Symposium on Computers and Communications (January 2006)
14. Farella, E., Pieracci, A., Brunelli, D., Acquaviva, A., Benini, L., Ricco, B.: Design and implementation of WiMoCA node for a body area wireless sensor network, pp. 342–347 (2005 Aug)
15. Barba, R., de Madrid, A.P., Boticario, J.G.: Development of an inexpensive sensor network for recognition of sitting posture. Int. J. Distrib. Sens. Netw. 28040 Madrid, Spain (2015 June)
16. Arora, K., Gupta, P., Chopra, S., Pathak, N.: Posture monitoring belt. Int. J. Innov. Res. Stud. **4**(Issue 6) (2015 June)
17. Wang, Q., Chen, W., Timmermans, A.A.A., Karachristos, C., Martens, J.B., Markopoulos, P.: Smart rehabilitation garment for posture monitoring. In: Annual International Conference for the IEEE Engineering in Medicine and Biology Society (2015 Aug)
18. Samiei-Zonouz, R., Memarzadeh-Tehran, H., Rahmani, R.: Smartphone-centric human posture monitoring system. In: IEEE Canada International Humanitarian Technology Conference (IHTC) (2014)

Sensor-Based Traffic Control System

Roopa Ravish, Datthesh P. Shenoy and Shanta Rangaswamy

Abstract An effective road traffic control system ensures continuous movement of traffic and helps in preventing crashes or accidents. Better traffic management requires traffic signal control based on vehicle density. One such technique proposed in this paper finds the solution to traffic flow control, depending on the number of vehicles on the lane. It has two separate systems to control the traffic flow. One of the systems first collects the vehicle density (data) on the individual lanes using ultrasonic sensors. Second system uses this data in order to control traffic lights. By sharing this data, using NRF24L01 transceiver module, the system handles the traffic lights. The LEDs in the traffic lights are then appropriately controlled and powered based on the data received. With separate systems, maintenance of the components is simpler and it can also be observed that the lack of physical contact between the systems enables us to modify the implementation in one of them without any requirement of changes in the other. This technique uses an Arduino Uno and an Arduino Mega as the base for the two different systems. This solution is built to be compatible with subsequent improvements and can be extended to an IoT model.

Keywords Traffic control · Traffic density · Traffic flow control using ultrasonic sensors and NRF24L01 · Arduino microcontroller

R. Ravish (✉) · D. P. Shenoy
Department of Computer Science and Engineering, PES University, Bangalore, India
e-mail: rooparavish@pes.edu

D. P. Shenoy
e-mail: datthesh.shenoy@gmail.com

S. Rangaswamy
Department of Computer Science and Engineering, Rashtreeya Vidyalaya College of Engineering, Bangalore, India
e-mail: shantharangaswamy@rvce.edu.in

© Springer Nature Singapore Pte Ltd. 2020
J. K. Mandal and S. Mukhopadhyay (eds.), *Proceedings of the Global AI Congress 2019*, Advances in Intelligent Systems and Computing 1112,
https://doi.org/10.1007/978-981-15-2188-1_17

1 Introduction

According to F. D. Hobbs et al., traffic control is a critical element in the safe and efficient operation of any transportation [1]. Traffic congestion is a major problem faced in metropolitan cities these days. It appears when too many vehicles attempt to use a common transportation infrastructure with limited capacity [2]. Therefore, building an efficient traffic control system would avoid people getting stranded for long hours. An efficient system would also result in lower fuel cost and a lower trip time. Lower trip times would, in turn, result in higher productive man hours.

In olden days, traffic personnel were deployed to control the traffic flow. Due to lesser number of vehicles on road in those days, the method was efficient. With the increase in the number of vehicles, the solution shifted to legacy systems which were based on the logic of static transitions after a fixed duration irrespective of any other parameters. In 1918, New York city was the first city where the three-colour traffic signalling system was introduced. This model had a manual control from a tower in the middle of the street [3]. But times have changed. The number of vehicles on road has increased. Following a traditional system will result in longer wait times leading to unnecessary longer trip times and frustration amongst the motorists. This demands newer solutions for ease in the flow of traffic.

In the proposed method, traffic is controlled based on traffic density. It is usually observed that particular lanes have a significantly higher number of vehicles during a particular time period. For example, the lanes leading to schools, universities and companies have a significantly high number of vehicles during the peak hours early in the morning when compared to the lanes on the route back home. During late evening, vice versa is true. Thus, giving a higher priority to lanes with higher number of vehicles would automatically result in a free-flowing traffic in the peak hours. It would also result in a lesser average trip time. Safety of the motorists being the main objective, vehicles on the lesser priority lanes are expected to wait longer than they would usually.

To support the solution of resolving to variable traffic flow instead of conventional traffic control system, author Suresh Kumar et al. [4] states, 'The conventional traffic control system in India is inefficient due to a change in the traffic density on the roads and the failure of signals and lack of intelligence for traffic signals has led to traffic congestion'. Nowadays, the production of automated vehicles has also increased. It is considered to be the next technological rage in the field of transportation [5]. This would mean that an efficient traffic control system would result in numerous benefits provided the vehicles would adhere to the traffic rules unlike people in recent times.

With the concept of having variable traffic flow, we obtain a viable solution for some current issues.

- Save on foreign exchange by reducing the crude oil import.
- Prevent air pollution by reducing the vehicle emissions.
- Avoid road rage resulting due to frustration of waiting for long durations.

The rest of the paper is divided into the following sections. Section 2 comprises the literature review. Section 3 briefly discusses the specifications of the components used in the proposed solution. Section 4 gives the implementation details of the proposed solution. Section 5 is about the experimental set-up. To sum up the entire project, Sect. 6 contains the circuit diagrams of the working model and the explanation of the obtained result. It also discusses the potential challenges, real-time and possible solutions to overcome the difficulties. Section 7 concludes the work.

2 Literature Review

Author Nellore et al. [6] classify the road traffic into recurring and non-recurring congestion. Recurring congestion can be addressed easily as traffic pattern maintains regularity with respect to time and place. However, the non-recurring congestion is difficult to identify as traffic pattern is not fixed and is unpredictable. Non-recurring congestion occurs due to unexpected situations such as climatic variation, festivals and road accidents.

In order to control the traffic, it is necessary to detect traffic density. Author Jain et al. [7] describe various traffic monitoring systems. One such technique is Situ traffic detector technology. Based on installation of detector at or below the road surface, this technique can be further divided into intrusive and non-intrusive technology. Intrusive technique causes disruption to the traffic, whereas non-intrusive causes little or no disruption. In order to monitor the traffic, author suggests sensor technique, in which various sensors are attached to the moving vehicles. These moving devices (vehicles) are made to interact with central monitoring traffic system. Based on the sensor data, central monitoring system takes control over traffic movement. In another technique, live video stream is feed from cameras at a junction to collect traffic data. Using video and image processing, traffic density can be calculated.

Author Parekh et al. [8] propose a solution on similar lines for detecting the traffic density using multiple IR sensors by categorising the vehicle density on the lanes into three different levels. By dividing the lanes into blocks and placing the sensor grids under the roads pave way for erroneous results to creep in. An article in BBC News [9] accounted badly maintained roads as one of the reasons for the high number of accidents. With the roads in that state, relying on sensor grids under the roads would not only expose the grids to possible damages but will also lead to erroneous results when the standards are not followed. Any modifications in the grids would mean a high maintenance cost given the fact that the work on and due to the top layer would also account for the cost.

Sundar [10] also proposes a technique using radio-frequency identification (RFID) sensors. It equips every vehicle with a RFID tag. When in the RFID reader's range, the signal indicating its presence is sent. This enables the system to track the number of vehicles and hence the congestion volume. Based on the calculated values, it sets the lights accordingly. The main highlight of the solution is its ability of vehicle clearance during emergencies. This is achieved by a ZigBee transmitter and receiver at the vehicle and traffic junction, respectively. To add to this, it is also designed to track down stolen vehicles by matching the missing vehicles' RFID tag with the

RFID tags at the signal and if found alerts the respective authorities. The major challenge in this solution is to issue a unique RFID tag for every vehicle on-road, including the existing ones and also ensure a non-tamperable RFID tag.

Misbahuddin et al. [3] propose IoT-based traffic management solutions for smart cities. The proposed technique controls traffic dynamically through smart phones operated by onsite traffic officer. Single-board computer such as Raspberry Pi adopted with networking feature is used to interface with external circuit. The technique can be further improved if traffic data is automatically passed to the Raspberry Pi unit, controlling lights at an intersection so that authorities can make decision quickly. Using a mobile agent, under a vehicular ad hoc network (VANET) automatically control of traffic is proposed by Rath et al. [11]. The function of mobile agent-based improved traffic control system (MITS) is triggered when any smart vehicle enters the VANET zone. MITS counts the number of vehicles and if exceeds the threshold value sensors triggers an alarm. This prevents further entry of vehicles in the network. In order to prevent congestion vehicles are also rerouted depending on their choice. Increase in vehicle ownership increases traffic management problem.

Traditional method of traffic management includes monitoring of vehicle speed through cameras and pollution check. Increase in number of vehicles causes inconvenience to adopt above methods. Author Dandala et al. [12] suggest IoV-based traffic management system in order to control traffic, accident detection, theft avoidance, etc. However, author also mentioned drawback of IoV in traffic management with respect to security and network failure. Sensors are placed at fixed position in order to collect the data generated. These data are feed to central server in order to control and communicate with different devices. Large number of vehicles on one area causes jams. Author Kumar et al. [13] suggest this jam can be reduced if vehicles are guided with rerouting. The proposed method is demonstrated by dividing street maps into small distinct maps. In order to find optimal path, ant colony algorithm is then applied. Fuzzy logic-based traffic intensity calculation function is proposed to model heavy traffic.

Wiering [14] proposes a solution which uses reinforcement learning (RL) algorithms for learning traffic light controllers to minimise the overall waiting time of the vehicles. L. Pack Kaelbling et al. state 'Reinforcement learning is the problem faced by an agent that learns behaviour through trial-and-error interactions with a dynamic environment' [15]. The solution defines a waiting queue comprising of the vehicles that would be immediately affected by the traffic light state and aims to minimise the cumulative waiting time. It also provides two approaches to proceed with—traffic-light-based and car-based controllers. The advantage of going for RL is that the agent not only learns the policy by interacting with the environment that provides feedback, it also learns to update the action evaluation function after each performed action. The author has also made note of the difficulties. The evident one would be computing the total trip times for all the vehicles and even if computed, the designed system would not be able to communicate the information making the information unusable.

On the similar lines, a deep Q learning was also deployed in a solution given by Liang [16]. It aims on optimising the traffic light duration by taking real-time traffic information and dynamically adjusting the duration accordingly. To overcome the large number of states in the traffic control system in vehicular networks, the solution opts for a convolutional neural network (CNN) to approximate the Q value. The actions are modelled as a Markov decision process, and the rewards are the cumulative waiting time difference between two cycles. The authors also recommend a double-duelling deep Q network for handling complex traffic scenarios. They also claim a 20% reduction in the average waiting time. A major challenge that the model could face would be erratic traffic in real time for the model to respond to the conditions within the stipulated time.

Computer vision-guided traffic management is another emerging area of research. Kumaran et al. [17] propose a solution of temporally clustering optical flow features of moving vehicles using Temporal Unknown Incremental Clustering (TUIC) model. The technique proposed also claims an improvement the clustering performance when compared to using Gibbs sampling in the inference as proposed by author Santhosh et al. [18]. The developed algorithm, Throughput Average Waiting Time Optimisation, uses cluster count, which acts as a measure to predict traffic phase durations, hence maximises the throughput and minimises the average waiting time.

3 Component Specifications

The proposed solution uses two sensors, ultrasonic sensor and NRF24L01 transceiver module.

3.1 Ultrasonic Sensor

The HC-SR04 ultrasonic sensor shown in Fig. 1 [19] uses sound navigation ranging to determine the distance of an obstacle. It is not affected by sunlight or materials in black (acoustically).

In the proposed solution, HC-SR04 is used to find the distance of the vehicle closest to it. The sensor by default calculates the time taken for the ultrasonic wave generated to return back to sensor. With the available microcontroller Arduino, the distance of the obstacle can be calculated using Eq. (1).

$$\text{Distance (in cm)} = \text{duration} * 340/20{,}000 \tag{1}$$

where 340 m/s is the speed of sound in air.

Distance is halved because the resulting distance would be sum total of the to and fro distance travelled by the wave.

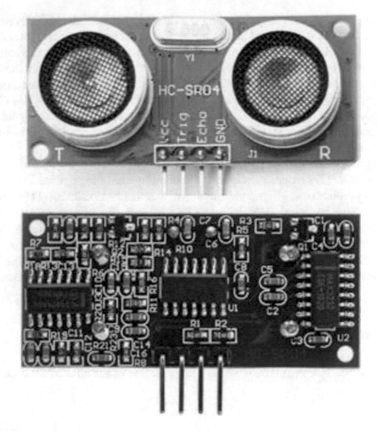

Fig. 1 Ultrasonic sensor

3.2 NRF24L01

The NRF24L01 transceiver module shown in Fig. 2 [20] is used to wirelessly communicate between two or more Arduino boards.

This is used to wirelessly communicate between the two Arduino boards here. The power is set to low for demonstration purpose, but can be set to max to obtain a better range. In the proposed solution, it is used to transmit and receive a structure containing the lane values and their respective waiting times which sum up to a total of 48 bytes for data. Therefore, even a 250 Kbps speed would work smoothly when it comes to information transfer between the two boards. Out of the available 125, any arbitrary channel can be chosen. The code uses the channel 115.

Fig. 2 NRF24L01 transceiver module

4 Implementation

The implementation of the proposed solution can be divided into three subsections, namely system for computation, system for traffic lights and traffic lights. System for computation computes the traffic density, system for traffic lights describes the transitions, and the traffic light subsection describes the states in the proposed solution.

4.1 System for Computation

This system consists of an Arduino powering multiple ultrasonic sensors as shown in Fig. 3. For demonstration, the proposed solution has one ultrasonic sensor, and the other three values are simulated. It works on logic of calculating the sum total of the distances from the vehicles nearest to the ultrasonic sensors as shown in Eq. (2). To avoid vehicles lining up on one side, ultrasonic sensors are used on either side of the lane with a reasonable offset such that there are very low chances of interference. This helps in obtaining a better value to gauge the density. It is evident that in a traffic condition like that of India, vehicles are tightly packed rather than dispersed. With this assumption, the more the number of vehicles in a particular lane, it would result in vehicles getting closer to the multiple ultrasonic sensors. This would lead to a lesser value as the sum total of all the ultrasonic sensors' output values in that particular lane when compared to a lane with lesser traffic. The reason being, there would be at least some ultrasonic sensors resulting in a value equal to the lane width (max distance as output) which would evidently be higher than the resulting sum of the other lanes. Before transmitting the value to the other system, it also checks the existence of lanes with a waiting time period beyond a certain threshold and if true, resets the corresponding value (sum total of the distances from the ultrasonic sensors) to zero to provide it the maximum priority for traffic flow. With this, the solution also takes care of traffic flow in the lanes with a very low traffic density at regular time intervals.

Fig. 3 Positing of the
ultrasonic sensors

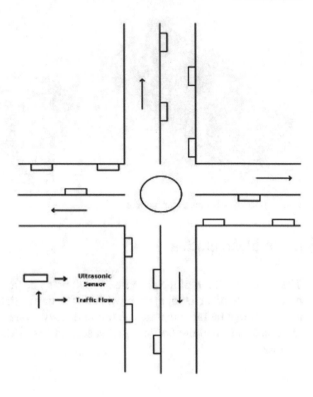

$$\text{Lane Sum} = \sum_{i=1}^{n} (\text{Ultrasonic sensor } i\text{'s value}) \qquad (2)$$

4.2 System for Traffic Lights

Based on the input from the computation system, this system runs a comparison
loop to check the lane with the maximum traffic, i.e. the lane with the lowest value.
After obtaining the lane number, it changes the traffic lights accordingly. It provides
a green signal for the lane with lowest value and retains the traffic on other lanes.

4.3 Traffic Lights

The traffic lights are modelled using LEDs. For demonstration, only red and green
LEDs are used. The start state is taken to be GRRR (green–red–red–red) corresponding to the lanes numbered 1–4. The transition is defined by setting the green LED to

active high and red LED to active low on the lane with maximum traffic and vice versa on the lanes with lesser traffic. It is assumed that the model would be restarted periodically because the return value of the millis() function used, returns the milliseconds passed since the Arduino board began running the current program. It would exceed the data structure limit after 50 days [21] and could result in erroneous results if it is not reset at appropriate time intervals.

5 Experimental Set-up

This section deals with the experimental set-up of the proposed solution.

5.1 Set-up Components

- Set up the circuit according to the circuit diagram as shown in Figs. 4 and 5. If another Arduino Mega is available, multiple ultrasonic sensors can be used instead of simulating the values.
- For multiple ultrasonic sensors, use appropriate free ports and ensure appropriate modification in the code.
- If the boards used are different, ensure appropriate connection of the NRF modules with the recommended pin numbers and modify the code appropriately.

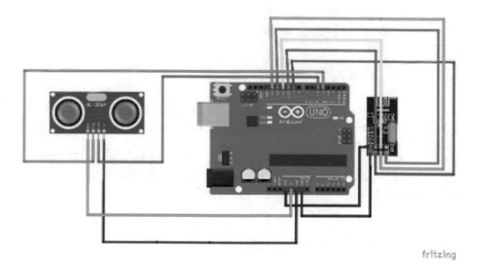

fritzing

Fig. 4 Circuit diagram of the computational part (transmitter)

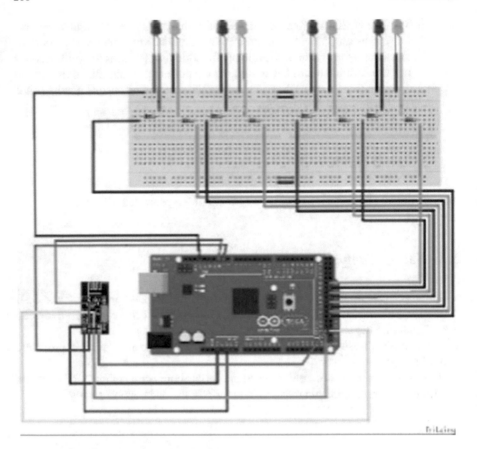

Fig. 5 Circuit diagram for powering the traffic lights (receiver)

5.2 Arduino Set-up

- Download the latest Arduino IDE on both the computer systems.
- Include the RF24-master library file. Zipped folder is available in the Google drive link provided for the code.
- If RF24-master is not available, any equivalent suitable library can also be included provided the connections and the code are modified appropriately.
- Set the appropriate boards on both the systems

5.3 Proposed Solution Execution

- Download the transmitter and receiver codes from the link provided, and run it on the respective boards.

- In case multiple ultrasonic sensors are used instead of simulation, a predesigned function called 'ultrasound' is available, which returns the sum total distances obtained from two ultrasonic sensors.
- Validate the results by checking the data on the COM monitor.

6 Results and Discussion

The demonstration model, as depicted in Fig. 6, has four lanes, namely 'A', 'B', 'C' and 'D'. The lane 'A' had the lowest lane sum value implying highest traffic density as shown in Fig. 7. The green LED corresponding to the lane 'A' was turned on, and the red LED on all the other lanes was turned on which was in accordance with the

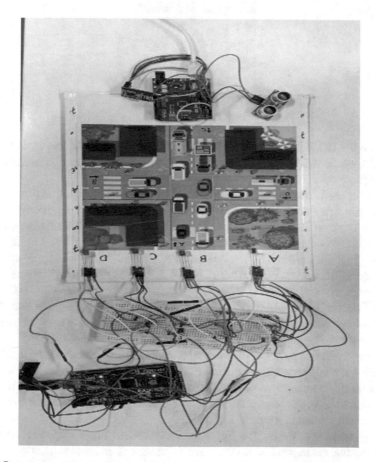

Fig. 6 Output state corresponding to the latest traffic density (in accordance with the output on the COM monitor)

Fig. 7 COM monitor output

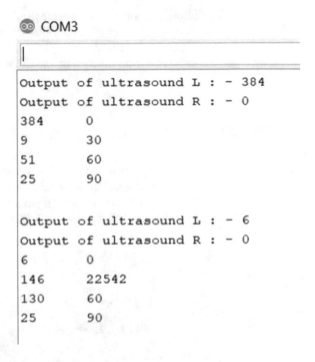

```
⊙ COM3

|

Output of ultrasound L :  - 384
Output of ultrasound R :  - 0
384        0
9          30
51         60
25         90


Output of ultrasound L :  - 6
Output of ultrasound R :  - 0
6          0
146        22542
130        60
25         90
```

output on the COM monitor as shown in Figs. 6 and 8. It is evident from the time field adjacent to the lane sum value which stores the most recent time of start of traffic flow.

The proposed solution consists of two systems—one for computation and the other for powering the traffic lights based on the input received. The computational system is based on the logic of prioritising the lanes based on the traffic density. This can be achieved using ultrasonic sensors, cameras or even satellites. In due course, if the Original Equipment Manufacturers (OEMs) manufacture vehicles with unique RFID tags, RFID readers can be used. This would also allow us to prioritise the lanes during emergencies by having specific RFID formats for the vehicles with higher priority (viz. ambulances). This would also ease the planning of green corridors to save lives. For demonstration purposes, ultrasonic sensors were used to gauge the traffic density.

For demonstration purpose, the internal clock of the Arduino was used to compute the waiting time. It brings in a restriction of time, wherein the value would overflow for the data structure used and would go back to zero after approximately 50 days. In real-time scenarios, a real-time clock module can be used to overcome the difficulty of resetting the board at frequent intervals.

The system for powering the lights uses an Arduino Mega to power the LEDs. The transitions are dependent on the input from the computational system.

In real-time scenarios, the assumption is that the Arduino boards can be powered through solar energy as well as regular electricity. This would ensure a hassle-free model which would work in extreme conditions as well. There is also a possibility of

Fig. 8 COM monitor post-traffic flow on lane A

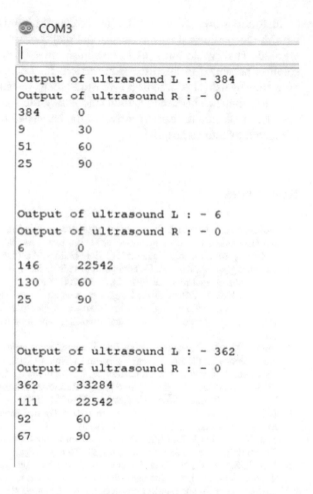

a mechanical damage. To overcome this, the solution can be programmed to shift to the traditional model during unforeseen circumstances and revert after it is handled. There is also a possibility of water clogging resulting in the malfunctioning of the ultrasonic sensors. A fall-back method can also be deployed to ensure free-flowing traffic.

7 Conclusion and Future Work

In order to control the traffic flow, sensor-based technique can be the better solution. Proposed technique first collects vehicle data on the road and then suggests appropriate signal to be turned ON. This is implemented using NRF24L01 transceiver module, LEDs and Arduino microcontroller.

In future work, we would like to improve our implementation in accordance with the technological advancements and their compatibility with all the vehicles on-road. That would allow us to recognise emergency conditions better than the solution presented in the discussion section. It would also help in improving the efficiency by using better sensors. We would focus on extending the proposed model to an IoT model. This would help in automating the signals appropriately for green corridors to ensure the earliest arrival of the harvested organs meant for transplants to reach the destined hospital.

References

1. Hobbs, F.D., Jovanis, P.P.: https://www.britannica.com/technology/traffic-control. Encyclopæ-dia Britannica, Inc, 08 September 2000. [Online]. Available: https://www.britannica.com/technology/traffic-control#accordion-article-history. Accessed 05 June 2019
2. Papageorgiou, M., Kiakaki, C., Dinopoulou, V., Kotsialos, A., Wang, Y.: Review of road traffic control strategies. Proc. IEEE **91**(12), 2043–2067 (2003)
3. Misbahuddin, S., Zubairi, J.A., Saggaf, A., Basuni, J., A-Wadany, S., Al-Sofi, A.: IoT based dynamic road traffic management for smart cities. In: 2015 12th International Conference on High-Capacity Optical Networks and Enabling/Emerging Technologies, HONET-ICT 2015 (2016)
4. Suresh Kumar, S., Rajesh Babu, M., Vineeth, R., Varun, S., Sahil, A.N., Sharanraj, S.: Autonomous Traffic Light Control System for Smart Cities, pp. 325–335 (2019)
5. Banerjee, S., Chakraborty, C., Chatterjee, S.: A survey on IoT based traffic control and prediction mechanism. In: Intelligent Systems Reference Library, vol. 154, pp. 53–75. Springer Science and Business Media Deutschland GmbH (2019)
6. Nellore, K., Hancke, G.P.: A Survey on Urban Traffic Management System Using Wireless Sensor Networks, vol. 16. MDPI AG (2016)
7. Jain, N.K., Saini, R.K., Mittal, P.: A review on traffic monitoring system techniques. In: Soft Computing: Theories and Applications, pp. 569–577. Springer, Berlin (2019)
8. Parekh, S., Dhami, N., Patel, S., Undavia, J.: Traffic signal automation through IoT by sensing and detecting traffic intensity through IR sensors. In: Information and Communication Technology for Intelligent Systems, pp. 53–65. Springer, Berlin (2019)
9. BBC, "BBC News," BBC, 10 June 2016. [Online]. Available: https://www.bbc.com/news/world-asia-india-36496375. Accessed 5 June 2019
10. Sundar, R., Hebbar, S., Golla, V.: Implementing intelligent traffic control system for congestion control, ambulance clearance, and stolen vehicle detection. IEEE Sens. J. **15**(2), 1109–1113 (2015)
11. Rath, M., Pati, B., Pattanayak, B.K.: Mobile agent-based improved traffic control system in VANET. In: Studies in Computational Intelligence, vol. 771. Springer, Berlin, pp. 261–269 (2019)
12. Dandala, T.T., Krishnamurthy, V., Alwan, R.: Internet of Vehicles (IoV) for traffic management. In: 2017 International Conference on Computer, Communication and Signal Processing (ICCCSP) (2017)
13. Kumar, P.M., Devi, U., Manogaran, G., Sundarasekar, R., Chilamkurti, N., Varatharajan, R.: Ant colony optimization algorithm with internet of vehicles for intelligent traffic control system. Comput. Netw. **144**, 154–162 (2018)
14. Wiering, M.A., Veenen, J., Vreeken, J., Koopman, A.: Intelligent traffic light control. Utrecht University: Information and Computing Sciences (2004)
15. Kaelbling, L.P., Littman, M.L., Moore, A.W.: Reinforcement learning: a survey. J. Artif. Intell. Res. **4**, 237–285 (1996)

16. Liang, X., Du, X., Wang, G., Han, Z.: A deep reinforcement learning network for traffic light cycle control. IEEE Trans. Veh. Technol. **68**(2), 1243–1253 (2019)
17. Kumaran, S.K., Mohapatra, S., Dogra, D.P., Roy, P.P., Kim, B.-G.: Computer vision-guided intelligent traffic signaling for isolated intersections. Expert Syst. Appl. (2019)
18. Santhosh, K.K., Dogra, D.P., Roy, P.P.: Temporal unknown incremental clustering model for analysis of traffic surveillance videos. IEEE Trans. Intell. Transp. Syst. **99**, 1–12 (2018)
19. Tutorialspoint (2019) Arduino—Ultrasonic Sensor. https://www.tutorialspoint.com/arduino/arduino_ultrasonic_sensor.htm. Accessed 05 June 2019
20. How to Mechatronics (2019) Arduino wireless communication—NRF24L01 tutorial. https://howtomechatronics.com/tutorials/arduino/arduino-wireless-communication-nrf24l01-tutorial/. Accessed 05 June 2019
21. Arduino (2019) millis(). https://www.arduino.cc/reference/en/language/functions/time/millis/. Accessed 5 June 2019

Addressing Grain-Matrix Differentiation in Sedimentary Rock Photomicrographs in the Light of Brightness Perception Modelling

Rajdeep Das, B. Uma Shankar, Tapan Chakraborty and Kuntal Ghosh

Abstract Brightness perception in real-life situations and subsequent image segmentation based on it is a complex phenomenon. The human visual system (HVS), apparently, does this effortlessly in the case of natural images. For specialized segmentation tasks, as in medical imaging, this is performed by the trained visual system of the specialist (radiologist, for instance, in medical imaging). In the present work, we shall concentrate on one such specialized task, viz. analysis of Photomicrograph of rock thin sections using petrological microscope, in the light of the HVS. For this, a new neural network model for the extended classical receptive field (ECRF) of Parvo (P) and Magno (M) cells in mid-level vision is elaborated at the outset. The model is based upon various well-known findings in neurophysiology, anatomy and psychophysics in HVS, especially related to the role parallel channels (P and M) in the central visual pathway. These two channels are represented by two different spatial filters that validate the reports of several psychophysical experiments on the direction of brightness induction. The mechanism of selecting the preferred channel for each of the stimuli consists of an algorithm that depends upon the output from an initial M channel filtering as captured in the visual cortex. We assume that the visual system of the geologist is training itself in the same way through such filtering processes in mid-level vision and identifying the important information in various situations in optical mineralogy and petrography. In the present work, the proposed model is applied in a simplified form on one such situation dealing with clast–matrix segregation from photomicrograph of sedimentary rocks, and is found to yield a promising result.

R. Das · B. U. Shankar · K. Ghosh (✉)
Machine Intelligence Unit, Indian Statistical Institute, Kolkata 700108, India
e-mail: kuntal@isical.ac.in

R. Das
e-mail: rajdeep0129@gmail.com

B. U. Shankar
e-mail: uma@isical.ac.in

T. Chakraborty
Geological Studies Unit, Indian Statistical Institute, Kolkata 700108, India
e-mail: tapan@isical.ac.in

© Springer Nature Singapore Pte Ltd. 2020
J. K. Mandal and S. Mukhopadhyay (eds.), *Proceedings of the Global AI Congress 2019*, Advances in Intelligent Systems and Computing 1112,
https://doi.org/10.1007/978-981-15-2188-1_18

Keywords Brightness perception · HVS · Figure–ground · Optical mineralogy ·
Petrology · Sedimentary rock · Clast–matrix · Clustering · Dice score

1 Introduction

The detritus formed through weathering of the pre-existing rock material is entrained,
transported and deposited by glaciers, rivers, wind or ocean waves to form clastic
sedimentary rocks. They can be deposited on the land surfaces or on the seafloor.
The sedimentary rocks cover almost sixty-five per cent of the exposed earth's surface
though the same is not true for the earth's crust which is dominantly made up of
igneous and metamorphic rock with a comparatively small proportion (only about
five per cent) of sedimentary rock. The sedimentary rocks are deposited layer by layer,
forming what is usually referred to as bedding. Fossil fuels (coal, petroleum, etc.) are
hosted by sedimentary rocks. That apart, most of the underground water is contained
in sedimentary rock layers. Thus, the study of the sedimentary rocks is crucial and
imperative in understanding the origin and location of these resources. Last but not
least, the study of the succession of sedimentary rock strata is the main source of our
knowledge of the surface processes and climatic conditions prevailing in the past.
Evolution of life itself can only be studied from the study of the sedimentary rocks
that contain the fossil record of the past life forms.

In this study we investigate the sedimentary rocks from an image processing perspective. The digital images that have been processed are generated through microscopic study of rock thin sections, which are slices of about 0.03 mm thickness,
mounted on glass slides. A chip of rock is, at first, cut out of a sample. One side of
the rock slice is then well-polished and carefully glued to a clean glass slide using
a transparent adhesive. The rock chip is then finally ground down to the desired
final thickness of about 0.03 mm. The colour and intensity of a particular point in a
rock thin-section under microscope depend on the optical property of crystal at that
point, its orientation and optical phenomena, viz. refraction, dispersion, absorption
and birefringence, that may occur at that point. The petrologic microscope consists
of two Nicol prisms having their polarizing planes oriented perpendicular to one
another. When only an isotropic material such as air, water or glass exists between
the polarizers, all light is blocked. However, most crystalline materials and minerals
change the vibration direction of light passing through them, allowing some of the
altered light to pass. Using one polarizer makes it possible to observe certain optical
properties (e.g. refractive index of the minerals) in plane-polarized light; using two
polarizers allows observation of certain other properties (e.g. birefringence) and is
referred to as the cross-polarized view. A look at the cross-polar images captured
through the microscope (shown and explained later in Fig. 7) may reveal the motivation for image processing and analysis tasks for such images through brightness
perception-based modelling, even to a non-expert. Depending upon the orientation
and type of the various crystals present in such slides, the colour and intensity of
the different segments of the picture, which a geologist is interested in extracting,

vary widely from very dark to very bright although such apparent variation may not actually be meaningful and hence may mislead. For instance, this is especially true in the matter of identifying the larger rock/mineral fragments called the grains (see Fig. 7). The matrix of the sedimentary rocks is fine-grained sedimentary material, such as clay or silt, that surrounds or occurs in the interstices of larger grains. To segment out the grains, for example to count such grains through an automated algorithm, we need computer vision algorithm/s that, akin to the trained geologists' eye, does not get deceived by the wide variation in brightness information. Hence, through image processing, such segmentation can also be achieved efficiently and accurately, which for the human being is a time-consuming and tedious job. In other words, the automatic detection and analysis of the photomicrographs can be efficient, fast, and accurate too, provided the brightness modelling is done competently.

The perceived brightness of any surface depends on the brightness of the surfaces that surround it, a phenomenon termed as brightness induction. Several studies by Helson [1] and Jameson and Hurvich [2] reveal that brightness contrast and brightness assimilation are two opposite aspects of brightness induction. The former aspect is believed to be exhibited by filtering out the low-frequency components of the stimulus and the later aspect by filtering out the high-frequency components. This Blakeslee and McCourt [3] demonstrated through their oriented difference-of-Gaussians (ODOG) filter which is considered as the state-of-the-art model of brightness perception in images.

The two typical examples of brightness contrast and assimilation have been shown in Fig. 1. In this work, we shall try to understand how the parallel channels, especially the magnocellular and the parvocellular channels in the central visual pathway can

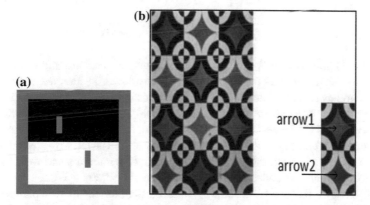

Fig. 1 Examples of **a** brightness contrast and **b** brightness assimilation. In **a**, the two test patches of equal intensity appear to be of different brightness based on the surroundings in such a way that one on darker background appears brighter than the other one on lighter background. In **b**, the brightness induction is opposite. The surrounding darkness or lightness has been assimilated into the centremost patches (marked by Arrow 1 and Arrow 2), thus representing a phenomenon opposite to contrast

play a significant role in perceiving the above-mentioned opposite aspects of brightness induction in mid-level vision. To this end, we examine a few cases of optical illusion, related to brightness perception, with our proposed model, and then investigate if this domain knowledge may provide clues to bio-inspired figure–ground grouping tasks for real-life problems in some geological studies.

2 Materials and Methods

The present model proposes that mid-level vision in human visual system (HVS) occurs in a two-step process: (a) *Vision at a glance* that uses incoming signals of the magnocellular pathway, and (b) *Vision with scrutiny* that uses the previous step to prioritize areas of scrutiny and uses the parvocellular pathway to scrutinize those areas. This finally leads to the figure–ground segregation in the brain. In other words, the parvocellular (P) pathway is enhanced by the presence of attention and has higher spatial resolution but lower temporal resolution. The magnocellular (M) pathway, on the other hand, is suppressed by the presence of attention. It has higher temporal resolution but lower spatial resolution. The M & P channels have been modelled using a 3-Gaussian feed-forward neural network model, called the extended classical receptive field filter (ECRF) [4, 5], in contrast to the scores of Gaussians in the ODOG filter mentioned earlier [6]. The ECRF filter that is composed of three Gaussian functions of varying widths representing the classical excitatory centre (σ_1), the classical inhibitory surround (σ_2) and the disinhibitory extended surround (σ_3) regions of visual receptive field, can be mathematically represented as follows:

$$\text{ECRF}(\sigma_1, \sigma_2, \sigma_3, x, y) = A_1 e^{-\frac{r^2}{2\sigma_1^2}} - A_2 e^{-\frac{r^2}{2\sigma_2^2}} + A_3 e^{-\frac{r^2}{2\sigma_3^2}}$$

where, A_1, A_2 and A_3 signify the weights to the centre, surround and extended surrounds. This neural network model has also recently been considered dynamically by other researches to make mid-level representations and corresponding applications [7, 8]. We further propose that the two spatial filters which model the mid-level neuronal networks as ECRF described by the above equation can represent the extended classical receptive fields of the P and M neurons in mid-level. To substantiate this claim, let us first put forth the corresponding sampling intervals and weights to the centre, surround and extended surrounds for these two filters in Table 1.

Table 1 shows that we have actually reduced the number of free parameters in the ECRF equation. The ratio of the three variances of the Gaussians is fixed [5]. The variation in sampling density effectively changes the size of the ECRF so that at lower sampling density, the ECRF of the M filter has lower spatial resolution than the P, as already mentioned for the model based on biological findings. In the value of parameters, it may be noted that inhibitory surround's contribution in the model is double for M as compared to P, like the sampling interval also, while the disinhibition is eightfold more.

Table 1 Spatial filters representing the two (M and P) parallel visual channels

Filter Parameters Mid-level Neurons	A_1	A_2	A_3	Sampling interval
The Magno neurons	10	1	0.08	0.5
The Parvo neurons	10	1	0.01	0.25

Let us next, take a glance at the frequency responses of these two filters when expressed in one dimension (Fig. 2). We find that they bear definite analogy, however crude, to the nature of the spatial frequency-dependent contrast sensitivity functions (CSF) in the central visual pathway as obtained from physiological and behavioural experiments. The solid curve in Fig. 2, whose corresponding spatial filter represents the lower row in Table 1, is analogous to the "parvo alone" contrast sensitivity function curve experimentally obtained by lesion studies conducted in LGN on macaque monkeys (Macaca nemestrina) (Fig. 2a in Merigan and Maunsell [9]). Though the analogy is crude, it should still be borne in mind that neurobiological responses are different under suprathreshold conditions [10]. We are merely trying to compare the nature of the curves. Similarly, the dotted curve in Fig. 2, whose corresponding spatial filter represents the upper row in Table 1, has an intermediate kink followed by a

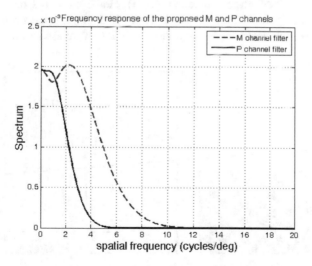

Fig. 2 Frequency response of the proposed P and M filters. The proposed M channel filter (dotted line) exhibits an intermediate kink and a slower fall-off compared to the P (solid line). This M channel behaviour is found to bear similarity to the mid-level neuronal cells from lesion studies. The P channel exhibits similar characteristic as shown by mid-level P neurons

similar slower fall-off as in the "Magno alone" CSF curve of Merigan and Maunsell [9], though in Merigan and Maunsell's curve the kink is less prominent.

With these two filters in hand, we shall now try to explain the context-dependent opposing nature of brightness induction as evident in some related optical illusions in the literature and then apply these as bio-inspired filters in dealing with some difficult problems in image processing of rock samples.

3 Results and Discussion

The checkerboard illusion (De Valois and De Valois [11]) is known to produce two opposite brightness effects in low and high spatial frequency of the inducing gratings.

By matching the mean luminance, Blakeslee et al. [3] confirmed that higher spatial frequency gratings (Fig. 3) indeed produced brightness assimilation, whereas lowering the spatial frequency (Fig. 4) reversed the effect to brightness contrast. Simulation results based on spatial filtering by the proposed ECRF model of P LGN cells confirm that when spatial frequency of the checkerboard is high, the test patch surrounded by four black squares looks darker than the same sharing borders with four white squares, a fact that implies brightness assimilation (Fig. 3b). At lower spatial frequency of checkerboard, the same spatial filter (i.e. proposed P channel) produces brightness contrast effect, so that the test patch surrounded by four black squares appears brighter in our simulation results in comparison with the same sharing borders with four white squares (Fig. 4b). However, in the results presented above, no justification has been provided as to which channel has been selected for which of the stimuli, and hence, our attempt would now be to first throw some light

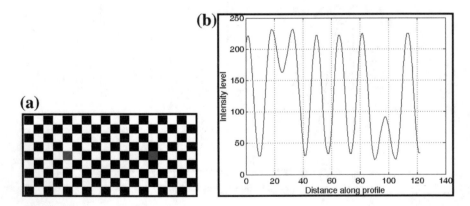

Fig. 3 **a** Checkerboard stimulus at 8 pixels width per column. **b** Horizontal intensity profile through the two test patches of the output image filtered by P channel, demonstrating brightness assimilation from the surrounding, with the left patch appearing brighter, as also observed in psychophysical experiments

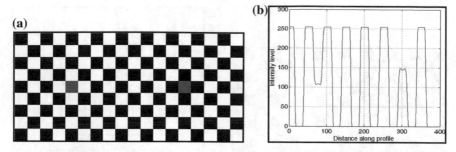

Fig. 4 **a** Checkerboard stimulus at 25 pixels checker dimension. **b** Horizontal intensity profile through the two test patches of the output image spatially filtered by P channel demonstrating brightness contrast with respect to surround, with the left patch appearing darker, as also observed in psychophysical experiments

on this aspect of channel selection. Based on the latencies of visual responses of neurons in different cortical areas, it has been argued that such characteristics of the M channel like faster conduction, high contrast sensitivity, poor chromatic selectivity, larger receptive fields and lower sampling rates makes it a probable candidate for a first-pass *vision at a glance* [6, 12]. The reverse hierarchy theory (RHT) of Ahissar and Hochstein [13] also envisages an initial *vision at a glance* to include results of automatic and implicit bottom-up processing, which makes the initial explicit perception introspectively direct without conscious antecedents, to be followed up later by *vision with scrutiny* for including more indirect conscious perceptual constructs. Our proposal to some extent also falls in line with these above-mentioned works. We propose that brightness perception is an integral part of the mechanism of attentive vision in the visual pathway, and this involves a final choice of channel in the mid-level vision based on the initial information extracted in the visual cortex through *vision at a glance* via the M pathway and subsequent feedback to the mid-level after computing certain features from this preliminary output possibly through the corticogeniculate pathway. To this effect, we propose an algorithm for attentive vision:

If (the initial M channel output identifies that the background around the test patches is *uniform* by virtue of satisfying *condition 1* ‖ (OR) *condition 2*),

/* *Uniform background* means M channel identifies

{

Condition 1. uniformity with respect to the absence of sharp intensity changes in the background outside the test patch, i.e. if the vertical and horizontal line profiles of the background above, below and on either side of the test patch do not contain contrast edges, it will be accepted as uniform background;

}

‖ *(OR)*

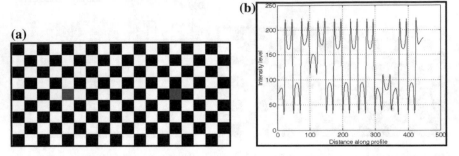

Fig. 5 **a** Checkerboard stimulus. **b** The corresponding intensity profile of the output image spatially filtered through the proposed M channel displays the sharp undershoots and overshoots in the vicinity of the test patches

```
{
```
Condition 2. uniformity with respect to the direction of Weber contrast * *across all the edges of the test patch, i.e. if such contrast of the test patch with its background is unidirectional (has the same sign) both across its horizontal and vertical edges, we call it high contrast surrounding or uniform background in both directions.*
```
}*/
{
```
P channel is invoked and the brightness percept is formed by P.

```
          }
```
Else
```
{
```
M channel produces the brightness percept.

```
             }
```
*Weber contrast =

$\frac{I-I_b}{I_b}$, *with I and I$_b$ representing the luminance of the region of attention (test patch) and the background luminance, respectively.*

This means that when any one of *condition 1* or *condition 2* is obeyed by the initial M output, then P channel is invoked and when both the conditions are violated or at least one is violated and the other remains inconclusive, then the visual system continues with the M channel. For instance, in Fig. 6a, i.e. the classical brightness-contrast effect, by virtue of both *condition 1* (neighbourhood is uniform) and *condition 2* (Weber contrast is uniform), the P channel is invoked, and the P channel correctly predicts the brightness direction as is evident from Fig. 6c. For Fig. 6b, i.e. the classical White effect, the output (Fig. 6d), from M channel filtering displays that both *condition 1* and *condition 2* for uniformity test are violated. The violation of *condition 1* is evident from the adjacent sharp undershoots and overshoots around the test patches (Fig. 6d) indicating the presence of contrast edges, while that of *condition 2*

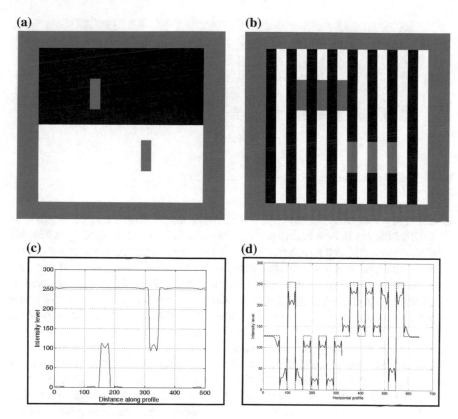

Fig. 6 **a** Classical example of brightness-contrast. **b** The White effect illusion. **c** Output of P channel over (**a**). **d** Output of M channel over (**b**)

is easy to see since the Weber contrasts across vertical and horizontal edges are of opposite sign. Hence, M remains the preferred channel.

Based on this algorithm, if we apply the initial M filter to any one of the checkerboard stimuli in Figs. 3 and 4, it is easy to see from Fig. 5b that the M output produces the sharp adjacent undershoots and overshoots and hence *condition 1* is clearly violated. On the other hand, it is easy to see that the Weber contrasts calculated across both horizontal and vertical edges of each test patch in the M filtered output are the same, i.e. *condition 2* is obeyed and P channel is thus preferred. The experimental findings [3] are indicated by the P outputs in Figs. 3b and 4b, respectively. The proposed algorithm is further justified in the light of the fact that the same channel produces opposite perception for such a difference in spatial frequency as observed in experiments.

Let us now try to apply this biologically inspired algorithm to some typical problems of processing rock thin-section images collected by the geologists of the Indian Statistical Institute, samples of which are shown in Fig. 7. The target here is to

grain

grain

grain

Fig. 7 Typical photomicrographs from sedimentary rocks

identify the grains from the background called the matrix, through image segmentation. This happens to be a highly relevant topic in the subject of geology where till date some works though have been done, yet no general solution to the grain-matrix segmentation problem has actually come out. References [14–19] are a few among examples of works done in this field. It is clear to see from the sample pictures in Fig. 7 that the brightness perception complexity as illustrated in Fig. 1 is very much an issue in these pictures, the possible reason why such a two-class segmentation still remains a difficult problem to solve. Neither the grain nor the matrix has any consistent greyscale range so that a suitable threshold for separation of the classes may be done through standard machine learning methods, and the same has been verified by us. We have demonstrated the ground truth for one such image in Fig. 8.

Under these limitations of the applicability of the classical image processing algorithms, we have, first of all, tried to keep things simple running a standard K-means clustering algorithm in the CIELAB colour space that was designed to be perceptually uniform with respect to human colour vision and see the results. It is well known that this colour space is exactly linearly proportional to its three colour contents. What we have done additionally is that we tried to mimic the visual system

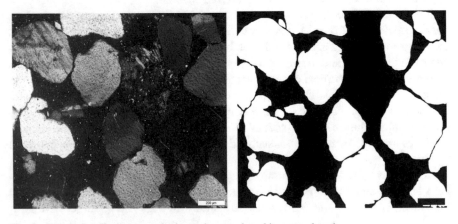

Fig. 8 A typical sedimentary rock photomicrograph and its ground truth

of the trained geologist working with a petrographic microscope in the light of the ECRF model described above and supply the model's output to our clustering algorithm. Since the region of interest is not predefined here, unlike the optical illusion figures, it is even more difficult to verify the conditions of uniformity and select the M or P filter accordingly to compute the final outputs, which in turn will act as inputs for K-means clustering. Therefore, we slightly depart from the model, and separately apply the P filter and the M filter on the original photomicrographs of sedimentary rock assuming that unlike natural image processing, the geologist uses both the M and P pathways in his/her visual system to train himself in grain identification. Then following the visual hierarchy, we already mentioned that we also apply the P filter on the M filter output. We refer to these three images in short as P image, M image and PM image. We used these three images and the original image as inputs to the K-means clustering algorithm which is described below and compared the grain-background similarities with ground truth as provided by the expert. The ground truth image has been built manually by filling the grain portions of a sedimentary rock thin-section image in white making the rest, i.e. matrix portion black. The performance of each of the filters (P, M and PM) is compared with the no-filter case in terms of dice score. The Sorensen–Dice coefficient, it is well known, is a basic and simple statistic used for comparing the similarity between ground truth and the segmented image:

$$\text{Dice score} = \frac{2|X \cap Y|}{|X| + |Y|}$$

where X is the set of all ground truth grain pixels and Y is the set of all grain pixels found by segmenting the image.

The working algorithm is as follows:

Input: RGB image

Targeted Output: Binary image (grain region in white (greyscale value 255) and background (matrix region) in black (greyscale value zero)) and the corresponding Dice score

Start:

1. *Give as input an image RGB image (original or M filter or P filter or MP filtered).*
2. *Transform it into CIELAB space.*
3. *Find 20^* clusters through K-means algorithm.*
4. *Order the 20 clusters using the norm over the cluster centres in LAB values.*
5. *Find the smallest frequency cluster (i.e. cluster with the smallest number).*
6. *This smallest number cluster is chosen as the threshold for dividing the 20 clusters into two groups (object class and background class).*
7. *The segmented results are obtained based on these two classes.*
8. *To evaluate the segmented result, it is compared with the ground truth using*

Dice score.

End.

Table 2 Dice scores for the original image and its filtered versions

Image type	Dice score
Original	0.6125
M image	0.6615
P image	0.6794
PM image	0.7117

*It has been verified by taking feedback from the expert that the image with 20 clusters highly resembles the original cross-polar image which the geologist uses for manual grain counting under the petrological microscope.

The Dice scores are provided in Table 2 for the four cases, viz. (a) original image, (b) M filtered image, (c) P filtered image and (d) PM filtered image.

It is interesting to note from Table 2 that first of all, each of the three filters, i.e. P filter, M filter and M and P filter applied sequentially produces larger dice scores compared to when no filter is applied on the original image before clustering. Among these, it is the PM filter that produces the highest dice score. This filter, as already explained, attempts to combine in our visual system, the effects of both contrast and assimilation of brightness in a targeted region of the image from its neighbourhood pixel values. It is possible that this is the reason why among the three bio-inspired filters, it is the PM filter which gives the best result when applied before clustering. However, instead of reading too much into it, this should still be considered as a preliminary result only that further encourages the search for a better combination of P and M. This may be through adaptive weighing as a future work, towards further improvement in the accuracy of grain identification.

4 Conclusion

The model of mid-level vision in human visual system (HVS) can explain both brightness contrast and brightness assimilation in different contexts. Checkerboard illusion and White effect illusion, two intriguing effects in brightness, can both be explained by the proposed model. We have applied the model to sedimentary rock thin-section images also presenting captivating brightness-contrast complexity, yet to be solved conclusively, assuming that the trained eye of the geologist is performing such clast–matrix separation tasks training the parallel channels in his/her central visual pathway. Our experiment shows promising result. However, it will be wrong to claim that we have applied the proposed algorithm on the rock thin sections in totality. This is because the best results are obtained by simply applying the P filter over the M filtered image before clustering. This, in fact, seems logical to some extent since any layperson, untrained in optical mineralogy is hardly expected to perform visual segmentation in such specialized pictures or under the microscope in the same way that an expert does. So, it is highly probable that the P and M pathways get differently trained in case of the specialists. The same is true for other specialized jobs like medical image segmentation. It is likely therefore that in a complex brightness perception scenario, as in the rock photomicrographs, the weights of M and P may have to be combined adaptively. In future, we shall try to learn features from the rock thin-section photographs in consultation with the geologists and do the figure–ground grouping task in an adaptive way. These modifications are also likely to improve the classification accuracy.

References

1. Helson, H.: Studies of anomalous contrast and assimilation. J. Opt. Soc. Am. **53**, 179–184 (1963)
2. Jameson, D., Hurvich, L.M.: Essay concerning color constancy. Ann. Rev. Psych. **40**, 1–22 (1989)
3. Blakeslee, B., McCourt, M.E.: A unified theory of brightness contrast and assimilation incorporating oriented multiscale spatial filtering and contrast normalization. Vis. Res. **44**, 2483–2503 (2004)
4. Ghosh, K.: A possible role and basis of visual pathway selection in brightness induction. Seeing and Perceiving **25**(2), 179–212 (2012)
5. Ghosh, K.: A neural network based model of M and P LGN cells. In: IEEE Proceedings, Bioinformatics and Systems Biology (BSB), International Conference on, pp. 1–5 (2016)
6. Bullier, J.: Integrated model of visual processing. Brain Res. Rev. **36**, 96–107 (2001)
7. Wei, H., Wang, X., Lai, L.L.: Compacts image representation model based on both nCRF and reverse control mechanisms. IEEE Trans. Neural Netw. Learn. Syst. **23** 150–162 (2012)
8. Wei, H.: A bio-inspired integration method for object semantic representation. J. Artif. Intell. Soft Comput. **6**, 137–154 (2016)
9. Merigan, W.H., Maunsell, J.R.H.: How parallel are the primate visual path ways? Ann. Rev. Neurosci. **16**, 369–402 (1993)
10. Bowker, D.O.: Suprathreshold spatiotemporal response characteristics of the human visual system. J. Opt. Soc. Am. **73**, 436–440 (1983)
11. De Valois, R.L., De Valois, K.K.: Spatial Vision. Oxford University Press, New York (1988)
12. Maunsell, J.H., Nealey, T.A., DePriest, D.D.: Magnocellular and parvocellular contributions to responses in the middle temporal visual area (MT) of the macaque monkey. J. Neurosci. **10**, 3323–3334 (1990)
13. Hotchstein, S., Ahissar, M.: View from the top: hierarchies and reverse hierarchies in the visual system. Neuron **36**, 791–804 (2002)
14. Fueten, F.: A computer controlled rotating polarizer stage for the petrographic microscope. Comput. Geosci. **23**(2), 203–208 (1997)
15. Goodchild, J.S., Fueten, F.: Edge detection in petrographic images using the rotating polarizer stage. Comput. Geosci. **24**, 745–751 (1998)
16. Lumbreras, F., Serrat, J.: Segmentation of petrological images of marbles. Comput. Geosci. **22**(5), 547–558 (1996)
17. Thompson, S., Fueten, F., Bockus, D.: Mineral identification using artificial neural networks and the rotating polarizer stage. Comput. Geosci. **27**, 1081–1089 (2001)
18. Izadi, H., Sadri, J., Mehran, N. A.: A new intelligent method for minerals segmentation in thin sections based on a novel incremental color clustering. Comput. Geosci. **81**, 38–52 (2015)
19. Jungmann, M., Pape, H., Wißkirchen, P., Clauser, C., Berlage, T.: Segmentation of thin section images for grain size analysis using region competition and edge-weighted region merging. Comput. Geosci. **72**, 33–48 (2014)

Identification of Interfaces of Mixtures with Nonlinear Models

Srirupa Das and Somdatta Chakravortty

Abstract The study attempts to identify the most appropriate unmixing model to detect endmember mixtures in a hyperspectral image scene. Least squares error-based linear spectral unmixing and gradient descent maximum entropy unmixing models have been used to identify linear mixtures in the datasets. Fuzzy C-Means-based unmixing algorithm extracts the interface of two major endmembers which is similar to the feature of the nonlinear spectral unmixing algorithms. Two popular bilinear models, Nascimento's model and Fan's model, have been studied. This study has been done on two hyperspectral datasets, i.e., homogeneous and heterogeneous. It is observed that the endmember interfaces detected by Fuzzy C-Means are more prominent in the dataset that has well-defined boundaries between endmembers, whereas Nascimento's and Fan's models show more accuracy in the dataset containing dense mixtures. The image derived fractional abundance values estimated by each model has been validated from ground abundance values after field visits.

Keywords Bilinear model · Endmember mixture · Linear Spectral unmixing · Fuzzy C-Means · Maximum entropy · Gradient descent · Nascimento's model · Fan's model

1 Introduction

Hyperspectral images are useful for the identification of objects as it has a wide range of spectral bands and depicts the signatures of different elements present in a land cover in a very detailed fashion. The images are course due to its spatial resolutions and cover a large area in a single image pixel. It is naturally quite difficult for a single

S. Das (✉) · S. Chakravortty
Department of Information Technology, Maulana Abul Kalam Azad University of Technology, West Bengal, Kolkata, West Bengal, India
e-mail: srirupadas000@gmail.com

S. Chakravortty
e-mail: csomdatta@rediffmail.com

© Springer Nature Singapore Pte Ltd. 2020
J. K. Mandal and S. Mukhopadhyay (eds.), *Proceedings of the Global AI Congress 2019*, Advances in Intelligent Systems and Computing 1112, https://doi.org/10.1007/978-981-15-2188-1_19

object to cover such a large land cover area and thus shows the signature of a mixed pixel [1, 2].

There are several unmixing techniques of mixed pixels. The techniques have been used to compute the proportions of endmember, which are in general in fractional form. These unmixing algorithms can be classified into three groups. The first works on endmember extraction. This includes spectral angle mapper [3], projection pursuit [4, 5], convex hull geometry [6–9], etc. The second group acts on estimating abundance on given endmembers, which includes maximum likelihood estimation [10], linear mixture analysis [11–13] Fuzzy C-Means [14, 15], to artificial neural networks [16]. The third group combines both first and second groups, including iterative LSE [2, 12, 13, 17] or ICA [18, 19], and NMF [20, 21].

Due to simplicity and flexibility, linear approaches are taken into account [22, 23]. In LSMM [24], unconstrained (LS) and fully constrained least squares (FCLS) [13] are commonly used.

In spite of practical advantages, the applicability of the linear approach is not universal. It has been shown in [25–35] that the nonlinear approaches may work better where multiple scattering of incident light takes place, like unmixing of mixed spectra of densely vegetated area [32]. Various nonlinear approaches are there for unmixing, such as ANN [36], FCM [15, 37, 38], bilinear model (BM) [30], support vector machine (SVM) [38, 39], and Gaussian mixture model (GMM) [40].

One of the most popular methods of linear spectral unmixing [41] is least squares error (LSE) [13]-based spectral unmixing that uses endmember detection algorithms for the identification of pure endmember signatures prior to unmixing. The gradient descent maximum entropy (GDME) [42] approach is another unmixing approach exploits the gradient descent approach to iteratively optimize the maximum entropy to find the fractional abundance values of the constituent endmembers of the mixed pixels. Nonlinear unmixing models [43] add nonlinear factors to incorporate the multiple scattering effects that are not considered in linear unmixing models. Nascimento's model [33] and Fan's model [28] are two popular bilinear models that take into account the amplitude of nonlinear interactions in the image scene under consideration. In this study, we have attempted to apply considered unmixing models on image scenes to identify and analyze the appropriate model suitable for the identification of endmember mixtures present in the specific land cover types.

Remaining part of the paper has been presented as Sect. 2 that holds the description of the considered unmixing methods such as LSU, GDME, FCM, Nascimento's model, and Fan's model; Sect. 3 describes the detailed study of the models in the hyperspectral scenes; and the conclusion of the study is discussed in Sect. 4.

2 Methodology

2.1 Least Squares-Based Linear and Nonlinear Spectral Unmixing

Least squares-based unmixing [13] can be broadly classified into linear least squares and nonlinear least squares. Least squares linear unmixing approach considers the singular scattering of radiation on endmembers. Equation (1) shows the LMM where the spectral response of a mixed pixel y is expressed as the linear combination of its constituent pure substances along with its fractional abundance values a.

$$y = \sum_{k=1}^{p} a_k s_k + n \tag{1}$$

Here p is the number of endmembers in the scene, n is the error considered of the model.

Nonlinear models which consider interactions up to second order are addressed as bilinear mixing models [26, 27]. Nascimento's model [33] and Fan's model are two popular bilinear models, and we have used in our study to observe the amplitude of nonlinear interactions in our considered image scene. Nascimento's model extends the linear model as follows:

$$y = \sum_{k=1}^{p} a_k s_k + \sum_{i=1}^{p-1} \sum_{j=i+1}^{p} \beta_{i,j} m_i \odot m_j + n \tag{2}$$

Here $\beta_{i,j}$ is the amplitude of interactions between two pure endmembers in the scene and $m_i \odot m_j$ is the Hadamard product of those pure endmember spectra. Equation (2) satisfie s the following constraints:

$$a_k \geq 0; \; k = 1, \ldots, p,$$
$$\beta_{i,j} \geq 0 i, ; \; j = 1, \ldots, p, i \neq j \tag{3}$$

$$\sum_{k=1}^{p} a_k + \sum_{i=1}^{p-1} \sum_{j=i+1}^{p} \beta_{i,j} = 1 \tag{4}$$

Here interaction terms are considered as additional pure endmembers, and abundance values are measured by transforming the bilinear equation to LMM, where the total number of endmembers are assumed to be $p*$ as in Eq. (5).

$$p* = \frac{1}{2} p(p + 1) \tag{5}$$

These abundance values are assumed to be an amplitude of interactions.

In Fan's model [28], the amplitude of interactions is calculated considering the abundance value of the pure endmember components involved in the mixture. In this case, the nonlinear factor can be expressed as,

$$\sum_{i=1}^{p-1} \sum_{j=i+1}^{p} a_i a_j m_i \odot m_j \tag{6}$$

where a_i and a_j are the fractional abundance values of ith and jth pure endmembers present in the mixture.

2.2 Gradient Descent Maximum Entropy Approach for Spectral Unmixing

This method exploits the gradient descent approach for iteratively optimizes the maximum entropy to find the abundance fractions of the constituent endmembers of mixed pixels. Here to initialize the GDME method the constituent endmembers are extracted through NFINDR [8]. The maximum entropy principle is used to formulate some minimization problem.

$$\begin{aligned}
\text{minimize} \quad & f_0(s) = \sum_{j=1}^{c} s_j \ln s_j \\
\text{subject to} \quad & h_0(s) = 1^T s - 1 = 0, \\
& h_i(s) = \sum_{j=1}^{c} a_{ij} s_j - x_i = 0, \quad i = 1, \ldots, l
\end{aligned} \tag{7}$$

where f_0 is the positive orthant, $R_{++}^c = \{s \in S\}$. The two equality corresponds to sum-to-one constraint $h_0(s)$, and $h_i(s)$ is the sensor measurement model.

2.3 Fuzzy C-Means-Based Spectral Unmixing

In FCM unmixing algorithm [14, 15, 44], the fuzzy membership values are treated as the fractional abundance values for the target endmember classes. It is an iterative process which terminates when objective function in Eq. (8) converges.

$$\text{£}(U, A) = \sum_{j=1}^{N} \sum_{i=1}^{K} (U_{ij})^m d^m (X_j, A_i) | \sum_{i=1}^{K} (U_{ij}) = 1, \forall j = 1, \ldots, N \tag{8}$$

Here X is the hyperspectral image of N pixels, K is the number of endmember classes with cluster center $A_i | i = 1, 2, \ldots, K$, and U_{ij} is the fractional abundance value of the ith endmember with respect to the jth pixel in X. Here, d is the Euclidean distance of the ith endmember with respect to the jth pixel in X and m is the fuzzifier ($m \geq 2$). Using Eqs. (9) and (10), cluster center A_i and fractional abundance value U_{ij} are obtained, respectively.

$$A_i = \frac{\sum_{j=1}^{N} U_{ij}^m X_j}{\sum_{j=1}^{N} U_{ij}^m} \tag{9}$$

$$U_{ij} = \frac{1}{\sum_{k=1}^{K} \left(\frac{d_{ij}}{d_{kj}}\right)^{2/(m-1)}} \tag{10}$$

3 Results and Discussions

3.1 Analysis with Standard Dataset of Samson [45]

In this paper, we have studied the results of the LSU, GDME, FCM, Nascimento's and Fan's models to identify the appropriate algorithms for the detection of different kinds of nonlinear mixtures in land cover types. The algorithms have been applied on the standard dataset of Samson, having 156 bands and 3 major endmembers, i.e., soil, vegetation, and water [45]. Besides pure endmembers, the interface between different endmembers such as soil–vegetation, vegetation–water and soil–water represents mixtures of respective endmembers. Moreover, as water approaches land it becomes shallow and generates mixed spectra of water influenced by the land.

Table 1 Comparative analysis between least squares and GDME-based spectral unmixing

Coordinate	Method	Endmembers		
(13,3)	Ground truth	0	0	1
	LSU	0	0.02	0.98
	GDME	0.0767	0.1189	0.8042
(67,53)	Ground truth	0.7974	0.1966	0.0061
	LSU	0.7973	0.1965	0.0060
	GDME	0.7411	0.1804	0.0783
(88,70)	Ground truth	0.7760	0.2065	0.0174
	LSU	0.7760	0.2065	0.0174
	GDME	0.7556	0.1653	0.0789

In comparison with the above-mentioned methods shown in Table 1, it is seen that the LSU [13] shows the closest similarity with ground abundance values for dominant endmembers, which is also better than the GDME [42].

The Fuzzy C-Means-based unmixing algorithm when applied on the Samson dataset leads to the formation of different clusters of different land cover types. This algorithm therefore also extracts the interface of two major endmembers as a third endmember, as the spectra of the interface are different from the pure constituents and hence form a separate cluster on the execution of the algorithm, which are shown in Table 2 and Fig. 1. From Fig. 1, it is observed that the FCM can be more suitable to identify the interfaces.

From Figs. 1 and 2, it is evident that the interface between endmembers is not as prominent in Nascimento's model as compared to Fuzzy C-Means.

From Table 2, it is observed that FCM considers the interface between adjacent endmembers as a new endmember. As a result, the abundance fractions of these interface endmembers show large values in regions of interaction. On the other hand, Nascimento's model calculates interactions between two endmembers as a Hadamard product between them, thus making it a derived endmember within the same pixel area. It reduces the abundance values of the constituent pure components of the mixture, generates a value for the derived endmember, and makes the whole sum to one. So the estimated abundance values for derived mixed endmembers are comparatively lesser than its pure constituents as well as the mixed endmember identified by Fuzzy C-Means. On the contrary, abundance values estimated by Fan's model show more similarity with the linear models in the image scenes where dense mixtures are not present and a well-defined boundary separates two pure regions.

3.2 Analysis with Real Hyperion Datasets of Henry Island

The hyperspectral image of Henry Island, Sundarbans, West Bengal, is used to analyze the efficiency of the unmixing models in areas where a similar type of mangrove endmembers is located in close association with each other. The models are applied on the vegetation cover of the image that has been retained by masking out the other land cover types. This image contains 155 spectral channels with 10 nm and 30 m spectral and spatial resolution, respectively. Henry Island is an appropriate location to study the intense mixture of mangrove endmembers. Seven dominant mangrove species of this island, namely *Excoecaria agallocha, Ceriops decandra, Phoenix paludosa, Avicennia alba, Avicennia marina, Bruguiera cylindrica, Avicennia officinalis,* have been considered in this study.

A field evaluation of the area under consideration has been done to distinguish the different species of mangrove, whose labeling has been carried out [46]. In these dense mangrove forests, various species of mangrove endmembers reside in close contact with each other resulting in multiple reflection of incident radiation. Figure 3a, b shows fractional abundance images of pure patches of *Excoecaria agallocha and Ceriops decandra* generated from Nascimanto's model, and Fig. 3c shows fractional abundance images of interactions of *Excoecaria agallocha* and *Ceriops decandra*

Table 2 Comparative analysis of fractional abundances of Nascimento's model, Fan's model, and Fuzzy C-Means model for Samson dataset

Coordinate	Algorithm	Abundance values of respective endmembers					
		Soil	Vegetation	Water	Soil–vegetation	Soil–water	Vegetation–water
(13, 3)	Ground truth	0	0	1	–	–	–
	FCM	0	0.11	0.55	0	–	0.3295
	Nascimento's model	0	0.0463	0.9534	0	0.0003	0
	Fan's model	0	0	0.9976	0	0.0023	0
(67, 53)	Ground truth	0.7974	0.1966	0.0061	–	–	–
	FCM	0.1229	0.0076	0.0105	0.8538	–	0.0051
	Nascimento's model	0.6674	0.1735	0.1574	0	0	0.0020
	Fan's model	0.99	0.002	0	0.001	0	0
(88, 70)	Ground truth	0.7760	0.2065	0.0174	–	–	–
	FCM	0.1269	0.0085	0.0003	0.8538		0
	Nascimento	0.5941	0.0939	0.3101	0	0	0.0026
	Fan's model	0.99	0	0	–	–	–

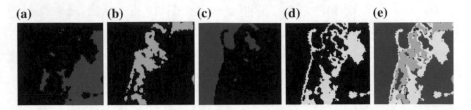

Fig. 1 Results of FCM. **a** Fractional abundance image of soil. **b** Fractional abundance image of vegetation. **c** Fractional abundance image of water. **d** Fractional abundance image of interface between water and vegetation. **e** Integrated fractional abundance image

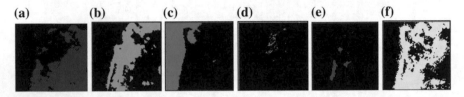

Fig. 2 Results of Nascimento's model **a** Fractional abundance image of soil. **b** Fractional abundance image of vegetation. **c** Fractional abundance image of water. **d** Fractional abundance image of soil–vegetation interactions. **e** Fractional abundance image of soil–water interactions. **f** Fractional abundance image of vegetation–water interactions

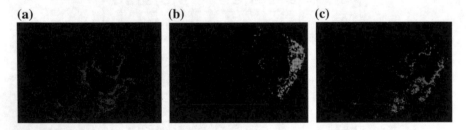

Fig. 3 a, b Fractional abundance images of pure patches of *Excoecaria agallocha and Ceriops decandra* generated from Nascimanto's model. **c** Fractional abundance images of interactions of *Excoecaria agallocha and Ceriopsdecandra* generated from Nascimanto's model

generated from Nascimanto's model. Application of FCM on the dataset shows that though it identifies interface between two endmembers as a third endmember, the dense mangrove mixtures do not reveal sharp demarcation among themselves. Therefore, a proper cluster cannot be formed by any particular endmember. It has been observed that FCM performs better clustering in the case of homogeneous mixtures than non-homogeneous mixtures.

Figure 4 displays the abundance images of mangrove endmembers in Henry Island after the application of FCM [47]. It is evident that the results do not have much accuracy with ground values when compared with Nascimento's model [33] or Fan's [28] model because dense mangrove forest is a proper example of non-homogeneous

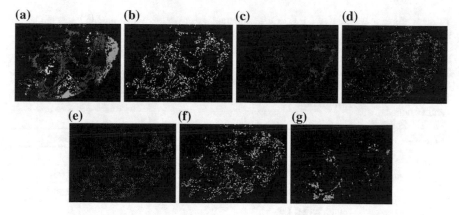

Fig. 4 Fractional abundance images of mangrove species generated from Fuzzy C-Means

mixture of different mangrove species. The endmember interfaces detected by this model are more prominent in the Samson dataset that has well-defined boundaries between endmembers as compared to the Henry Island dataset, which contains dense mangrove mixture.

Figure 7a shows the classification of the dominant endmembers present in the image derived by LSU, and Fig. 4b–g shows the endmember clusters generated by FCM. From Fig. 4, we can say that the derived clusters generated by FCM do not fully follow the distribution of dominant endmember classes derived through LSU. On the other hand in cases of dense mixtures, bilinear models are observed to give comparatively better results in finding the abundance values of the mixtures. A comparative analysis of abundance values estimated using Nascimento's and Fan's bilinear models is shown in Table 3, and fractional abundance values generated by FCM is depicted in Table 4.

4 Conclusion

In this paper, linear and nonlinear unmixing models have been applied on hyperspectral images to analyze and identify the appropriate model for the identification of endmember mixtures present in specific land cover types. Least squares-based linear spectral unmixing algorithm and gradient descent maximum entropy-based unmixing algorithm have been applied separately on the Samson dataset to unmix the mixed pixels and identify the proportions of constituent endmembers of that pixel. It is observed that LSU shows the closest similarity with ground abundance values for dominant endmembers as compared to GDME that also estimates major endmember abundances present in a mixture but with lower accuracy. An implementation of Fuzzy C-Means has also been analyzed on the same hyperspectral images. It has been observed that the endmember interfaces detected by Fuzzy C-Means are

Table 3 Comparative analysis of fractional abundances of Nascimento's model, Fan's model for Henry Island, Sundarbans dataset

Location	Endmembers	Nascimento's model	Fan's model	Ground truth
Latitude 21° 34′ 23.74″N and Longitude 88° 17′ 48.35″E (67, 86)	Species1	0.3710	0.1408	0.24
	Species2	0	0	0
	Species3	0	0	0
	Species4	0.0587	0.0259	0.04
	Species5	0	0	0
	Species6	0.0657	0.0313	0.05
	Species7	0.1299	0.0549	0.08
	Interaction 1 and 2	0	0	0
	Interaction 1 and 3	0	0	0
	Interaction 1 and 4	0.0713	0.0036	0.05
	Interaction 1 and 5	0	0	0
	Interaction 1 and 6	0.0799	0.0044	0.05
	Interaction 1 and 7	0.1579	0.0077	0.09
	Interaction 2 and 3	0	0	0
	Interaction 2 and 4	0	0	0
	Interaction 2 and 5	0	0	0
	Interaction 2 and 6	0	0	0
	Interaction 2 and 7	0	0	0
	Interaction 3 and 4	0	0	0
	Interaction 3 and 5	0	0	0
	Interaction 3 and 6	0	0	0
	Interaction 3 and 7	0	0	0
	Interaction 4 and 5	0	0	0
	Interaction 4 and 6	0.0126	0.0008	0.01
	Interaction 4 and 7	0.0250	0.0014	0.02
	Interaction 5 and 6	0	0	0
	Interaction 5 and 7	0	0	0
	Interaction 6 and 7	0.0280	0.0017	0.02
	RMSE	0.0306	0.0286	

more prominent in datasets that have well-defined boundaries between endmembers. In cases of dense mixtures, as in the Henry Island dataset, Nascimento's and Fan's models are observed to give comparatively better results where various endmembers reside in close contact with each other, resulting in multiple reflections of incident radiation between endmembers.

Table 4 Fractional abundances of Fuzzy C-Means model for Henry Island, Sundarbans dataset

FCM identified Endmembers	FCM generated abundance values
Endmember1	0.0071
Endmember2	0.0835
Endmember3	0.0341
Endmember4	0.3666
Endmember5	0.0158
Endmember6	0.4467
Endmember7	0.0460

Acknowledgements We acknowledge the Department of Science and Technology, Government of India, for providing necessary funds to the major research project "Development of Algorithms for Spectral Unmixing and Sub-Pixel Classification of Hyperspectral Image Data."

References

1. Keshava, N., Mustard, J.F.: Spectral unmixing. IEEE Signal Process. Mag. **19**(1), 44–57 (2002)
2. Meer, F.V.D.: iterative spectral unmixing (ISU). Int. J. Remote Sens. **20**(17), 3431–3436 (1999)
3. Yuhas, R.H., Goetz, A.F.H., Boardman, J.W.: Discrimination among semi-arid landscape end-members using the spectral angle mapper (sam) algorithm. In: Green, R. O. (eds.) Proc. Summaries 3rd Annu. JPL Airborne Geoscience Workshop, vol. 1, pp. 147–149 (1992)
4. Cross, A.M., Settle, J.J., Drake, N.A., Paivinen, R.T.M.: Subpixel measurement of tropical forest cover using AVHRR data. Int. J. Remote Sens. **12**(5), 1119–1129 (1991)
5. Healcy, G., Slater, D.: Models and methods for automated material identification in hyperspectral imagery acquired under unknown illumination and atmospheric conditions. IEEE Trans. Geosci. Remote Sens. **37**(6), 2706–2717 (1999)
6. Craig, M.D.: Minimum-volume transforms for remotely sensed data. IEEE Trans. Geosci. Remote Sens. **32**(3), 542–552 (1994)
7. Boardman, J.W., Kruse, F.A., Green, R.O.: Mapping target signatures via partial unmixing of AVIRIS data. In: Green, R. O. (eds.) Proc. Summaries 5th Annu. JPL Airborne Earth Science Workshop, vol. 1, pp. 23–26 (1995)
8. Winter, M.E.: N-findr: an algorithm for fast autonomous spectral end-member determination in hyperspectral data. In: Proc. SPIE Conf. Imaging Spectrometry V, pp. 266–275 (1999)
9. Nascimento, J.M.P., Dias, J.M.B.: Vertex component analysis: A fast algorithm to unmix hyperspectral data. IEEE Trans. Geosci. Remote Sens. **43**(4), 898–910 (2005)
10. Settle, J.J.: On the relationship between spectral unmixing and subspace projection. IEEE Trans. Geosci. Remote Sens. **34**(4), 1045–1046 (1996)
11. Settle, J.J., Drake, N.A.: Linear mixing and the estimation of ground cover proportions. Int. J. Remote Sens. **14**(6), 1159–1177 (1993)
12. Chang, C.I., Heinz, D.C.: Constrained subpixel target detection for remotely sensed imagery. IEEE Trans. Geosci. Remote Sens. **38**(3), 1144–1159 (2000)
13. Heinz, D.C., Chang, C.-I.: Fully constrained least squares linear spectral mixture analysis method for material quantification in hyperspectral imagery. IEEE Trans. Geosci. Remote Sens. **39**(3), 529–545 (2001)
14. Foody, G.M.: Approaches for the production and evaluation of fuzzy land cover classifications from remotely-sensed data. Int. J. Remote Sens. **17**, 1317–1340 (1996)

15. Bastin, L.: Comparison of Fuzzy C-Means classification, linear mixture modelling and MLC probabilities as tools for unmixing coarse pixels. Int. J. Remote Sens. **18**(17), 3629–3648 (1997)
16. Foody, G.M.: Thematic mapping from remotely sensed data with neural networks: MLP, RBF, and PNN based approaches. J. Geograph. Syst. **3**, 217–232 (2001)
17. Neville, R.A., Staenz, K., Szeredi, T., Lefebvre, J., Hauff, P.: Automatic endmember extraction from hyperspectral data for mineral exploration. In: Proceedings of 21st Canadian Symposium on Remote Sensing, Ottawa, ON, Canada, vol. 2, pp. 891–896 (1999)
18. Bayliss, J., Gualtier, J.A., Cromp, R.: Analyzing hyperspectral data with independent component analysis. Proc. SPIE **3240**, 133–143 (1997)
19. Nascimento, J.M.P., Dias, J.M.B.: Does independent component analysis play a role in unmixing hyperspectral data. IEEE Trans. Geosci. Remote Sens. **43**(1), 175–184 (2004)
20. Liou, C.Y., Yang, K.O.: Unsupervised classification of remote sensing imagery with non-negative matrix factorization. In: International Conference on ONIP (2005)
21. Paura, V.P., Piper, J., Plemmons, R.J.: Non negative matrix factorization for spectral data analysis. Linear Algebr. Appl. (2005)
22. Chang, C.I.: Hyperspectral imaging: techniques for spectral detection and classification. Springer Science & Business Media, Berlin, Heidelberg (2003)
23. Chen, J., Richard, C., Honeine, P.: Nonlinear unmixing of hyperspectral data based on a linear-mixture/nonlinear-fluctuation model. IEEE Trans. Signal Process. **61**(2), 480–492 (2013)
24. Zanotta, D.C., Haertel, V., Shimabukuro, Y.E., Renno, C.D.: Linear spectral mixing model for identifying potential missing endmembers in spectral mixture analysis. IEEE Trans. Geosci. Remote Sens. **52**(5), 3005–3012 (2014)
25. Dobigeon, N., Tits, L., Somers, B., Altmann, Y., Coppin, P.: A comparison of nonlinear mixing models for vegetated areas using simulated and real hyperspectral data. IEEE J. Sel. Top. Appl. Earth Obs. Remote. Sens. **7**(6), 1869–1878 (2014)
26. Altmann, Y., Dobigeon, N., Tourneret, J.Y.: Bilinear models for nonlinear unmixing of hyperspectral images. In: Paper presented at the Hyperspectral Image and Signal Processing: Evolution in Remote Sensing (WHISPERS), Lisbon (2011)
27. Altmann, Y., Halimi, A., Dobigeon, N., Tourneret, J.-Y.: Supervised nonlinear spectral unmixing using a polynomial post nonlinear model for hyperspectral imagery. In: Paper presented at the IEEE International Conference on Acoustics, Speech and Signal Processing, Prague (2011)
28. Fan, W., Baoxin, H., Miller, J., Mingze, L.: Comparative study between a new nonlinear model and common linear model for analysing laboratory simulated—forest hyperspectral data. Int. J. Remote Sens. **30**(11), 2951–2962 (2009)
29. Guilfoyle, K.J., Althouse, M.L., Chang, C.I.: A quantitative and comparative analysis of linear and nonlinear spectral mixture models using radial basis function neural networks. IEEE Trans. Geosci. Remote Sens. **39**(10), 2314–2318 (2001)
30. Halimi, A., Altmann, Y., Dobigeon, N., Tourneret, J.-Y.: Nonlinear unmixing of hyperspectral images using a generalized bilinear model. IEEE Trans. Geosci. Remote Sens. **49**(11), 4153–4162 (2011)
31. Heylen, R., Parente, M., Gader, P.: A review of nonlinear hyperspectral unmixing methods. IEEE J. Sel. Top. Appl. Earth Obs. Remote. Sens. **7**(6), 1844–1868 (2014)
32. Ray, T.W., Murray, B.C.: Nonlinear spectral mixing in desert vegetation. Remote Sens. Environ. **55**(1), 59–64 (1996)
33. Nascimento, J.M.P., Bioucas-Dias, J.M.: Nonlinear mixture model for hyperspectral unmixing. In: Bruzzone, L., Notarnicola, C., Posa, F. (eds.) Proceedings of the SPIE Image and Signal Processing for Remote Sensing XV, vol. 7477, Berlin, Germany (2012)
34. Altmann, Y., Dobigeon, N., Tourneret, J.Y. Bermudez, J.C.M.: A robust test for nonlinear mixture detection in hyperspectral images. In: Proceedings of the IEEE International Conference Acoustic, Speech and Signal Processing, pp. 2149–2153. Vancouver, Canada, (2013)
35. Somers, B., Cools, K., Delalieux, S., Stuckens, J., Zande, D., Verstraeten, W., Coppin, P.: Nonlinear hyperspectral mixture analysis for tree cover estimates in orchards. Remote Sens. Environ. **113**, 1183–1193 (2009)

36. Licciardi, G.A., Frate, F.D.: Pixel unmixing in hyperspectral data by means of neural networks. IEEE Trans. Geosci. Remote Sens. **49**(11), 4163–4172 (2011)
37. Dunn, J.C.: A fuzzy relative of the ISODATA process and its use in detecting compact well-separated clusters. J. Cybern. **3**(3), 32–57 (1973)
38. Ping-Xiang, L., Wu, B., Zhang, L.: Abundance estimation from hyperspectral image based on probabilistic outputs of multi-class support vector machines. In: Paper presented at the IEEE International Geoscience and Remote Sensing Symposium, Seoul (2005)
39. Tang, Y., Krasser, S., Yuanchen, H., Yang, W., Alperovitch, D.: Support vector machines and random forests modeling for spam senders behavior analysis. In: Paper presented at the IEEE Global Telecommunications Conference, New Orleans (2008)
40. Cheng, B., Zhao, C., Wang, Y.: Algorithm to Unmixing Hyperspectral Images Based on APSO-GMM. Paper presented at the 2010 First International Conference on Pervasive Computing, Signal Processing and Applications Signal Processing and Applications, Harbin (2010)
41. Hu, Y.H., Lee, H.B., Scarpace, F.L., Fort, J., Fish, S., Wilson, S., Werle, D.: Optimal linear spectral unmixing. IEEE Trans. Geosci. Remote Sens. **37**(1), 639–644 (1999)
42. Miao, L., Qi, H., Szu, H.: A Maximum entropy approach to unsupervised mixed-pixel decomposition. IEEE Trans. Image Process. **16**(4), 1008–1021 (2007)
43. Liu, W., Wu, E.Y.: Comparison of non-linear mixture models: sub-pixel classification. Remote Sens. Environ. **94**(2), 145–154 (2005)
44. Fisher, P.F., Pathirana, S.: The evaluation of fuzzy membership of land cover classes in the suburban zone. Remote Sens. Environ. **34**(2), 121–132 (1990)
45. http://www.escience.cn/people/feiyunZHU/Dataset_GT.html
46. Chakravortty, S., Sinha, D.: Development of higher-order model for nonlinear interactions in hyperspectral data of mangrove forests. Current Sci. **111**(6), 1055 (2016)
47. Bezdek, J.C., Ehrlich, R., Full, W.: FCM: the Fuzzy C-Means clustering algorithm. J. Comput. Geosci. **10**, 191–203 (1984)

Random Forest Boosted CNN: An Empirical Technique for Plant Classification

Somnath Banerjee and Rajendra Pamula

Abstract Plant identification and classification is one of the toughest jobs for in-experienced botanists. This paper proposes a hybrid model by combining random forest and along with the convolutional neural network (CNN) to classify the images. Our proposed method comprises of two phases; feature extraction using CNN and training the random forest model. The layers in convolutional neural network are useful to extract essential features. In this work, we have used PlantCLEF 2019 dataset (Amazonian Rainforest) to train and evaluate our model. We compare our method with the state-of-the-art methods for plant classification. The experimented method produces relatively higher accuracy than earlier methods.

Keywords Plant classification · CNN · Random forest · PlantCLEF 2019

1 Introduction

The current boom in the application of deep learning on computer vision has brought artificial intelligence in picture again. Problems of computer vision can be addressed using some novel algorithms of deep learning. Along with that, several popular machine learning algorithms [23, 27] boost the process of addressing problem of computer vision. Plant classification and identification [24] is one of the major problems in the field of computer vision due to its demand in agricultural [25] and environmental research [3]. In recent years, automated plant identification has helped Botanist to resolve the plant identification task [4].

S. Banerjee (✉) · R. Pamula
Indian Institute of Technology (ISM), Dhanbad, India
e-mail: mailto_somnath.16kt000102@cse.ism.ac.in

R. Pamula
e-mail: rajendra@iitism.ac.in

© Springer Nature Singapore Pte Ltd. 2020
J. K. Mandal and S. Mukhopadhyay (eds.), *Proceedings of the Global
AI Congress 2019*, Advances in Intelligent Systems and Computing 1112,
https://doi.org/10.1007/978-981-15-2188-1_20

1.1 Plant Classification

Plant classification is one of the difficult jobs for botanist. They usually identify plants by observing its flower, leaf, size, etc. For an inexperienced botanist, it is difficult to identify plants by observing such features. Automated plant classification is one of the major tasks in the domain of deep learning. In machine learning, plant classification and identification is considered as supervised problem. Multiple solution mechanisms for plant classification problems are already exist and were mentioned by Cope et al. [3]. Most of such algorithms are not applicable directly but need to train the classifier to distinguish the classes. For species classification, the training phase accumulates the design and analysis of images that have been uniquely considered as taxa and from now onward will be used as a maximum discrimination between trained species. In the next phase, the trained classifier will help the model to classify the unidentified species and classify the unknown plants.

1.2 Our Contribution

In this article, we prepare a hybrid model which can classify the plant images. We have compared and tested our approaches with popular existing approaches. In our experiment, we use PlantCLEF 2019 dataset which consists of 10k species of plants from Amazonian forest. According to PlantCLEF team, majority variation of plants in the world are there in Amazonian forest[1] ($\sim$369 K species). The main objective is to identify the Amazonian species from Amazonian rainforest. Our contribution basically here is of threefold.

1. Extraction of features using CNN.
2. Classification of images by feeding feature vectors to random forest.
3. Compare the results with existing models.

Toward the first objective, we convert the image into one-dimensional array. Then we identify top 500^2 important features of each image. We pass these features through two different algorithms. Our objective is to passing the plant images from Amazonian forest with species id for training our model and from the test data we have to classify the images with their proper species. The training images contain total 10,000 images of multi organ tree images from ImageCLEF2019. In literature, it is found that convolutional neural network is the most powerful algorithm for classification problems and along with that we are adding some machine learning boosting algorithms like RandomForest in order to get more optimized output. Our another target will be to compare some common boosting algorithms like Adaboost [16] with their performances in our article.

---

[1]https://www.imageclef.org/PlantCLEF2019.
[2]Our results hold for other feature size also.

1.3 Outline

The orientation of this paper is as follows. In Sect. 2 we have presented literature review. Section 3 describes the *PlantCLEF 2019* dataset used in this study. Section 4 presents the existing classification methods and our approach in detail. In Sect. 5, we analyze the results produced by different algorithms. Finally, we conclude in Sect. 6 and identify future work.

2 Related Work

Several earlier work has been done on identification of multi-organ of plants and its classification. Existing research mainly focuses on plant identification and classification [25]. Over the last few decades, researchers carried out their work on recognizing species of plant using solely single plant organs such as leaf, flower, etc. Most of the earlier work used characteristics of leaves to identify species. Characteristics of leaf such as shape [3, 24], texture [4], and vein structure [2, 26] are the important feature to identify the leaves of different plant. There are several research carried out on using flower [17, 18] to identify plant species. To overcome real problem where botanist identifies a plant by observing several organs of the plant from different point of view when other organs are not in season, computer vision researcher focus on automate the plant identification process to identify plant from different plant organ images. Existing research conduct feature extraction from images and classifying these images as two different stages and later they engineered the features. There are several approaches were adopted in order to automate plant identification task—(i) state-of-the-art ML-based approach, (ii) model-free approach and (iii) deep learning approach.

2.1 State-of-the-Art ML-Based Approach

Images comprise of billions of pixel values associated with color information. Such huge information is inefficient to be directly use by some machine learning algorithms. Direct use of such rich information toward any machine learning algorithm unable to capture important features of image and there is information loss. During last few years, researchers mainly focus on the detection, extraction and identification of important features and computation of feature vectors. Designing such mechanism is a problem specific task which means a particular model is solely implemented for a specific task such as plant leaf, flower identification. For example, [27] employ probabilistic neural network (PNN) normally uses image conversion to binary which is a unique mechanism to distinguish the plant leaves, contour detection, and extraction of 12 types of leaf-like geometric shapes and features. This approach has already done at its best by giving result of 90% by just evaluated on only 32 species. Only challenge here is this process will be able to deal on leaf images. Jin et al. [8] first introduced the method of leaf tooth after binaraization, segmentation, and contour corner detection. This experiment has given 76% accuracy but the main challenge here was it was only applicable on leaf tooth specific species [1]. The first stage of making the platform by means of extracting important features normally takes 90% of development time [1].

2.2   Model-Free Apporach

The main focus of model-free approach is to overcome the above limitation of ML-based approaches. Model-free approach does not provide solution for application-specific task but provide a generalized solution across different classes such as species and organs of plants. Model-free approach aims to identify characteristics of interesting features using some generalized algorithms, i.e., speeded-up robust features (SURF), scale-invariant feature transform (SIFT), and histogram of gradients (HOG). Normally this type of generic function captures important features and gradient orientation for plant identification higher studies [28]. Comparatively evaluate [21] model-free image classification pipeline for image identification. As per result, it is saying SURF detector with SIFT descriptor to be the best over detector descriptor combination. In case of encoding interest points Fisher kernel encoding is the best for classification.

2.3   Deep Learning Approach

Sun et al. [23] proposed a deep learning approach for identifying important feature of a leaf and found that different orders of leaf venation are more distinguishable than leaf shape features. To conduct large-scale plant identification, a novel method was proposed by building a hierarchical classifier [5]. The performance reported in hierarchical classifier is reliable but this performance is much affected by the selected best subset of hand crafted features which are task/dataset dependent. Recently, researcher adopt several deep learning approaches to automate plant species identification. In 2014, the deep learning network model with eight layers was first been introduced and was namely Alexnet and have won ImageNet Challenge (ILSVRC) [10]. The beauty of deep learning procedure is, normally it will extract the critical features of species and classify them properly by using its layered based structure. Over years, winning architecture is modified and more sofisticated mechanisms are appended around the basic winning architecture. Now on the history point of view the first classification by using CNN was [13] plant leaf identifier which was using Alexnet and was getting 99.5% accuracy within 44 species. Using six-layer architecture, we have got around 94.69% accuracy in lavia dataset [29]. Barré et al. [1] have done the same experiment with 17-layer structure and got little bit good result comparing to earlier one and was 97.9%. Eventually, [23] But eventually 26-layer structure which we call as ResNet have won the prize and got almost 99.65% accuracy in Flavia dataset. Again here have used CNN as extracting features and then SVM as classifier and have got around 95.34% on the Oxford Flowers 102 dataset [22].

3 Dataset

In our experiment, we use PLANTClcf-2019 dataset to evaluate our model. Provided by CrowdAI team, the PLANTClef-2019 dataset is an image dataset of Amazonian forest where 10K species of colorful images of different random shapes are classified into 3500 categories. So total 10,000 images they have provided for training the model. According to CROWDAI team, "The average number of images per species in that new dataset will be much lower than the dataset used in the previous editions of PlantCLEFs (about 10 vs. 100). Many species will contain only a few images and some of them might even contain only 1 image". The details of the dataset are given in Table 1.

4 Method

In this section, we describe our hybrid model RandomForest Boosted CNN model which is a combined method of convolutional neural network and random forest model. Along with that, we compare existing models with our model for performance analysis. Traditional neural network is used as main learner but the major drawback is high cost for training along which reduces the performance as well. In our model, we try mitigate these issues. We describe the existing classification models and proposed model in detail.

Table 1 Dataset description

Dataset	# Classes	# Images
PlantCLEF training data	10K	>10K
PlantCLEF test data	–	>5K

4.1 CNN Based Approach

Lee et al. [14] used a CNN-based model to solve this classification problem. They use 16-layered VGG-net which consists of four major components shared, organ, species and fusion layers. Firstly, they train the organ layer with organ classes. Then, shared and organ layer are used to train the species layer. At last, Both of them are combine to train the fusion layer. In order to get rich features, they use downsizing convolutional layer on the feature map before passing it through fusion layer. We regenerate their results on PlantCLEF 2019 dataset. Results are reported in Table 5.

4.2 SVM-Based Approach

Priya et al. [20] use state-of-the-art machine learning algorithm to classify the leaf images. They employ support vector machine [19] algorithm on the extracted digital morphological features (DMF). In our article, we use their approach on our Plant-CLEF data and test it. Results are reported in Table 5.

4.3 Random Forest BOOSTed CNN

Our approach comprises of two phases, i.e., CNN [11] and random forest [15]. The major objective behind the training of CNN is to optimize the process and improve performance. We extracted feature set by removing output layer of CNN. Later, we use these features for Softmax classifier [6] and pass the output to random forest model to get the final classification.

The Design of Random Forest BOOSTed CNN: Figure 1 shows the Random Forest Boosted CNN structure. The orange-colored cube shows training module of the CNN model, where Random Forest Boosted CNN is trained as a normal CNN. Sky blue module will be the training module of random forest model. Actually we are considering convolutional neural network for extracting the features from the images and post that will use random forest classifier as base classifier to classify the species. The training procedure of full model is described as flow charted version in Fig. 2.

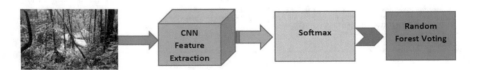

Fig. 1 Proposed random forest boosted CNN

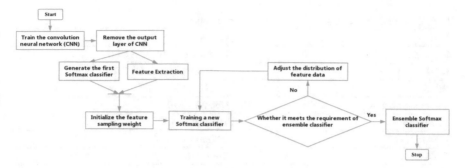

Fig. 2 Random forest boosted CNN training procedure

The Learning Phase of CNN: The training procedure of convolutional neural network model includes Adam optimation technique [9] for reduction of the gradient in learning. The complex problems of a partial optimum can be sorted out easily if we can store the old gradients somewhere and tweaking the parameters based on the historical records of CNN.

The Training of Random Forest Model: Training of random forest is easy process like, the trained convolution network is used to extract the features. In the process, first Softmax classifier will be considered as the initial base learner. The specific training process consists of few steps:

– Firstly, we remove the output layer from the convolutional neural network as in the output layer it will classify the features to the respective species probabilities. Now, we extracted the features from the data.
– We can consider the Softmax classifier as the output layer of the convolutional neural network, We have to capture the accuracy here and sampling weights of the extracted data.
– The weights which we obtained after sampling will be used for next base learner and storing the probability of correctness.
– Initiate the iterative training of several Softmax classifiers; in the end, we consider the voting output of Softmax classifier and will propagate that to random forest classifier for final classification.

5 Experimental Analysis

5.1 Lab for Experiment

Table 2 showing the technical configuration of the machine we have used for experiment. We have used Scikit-Learn and Tensor flow deep learning library for training.

Table 2 Hardware and software configuration of computer

Item	Content
Processor	Intel Core i7-7th Gen CPU
GPU	NVIDIA GeForce GTX 1050Ti
RAM	16 GB
Operating system	Ubuntu 18.10
Tensorflow	Tensorflow 1.0
Python	Python 3.6
Cuda	Cuda 8.0
Scikit-Learn	Scikit-learn 1.0

Table 3 Convolution neural network specification

Input: 10,000 * 3072		
Hidden1 layer	conv	Size 5 * 5; quantity: 64; method: same
	ReLU	Max(0,x)
	Max pooling	Size: 3 * 3; stride: 2
	Batch norm	alpha = 0.01/9, beta = 0.5
Hidden2 layer	conv	Size 5 * 5; quantity: 64; method: same
	ReLU	Max(0,x)
	Max Pooling	Size: 3 * 3; stride: 2
	Batch norm	alpha = 0.01/9, beta = 0.5
Hidden3 layer	Full connect	Weight size: [1200,380]
	ReLU	Max(0,x)
	Dropout	Probability of activation: 0.5
Hidden4 layer	Full connect	Size of weight: [380,190]
Output layer	Softmax	Size of weight: [190,10]

Tables 3 and 4 describes the configuration overview of the CNN model and the specifications of random forest, respectively.

5.2 Result Analysis

As mentioned in Table 1, the PlantCLEF-2019 dataset we have used to test the performance of CNN [14] approach, CNN + AdaBoost [12] approach, SVM approach [20] and CNN + Random Forest (our method). According to Table 5, convolutional neural network model provides 56% accuracy on our dataset whereas the accuracy increases to 59.6% when we use CNN + Adaboost. Another method which uses

Table 4 Configuration of random forest

Input	Use the feature extraction data of the convolution network: [190, 10]
Softmax1	Size of weight: [190,10]
Softmax2	Size of weight: [190,10]
Softmax3	Size of weight: [190,10]
Softmax4	Size of weight: [190,10]
Softmax5	Size of weight: [190,10]
Softmax6	Size of weight: [190,10]
Softmax7	Size of weight: [190,10]
Output	Results of weight voting of the categories

Table 5 Experimental configuration of PlantCLEF-2019 dataset

Classifier	Accuracy of testing (%)
CNN [14]	56
CNN + Adaboost [12]	59.6
SVM [20]	58
CNN + Random Forest (our method)	67

SVM mode shows relatively better results than CNN approach (58% accuracy) but not better than the CNN + Adaboost. In our model we get relatively better accuracy (67%) than other existing methods.

6 Conclusion and Future Scope

In this work, our main motivation is to improve the quality and performance of popular convolutional neural network (in addition with random forest classifier) in case of small sample data in each class of the dataset. Various existing methods are present in the literature for improving the performance. We observe ensemble learning provides relatively better performance than other methods. Also we observe that ensemble methods increase the performance by using the multiple weak learners and by voting the results of them. Our assumption was that the combination of the convolutional neural network and random forest may provide better accurancy. As per our experiments and results, we have shown that our model can improve the performance by 8%. In our analysis, we notice that the performance of the classification solely depends on the first stage which is the image feature extraction as well as performance can be improved by increasing the layers and using high end configuration.

Though convolutional neural network is a most successful model for image classification but it fully depends on the datasets. Due to the increasing quality of datasets and computer hardware, the performance of convolutional neural network is increasing day by day. Similarly, because of high-resolution images the training of model is getting difficult too and training time is also getting high. Even very high performance computers are taking ages to get trained. So the reducing time and cost will be always a open problem. Integrating two approaches generally can improve the performance but if we want to really get a optimum result then the best way will be reinforcement learning approach. As because, we have small dataset to train, as per our perception and literature survey reinforcement learning can give more optimum result. Our future plan is to apply generative adversarial network (GAN) [7] to classify plant images. By using GAN, we can train the model with small number of datasets but as because we will use the semi-supervised learning then it will be able to classify species more accurately.

References

1. Barré, P., Stöver, B.C., Müller, K.F., Steinhage, V.: Leafnet: a computer vision system for automatic plant species identification. Ecol. Inf. **40**, 50–56 (2017)
2. Bruno, O.M., de Oliveira Plotze, R., Falvo, M., de Castro, M.: Fractal dimension applied to plant identification. Inf. Sci. **178**(12), 2722–2733 (2008)
3. Cope, J.S., Corney, D., Clark, J.Y., Remagnino, P., Wilkin, P.: Plant species identification using digital morphometrics: a review. Expert Syst. Appl. **39**(8), 7562–7573 (2012)
4. Cope, J.S., Remagnino, P., Barman, S., Wilkin, P.: Plant texture classification using gabor co-occurrences. In: International Symposium on Visual Computing, pp. 669–677. Springer, Berlin (2010)
5. Deng, J., Dong, W., Socher, R., Li, L.J., Li, K., Fei-Fei, L.: Imagenet: a large-scale hierarchical image database. In: 2009 IEEE Conference on Computer Vision and Pattern Recognition, pp. 248–255. IEEE (2009)
6. Duan, K., Keerthi, S.S., Chu, W., Shevade, S.K., Poo, A.N.: Multi-category classification by soft-max combination of binary classifiers. In: International Workshop on Multiple Classifier Systems, pp. 125–134. Springer, Berlin (2003)
7. Goodfellow, I., Pouget-Abadie, J., Mirza, M., Xu, B., Warde-Farley, D., Ozair, S., Courville, A., Bengio, Y.: Generative adversarial nets. In: Advances in Neural Information Processing Systems, pp. 2672–2680 (2014)
8. Jin, T., Hou, X., Li, P., Zhou, F.: A novel method of automatic plant species identification using sparse representation of leaf tooth features. PloS ONE **10**(10), e0139482 (2015)
9. Kingma, D.P., Ba, J.: Adam: a method for stochastic optimization. arXiv preprint arXiv:1412.6980 (2014)
10. Krizhevsky, A., Sutskever, I., Hinton, G.: Imagenet classification with deep convolutional neural. In: Neural Information Processing Systems, pp. 1–9 (2014)
11. Krizhevsky, A., Sutskever, I., Hinton, G.E.: Imagenet classification with deep convolutional neural networks. In: Advances in Neural Information Processing Systems, pp. 1097–1105 (2012)
12. Lee, S.J., Chen, T., Yu, L., Lai, C.H.: Image classification based on the boost convolutional neural network. IEEE Access **6**, 12755–12768 (2018)
13. Lee, S.H., Chan, C.S., Wilkin, P., Remagnino, P.: Deep-plant: Plant identification with convolutional neural networks. In: 2015 IEEE International Conference on Image Processing (ICIP), pp. 452–456. IEEE (2015)

14. Lee, S.H., Chang, Y.L., Chan, C.S., Remagnino, P.: Plant identification system based on a convolutional neural network for the lifeclef 2016 plant classification task. In: CLEF (2016)
15. Liaw, A., Wiener, M., et al.: Classification and regression by randomforest. R News **2**(3), 18–22 (2002)
16. Moghimi, M., Belongie, S.J., Saberian, M.J., Yang, J., Vasconcelos, N., Li, L.J.: Boosted convolutional neural networks. In: BMVC, pp. 24–1 (2016)
17. Nilsback, M.E., Zisserman, A.: A visual vocabulary for flower classification. In: 2006 IEEE Computer Society Conference on Computer Vision and Pattern Recognition (CVPR'06). vol. 2, pp. 1447–1454. IEEE (2006)
18. Nilsback, M.E., Zisserman, A.: Automated flower classification over a large number of classes. In: 2008 Sixth Indian Conference on Computer Vision, Graphics & Image Processing, pp. 722–729. IEEE (2008)
19. Patil, B., Pattanshetty, A., Nandyal, S.: Plant classification using SVM classifier (2013)
20. Priya, C.A., Balasaravanan, T., Thanamani, A.S.: An efficient leaf recognition algorithm for plant classification using support vector machine. In: International Conference on Pattern Recognition, Informatics and Medical Engineering (PRIME-2012), pp. 428–432 (March 2012). 10.1109/ICPRIME.2012.6208384
21. Seeland, M., Rzanny, M., Alaqraa, N., Wäldchen, J., Mäder, P.: Plant species classification using flower images– a comparative study of local feature representations. PLoS ONE **12**(2), e0170629 (2017)
22. Simon, M., Rodner, E.: Neural activation constellations: unsupervised part model discovery with convolutional networks. In: Proceedings of the IEEE International Conference on Computer Vision, pp. 1143–1151 (2015)
23. Sun, Y., Liu, Y., Wang, G., Zhang, H.: Deep learning for plant identification in natural environment. In: Computational Intelligence and Neuroscience, vol. 2017 (2017)
24. Wäldchen, J., Mäder, P.: Plant species identification using computer vision techniques: a systematic literature review. Arch. Comput. Methods Eng. **25**(2), 507–543 (2018)
25. Wäldchen, J., Rzanny, M., Seeland, M., Mäder, P.: Automated plant species identification–trends and future directions. PLoS Comput. Biol. **14**(4), e1005993 (2018)
26. Wilf, P., Zhang, S., Chikkerur, S., Little, S.A., Wing, S.L., Serre, T.: Computer vision cracks the leaf code. Proc. Natl. Acad. Sci. **113**(12), 3305–3310 (2016)
27. Wu, S.G., Bao, F.S., Xu, E.Y., Wang, Y.X., Chang, Y.F., Xiang, Q.L.: A leaf recognition algorithm for plant classification using probabilistic neural network. In: 2007 IEEE International Symposium on Signal Processing and Information Technology, pp. 11–16. IEEE (2007)
28. Xiao, X.Y., Hu, R., Zhang, S.W., Wang, X.F.: Hog-based approach for leaf classification. In: International Conference on Intelligent Computing, pp. 149–155. Springer (2010)
29. Zhang, C., Zhou, P., Li, C., Liu, L.: A convolutional neural network for leaves recognition using data augmentation. In: 2015 IEEE International Conference on Computer and Information Technology; Ubiquitous Computing and Communications; Dependable, Autonomic and Secure Computing; Pervasive Intelligence and Computing, pp. 2143–2150. IEEE (2015)

Edge–Texture-Based Characteristic Attribute on Local Radius of Gyration Face for Human Face Recognition

Pinaki Prasad Guha Neogi

Abstract It is a well-known fact that most of the edges in an image can be spotted on the facial segments and they have a place on the image's high-frequency components. Besides edges, another crucial feature for face matching is the texture. Therefore, both edges and texture can have significant contribution in extracting facial attributes and thus in human face recognition. This paper puts forward a novel edge–texture characteristic attribute for human face recognition based on the concept of radius of gyration face, which is invariant to changes in illumination, rotation and noise. The supremacy of the proposed approach in human face recognition is exhibited when its recognition accuracy is compared with other recent state-of-the-art techniques over challenging databases like CMU-PIE database, Extended Yale B database, AR database and CUFS database under varying conditions of illumination, noise, rotation and face sketch recognition.

Keywords Human face recognition · Radius of gyration · Edge · Texture · Illumination · Rotation · Noise

1 Introduction

In spite of the fact that very often people forget the name of a person, it is less likely for them to forget the face. Only a glimpse of an individual's face unveils a great deal of information about the person to whom it belongs to. Albeit all human beings have a similar arrangement of facial components, like nose, eyes, eye-brows and lips, each of them exhibits distinguishing attributes. Human beings are astonishingly quick and precise in identifying faces—a lot quicker and more exact than machines are—however, it is practically inconceivable to recognize hundreds or thousands of images within a stipulated time period and here arises the necessity to develop a system that should be able to accurately coordinate a picture to its proper class. Indeed, human face recognition is a commoving domain of research with an enormous scope of latent

P. P. Guha Neogi (✉)
Department of CSE, Meghnad Saha Institute of Technology, Kolkata, India
e-mail: ratul.ng@gmail.com

© Springer Nature Singapore Pte Ltd. 2020
J. K. Mandal and S. Mukhopadhyay (eds.), *Proceedings of the Global AI Congress 2019*, Advances in Intelligent Systems and Computing 1112, https://doi.org/10.1007/978-981-15-2188-1_21

applications, starting from ubiquitous surveillance to individual security frameworks. The domain of biometric pattern recognition is becoming progressively significant with the passage of time, specifically in fields like security, crime investigation, individual recognition proof and validation. The recent years have experienced such an expansion in societal fraud that it has turned out to be quite natural for mankind to lose reliance on empirical proof. In any case, well-grounded individual identification information has turned out to be increasingly challenging to acquire. Although evidences like fingerprint and blood sample provide biological information which are of very little significance, yet they have been utilized for decades as vital forensic evidences for finding out the criminals. But with the advancement in the field of technology, there have been numerous aspects of individual recognition that can be explored, human face recognition being one of the most significant one.

A lot of work has been done in the field of human face recognition. Among various handcrafted local descriptors for face recognition, local binary pattern (LBP) has reached the highest level of popularity due to its excellent performance and less computational cost. But, LBP is noise-sensitive and it also has a high-dimensional attribute that has a detrimental effect on its conduct. To overcome the limitation of LBP, Ren et al. [1] put forward a more proficient variant of LBP, called noise-resistant local binary pattern (NRLBP). Lai et al. [2] presented an approach called MSLDE, based on multi-scale logarithm difference edge-mapping for face recognition against varying lighting conditions, but does not say anything about rotational changes and effect of noise. Barnouti et al. [3] put forward a face detection methodology utilizing Viola–Jones with PCA-LDA (VJ-PL) and square Euclidean distance. Utilizing the distinctiveness of Gabor features, Soula et al. [4] proposed a novel kernelized face recognition (NKFR) methodology. Gumede et al. [5] put forward a hybrid component-oriented face recognition (HCFR) system, as an alternative to the prevailing human face detection approaches. But primarily, the performance of most of these algorithms has been evaluated on face pictures gathered under very much controlled studio conditions. And as a consequence, the greater part of these methodologies face trouble on account of dealing with characteristic pictures, which are subjected to illumination, rotation and noise and the outcome debases radically with the variation of these parameters. Thus, the need arises for a system of face recognition and classification that is robust against variation of rotation, noise and illumination.

2   Concept of Radius of Gyration

Mass moment of inertia can be characterized as a quantifiable feature or attribute that determines the ability of an arrangement of rigid bodies to resist the variations in rotational velocity about the pivot or hinge. The estimation of mass moment of inertia about the rigid body's centre of mass is predominantly reliant upon the basic structure and mass of the body; additionally, it is affected by the orientation of the rotational axis of the rigid body. About the centre of mass of the inflexible body,

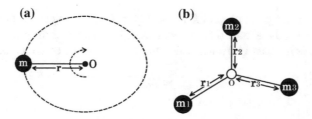

Fig. 1 **a** Moment of inertia of a system containing a mass m and **b** moment of inertia of a system containing three different masses m_1, m_2 and m_3 at a distance of r_1, r_2 and r_3 from the axle of rotation

the moment of inertia (I_o) is applicable for its rotation about not only fixed but also translating axis. So as to cognize the manner in which the moment of inertia is depicted in a quantitative frame, a very basic and comprehensible arrangement is depicted in Fig. 1a, where a rigid body having a mass m is rotated about a riveted point O, with the help of a string of length r.

For this arrangement, the mathematical illustration of the moment of inertia can be presented as follows:

$$I_o = m \cdot r^2 \tag{1}$$

If instead of a single rigid body we take into consideration a system of inflexible bodies, then the arrangement will be as depicted in Fig. 1b, which shows a system of three different bodies, having masses m_1, m_2, m_3, respectively, and are rotating about a common axis O, with a separation of r_1, r_2 and r_3 units from the axis, respectively. Assumption is being made that all the masses m_1, m_2, m_3 are either rotating the anticlockwise direction or in the clockwise direction. For such an arrangement, the mathematical illustration of the moment of inertia can be presented as shown below:

$$I_o = m_1 \cdot r_1^2 + m_2 \cdot r_2^2 + m_3 \cdot r_3^2 \tag{2}$$

In a similar fashion, for an arrangement comprising x number of such objects rotating in the same direction about a common axis, having masses m_1, m_2, m_3, …, m_{x-1}, m_x, respectively, and are at a distance of r_1, r_2, r_3, …, r_{x-1}, r_x, respectively, from the common axis, the moment of inertia can be represented as follows:

$$I_o = \sum_{i=1}^{x} m_i \cdot r_i^2 \tag{3}$$

On the other hand, the radius of gyration is basically denoted as the distance of that very point from the axis of rotation where the whole mass of a specific system is concentrated, so that around the axle of rotation, the moment of inertia (I_o) remains the same. In precis, gyration is merely the dissemination of the components of any rigid body. Generally, the symbol K is employed for denoting the radius of gyration of a body. For a body of mass (M), it is also possible to express the moment of Inertia (I_o) in terms of its mass (M) and the radius of gyration (K) as shown below:

$$I_o = MK^2 \tag{4}$$

$$\Rightarrow K = \sqrt{\frac{I_o}{M}} \tag{5}$$

3 Proposed Method

3.1 Local Radius of Gyration Face (LRGF) Representation

The most fundamental and elementary units for an image are the pixels that comprise it and each of these pixels is comparable to an individual body or a charged particle in space. Under such circumstances, it is taken into consideration that these individual pixels would demonstrate physical properties identical to those exhibited by singular bodies in an organization of inflexible mass, e.g. moment of inertia, centre of mass and radius of gyration.

If we consider a 3×3 block of pixels in an image, as depicted in Fig. 2, then, taking each of the pixels as an individual body, each having a mass equivalent to its intensity value, we can reckon the centre of mass of the arrangement, as follows:

$$\vec{C} = \frac{\sum_{i=1}^{9}(m_i \vec{v_i})}{\sum_{i=1}^{9}(m_i)} \tag{6}$$

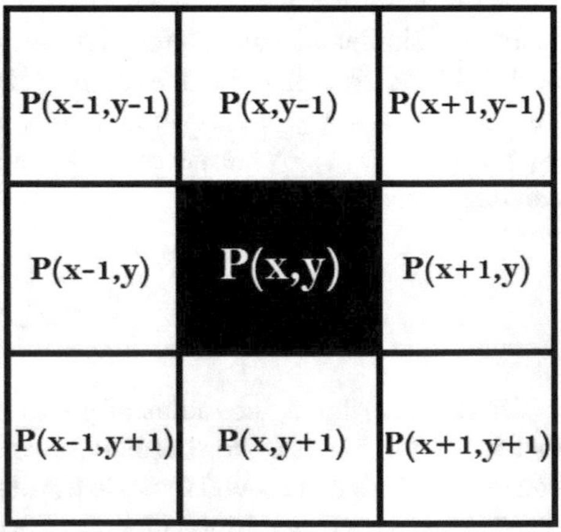

Fig. 2 Pixel orientation in a 3×3 block

$P(x-1,y-1)$	$P(x,y-1)$	$P(x+1,y-1)$
$P(x-1,y)$	$P(x,y)$	$P(x+1,y)$
$P(x-1,y+1)$	$P(x,y+1)$	$P(x+1,y+1)$

– where for each pixel i, the mass is denoted as m_i and its position vector is represented as $\vec{v_i}$. The denominator denotes the summation of the masses of each of the nine individual pixels, i.e. the total mass of the system. Since an analogy has been drawn between the pixel intensities and their masses, the relation (6) can also be restructured as shown below:

$$\vec{C} = \frac{\sum_{i=1}^{9}(I_i \vec{v_i})}{\sum_{i=1}^{9}(I_i)} \tag{7}$$

$$= \frac{1}{I} \sum_{i=1}^{9}(I_i \vec{v_i}) \tag{8}$$

– where the term $I = \sum_{i=1}^{9}(I_i)$ denotes the summation of the intensities of each of the nine pixels in the block, i.e. the block's aggregate intensity value.

Considering the centre of mass of a block of pixel as the axis of rotation, the local radius of gyration face (LRGF) [6] is found out. For this purpose, the block's moment of inertia is computed employing relation (3), where the mass m_i for a specific pixel i is taken as the intensity value of the pixel and r_i corresponds to the distance of the pixel i from the centre of mass of the block. Following this, for every individual pixel (i, j), the radius of gyration $K_{i,j}$ is reckoned according to relation (5), where the block's moment of inertia is divided by the mass (intensity value) of that pixel (i, j) and taking its square root. These values of the radius of gyration for the individual pixels constitute the local radius of gyration face (LRGF), which has been proved to be invariant to variations in noise, rotation and illumination [6]. Figure 3 shows the local radius of gyration face for a sample image that has been acquired from the Extended Yale B database.

Fig. 3 **a** Sample face image from the Extended Yale B database and **b** local radius of gyration face (LRGF) for Fig. 3a

(a)　　　　　　　　　**(b)**

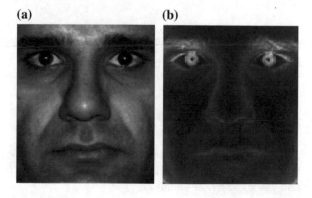

3.2   LRGF Edge–Texture Characteristic Face Representation

A binary texture image is created from the radius of gyration face by following the condition that if in a 3×3 neighbourhood, the radius of gyration value of a specific pixel is greater than or equal to the radius of gyration value of the central pixel, then the intensity of the pixel is set to 1, else 0. At the same time, an edge image is also created from the radius of gyration face using Canny edge detection [7] technique.

In order to develop the edge–texture characteristic image, six different parameters have been taken into consideration: $P\_Edge(i, j)$, i.e. the probability of a specific pixel (i, j) in the Canny edge image to be of value 1; $P\_Text(i, j)$, i.e. the probability of a specific pixel (i, j) in the binary texture image to be of value 1; $Nearest\_Edge\_of\_Each\_Edge(i, j)$ indicating the distance of the specific pixel (i, j) having a value equals 1 in the edge matrix from the nearest pixel in a 3×3 block, whose value is also 1 in the edge matrix; $Nearest\_Texture\_of\_Each\_Edge(i, j)$ indicating the distance of the specific pixel (i, j) having a value equals 1 in the edge matrix from the nearest pixel in a 3×3 block having value as 1 in the binary texture matrix; $Nearest\_Texture\_of\_Each\_Texture(i, j)$ indicating the distance of the specific pixel (i, j) having a value equals 1 in the binary texture matrix from the nearest pixel in a 3×3 block whose value is also 1 in the binary texture matrix; $Nearest\_Edge\_of\_Each\_Texture(i, j)$ indicating the distance of the specific pixel (i, j) having a value equals 1 in the binary texture matrix from the nearest pixel in a 3×3 block having value as 1 in the edge matrix. Employing all these six parameters, the characteristic edge–texture matrix ($\mathbb{M}$) is formed according to the following relation:

$$
\begin{aligned}
\mathbb{M}(i, j) = \{ & (P\_Edge(i, j) \times Nearest\_Edge\_of\_Each\_Edge(i, j)) \\
+ & (P\_Edge(i, j) \times Nearest\_Texture\_of\_Each\_Edge(i, j)) \\
+ & (P\_Text(i, j) \times Nearest\_Texture\_of\_Each\_Texture(i, j)) \\
+ & (P\_Text(i, j) \times Nearest\_Edge\_of\_Each\_Texture(i, j)) \} \div 4
\end{aligned}
$$

$$(9)$$

The values in the characteristic edge–texture matrix ($\mathbb{M}$) are utilized for creating the feature vector of the image.

4   Similarity Measure

For the similarity measure between two images, the Jaccard similarity [8] method has been taken into account. The Jaccard distance $JD(x, y)$ between two vectors x and y is given by the equation:

$$JD(x, y) = \left(1 - \frac{\sum_{i=1}^{\text{No. of Pixels}}(x_i \ \& \ y_i)}{\sum_{i=1}^{\text{No. of Pixels}}(x_i y_i)} \right) \qquad (10)$$

The query image x is classified into that specific class with whose gallery image y the $JD(x, y)$ value is minimum.

5 Experimental Set-Up and Result

The performance of the proposed method in face detection has been evaluated on a number of standard and challenging databases. Challenging scenarios were established for the databases upon which the efficiency of the method being put forward sustained itself. To study the recognition accuracy with variation of illumination, the proposed method has been evaluated on the CMU-PIE database [9] and the Extended Yale B database [10], the outcome of which has been tabulated in Table 1. Figure 4 indicates some sample face images from the Extended Yale B database at various illumination conditions and their corresponding radius of gyration face, texture image, edge image and characteristic edge–texture images. To test the robustness against rotational changes, the AR database [11] has been taken into consideration. The reference images are the original images taken from the AR database (angle of rotation $= 0°$) and the test images have been subjected to rotations, with angle of rotation as $10°$, $20°$ and $30°$, as shown in Fig. 5. The recognition accuracy for different angle of rotation is shown in Table 2. To test the robustness against noise also, the AR database has been considered. The reference images are the original images taken from the AR database ($\sigma = 0.00$) and the test images have been added with Gaussian noise with standard deviations (σ) as ($\sigma = 0.01$, $\sigma = 0.02$, $\sigma = 0.03$, $\sigma = 0.04$), as depicted in Fig. 6. The recognition accuracy for variations in noise is tabulated in Table 3. Besides, the proposed approach has also been evaluated on the CUHK Face Sketch (CUFS) [12] database to study its efficiency in heterogeneous face recognition (HFR). Figure 7 shows some sample input and output images of the proposed method on the CUFS database, and Table 4 shows the recognition accuracy on CUFS

Table 1 Recognition accuracy (%) with variations in illumination

Approaches	Recognition accuracy (%)	
	CMU-PIE database	Extended Yale B database
NKFR	90.20	89.91
MSLDE	94.92	93.00
VJ-PCA-LDA	92.17	90.08
HCFR	95.09	92.88
NRLBP	94.84	94.18
Proposed method	97.16	98.04

Fig. 4 **a** Sample face images of a subject from the Extended Yale B database under varying illumination conditions and **b** the corresponding RG faces, **c** corresponding texture images, **d** corresponding edge images and **e** characteristic edge–texture images

Fig. 5 **a** Original image of an individual from the AR database and its corresponding images with different angle of rotation (10°, 20° and 30°), **b** the corresponding RG faces of the images shown in Fig. 5a, **c** corresponding texture, **d** corresponding edge images and **e** characteristic edge–texture images

θ=0°   θ=10°   θ=20°   θ=30°

Table 2 Recognition accuracy (%) on AR database with variations in rotation

Approaches	Recognition accuracy (%)			
	AR database (0° rotation)	AR database (10° rotation)	AR database (20° rotation)	AR database (30° rotation)
NKFR	87.20	81.52	76.82	66.76
MSLDE	88.40	83.98	77.40	67.09
VJ-PCA-LDA	88.40	85.12	81.76	79.68
HCFR	93.75	91.68	88.18	84.42
NRLBP	92.80	92.08	87.66	86.88
Proposed method	96.08	94.06	92.14	90.02

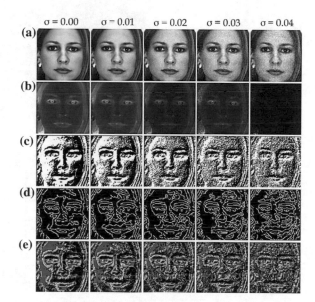

Fig. 6 **a** Original image from the AR database and its corresponding images with noise variations ($\sigma = 0.01$, $\sigma = 0.02$, $\sigma = 0.03$, $\sigma = 0.04$), **b** the corresponding RG faces of the images shown in Fig. 6a, **c** corresponding texture images, **d** corresponding edge images and **e** characteristic edge–texture images

Table 3 Recognition accuracy (%) on AR database with variations in noise

Approaches	Recognition accuracy (%)				
	AR database ($\sigma = 0.00$)	AR database ($\sigma = 0.01$)	AR database ($\sigma = 0.02$)	AR database ($\sigma = 0.03$)	AR database ($\sigma = 0.04$)
NKFR	87.20	83.42	78.27	72.22	65.78
MSLDE	88.40	85.84	80.02	76.75	69.64
VJ-PCA-LDA	88.40	85.84	83.62	80.42	77.20
HCFR	93.75	90.02	87.32	93.96	81.83
NRLBP	92.80	92.08	90.04	87.44	86.18
proposed method	96.08	93.75	92.22	91.08	90.02

Fig. 7 a Sample photo-sketch pair images from the CUFS database, **b** the corresponding RG faces of the images shown in Fig. 7a, **c** corresponding texture images, **d** corresponding edge images and **e** characteristic edge–texture images

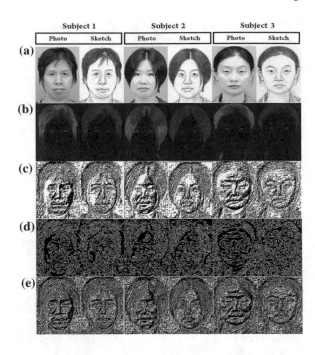

Table 4 Recognition accuracy (%) on CUHK Face Sketch (CUFS) database

Approaches	Recognition accuracy (%)
NKFR	88.28
MSLDE	94.08
VJ-PCA-LDA	87.34
HCFR	94.42
NRLBP	93.26
Proposed method	100.00

database. The performance of the proposed approach has also been compared with other recent state-of-art methods, like NKFR, MSLDE, VJ-PL, HCFR and NRLBP, which clearly exhibits the supremacy of the proposed approach.

6 Conclusion

This paper puts forward the proposal of a novel edge–texture-oriented modality-invariant attribute representation for face recognition based on the radius of gyration face, which is invariant to changes in illumination, rotation and noise at the same time. It is seen that the proposed method demonstrates momentous execution accuracy when the outcome is compared to other prevalent techniques over challenging databases like the CMU-PIE, Extended Yale B and AR database. Besides, this also exhibits a near perfect accuracy in case of CUFS photo-sketch pair dataset, thus proving its credibility in heterogeneous face recognition.

References

1. Ren, J., Jiang, X., Yuan, J.: Noise-resistant local binary pattern with an embedded error-correction mechanism. IEEE Trans. Image Process. **22**(10), 4049–4060 (2013)
2. Lai, Z., Dai, D., Ren, C., Huang, K.: Multiscale logarithm difference edgemaps for face recognition against varying lighting conditions. IEEE Trans. Image Process. **24**(6), 1735–1747 (2015)
3. Barnouti, N.H., Matti, W.E., Al-Dabbagh, S.S., Naser, M.A.: Face detection and recognition using Viola-Jones with PCA-LDA and square euclidean distance. Int. J. Adv. Comput. Sci. Appl. (IJACSA) **7**(5), 371–377 (2016)
4. Soula, A., Said S.B., Ksantini, R., Lachiri, Z.: A novel kernelized face recognition system. In: 4th International Conference On Control Engineering & Information Technology (CEIT), Hammamet, pp. 1–5. https://doi.org/10.1109/ceit.2016.7929118. (2016)
5. Gumede, A., Viriri, S., Gwetu, M.: Hybrid component-based face recognition. In: Conference on Information Communication Technology and Society, Umhlanga, pp. 1–6 (2017)
6. Kar, A., Neogi, P.P.G.: Appl Intell. (2019). https://doi.org/10.1007/s10489-019-01545-x
7. Chitra, A.D., Ponmuthuramalingam P.: An approach for canny edge detection algorithm on face recognition. Int. J. Sci. Res. **4**(11) Paper ID: NOV151012 (2015)
8. Marzieh, O., Morteza, M.Z.: Pairwise document similarity measure based on present term set. J. Big Data (2018). https://doi.org/10.1186/s40537-018-0163-2
9. Sim, T., Baker, S., Bsat, M.: The CMU pose, illumination, and expression (PIE) database. In: Proceedings of Fifth IEEE International Conference on Automatic Face and Gesture Recognition, pp. 46–51. IEEE (2002)
10. Lee, K.C., Ho, J., Kriegman, D.: Acquiring linear subspaces for face recognition under variable lighting. IEEE Trans. Pattern Anal. Mach. Learn. **27**(5), 684–698 (2005)
11. Martinez, A.M., Benavent, R.B.: The AR Face Database, CVC Technical Report #24, June (1998)
12. Zhang, W., Wang, X., Tang, X.: Coupled information-theoretic encoding for face photo-sketch recognition. In: Proceedings of IEEE International Conference on Computer Vision and Pattern Recognition, pp. 513–520 (2011)

Centroid-Based Hierarchy Preserving Clustering Algorithm Using Lighthouse Scanning

Soujanya Ray and Anupam Ghosh

Abstract Clustering is an important topic in machine learning and data mining. But determining the number of clusters in a data set, a quantity that is often labelled as 'k' as in a k-means algorithm, is a distinctive problem from the process of actually solving the clustering problem. Another challenge faced by the existing clustering algorithms is the final result and the converge time largely depends on the predicted position of the centroids. It is an interesting problem because if the number of clusters and the centroid positions could be roughly estimated, then the convergence time of any clustering algorithm could be decreased and the resulting redundancy could be preserved. In this paper, we implemented our novel mathematical idea of two-way lighthouse scan, on existing k-means algorithm, to predict the quantity and position of the clusters. The recursive implementation of the algorithm decreases the number of iterations of any centroid-based clustering algorithm while preserving the rank of the clusters.

Keywords Clustering · Non-hierarchical clustering · Unsupervised learning · Recursive clustering · Lighthouse scanning

1 Introduction

Clustering is a very primary, yet it is an essential and powerful method in the domain of unsupervised machine learning. Clustering is used on unlabelled data to recognise similar or dissimilar entities in a pool of entities. The use and application of this kind of algorithm increased with the increasing computational power. The main idea behind clustering is to group entities with similar features together. This is done by plotting the features of the unlabelled data in a multidimensional space. The various

S. Ray (✉) · A. Ghosh
Department of Computer Science and Engineering, Netaji Subhash Engineering College, Kolkata, India
e-mail: shaun.ray.1996@gmail.com

A. Ghosh
e-mail: anupam.ghosh@rediffmail.com

© Springer Nature Singapore Pte Ltd. 2020
J. K. Mandal and S. Mukhopadhyay (eds.), *Proceedings of the Global AI Congress 2019*, Advances in Intelligent Systems and Computing 1112, https://doi.org/10.1007/978-981-15-2188-1_22

types of clustering algorithms use different kinds of mathematical functions and methods to group these entities together. These clustering algorithms can be broadly classified into hierarchical clustering [1] and non-hierarchical clustering [2]. Both of these techniques suffer from few distinctive problems.

Hierarchical agglomerative clustering algorithms mostly have a complexity of $O(n^3)$ and use a space of $O(n^2)$, this makes it slow even for medium-sized data sets. Few algorithms such as SLINK [3] and CLINK [4] tackle this problem, and using a heap data structure further brings down the complexity to $O(n^2 \log n)$. The hierarchical divisive algorithm with an exhaustive search has $O(2^n)$. The method of calculation of Proximity of the Nearest Neighbours differs from algorithm to algorithm and thus is not definitive, but we are interested in non-hierarchical clustering processes. Non-hierarchical clustering algorithms (in particular, k-means, k-medoids and expectation maximisation [5–7]), use a parameter commonly referred to as 'k' that specifies the number of clusters to detect. These algorithms require the number of clusters—k, as an input. Provided the k is available, the algorithm proceeds to group the data points into clusters. The correct choice of k is often ambiguous, with interpretations depending on the shape, scale, variety of the distribution of points in a data set and the desired clustering resolution of the user. This is a problem. It is difficult to determine the number of clusters in a comparatively larger data set. Another distinct problem in non-hierarchical clustering is the pattern of clustering is highly dependent on the position of the initially chosen centroids. Few existing methods try to solve this problem by taking a snippet of the data set and then using hierarchical algorithms to determine the number of clusters in the first place [16]. Another way to tackle this issue is by using spatially constrained spectral clustering [17]. Then with this information, some non-hierarchical algorithm is used for better and desired result.

But in our paper, we have devised and implemented the novel idea of one-way and two-way lighthouse scans—LiHoS and breaking down the entire clustering task into recursive clustering to decrease the total number of iterations of any non-hierarchical clustering algorithm. We used the k-means as a proof of concept, but any centroid-based clustering algorithm can be used. Recursive clustering retains the information of the data distribution that is rank of data points through all the clusters. This method provides building and understanding actual relationships between clusters and solves real-world problems. During the recursion, the algorithm preserves this information in the form of a hierarchy tree.

Related works

In his fascinating dissertation [8], Mingjin Yan discussed another method to solve the problem of clustering without the knowledge of the number of clusters. Altaf R. and Vincent N. in their paper narrowing the modelling gap: a cluster-ranking approach to coreference resolution [9] explained how clusters can be ranked and how ranking clusters can extract more information from the data set. Hervé Cardot, Peggy Cénac, Jean-Marie Monnez in their paper a fast and recursive algorithm for clustering large data sets with k-median [10] showed how recursive clustering work and analysed the algorithm. Shi Na, Liu Xumin, Guan Yong in their paper research on k-means clustering algorithm: an improved k-means clustering algorithm [11] improved on the k-means using a nominal data structure to store some information in every iteration. That led to improve the clustering results.

Motivation

The clustering algorithms had no measure of number of clusters and lost the hierarchy of the clustering while using a non-hierarchical algorithm. Our work helps to preserve both the quality and the type of clustering. This ensures better data analysis and interpretation. Moreover, the hierarchy of the data is interpreted. This hierarchy can be used for identifying the rank of any data point sample with respect to the entire dataset. This hierarchy preservation through a centroid-based algorithm provided the motivation to pursue this work. In addition to this, the future work is also under process.

2 Methodology

Proposed Algorithm

We considered the entire data set as one single cluster and then we used the cluster properties to determine the threshold values. These threshold values would serve as the convergence criteria of the algorithm.

1. Set data as

$$D = \overrightarrow{X}_{ij} \quad \text{where} \quad \overrightarrow{X}_{ij} =$$

$$\begin{matrix} \overrightarrow{X_1} = \\ \overrightarrow{X_2} = \\ \overrightarrow{X_n} = \end{matrix} \begin{bmatrix} x_{1,1} & x_{1,2} & \cdots & x_{1,m} \\ x_{2,1} & x_{2,2} & \cdots & x_{2,m} \\ \vdots & \vdots & \cdots & \vdots \\ x_{n,1} & x_{n,2} & \cdots & x_{n,m} \end{bmatrix}$$

where m is the total dimension of the data and n is the total count of data.

2. The centroid of the entire data set is calculated.

$$Centroid \leftarrow cent(D)$$

3. Calculate the average distance circle.

$$avg\_dist \leftarrow avg\_dist\_circle\,(cent,\ D)$$

4. Generate the cluster table

$$cst \leftarrow LiHoS(centroid, \ D, \ avg\_dist)$$

5. Calculate the threshold for the data set.

$$Threshold \leftarrow threshold(cst)$$

6. Assign cluster centroids outside and inside the average distance circle.

$$Predicted\_centroids \leftarrow assign\_cen(cst, \ Threshold)$$

7. Run K-means on the assigned centroids.

$$D[] = K - means(D, \ Predicted\_centroids)$$

8. Each new cluster formed is considered as an independent data set. Now, the flow is done recursively

recursive_clustering(D):
while (threshold $\neq -1$)
  centroid $\leftarrow$ cent(D)
  avg_dist $\leftarrow$ avg_dist_circle(cent, D)
  cst $\leftarrow$ gen_clsuter_table(centroid, D, avg_dist)
  threshold $\leftarrow$ threshold(cst)
  predicted_centroids $\leftarrow$ assign_cen(cst, threshold)
  D[] $\leftarrow$ K_means(D, predicted_centroids)
  recursive_clustering(D[])

Parameters and Functions

cent (D):
  for j = 1 to m

$$\vec{X}_{\bar{n},j} \leftarrow \frac{\sum_{i=1}^{n} X_i}{n}$$

return $\left[x_{\bar{n},1}, x_{\bar{n},2}, \ldots, x_{\bar{n},m} \right]$

This function returns the coordinates for the centroid of the data set using the mean method.
avg_dist_circle(cent, D):

$$return \ \frac{\sum_{i=1}^{n} dist\left(cent, \vec{X}_i\right)}{n}$$

/*where dist(a, b) returns the euclidean distance between point a and point b*/

LiHoS (centroid, D, avg_dist):

cluster_table[s_index][in, out]

for i ← 0 to n

   s_index ← slope (centroid, $\overrightarrow{X_i}$)

   if $\overrightarrow{X_i} > avg\_dist$

     cluster_table[s_index][out] ← cluster_table[s_index][out] +
(dist(centroid, $\overrightarrow{X_i}$)-avg_dist)$^{-1}$

    else

     cluster_table[s_index][in] ← cluster_table[s_index][in] + (avg_dist -
dist(centroid, $\overrightarrow{X_i}$)$^{-1}$

return cluster_table

To generate the cluster table, the data points are scanned from the circumference of the average distance circle, both inside and outside. The index to the cluster table angles from 0 to 360. We fill the cluster table values with respect to the data points' slope, as an index, to an assumed normal. This method of scanning is novel and we call it two-way lighthouse scan. The inverse of the distance is calculated for all the points and the distances' angle. Then that angle value of the cluster table is updated by the mean of the calculated number and previously stored number. This is done to find a canopy-like cover [13].

The cluster table is the data set that LiHoS returns. It has 360 rows and 3 columns. So, the size of the cluster table is fixed irrespective of the data set used. Cluster table will help us to assign centroids to the supposed clusters determined by the scan line. This is plotted on a graph to give a visualisation of the possible clusters. The two lines plotted are the inner cluster measurement and outer cluster measurement (Table 1).

assign_cen (cst, threshold):

for i in cst[inner]

 if (i > threshold)

   inner_centroids ← cst#index

for j in cst[outer]

 if (j > threshold)

   outer_centroid ← cst#index

/* selects every bump that is above the threshold value*/

Based on the cluster table, we searched for the maximum surge of data points in the inner and outer sections of the cluster table. Each surge is measured and compared to the threshold that was determined. This measurement results in n number of inner clusters and m number of outer clusters. The quantity m and n are different for every

Table 1 A sample of the cluster table

S_Index	In	Out
1	0	1.23
2	2.46	1.35
:	:	:
359	4.65	0
360	4.25	0

recursive iteration. These surges are then allocated as clusters. This results in m outer clusters (outside the perimeter of the average distance circle) and n inner clusters (inside the average distance perimeter). In total, for the first iteration, we are left with $m + n$ cluster.

Recursive Clustering

Using k-means or k-medoids, we implemented recursive clustering. This takes place after we allocate the centroids in the first run. After the centroids are allocated, we converge the centroids to cluster using k-means. After the k-means converges, we get an independent cluster of clusters. The idea is to recursively use the entire process into these independent clusters of clusters. This would go on recursively until the convergence criteria are reached. This property is by default preserved in the state-of-the-art deep embedded clustering [12] which uses neural networks. This similarity relationship has been also explored [14].

As stated earlier, recursive clustering preserves the hierarchy of the data and ranks the cluster. This helps in narrowing the modelling gap. In their paper, A. Rahman and V. Ng used ranking methodology for an approach to coreference resolution. The ranking would result in understanding the pattern and relation of clusters with one another. But clusters can also be ranked based on their dissimilarity measure [15]. Hierarchical rank of clusters can be explained using an analogy of classification of class of vehicles (Fig. 1).

Threshold (cst):

> *packing ← maximum (cst[inner]) - maximum(cst[outer])*
> *if (avg_dist ~ packing)*
> *set threshold ← -1*
> *else*
> *set threshold ← avg_dist - packing*
> *return threshold*

The threshold is a function of the inverse of the mean distance from the *average_dist_circle*. For our test runs, the value of 0.5 gave the best results. The threshold of cluster allocation holds great influence over the quality of the clusters that are being formed. A threshold value of more than 0.5 resulted in loosely packed clusters and more recursive runs. This resulted in bottom-heavy cluster ranking. Bottom-heavy cluster ranking refers to the hierarchy tree of cluster ranking with less branches on the top and more branches at the bottom. This classification gives a better sense to specification. On the other hand, a value of less than 0.5 resulted in tight clusters with top-heavy ranking. Top-heavy ranking refers to the hierarchy tree with more branches at the top and less branches at the bottom. This classification gives a better sense to generalisation.

Computational Complexity

Below is the calculation of the complexity.

Number of recursive runs $= m$

Number of recursive runs $= i$

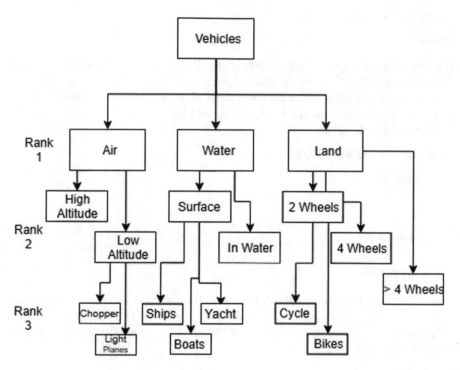

Fig. 1 The ranked representation of vehicles

Number of clusters formed $= m * i$
Then the total number of clusters k can be given by

$$k = m * m * i$$

Time

The computational complexity of each function is calculated. Let the total number of data points be n.

1. cent $(D) = O(n)$
2. avg_dist_circle$(D, cent) = O(n)$
3. LiHoS$(cent, D, avg\_dist) = O(n)$
4. threshold$(cst) = O(n)$
5. assign_cent$(cst, threshold) = O(n)$
6. Regular K-means $= O(n)$

Now, calculating the recursive time complexity of the recursive run.

Let the total number of cluster formed after the algorithm converges be k (total number of leaf nodes of the recursion tree). And the total recursion runs be h (the height of the recursion tree). Therefore

7. recursive_clustering(D[]) = $O(k * h)$.

Space

The cluster table's space is independent of the size of the data set. It takes a constant space that is 360×3. Every recursive cluster break into m clusters will take m $\times$ 360×3 space and the previous table will be freed up. When the algorithm converges with k clusters, the max space complexity will be $k \times 360 \times 3$. The recursive clustering takes $O(n)$ space.

Flowchart to the Algorithm

See Fig. 2.

3 Results

Data Set Description

The data set used in this project was used from Kaggle and KD-Nuggets. These data sets were specifically designed for analysis of different kinds of clustering algorithms. There were four variants of clusters used. Skew clusters and indefinite-shaped cluster were intentionally placed.

Analysis of the Result

The algorithm clustered four of the Kaggle cluster analysing data sets. This picto-graphic analysis shows a specific run of one of the four datasets. In Fig. 3, the data points are plotted on the given space. The red circle represents the average distance circle of the first recursive iteration. The centroid is in red dot.

Next, the algorithm runs the scan line to build the cluster table. The cluster table is a 360×3 data structure with the angular density values inside and outside the average distance circle, respectively. The plot in Fig. 4 represents the cluster patches in the form of bumps. The algorithm compares these bumps to the threshold in order to predict the number of possible clusters. The inner scan is plotted in red and outer plot in green, respectively.

Using the threshold criteria, the algorithm predicted the number of clusters and their probable position with respect to the average distance circle and the centroid. The algorithm assigns centroids to all these measured probable clusters. In the given example plot in Fig. 5, the green dots refer to the predicted cluster centroids outside the perimeter and the red dots inside the perimeter.

After predicting the number of clusters and their probable position, K-means to converge the algorithm was used. Any hierarchical clustering algorithm can be used according to the choice. Figure 6 shows the first complete run. The iterations of the k-means decrease drastically due to pre-defined territory and centroids.

After the first iteration, few clusters have attained the threshold and get saturated, thus the algorithm will declare them clusters and not use them in upcoming recursive

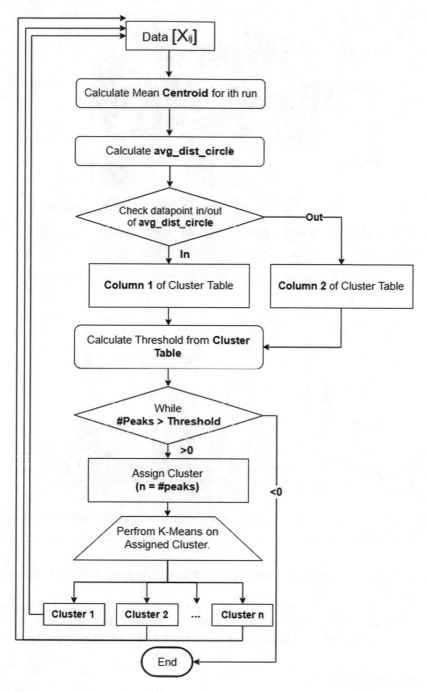

Fig. 2 The flowchart to the algorithm

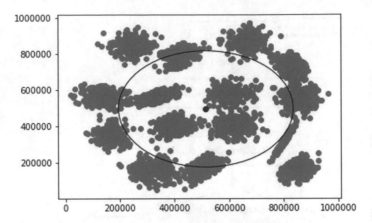

Fig. 3 The centroid and the average distance circle after the first iteration

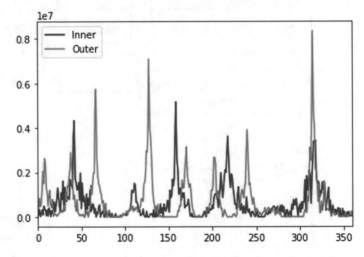

Fig. 4 The cluster table plot. The Y-axis denotes the inverse square distances and the X-axis denotes the respective angles

runs. The few clusters that did not attain the threshold proceed for the first recursive iteration as shown in Fig. 7.

After the first recursive iteration, the remaining clusters were assigned and attained the threshold value for the data set as shown in Fig. 8. These clearly show the algorithm's accuracy in predicting the number and position of the clusters in the data set.

Validation of the result

The following tables and graphs show the specificity, sensitivity, precision and F-score data sets. Note that we have averaged the values over clusters. The two measured

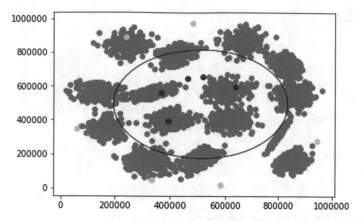

Fig. 5 Assigning the possible centroids outside (green) and inside (red) of the average distance circle

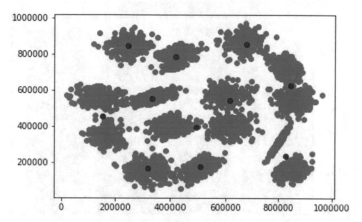

Fig. 6 The converged centroids

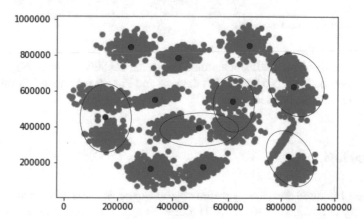

Fig. 7 The second iteration and the average distance circles of the local rank two clusters

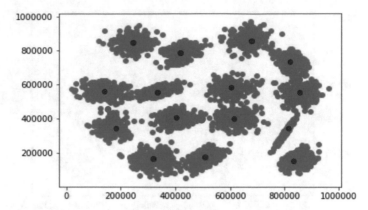

Fig. 8 Second round of recursive clustering

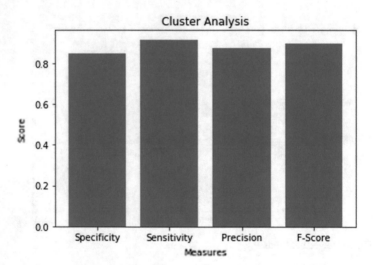

Fig. 9 Comparison of the first dataset

data sets, D1 data set has 800 samples and D2 data set has 5000 samples. Figure 9 represents the validation score for the first data set and Fig. 10 represents the validation score for the second data set (Table 2).

4   Conclusion

For any unstructured data, the algorithm was able to achieve clustering with comparatively very less recursive iterations. Under each recursive iteration, the supporting algorithm that was used was K-means. But when k-means was provided, the existing

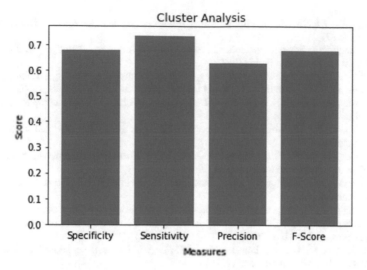

Fig. 10 Comparison of the second dataset

Table 2 Comparison of specificity, sensitivity, precision and F-score of first and second data set, respectively

	Specificity	Sensitivity	Precision	F-score
D1	0.687	0.737	0.628	0.678
D2	0.849	0.915	0.875	0.894

cluster data used less number of iterations than without our cluster allocation algorithm for the same data set. The recursive clustering helped to preserve the rank of the clusters and form a hierarchy of the data set with the clusters as the classes.

In addition to this as a future scope of work, we would try to improve the threshold value and its relation to the data set. Improvements can be made on the process of calculating the average distance circle. The better-approximated shape, the better LiHoS works in distinguishing clusters. In our method of assigning the cluster centroid based on the cluster, table uses Euclidean distances. Other distances such as Manhattan distances or Mahalanobis distance could be compared to the current results.

References

1. Johnson, S.C.: Hierarchical clustering schemes. In: Psychometrika, vol. 3. Springer, US (1967)
2. Chandrasekharan, M.P., Rajagopalan, R.: An ideal seed non-hierarchical clustering algorithm for cellular manufacturing. Int. J. Product. Res. IFPR (1984)
3. Sibson, R.: SLINK: an optimally efficient algorithm for the single-link cluster method (PDF). Comput. J. British Computer Society (1973)

4. Defays, D.: An efficient algorithm for a complete-link method. Comput. J. British Computer Society (1977)
5. MacQueen, J.B.: Some Methods for Classification and Analysis of Multivariate Observations. University of California Press (1967)
6. Kaufman, L., Rousseeuw, P.J.: Clustering by means of medoids. In: Dodge, Y. (ed.) Statistical Data Analysis Based on the—Norm and Related Methods, North-Holland (1987)
7. Dempster, A.P., Laird, N.M., Rubin, D.B.: Maximum likelihood from incomplete data via the EM algorithm. J. R. Stat. Soc., Ser. B (1977)
8. Yan, M., Ye, K.: Determining the number of clusters using the weighted gap statistics. J. Int. Biometric Soc. (2007)
9. Rahaman, A., Ng, V.: Narrowing the modeling gap: a cluster-ranking approach to coreference resolution. J. Artif. Intell. Res. (2011)
10. Cardot, H., Cenac, P., Monnez, J.M.: A fast and recursive algorithm for clustering large datasets with k-medians (2011)
11. Na, S., Xumin, L., Yong, G.: Research on k-means clustering algorithm: an improved k-means clustering algorithm. In: Third International Symposium on Intelligent Information Technology and Security Informatics (2010)
12. Ren, Y., Hu, K., Dai, X., Pan, L., Hoi, S.C.H., Xu, Z.: Semi supervised deep embedded clustering. Neurocomputing **325** (2019)
13. Li, J., Zhang, K., Yang, X., Wei, P., Wang, J., Mitra, K., Ranjan, R.: Category preferred canopy-k-means based filtering algorithm. Futur. Gener. Comput. Syst. **93** (2019)
14. Huang, S., Kang, Z., Tsang, I.W., Xu, Z. Auto-weighted multi-view clustering via kernelized graph learning. Pattern Recogn. **88** (20190
15. Pimentel, B.A., de Carvalho, A.C.P.L.F.: A new data characterisation for selecting clustering algorithms using meta-learning. Inf. Sci. **427** (2019)
16. Wohwe Sambo, D., Yenke, B.O., Forster, A., Dayang, P.: Optimized clustering algorithms for large wireless sensor networks: a review. Sens. Netw **19** (2019)
17. Yuan, S., Tan, P.N., Cheruvelil, K.S., Collins, S.M., Soranno, P.A.: Spatially constrained spectral clustering algorithms for region delineation (2019)

Design and Implementation of an Automatic Summarizer Using Extractive and Abstractive Methods

Adhiraj Chattopadhyay and Monalisa Dey

Abstract Communication via a natural language requires two fundamental skills: producing 'text' (written or spoken) and understanding it. Here, the two terms, natural language processing (NLP) and natural language generation (NLG) become important. NLU, also known as natural language understanding, is where the system understands to disambiguate the input sentences (in some human or natural language) to produce the machine representation language. NLG, on the other hand, is the technique of generating natural language from a machine representation system (a database/logical form). An example of a simple NLG system is the Pollen Forecast for Scotland system (Turner et al., in generating spatio-temporal descriptions in pollen forecasts 2006, [1]) that could essentially be a template. NLG system takes as input six numbers, which predicts the pollen levels in different parts of Scotland. From these numbers, a short textual summary of pollen levels is generated by the system as its output. In this paper, our aim is to test and/or improve the preexisting algorithms for our NLG system and if required, to develop our own algorithms for the said system. The paper starts, first defining the stages and components of the NLG task and their distinctive roles in accounting for the coherence and appropriateness of natural texts. It then sets out the principal methods that have been developed in the field for building working computational systems. Thereafter, the attempts to define a new method for the application being developed are shown. Finally, the problem faced in developing an NLG system and potential applications are discussed.

Keywords Natural language processing (NLP) · Natural language generation (NLG) · Natural language understanding (NLU) · Algorithms

A. Chattopadhyay (✉) · M. Dey
Institute of Engineering and Management, Saltlake Sector-V, Kolkata, India
e-mail: admin@iemcal.com

© Springer Nature Singapore Pte Ltd. 2020
J. K. Mandal and S. Mukhopadhyay (eds.), *Proceedings of the Global AI Congress 2019*, Advances in Intelligent Systems and Computing 1112,
https://doi.org/10.1007/978-981-15-2188-1_23

289

1   Introduction

Our aim is to summarize, extractively and abstractively a collection of over 200 research papers [2]. NLP is a machine learning technique to analyse documents with human language and make them machine readable. Summarization is a process of presenting a particular content (stories, articles) in a shorter form while preserving all of the points conveyed in the original content. The task of summarization comes very easily to us. However, the amount of data that we are having to deal with keeps increasing with time. Most of the data is unstructured, i.e. the data does not have a defined format (csv file or it does not follow a defined structure). To use human labour to format all unstructured data and then perform some operation on them would be a terrible waste of human resources. Hence came the idea to automate these tasks. Making computers perform tasks like summarization is difficult. However, different algorithms have been developed to accomplish this task in different ways [3]. Presently, there exist two ways of summarization: (1) extractive summarization and (2) abstractive summarization. In the upcoming sections, we have dealt with both these methods. Extractive summarization ranks the vectorized words/sentences based on word count. Extracting the top-ranked sentences and then stacking them together to create a summary, abstractive summarization uses neural networks to generate new words from its own vocabulary on being fed some input. Most summarizers are designed as extractive ones as they are easier to train and build. However, research and work have improved various models for abstractive summarization. In the upcoming section, terminologies related to machine learning like supervised and unsupervised learning and those algorithms specific to NLP would be discussed. This would be followed by the section which discusses, in some detail, how these terminologies are relevant for our two summarization techniques.

2   Background

Machine learning and deep learning techniques are widely used in natural language processing. For example, TextRank algorithm used to rank the vectorized sentences in a document is an example of unsupervised learning. For abstractive summarization, SeqtoSeq models are used. To implement them, special types of neural networks (RNN—recursive neural networks) are used.

2.1   Machine Learning

Machine learning aims to train machines to perform human tasks through a lot of data. This happens in two phases. The data set given is split into training data and test data. The machine learning algorithm is thereafter trained on the training data to obtain a fairly high level of accuracy. After this, the proficiency of the trained algorithm is tested on the test data.

Machine learning can be classified into three categories:

Supervised Learning Here, the machine learning algorithm is trained on a data set with its input and output clearly labelled. The learning process is to find patterns/regularities among the data. Supervised learning can be further categorized into classification and regression problems. Supervised learning algorithms include logistic regression, neural networks (NNs), support vector machines (SVMs), decision trees (DTs) and k-nearest neighbours (k-NN).

Unsupervised Learning Here, the machine learning algorithm is trained on a data set where the data is not properly labelled. For example, considering the problem of arranging webpages in the Internet and finding out if one webpage could be accessed from another, the machine learning algorithm is given a bunch of webpages. It finds out if other webpages are referenced in a webpage and how many times. Webpages can be accessed from another webpage if they are referenced in that webpage. This is the PageRank algorithm. A derivative of this algorithm, the TextRank algorithm, is used in the classification of NLP to rank sentences in the documents. Thus, the task of unsupervised learning algorithms is to find (hidden) patterns among the data. An example of unsupervised learning is clustering. Representative unsupervised learning algorithms are k-means, auto-encoders and generative adversarial networks (GAN).

Reinforcement Learning RL aims at choosing the most suitable action at a specific state in an environment to maximize the cumulative reward. Classic RL algorithms are Q-Learning, Monte Carlo RL, SARSA, etc.

Deep Learning These are also autonomous learning algorithms which give better performance than classical machine learning algorithms. Deep learning is a subset of machine learning. Neural neworks are deep supervised learning algorithms. Neural networks can be categorized as artificial neural networks (ANN), recurrent neural networks (RNN) and convolutional neural networks (CNN). For abstractive summarization, we have worked with a special kind of RNN called long short-term memory (LSTM).

2.2 Computational Linguistics

There are three categories to computational linguistics: natural language processing (NLP), natural language understanding (NLU) and natural language generation (NLG). NLP is the preliminary task, processing the data so that it can be utilized with algorithms. NLU deals with understanding the context of each sentence and word to make meaning of the data provided. For example, depending on the sentences that precede and supersede it, 'A Clammer' might mean a sound or an argument. Understanding the context of a text is important in problems of classification and summarization as it would help the program construct meaningful sentences. NLG is the generation of natural language by the program. This aspect is what will be dealt with in the paper.

We will now move to understand automatic summarization and its different categories, namely extractive summarization and abstractive summarization.

3 Automatic Summarization

Extractive summarization and abstractive summarization (more complex) are two types of automatic summarizations.

3.1 Extractive Summarization

See Fig. 1.

Data Preprocessing After our program imports and reads the data, it has to be cleaned. There are a number of aspects to data cleaning. They are discussed one by one.

Normalization It is the process of making our text data more uniform. It includes the removal of stop words, punctuation symbols and specific patterns (formatted dates, e-mail ids) from the text. It also includes lower-casing all of the text. Some examples of stop words are a, an and the. They are most numerous in any text. Many algorithms select out sentences for making up the summary based on word count. Hence, it is important to remove them during data preprocessing as otherwise they would most likely make up most of the summary. Punctuation marks like ! , . ? are removed at this stage or they are treated as words and ranked. Before moving forward, all the words are converted to lower case. To remove different patterns present, regex is used to remove patterns deemed unnecessary.

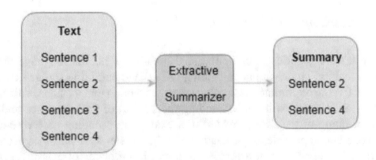

Fig. 1 Extractive method

Form	Morphological information	Lemma
studies	Third person, singular number, present tense of the verb study	study
studying	Gerund of the verb study	study
niñas	Feminine gender, plural number of the noun niño	niño
niñez	Singular number of the noun niñez	niñez

Fig. 2 Example of lemmatization

Lemmatization Words (long words) are converted to their root words. Running becomes Run. Lemmatization takes into consideration the morphological analysis of the words. To do so, it is necessary to have detailed dictionaries which the algorithm can look through to link the form back to its lemma (Fig. 2).

Tokenization In the last step of data cleaning, the data is now cut up into pieces formatted in a way to be vectorized. According to one's needs, the document may be tokenized into individual sentences or words.

Vectorization Vectorization is the method of transforming human readable text data to machine readable vectors (numeric data).

Bag of words Bag of words is a vectorization technique where a fixed length vector is defined with each entry corresponding to a word in our predefined dictionary of words. The size of the vector equals the size of the dictionary. Then, for representing a text using this vector, we count how many times each word of our dictionary appears in the text, and we put this number in the corresponding vector entry. The problem with this method is that it does not capture the semantics of the text.

Vectorization with deep learning The efficiency of different models has increased significantly after deep learning techniques were used. The reason for this is that they use neural networks which have a greater capacity to learn. Word embeddings are models that learn to map a set of words in a vocabulary to vectors of numeric data. Pre-trained word embedding models are used like GloVe [4] and Word-to-vec [5] for vectorization. They preserve the semantics of our text data better than bag of words models.

Similarity Matrix A matrix is created (word vs. word) indicating how close a word is to another. This is normally done by calculating the distances between them. The distance between them can be calculated using Euclidean distance;

$$\text{dist}(d) = |a - b|^2 = (a1 - b1)^2 + ... + (an - bn)^2 \qquad (1)$$

(a, b are the positions of the individual words. n denotes the number of words) or it could also be calculated using cosine distance where basically the cosine of the angle between the word vectors is calculated;

$$\text{dist}(d) = \cos x = 1 - (a.b)/(|a|.|b|) \tag{2}$$

Any of the above methods can be used as per our convenience. At different problems, different methods yield better results.

Ranking vectorized sentences After vectorization and creation of the similarity matrix, we must rank them.

Summarization Finally, the top-ranking sentences are selected out from our matrix, and they are used to create our summary. However, a problem with this method is that sometimes the summaries generated might not be very coherent (meaningful). So, for this reason, even if extractive summarization works most of the time, a better approach towards the problem is abstractive summarization using deep learning techniques.

3.2 Abstractive Summarization

In this approach, new sentences are generated from the original text. This is in contrast to the extractive approach seen earlier where we used only the sentences that were present. The sentences generated through abstractive summarization might not be present in the original text! (Fig. 3).

To improve soundness and readability of automatic summaries, it is necessary to improve the topic coverage by giving more attention to the semantics of the words and to experiment with re-phrasing of the input sentences in a human-like fashion. Recently, NLP has seen a rise of deep learning-based models that map an input sequence into another output sequence, called sequence-to-sequence models [6] that have been successful in such problems as machine translation [7], speech recognition [8] and video captioning [9]. However, despite the similarities, abstractive summarization is vastly different from machine translation: in summarization, the target text is typically very short and does not depend very much on the length of

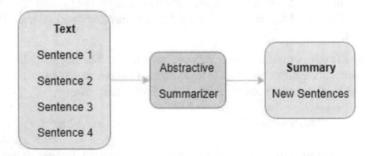

Fig. 3 Abstractive method

the source. More importantly, in machine translation, there is a strong one-to-one word level alignment between source and target, but in summarization, it is less straightforward.

Our approach to abstractive summary generation has also been with the use of sequence-to-sequence models. In general, to map an input sequence to an output sequence, recurrent neural networks (RNA) are used. However, when the task is to learn a very long sequence of words/sentences, RNAs cannot remember too far back in this sequence. This is known as the 'vanishing gradient' problem. To overcome this problem, long short-term memory (LSTM) networks are used. These have an 'attention layer'. With the help of this attention layer, the LSTM can decide which data it wants to remember and which to ignore and forget. This increases its efficiency in processing sequences. In case of summary generation, the output gives a shorter text than the input, i.e. the sequence mapping would be 'many-to-one'. When there is a large text data, processing all of it as a sequence using one LSTM network would be inefficient. So, a stack of LSTMs is used at input (encoder) and output (decoder). For this particular problem, the decoder has lesser number of LSTMs than does the encoder.

Figure 4 shows the architecture of an encoder–decoder set-up. 'S', in between, represents the hidden layer. The Softmax layer is an activation function that outputs decimals between (0–1). This ensures that an enormously large number is not outputted. The encoder is a stack of several RNA/LSTMs, each of which accepts a single element of the input sequence, processes it and propagates it forward to the hidden layer. The decoder is a stack of several RNA/LSTMs where each predicts an output at each timestamp. It passes on the output to the Softmax layer.

The above encoder–decoder architecture with attention has been attempted by us for abstractive summary generation.

Some other techniques are using transformer networks with attention [10], using pre-trained transformer models like Google's bidirectional encoders representations from transformers (BERT) [11], GPT [12], Microsoft's masked seq to seq pre-training for language generation (MASS) [13] and XLNet (an improvement on BERT) [14].

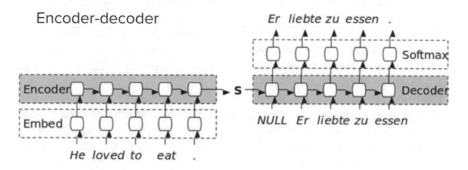

Fig. 4 Encoder–decoder architecture

4 Our Work

So far we have explained various concepts related to machine learning, deep learning and to our task of summarization in the domain of NLP. In this section, we have explained how we have gone forward with extractive and abstractive summarization, our results, problems faced and future scope of this work.

In both our works with extractive and abstractive summarizations, we have used GloVe word embeddings instead of creating word vectors with bag of words methods. This is because pre-trained word vectors give better results. For abstractive summary generation, which is still under development, we attempted to use LSTM encoder–decoder model with attention, of which we talked about, in the previous section, to generate the summaries.

4.1 Data Set

For our data set, we collected about 200 research papers from the arXiv website. The topics of these papers were mostly on math, computer science and the general sciences. Then to have the information and the content of all the papers in one place, we convert them to a csv file with the headers (paper id, content). This csv file was used as the structured data to generate our summaries. This is an example of multi-document summary generation [15].

All our code is written using the Python3 language. For our NLP tasks, we have tried out using both NLTK and spaCy libraries in Python, separately, for our task. spaCy is more specialized than NLTK, so we attempted to compare and contrast the results. Differences between NLTK and spaCy

- NLTK provides a plethora of algorithms to choose from a particular problem which is great for researches. spaCy keeps the best algorithm for a problem in its toolkit and keeps updating it. Hence, it is very development specific.
- NLTK supports various languages; whereas, spaCy has statistical models for seven languages (English, German, Spanish, French, Portuguese, Italian and Dutch).
- NLTK is a string processing library. It takes strings as input and returns strings or lists of strings as output; whereas, spaCy uses object-oriented approach. When we parse a text, spaCy returns document object whose words and sentences are objects themselves.
- spaCy has support for word vectors; whereas, NLTK does not.
- As spaCy uses the latest and best algorithms, its performance is usually better than NLTK. As we can see below, in word tokenization and POS-tagging, spaCy performs better, but in sentence tokenization, NLTK outperforms spaCy. Its poor performance in sentence tokenization is a result of differing approaches: NLTK attempts to split the text into sentences. In contrast, spaCy constructs a syntactic

tree for each sentence, a more robust method that yields much more information about the text.

Towards the end of this section, a comparison chart has been provided for NLTK and spaCy.

4.2 Extractive Summarization

The general algorithm that we follow to generate extractive summaries (Fig. 5) is as follows:

- The necessary libraries are imported to be able to parse csv files, perform numeric operations, to perform the necessary NLP tasks and to plot graphs for performance analysis and better visualization.
- The contents of the different research papers are accessed by our program as it iterates the csv data file row by row.
- Sentence vectorization of each of the texts is performed.
- The vector representation (word embeddings) for each and every sentence is found.
- Similarities between sentence vectors are then calculated and stored in a matrix.
- The similarity matrix is then converted into a graph, with sentences as vertices and similarity scores as edges, for sentence rank calculation.
- Finally, a certain number of top-ranked sentences form the final summary.

Results and Comparison
In general, we were able to generate summaries using both NLTK and spaCy. After generating the summaries, we formatted them for each paper in JSON format with the headers listed below:

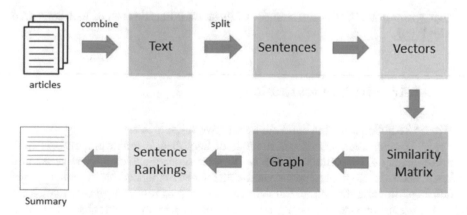

Fig. 5 Process for extractive summarization

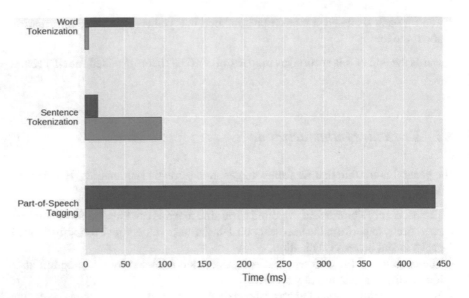

Fig. 6 Comparison of NLTK and spaCy (blue = NLTK, green = spaCy)

- Paper id
- Paper name
- Paper content
- Extractive summary

Some problems faced were that, though the summaries were created, some of them did not make much sense. This appears to be the downside of the methods of extractive summary generation.

SpaCy versus NLTK The performance of these NLP libraries can be best compared with Fig. 6.

5   Abstractive Summarization

We use SeqtoSeq model to build our abstractive summarizer.
Under Development: Our initial attempts at SeqtoSeq abstractive summarization, unfortunately, did not quite work. Hence, the work on abstractive summary generation is still under development. For that reason, in the rest of this section, the concepts and details of the steps that we are following to get our final results have been explained.

We use the encoder–decoder model with attention to obtain our abstractive summaries. The encoder and decoder are made up of sequence of LSTM networks. Figure 7 is a typical SeqtoSeq model architecture.

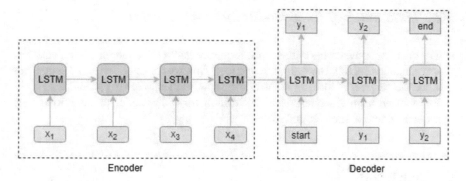

Fig. 7 Encoder–decoder model

The general algorithm that was followed for abstractive summary generation is as follows:

- The necessary libraries were loaded just like in extractive summary generation.
- The data was then loaded. Program iterates through the csv file data and performs the following function.
- Data preprocessing;

  • Everything was converted to lowercase.
  • HTML tags were removed.
  • Lemmatization.
  • Removal of ('s).
  • Any text inside the parenthesis ( ) was removed.
  • Punctuations and special characters were eliminated.
  • Stop word and short words were removed.

- The text of each research paper was vectorized with the help of word embeddings.
- The vectorized data was divided into training and test data.
- The model for the abstractive summary generation was defined.

  • The LSTM feed-forward encoder was defined.
  • Back-propagation for the encoder was defined.
  • The attention mechanism was defined.
  • The LSTM decoder with attention was defined.
  • Parameters of the neural network (cross-entropy, cost-function) were defined and initialized.
  • Accuracy parameters were defined and initialized.
  • Optimizer was defined.

- Thereafter, this model was trained, and its results were evaluated.

Following this algorithm, however, our program was confronted with a number of errors. Thus, we could not complete abstractive summary generation. As it is still under development, it falls under the future scope of this entire project.

5.1 Limitations of Encoder–Decoder Architecture

– The encoder converts the entire input sequence into a fixed length vector, and then the decoder predicts the output sequence. This works only for short sequences since the decoder is looking at the entire input sequence for the prediction.
– Problem with long sequences; it is difficult for the encoder to memorize long sequences into a fixed length vector.

6 Conclusion

In this paper, we presented our attempt to create a corpus from the data set for summarization.

We then presented our efforts at extractive and abstractive summarizations.

We were successful at creating extractive summaries. Limitations to these summaries were coherence, i.e. a lot of the summaries did not make concrete sense. Our initial attempts at abstractive summarization did not work; hence, the work is currently under development.

7 Future Scope

Our encoder–decoder with attention abstract summary generator is still under active development. We hope to successfully program it within some time.

A different approach could be taken for this problem by mixing the techniques of both abstractive and extractive summarization. This method can be accomplished by pointer-generated networks [16]. These networks combine the extractive and abstractive model by switching probability. This is a hybrid network that can choose to copy words from the source (like extractive summarization) via pointing while retaining the ability to generate words from the fixed vocabulary (abstractive summary generation). This method could yield better results, subject to implementations.

Abstract summarization, these days, is being used more and more. Many new techniques are being developed for it. Google came up with BERT (bidirectional encoder representation from transformers) then there is Transfer learning where while training a network, if there exists a similarly trained network, then in place of training this network from scratch, the training of that network is transferred to this network.

References

1. Turner, R., Sripada, S., Reiter, E., Davy, I.P.: Generating spatio-temporal descriptions in pollen forecasts. Published in EACL(2006)
2. Nikolov, N.I,, Pfeiffer, M., Hahnloser, R.H.R.: Data-driven summarization of scientific articles (2018). In Arxiv preprint arXiv:1804.08875v1 [cs.CL]
3. Gatt, A., Krahmer, E.: Survey of the state of the art in natural language generation: core tasks, applications and evaluation (2018). In Arxiv preprint arxiv:1703.09902v4 [cs.CL], published in J. AI Res. **60** (2017)
4. Pennington, J., Socher, R., Manning, C.D.: GloVe: Global vectors for word representation. In: Proceedings of the 2014 Conference on Empirical Methods in Natural Language Processing (EMNLP), pp. 1532–1543, October 25–29, 2014, Doha, Qatar. Association for Computational Linguistics (2014)
5. Mikolov, T., Chen, K., Corrado, G., Dean, J.: Efficient estimation of word representations in vector space (2013). In Arxiv preprint arXiv:1301.3781v3 [cs.CL]
6. Shi, T., Keneshloo, Y., Ramakrishnan, N., Reddy, C.K.: Neural abstractive text summarization with sequence-to-sequence models (2018). In Arxiv preprint arXiv:1812.02303v2 [cs.CL]
7. Bahdanau, D., Cho, K.H., Bengio, Y.: Neural Machine Translation by jointly learning to align and translate. In Arxiv preprint arXiv:1409.0473v7 [cs.CL] (2016)
8. Bahdanau, D., Chorowski, J., Serdyuk, D., Brakel, P., Bengio, Y.: End-to-end attention based large vocabulary speech recognition. In: 2016 IEEE International Conference on Acoustics, Speech and Signal Processing (ICASSP) (2016)
9. Venugopalan, S., Rohrbach, M., Donahue, J., Mooney, R., Darrell, T., Saenko, K.: Sequence to sequence–video to text. In: The IEEE International Conference on Computer Vision (ICCV) (2015)
10. Vaswani, A., Shazeer, N., Parmar, N., Uszkoreit, J., Jones, L., Gomez, A.N., Kaiser, Ł., Polosukhin, I.: Attention is all you need. In: 31st Conference on Neural Information Processing Systems (NIPS 2017) (2017)
11. Devlin, J., Chang, M.-W., Lee, K., Toutanova, K.: BERT: pre-training of deep bidirectional transformers for language understanding. In: Proceedings of NAACL-HLT 2019, pp. 4171–4186. Minneapolis, Minnesota, June 2–7, 2019. Association for Computational Linguistics (2019)
12. Liao, Y., Wang, Y., Liu, Q., Jiang, X.: GPT-based generation for classical Chinese poetry (2019). In Arxiv preprint arXiv:1907.00151v4 [cs.CL]
13. Song, K., Tan, X., Qin, T., Lu, J., Liu, T.-Y.: MASS: masked sequence to sequence pre-training for language generation (2019). In Arxiv preprint arXiv:1905.02450v5 [cs.CL]
14. Yang, Z., Dai, Z., Yang, Y., Carbonell, J., Salakhutdinov, R., Le, Q.V.: XLNet: generalized autoregressive pretraining for language understanding. Arxiv preprint (2019). https://arxiv.org/abs/1906.08237
15. Liu, P.J., Saleh, M., Pot, E., Goodrich, B., Sepassi, R., Kaiser, Ł., Shazeer, N.: Geneating wikipedia by summarizing long sequences. In: ICLR 2018 Conference (2018)
16. See, A., Liu, P.J., Manning, C.D.: Get to the point: summarization with pointer-generator networks. In: Proceedings of the 55th Annual Meeting of the Association for Computational Linguistics (Volume 1: Long Papers) (2017)

Context-Aware Conversational Agent for a Closed Domain Task

Srimoyee Bhattacharyya, Soumi Ray and Monalisa Dey

Abstract With the ever-growing complexity in automated business processes and systems, manual navigation through the large volumes of data and hyperlinks on applications has become impractical and inefficient. Thus, conversational agents for closed domain tasks have received considerable interest in recent years. One of the fundamental characteristics of a conversational agent is its ability to interpret a statement based on the context of the entire conversation or in the context of a specific task. Keeping this view as the central focus, we study the scope of the Rasa chatbot development framework for identifying user intent and predicting the agent's future actions. We have presented a processing pipeline that allows optimized and accelerated data generation, action path definition and evaluation of the different components to demonstrate the generalization ability of the chatbot as a whole. The proposed application demonstrates fairly good performance in identifying user intent and completing the user's intended action.

Keywords Conversational agents · Chatbots · Deep learning for dialogue agent

1 Introduction

Many recent automated technical or business processes or systems have attained a level of complexity that makes the navigation of said systems through regular user interfaces highly complicated as well as time consuming. Especially in the case of

S. Bhattacharyya
University of Massachusetts, Amherst, US
e-mail: srimoyeebhattacharyya@gmail.com

S. Ray (✉)
Institute of Engineering and Management, Kolkata, India
e-mail: raysoumi1001@gmail.com

M. Dey
Jadavpur University, Kolkata, India
e-mail: monalisa.dey.21@gmail.com

© Springer Nature Singapore Pte Ltd. 2020
J. K. Mandal and S. Mukhopadhyay (eds.), *Proceedings of the Global AI Congress 2019*, Advances in Intelligent Systems and Computing 1112, https://doi.org/10.1007/978-981-15-2188-1_24

banking systems, the multiple offers and services available make the situation inefficient for the user with respect to effort and time required to find the appropriate information. Conversational agents are fast becoming a basic addition to such systems to direct the user quickly and effectively to the right resources. Additionally, when performing banking tasks such as processing a transaction or searching for loan information, the agent needs to maintain the context of the conversation across multiple discourse stages. The overall motivation of this project is to evaluate the efficiency of generating training and testing data sets and using the Rasa framework to define appropriate processing pipelines in order to develop an end-to-end dialogue system to assist in or perform a closed domain task like banking.

2   Related Work

The conversational agent implemented here is developed using the Rasa framework [1] which is composed of Rasa NLU for text interpretation and Rasa Core for dialogue management. Braun et al. [2] show that Rasa NLU's performance compares favourably to various closed-source solutions. Custom entities for the banking domain in our implementation are recognized using a conditional random field [3]. The approach to dialogue management in Rasa Core is most similar to the Hybrid Code Network (HCN) approach [4], as opposed to the data-intensive end-to-end learning of recurrent neural networks (RNNs), as in [5, 6], where the system uses dialogue transcripts to jointly learn natural language understanding, dialogue state tracking, dialogue act management and response generation. The language understanding module in Rasa is completely decoupled from its dialogue management module, thereby enabling the reuse of a trained dialogue management model with multiple language models.

In the current implementation, response generation involves handcrafted templates for each response type for limited system actions in closed domain tasks. This is currently easier and more reliable than language generation using (for example) a neural network to generate grammatically coherent and semantically correct responses as in [8]. The Rasa framework supports training the dialogue policy interactively, by providing real-time feedback and corrections for the predictions of both the NLU and the dialogue management system. This is similar to the machine teaching approach [7] used by Williams, where end-to-end dialogue systems are interactively trained. However, in the work, the policy is directly exposed to the surface form of the user utterances, whereas in Rasa Core the dialogue policy only receives the recognized intent and entities.

3   Approach and Methodology

3.1   *Introduction to Rasa Components*

3.1.1   Rasa NLU:

    3.1.1.1   ner_crf component is used for named entity recognition for custom entities.

    3.1.1.2   spacy component used for

        3.1.1.2.1   user input tokenization.

        3.1.1.2.2   intent featurization.

    3.1.1.3   sklearn component is used for intent classification of user input.

3.1.2   Rasa Core:

    3.1.2.1   Dialogue state tracker stores and maintains the state of the dialogue with a single user.

    3.1.2.2   Agent allows you to train a model, load and use it.

    3.1.2.3   Policies use the current state of the conversation (provided by the tracker) to choose the next action to take.

3.2   *Overview of Approach*

3.2.1   NLU data file is built with examples using labelled intents and entities in the JSON format using Chatito framework. Dialogue data set, i.e. stories, file is built which contains training data samples for dialogue system.

3.2.2   Domain file is built which contains the following: slots, entities, intents, templates, actions.

3.2.3   Actions are defined to perform tasks like login, loaninfo, etc.

3.2.4   Using Rasa NLU:

    3.2.4.1   Intent of request submitted by the user is classified.

    3.2.4.2   Named entity recognition is performed.

    3.2.4.3   Interpreter is trained to predict intents and entities.

3.2.5   Using Rasa Core:

    3.2.5.1   Appropriate actions are performed based on classified intents and extracted entities.

    3.2.5.2   Agent is trained to predict actions.

3.3   Data Set Structure and Description

NLU Data Set. The following is an example of an annotated user input text containing intent and entity:

```
{"text": "log me in with username as1202 and pwd 2222",
 "intent": "login",
 "entities": [
   {
     "end":  30,
     "entity": "user_id",
     "start": 24,
     "value": "as1202"
   },
   {
     "end": 43,
     "entity": "pin",
     "start": 39,
     "value": "2222"
   }
 ]
}
```

Dialogue Data Set. The following is an example of an annotated sequence of user intents and agent actions that make up a conversation (called a 'story' in Rasa).

```
## story user_greet
* greet
  - utter_greet
> check_user_option

## story loan_info
> check_user_option
* loaninfo{"loan_type": "education"}
  - action_loaninfo
* loaninfo{"loan_type": "education", "search_key": "eligible"}
  - action_loaninfo
  check_user_happy

## story close_conversation
> check_user_happy
* thank
  - utter_happytohelp
* bye
  - utter_goodbye
```

3.4 Data Set Generation

For building chatbots using commercial models, open-source frameworks or writing an own natural language processing model, training and testing examples are required. We used Chatito which assists in generating data sets for training and validating chatbot models using a simple DSL.

Chatito is a domain-specific language designed to simplify the process of creating, extending and maintaining data sets for training NLP models for text classification, named entity recognition, slot filling or equivalent tasks.

```
The syntax of the language used is as follows:
%[some intent]('training': '1')
    @[some slot]
@[some slot]
    ~[some slot synonyms]
~[some slot synonyms]
    synonym 1
    synonym 2
```

In this example, the generated data set will contain the entity_synonyms of 'synonym 1' and 'synonym 2' mapping to 'some slot synonyms'.

Following the above principles and syntax, an example of the language and its generated output is as follows:

```
%[greet]('training': '4', 'testing': '4')
    ~[hi] @[name?] ~[whatsUp?]
~[hi]
    hi
    hey
@[name]
    Janis
    Bob
~[whatsUp]
    whats up
    how is it going
```

The above DSL code generates four training examples and four testing examples for the greet intent which are as follows.

Training data set:

```
{"rasa_nlu_data": {
  "regex_features": [],
  "entity_synonyms": [],
  "common_examples": [
    {"text": "hi Bob",
      "intent": "greet",
      "entities": [
        {"end": 6,
          "entity": "name",
          "start": 3,
          "value": "Bob"
        }
      ]
    },
    {"text": "hey",
      "intent": "greet",
      "entities": []
    },
    {"text": "hi Janis how is it going",
      "intent": "greet",
      "entities": [
        {"end": 8,
          "entity": "name",
          "start": 3,
          "value": "Janis"
        }
      ]
    },
    {"text": "hi whats up",
      "intent": "greet",
      "entities": []
    }
  ]
  }
}
```

```
{
  "rasa_nlu_data": {
    "common_examples": [
      {
        "text": "hey Bob whats up",
        "intent": "greet",
        "entities": [
          {
            "end": 7,
            "entity": "name",
            "start": 4,
            "value": "Bob"
          }
        ]
      },
      {
        "text": "hey Janis",
        "intent": "greet",
        "entities": [
          {
            "end": 9,
            "entity": "name",
            "start": 4,
            "value": "Janis"
          }
        ]
      },
      {"text": "hey Bob",
        "intent": "greet",
        "entities": [
          {
            "cnd": 7,
            "entity": "name",
            "start": 4,
            "value": "Bob"
          }
        ]
      },
      {"text": "hi how is it going",
        "intent": "greet",
        "entities": []
      }
    ]
  }
}
```

3.5 Domain File Description

The domain file defines the 'universe' our bot is supposed to function in. The domain file contains a list of items of each constituent component as described in Table 1.

Table 1 Components of the domain file

Components	Description
Intents	Type of things the user is expected to say For example: login, greet, etc.
Entities	Pieces of information that need to be extracted from requests by user. For example: user_id, loan_type, etc.
Actions	Things that the bot can say or do For example: ActionLogin, etc.
Slots	Locations where the entities are stored or information to keep track of during a conversation For example: user_id, acc_num, etc.
Templates	Template strings for the things that the bot can say For example: utter_greet, utter_goodbye, etc.

Table 2 Components of the action file

Components	Description
ActionLogin	Allows the user to login with the correct credentials
ActionLogout	Allows the user to logout by deleting session information
ActionLoanInfo	Allows the user to get the desired information (loan eligibility criteria, interest rates, etc.) about a particular loan type
ActionShowBalance	Allows the user to check his/her balance amount in the account

3.6   Action File Description

At each iteration, Rasa Core predicts which action to take from a predefined list. An action can be a simple utterance, i.e. sending a message to the user, or it can be an arbitrary function to execute. The actions.py file contains the description of all the actions used in the code. Sample actions in the banking domain are as follows (Table 2).

3.7   Policy File Description

The job of a policy is to select the next action to be executed with the help of the tracker object. The policy file contains the following ensemble of policies that were used in the making of the bot. The final action was chosen by the ensemble system by pooling the individual predictions of each policy and then selecting the prediction that had the highest confidence score (Table 3).

Table 3 Components of the policy file

Components	Description
KerasPolicy	Trains a deep learning model on the featurized dialogue data set in order to predict the next action
MemoizationPolicy	Exactly memorizes the dialogue data set in order to make predictions with confidence of 1 if the user input exactly matches a training sequence
FormPolicy	Predicts the next action based on the fields in a predefined form that needs to be filled before the corresponding form action can be triggered
FallBackPolicy	Returns a default action if the intent classification or next action prediction modules return results with low confidence

Table 4 Architecture of the dialogue prediction system

Layer type	Output shape	Number of parameters
Masking	(5, 45)	0
LSTM	(32)	9984
Dense	(24)	792
Activation	(24)	0

3.8 Dialogue Prediction System Architecture

The KerasPolicy defines a deep learning LSTM model which is trained on the featurized representations of the dialogue data set, i.e. Rasa stories. Hence, the model uses both the previous user intents and the previous dialogue system actions to predict the next system action. The architecture of the model is as follows (Table 4).

3.9 Processing Pipeline

Incoming messages are processed by a sequence of components. These are executed one after another in a so-called processing pipeline. There are components for entity extraction, intent classification, pre-processing and others. Each component processes the input and creates an output, which can be used by any component that comes after this component in the pipeline (Table 5).

When a user enters a message, it goes through the components of the pipeline used in our project in the following manner and produces a corresponding output (Fig. 1).

3.10 Visualization of Dialogue States

See (Fig. 2.

Table 5 Components of the processing pipeline

Component	Output	Description
Spacy NLP	Nothing	Initializes spacy structures for pipelines that use spacy components
Spacy tokenizer	Nothing	Generates tokens using the spacy tokenizer
Spacy featurizer	Used as input to intent classifiers	Creates word vector representation as features for intent classification
sklearn intent classifier	Intent and intent rankings	Trains a linear SVM, optimized using a grid search
CRF entity extractor	Entities	Conditional random fields for named entity recognition

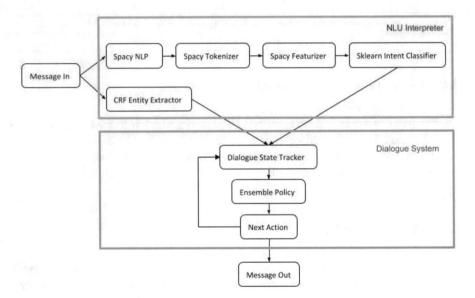

Fig. 1 User input processing pipeline

4 Results

4.1 Experimental Set-Up

The performance of the conversational agent is measured through both the cross-validation evaluation of the training data set for intent and entity recognition and the analysis of the classification report on the test data set for the following steps: intent

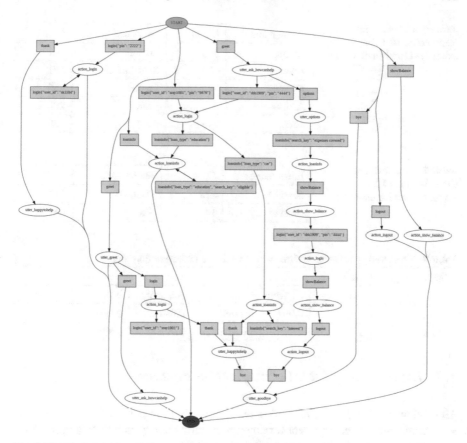

Fig. 2 Graph showing multiple conversation paths between user and agent

identification (SVM linear classifier), entity extraction (conditional random fields) and next dialogue action prediction (custom Keras model).

Data set statistics. The training and test data sets contain 1425 and 1270 surface forms of user utterances, respectively, which have been annotated with intents and optional entities.

The test data set annotated with intents and entities has the following statistics (Tables 6 and 7).

4.2   Cross-Validation Evaluation (Number of Folds = 10)

See Table 8.

Table 6 Counts of intent examples: total 1270 examples (10 distinct intents)

Intent name	Count	Intent name	Count
Login	300	Logout	61
Loan info	200	Show balance	51
Options	100	Thank	50
Affirm	200	Deny	200
Greet	48	Bye	60

Table 7 Counts of entity examples: total 1021 examples (4 distinct entities)

Intent name	Count	Intent name	Count
Search key	277	User_ID	297
Loan type	148	Pin	299

Table 8 Cross-validation results of the natural language understanding stage

Pipeline stages	Train			Test		
	Accuracy	F1 score	Precision	Accuracy	F1 score	Precision
Intent	0.997	0.997	0.997	0.980	0.980	0.982
Entities	0.998	0.998	0.998	0.995	0.995	0.995

4.3 Evaluation of Intent Classification Pipeline

The following classification report demonstrates that the performance of the conversational agent is near perfect for domain-specific intents such as 'loaninfo' (user asks for information about loans available) and 'showbalance' (user wants to view account balance) but shows lower F1 scores for generic intents such as 'greet' and 'bye' which have broader scope (Table 9; Figs. 3, and 4).

4.4 Evaluation of Entity Extraction: Conditional Random Fields

Entity evaluation departs from BILOU-based scoring to follow a more lenient tag-based approach that rewards partial extraction and does not punish the splitting of entities (Table 10).

Table 9 Intent classification report: (accuracy: 0.962992125984252)

Intents	Precision	Recall	F1 score	Support
Affirm	0.94	0.99	0.97	200
Bye	1.00	0.57	0.72	60
Deny	1.00	1.00	1.00	200
Greet	0.63	1.00	0.77	48
Loaninfo	1.00	1.00	1.00	200
Login	0.99	1.00	1.00	300
Logout	0.98	0.95	0.97	61
Options	1.00	1.00	1.00	100
Showbalance	1.00	0.90	0.95	51
Thank	0.95	0.76	0.84	50
Avg./total	0.97	0.96	0.96	1270

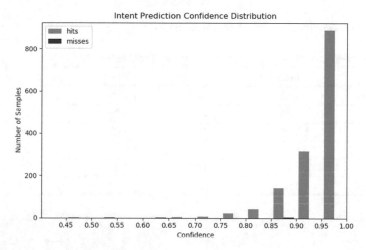

Fig. 3 Intent prediction confidence distribution

4.5 Evaluation of the Dialogue Prediction Model

The dialogue model takes as input the intents and entities extracted by the NLU segment of the processing pipeline, and proceeds to predict the next action based on these inputs and the memory of previous actions. The training is accomplished by defining sequences of steps that the conversational agent should undertake to achieve a successful user interaction. These are known as 'stories'. The evaluation script takes a set of stories and attempts to predict the sequence of steps (called 'actions') based on the previous steps or actions that it encounters. Thereafter, the confusion matrix is composed based on total counts of actions present in the test stories (Fig. 5).

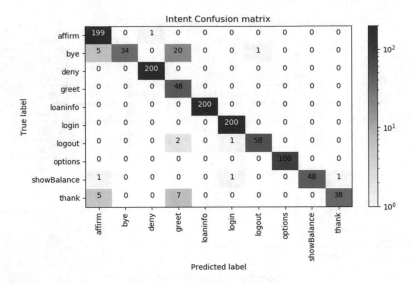

Fig. 4 Intent confusion matrix without normalization

Table 10 Entity classification report (accuracy: 0.9987156434626252)

Entity	Precision	Recall	F1 score	Support
loan_type	1.00	1.00	1.00	148
no_entity	1.00	1.00	1.00	6765
Pin	1.00	1.00	1.00	299
search_key	0.99	0.97	0.98	277
user_id	1.00	1.00	1.00	297
Avg./total	1.00	1.00	1.00	7786

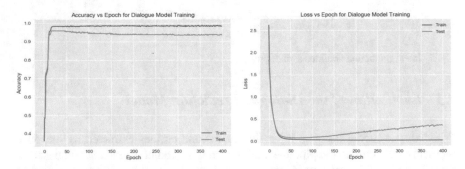

Fig. 5 Accuracy (left) and loss (right) versus epoch for dialogue model training

5 Discussions

An effective conversational agent for the banking domain has been successfully built. What makes this chatbot stand out from other existing chatbots is its ability to be aware of the context of the current conversation and respond accordingly. The performance of the agent built is experimentally found to be good and comparable with existing agents. However, methods to increase its robustness have been discussed in future scope (Fig. 6).

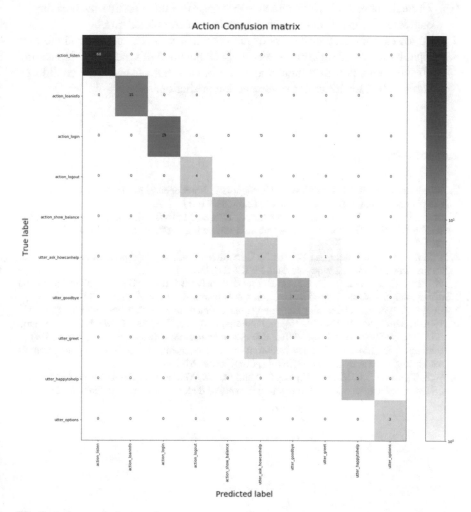

Fig. 6 Action confusion matrix

The following aspects with regard to development of the bot can be explored in future:

5.1 To add more features to the bot to increase its scope. For example:

   5.1.1 The bot may contain options for user registration.
   5.1.2 The bot may contain options to apply for loan.

5.2 To enhance training data in order to ensure that the bot is effective at carrying out a smooth conversation with the user despite receiving out of context responses or change of context.
5.3 To create more stories by interactive learning to ensure effective and meaningful conversations that lead to higher percentage of completed tasks.
5.4 To increase the responsiveness of the bot when it fails to understand the user input by incorporating multi-level fallback intents (with the last level handing off the interaction to a human agent) when the bot receives low confidence during intent identification or next action prediction.

References

1. Bocklisch, T., Faulkner, J., Pawlowski, N., Nichol, A.: Rasa: open source language understanding and dialogue management. ArXiv, abs/1712.05181 (2017)
2. Braun, D., Hernandez-Mendez, A., Matthes, F., Langen, M.: Evaluating natural language understanding services for conversational question answering systems. In: SIGDIAL Conference (2017)
3. Lafferty, J.D., McCallum, A., Pereira, F.: Conditional random fields: probabilistic models for segmenting and labeling sequence data. In: ICML (2001)
4. Williams, J.D., Asadi, K., Zweig, G.: Hybrid code networks: practical and efficient end-to-end dialog control with supervised and reinforcement learning. ArXiv, abs/1702.03274 (2017)
5. Bordes, A., Weston, J.: Learning end-to-end goal-oriented dialog. ArXiv, abs/1605.07683 (2017)
6. Wen, T., Gasic, M., Mrksic, N., Rojas-Barahona, L.M., Su, P., Ultes, S., Vandyke, D., Young, S.J.: A network-based end-to-end trainable task-oriented dialogue system. In: EACL (2017)
7. Williams, J.D., Liden, L.: Demonstration of interactive teaching for end-to-end dialog control with hybrid code networks. In: SIGDIAL Conference (2017)
8. Wen, T., Gasic, M., Mrksic, N., Su, P., Vandyke, D., Young, S.J.: Semantically conditioned LSTM-based natural language generation for spoken dialogue systems. In: EMNLP (2015)

An End-to-End Approach for Benchmarking Time-Series Models Using Autoencoders

Abhirup Das, Shramana Roy, Soham Chattopadhyay and Soumik Nandi

Abstract Data generation and augmentation have become an important process in the area of model training and validation by allowing us to measure a model's robustness in comparison with others. Training on sparse and small datasets can reduce the effectiveness of a model as well as make it prone to overfitting. Thus, it is important to be able to validate its accuracy on data that has the same inherent characteristics as the original but includes subtle variances potentially present in real-world data. Our novel contribution makes use of a generative modeling technique to encode such discriminative features and learn their representation through the use of an encoder and generate new data through the decoding of such encodings. We propose to train the decoder through the use of a differentiable version of the popular Dynamic Time Warping (DTW) algorithm to formulate the objective loss of our end-to-end data generation model. The model will be able to generate suitable validation data for testing existing datasets, especially low resource and scarce ones.

Keywords Time series · Autoencoder · Dynamic Time-warping · VAE · LSTM

1 Introduction

Synthetic data is "any production data applicable to a given situation that is not obtained by direct measurement" according to the McGraw-Hill Dictionary of Scientific and Technical Terms. They are generated to meet certain specific needs. Sparse data can be one of the vital reasons for the algorithm to overfit. To avoid overfitting, models need to be trained over extensive datasets that make allowances for irregularity and variations in order for real-life applications. One way to address this problem is to increase the training set by generating synthetic examples and is called data augmentation.

Generating time-series data is a challenging work and very different from traditional data generation. General data augmentation algorithm does not work for

A. Das (✉) · S. Roy · S. Chattopadhyay · S. Nandi
IEM, Kolkata, India

© Springer Nature Singapore Pte Ltd. 2020
J. K. Mandal and S. Mukhopadhyay (eds.), *Proceedings of the Global AI Congress 2019*, Advances in Intelligent Systems and Computing 1112,
https://doi.org/10.1007/978-981-15-2188-1_25

sequential dataset. Normally, we can use flipping, rotating (orientation change), cropping, changing location (translation), scaling, adding Gaussian noise to the existing datasets to have additional synthetic modified data. But for time-series dataset changing in sequence of data will lead to the wrong prediction.

The general idea behind augmentation is to decrease the errors of the model overall training set that occurs due to variance or less data. We can influence the variance by adding or removing bias. In the case of [1], a small number of representative features are augmented with other noisy ones with the aim of benchmarking proposed feature-selection algorithms and models. Augmentation with synthetic example is often used to handle datasets with skewed classes (meaning that the class distribution is not uniform or comparable throughout all the classes in the training dataset). Data augmentation is used vastly in the field of computer vision due to the high variance in images. In the case of sparse datasets, it is often used to extend existing datasets and falls under domains such as feature engineering and data mining.

In this paper, we propose an architecture, which is an aggregation of LSTM and variational autoencoder (LSTM-VAE), for the generation of a modified time-series data maintaining the statistical and periodical characteristics/features extracted from the existing time-series data. The encoded/hidden representation of the data once trained can be used for reconstruction of a synthetic dataset, and we propose to use soft-DTW as loss function in our model, which outperforms for time-series data.

2 Related Works

Existing literature on dataset generation and data augmentation includes works such as that by Forestier et al. [2] which makes use of the averaging algorithm using DBA [3] to reduce the variance in time-series sequences allowing models to be trained on such synthetic samples. Other approaches such as [4] involve the usage of deep-residual networks to augment sparse datasets. In the field of event-analysis and time-series forecasting, much research includes the use of autoencoders to extract important features from given data, while making use of LSTMs for the purpose of capturing temporal patterns within these features. Works such as [5] use twin-model approach to effectively capture encoded features while using the LSTM-based decoder for forecasting. A certain work by Naul et al. [6] uses LSTM autoencoders (LAEs) for generating evenly sampled data from uneven input data to train their model.

Extending the boundaries of literature, LAE has also been used for dialect analysis, mentioned among the works of Rama and Çöltekin [7], to calculate the distances between two phoneme sequences. Keeping up the analogy of the dialect sequence with the time-series sequence, phonemes can be characterized as the distinguishing features of a time-series data, enabling the model to figure out the characteristics of a sequence and maintain it for the reconstruction of sequential data.

Sentence generation using conventional RNN's have also been modified by the works of Bowman [8]. An RNN-based variational autoencoder generative model has been incorporated to generate distributed latent representations of entire sequences. This factorization allows it to explicitly model holistic properties of time-series data, such as statistic and periodic properties.

Recent work on generative text modeling has found that variational autoencoders (VAE) with LSTM decoders perform worse than simpler LSTM language models [8]. Hence, a new approach has been implemented, which has been mentioned in the works of Yang [9]. They proposed a new type decoder, i.e., a dilated CNN, which was found to achieve better results than the conventional LSTM decoders. Further implementation of this concept has been reflected in the works of Lin [10] a neural machine transmission (NMT) has been proposed that decodes the sequence with the guidance of its structural prediction of the context of the target sequence. This approach can be used for synthetic generation of sequential data for scarce dataset.

In the field of forecasting, a good improvement has been achieved by using Dynamic Time Warping (DTW) along with nonparametric classifiers such as nearest neighbors. In this paper, we propose a mechanism of generating synthetic time-series data from a set of given time-series data using LSTM autoencoders, with soft-DTW as an objective loss. Soft-DTW is a differentiable loss function for time-series data, which computes the soft-minimum of all alignment costs. We also propose to tune the parameters of a machine that outputs time series by minimizing its fit with ground-truth labels in a soft-DTW sense.

3 Fundamental Concepts

3.1 Time-Series Data

Time series refers to a series of data points indexed (or listed or graphed) in temporal order. It is a sequence taken at successive equally spaced points in time.

Most time-series patterns can be described in terms of two basic classes of components: trend and seasonality. *Trend* represents a general systematic linear or (most often) nonlinear component that changes over time and does not repeat or at least does not repeat within the time range captured by our data. *Seasonality* may have a formally similar nature; however, it repeats itself in systematic intervals over time.

Stationary time series is one whose statistical properties such as mean, variance, autocorrelation, etc. are all constant over time. Most statistical forecasting methods are based on the assumption that the time series can be rendered approximately stationary (i.e., "stationarized") through the use of mathematical transformations. Since we are interested in the generation of synthetic time-series dataset, these characteristics are the important aspects to be taken care of along with the statistical components of a particular dataset. There are various methods for the analysis of a time-series dataset and further for forecasting using ARIMA, GARCH models. In order to find the similarity between various datasets, we have an algorithm known as Dynamic Time Warping (DTW).

The procedure of generation of the time-series data needs to preserve these aspects and should be able to generate data retaining the same. This fits our objective of synthesizing data retaining such salient characteristics but containing variation unseen in the original input dataset. This will allow us to benchmark and thus validate the effectiveness of a trained model on realistic, noisy data.

3.2 LSTM-VAE Model

Autoencoders, unlike standard neural networks that learn the value of the weights for the purpose of classification, do not require any output label. Rather, they learn to reconstruct the input. Typical autoencoders feature a hourglass architecture where the lower half of the hourglass architecture is used for learning a hidden representation, whereas the upper half of the architecture learns to reconstruct the input through backpropagation from the encoded /hidden representation. This enables the network to remove the redundant features for the regeneration of the input.

3.2.1 LSTM Autoencoders

An LSTM autoencoder is an implementation of an autoencoder for sequence data using an Encoder–Decoder LSTM architecture. Once fit, the encoder part can be used for feature extraction purposes and thus can be used for further supervised learning models, while the decoder part of the model can be used to regenerate a sequence data from the compressed features maintaining the characteristics of the input data.

The LSTM autoencoder has two parts: encoder and decoder. The encoder part transforms the input sequence $(x_1, \ldots x_T)$ to a hidden representation $h \in R^k$ where, k is a predetermined dimensionality of h. The decoder is another LSTM layer of length T. The $h_t = h$ representation at each time step t is fed to a softmax function $\frac{e^h_{tj}}{\sum e^h_{tk}}$ that outputs a $x_t \in R^{|P|}$ length probability vector where P is the set of the distinct units in a data or the group of data under investigation.

Since our point of consideration is a time-series dataset, i.e., a sequential dataset, we need a model which can sequentially encode the data preserving the characteristics of a time-series data and regenerating it with imputation of some differences with the actual data. Recurrent neural networks, such as the long short-term memory, or LSTM, are specifically designed to support input time series data. They are capable of learning the complex dynamics within ordered input sequences as well as use an internal memory to memorize or use the information across long input sequences.

3.2.2 Sampling from Encoding Layer

Vanilla autoencoders do not make use of an explicit global representation for the purpose of generation [8]. In order to fulfill our purpose of generating data that retains characteristics of input, it is important to incorporate the distributed latent representations of all input time sequences. This will allow the generative algorithm to explicitly model holistic properties such as stationarity, trend and seasonality and high level representative features.

Thus, we propose to make use of variational autoencoders (VAEs) in order to build a continuous latent representation that the decoder can sample from to generate time-series sequences. A VAE trains the encoded layer z (see Fig. 2) to act as the Gaussian prior distribution and enforces a regular geometry over these encodings. It learns code not as single points, but as ellipse-like regions in latent space, ensuring that the generated samples fill the space rather than replicate the training data as isolated time-series sequences.

If trained with a standard autoencoders reconstruction objective, the latent representation z would be encoded deterministically by making variances in encoder $q(z|x)$ extremely small by training [11]. Instead the VAE encourages the posterior distribution to be close to a prior $p(z)$, which is generally as Gaussian with $\mu = 0$ and $\sigma = 1$ (see architecture in Fig. 3).

$$L_{\text{vae}} = -KL(q(\frac{z}{x})||p(z)) + E_{q(\frac{z}{x})}[\log p(\frac{x}{z})] \ \leq \log p(x) \qquad (1)$$

This objective sets a valid lower bound on the true log-likelihood, making VAE suitable for generative modeling. Thus, the decoder model is trained to be able to decode time sequences from the encoded representation that has a reasonable probability under the prior.

3.2.3 Convolutional Decoders

Instead of using an LSTM decoder as described in Fig. 1, we make use of an intermediate dilated convolutional layer as our decoder (see Fig. 2). The convolutional layer does not directly decode the encoding; rather, it provides global information of dataset-specific contexts to the LSTM decoder such that the final generated output is able to retain the characteristics observed in input while sampling from prior $p(z)$. The effectiveness of such decoders has been observed in tasks such as sentence generation [10].

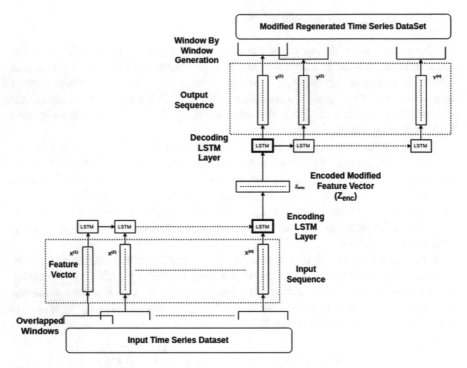

Fig. 1 A vanilla LSTM autoencoder architecture

3.3 Soft-DTW Loss

DTW or Dynamic Time Warping is an algorithm to measure the similarity between two temporal sequences, which may vary in speed. For instance, similarities between the walking of two persons can be measured by DTW, even if one person was walking faster than the other, or if there were accelerations and decelerations during the course of an observation. Unlike the traditional Euclidean distance similarity, DTW is robust because it can compare time series of variable size and time shifts. To compute DTW, one typically solves a minimal-cost alignment problem between two time series using dynamic programming. Given two time series $X = (x_1, \ldots, x_n) \in R^n$ and $Y = (y_1, \ldots, y_m) \in R^m$, DTW can be calculated as:

$$DTW(X, Y) = \min_{A \in A_{n,m}} < A, \Delta(X, Y) > \tag{2}$$

where, $A_{n,m} \subset 0, 1^{n \times m}$ represents the set of (binary) alignment matrices, i.e., paths on an $n \times m$ matrix that connects the top-left $(1,1)$ entry to the bottom-right (n, m) entry.

The paper [12] defines a single barycenter time series t under a set of normalized weights $\lambda_1, \ldots \lambda_N$ for a family of time sequences of size N such that $\sum_{i=1}^{N} \lambda_i = 1$.

Fig. 2 Proposed mechanism

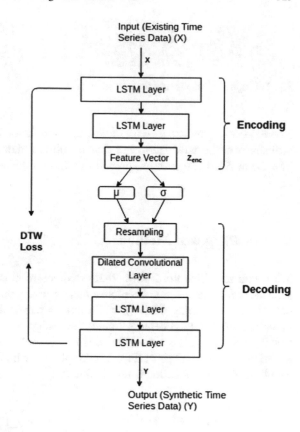

Soft-DTW loss aims to approximately solve the following minimization problem:

$$\min_{t \in R^n} \sum_{i=1}^{N} \frac{\lambda_i}{m_i} \mathrm{DTW}(t, y_i) \tag{3}$$

where m_i is the length of each such time series y_i.

3.4 Kullback–Liebler Divergence

The KL divergence is used to compare a probability distribution p with approximating distribution q as given:

$$KL(p\|q) = \sum_{i=1}^{N} p(x_i)(\log p(x_i) - \log q(x_i)) \tag{4}$$

Fig. 3 Usage of synthetic data for validation

The KL value provides us with the expectation of the log distribution between the probability of the data in the original distribution and the approximating distribution. The use of KL divergence has proved suitable for generative modeling, as observed in [8].

4 Our Proposed Method

We intend to utilize the LSTM-VAE architecture in order to generate time-series sequences that have the same representative characteristics seen in input training data (in given training dataset), while having substantial variations unseen in the latter. Our contribution centers around the design of a novel objective function for training our model; it is a modification of the vanilla autoencoder loss which makes use of reconstruction loss in Eq. (1) and replace it with the soft-DTW loss in Eq. (3). The final objective function is thus defined as

$$L_{\text{LVAE}} = -\gamma_1 KL(q(\frac{z}{x})||p(z)) + \gamma_2 \sum_{i=1}^{N} \frac{\lambda_i}{m_i} \text{DTW}(t, y_i) \tag{5}$$

Through experimentation, we find values of γ_1, γ_2 that provides the best results. The generative model trained as such will be sensitive to and be able to capture characteristics such as mean, variance, autocorrelation, trends, linearity, stationarity, seasonality, entropy of the given time series and introduce variations by sampling from the learnt latent space (Fig. 3).

5 Conclusion

This approach is mainly focused on the generation of synthetic time-series data that retains the statistical characteristics of the original dataset. The emphasis on the preservation of the underlying characteristics of a particular data is of utmost importance, since the synthesized data with added variances would be used for benchmarking the predictive learning models, i.e., by evaluating whether the model, after being trained on the synthesized data, is effective on real-world data that is more irregular that the time-series data samples it has already seen. The generated dataset is thus obtained from the original data, without spending any additional resources. This method can become an efficient and effective method for validating time-series classification and forecasting models in areas where data is limited.

References

1. Guyon, I.: Design of Experiments for the NIPS 2003 Variable Selection Benchmark (2003 July)
2. Forestier, G., Petitjean, F., Dau, H.A., Webb, G.I., Keogh, E.: Generating synthetic time series to augment sparse datasets. In: 2017 IEEE International Conference on Data Mining (ICDM), pp. 865–870. IEEE (2017 Nov)
3. Petitjean, F., Ketterlin, A., Ganc-arski, P.: A global averaging method for dynamic time warping, with applications to clustering. Pattern Recognit. **44**(3), 678–693 (2011)
4. Fawaz, H.I., Forestier, G., Weber, J., Idoumghar, L., Muller, P.A.: Data augmentation using synthetic data for time series classification with deep residual networks (2018). arXiv preprint arXiv:1808.02455
5. Laptev, N., Yosinski, J., Li, L.E., Smyl, S.: Time-series extreme event forecasting with neural networks at Uber Sydney, Australia (2017)
6. Naul, B., Bloom, J.S., Pérez, F., van der Walt, S.: A recurrent neural network for classification of unevenly sampled variable stars. Nat. Astron. **2**(2), 151 (2018)
7. Rama, T., Çöltekin, Ç.: LSTM autoencoders for dialect analysis Osaka, Japan (2016 December 12)
8. Bowman, S.R., Vilnis, L.: Generating Sentences from a Continuous Space (2016 May 12)
9. Yang, Z., Hu, Z., Ruslan, S.: Taylor Berg-Kirkpatrick: Improved variational autoencoders for text modeling using dilated convolutions, Sydney, Australia (2017)
10. Lin, J., Sun, X., Ren, X., Ma, S., Su, J., Su, Q.: Deconvolution-Based Global Decoding for Neural Machine Translation, Santa Fe, New Mexico, USA (2018 Aug 20–26)
11. Raiko, T., Berglund, M., Alain, G., Dinh, L.: Techniques for learning binary stochastic feed-forward neural networks (2014). arXiv preprint arXiv:1406.2989
12. Cuturi, M., Blondel, M.: Soft-DTW: a differentiable loss function for time-series. In: Proceedings of the 34th International Conference on Machine Learning, vol. 70, pp. 894–903. JMLR. org (2017 Aug)

Load Balancing for Solving of Supplementary Machines Over Jobs with Max–Min Algorithm

Ranjan Kumar Mondal, Enakshmi Nandi, Payel Ray and Debabrata Sarddar

Abstract Cloud computing is living in computing systems to make accessible the completion of tasks. It is a web distributed system. So, there are a few million machines attached to the web to supply numerous types of adjustments to give users. Restrained figures of machines perform a huge number of tasks at the same time. So, it is inflexible to perform each task at the same time. A few machines carry out each task, so there are required to balance total loads. Load balance reduces the execution time and executes all tasks also. There are not convenient always to stay the equal number of machines to perform an equal number of tasks. The tasks to be performed in the cloud would be not as much of the associated machines for a time. That is the overload machines perform a less number of tasks at a time. So, we are going to present in this paper an algorithm with minimization execution time with throughput. We are appropriate at this point a Max–Min method balances total loads in the computing arena. The Max–Min technique will assist us to reduce the problem.

Keywords Min–Min algorithm · Task scheduling · Parallel computing

1 Introduction

It is known that cloud computing is a web-based package. It presents the apps conveyed as services over the networks and software in the data centers offering those

R. K. Mondal (✉) · E. Nandi · P. Ray · D. Sarddar
Department of Computer Science & Engineering, University of Kalyani, Kalyani, India
e-mail: ranjan@klyuniv.ac.in

E. Nandi
e-mail: nandienakshmi@gmail.com

P. Ray
e-mail: payelray009@gmail.com

D. Sarddar
e-mail: dsarddar1@gmail.com

© Springer Nature Singapore Pte Ltd. 2020
J. K. Mandal and S. Mukhopadhyay (eds.), *Proceedings of the Global AI Congress 2019*, Advances in Intelligent Systems and Computing 1112,
https://doi.org/10.1007/978-981-15-2188-1_26

329

services. It has traveled the applications and records from the machine into data centers [1]. Cloud computing has customized IT companies provided work to be set to software. Since cloud computing has been running in its developing stage, there is a lot of complications stay on in cloud computing. There is an example [2–4] to certify skilled access control. Network-level immigration for the constraint of a small amount cost and time to transfer work for data transition period provides the right protection to the data. Data defines that the revelation of perceptive data is practicable, and the momentous difficulty of cloud computing is load balancing.

In usual data centers, materials are tied to exact servers that are over-provisioned to handle the upper-bound workload. This type of configuration builds data centers pricey to sustain with wasted energy and floor space, short material operation, and important organization overhead. With VM machinery, cloud data centers turn into more elastic, secure, and supply improved support for the on-demand provision. It conceals server heterogeneity, allows server consolidation, and picks up server operation. A node is proficient in hosting numerous virtual machines with potential dissimilar material specifications and changeable workload types. Servers hosting separated virtual machines with up-and-down and unpredictable workloads may reason a material practice unbalance, which effects in operation deterioration and destruction of service level agreements (SLAs). Unbalance material usage can be monitored in cases; for instance, a virtual machine is running a computation-intensive relevance while with small memory necessity.

Cloud data centers are decidedly energetic and irregular due to

(1) Irregular material practice patterns of users continually requesting virtual machines,
(2) Fluctuating material usages of VMs,
(3) Unstable rates of arrivals and exit of data center end-users, and
(4) The process of nodes to handle singular load level may differ very much.

These states of affairs are effortless to trigger unbalanced loads in the cloud center, and they can direct to operation dreadful conditions and SLA violations, entailing a load balancing mechanism to moderate this problem.

Load balancing in the cloud arena is a means that shares out the overload lively local loads ideally balanced among all the machines [5]. It is applied to conquer both enhanced end-user fulfillment and higher material operation, certifying that no particular device is besieged, thus improving the overall system operation. For virtual machines scheduling with balancing purpose in cloud computing, it plans to assign VMs to appropriate hosts and poise the material operation among all nodes [6].

Appropriate balancing algorithms can assist to operate the accessible materials optimally, thereby to reduce the material operation. It also facilitates to put into practice failover, to make possible scalability and sinking response time.

2 Related Works

It is said that cloud computing supplies an assemblage of facilities to the purchaser, as, sharing of multimedia, software, and storage [7]. Each machine executes a task/subtask in a cloud background [8]. The MCT method allocates jobs to the machines having the expected lowest completion time of this job over other machines [9]. The MM scheduling method presumes the similar scheduling method as the MCT method to assign a job, the machine to complete this job with the least amount of completion time over other machines [10]. The LBMM scheduling method implements MM scheduling method and load balancing strategy [11]. Our proposed applies Hungarian method where works are larger than each machine [12]. Load balancing of the unbalanced cost matrix is equivalent to the previous method.

3 Problem Definition

There are numerous machines in a cloud society. That is to say, every one machine has the dissimilar potential to perform the responsibility; hence, considering the central processing unit of the device is not as much as necessary while the machine is favored to execute a task or subtask. So, to accept a material machine to perform a job is extremely considerable in cloud surroundings [13] due to the task that has unusual attributes for the *consumer* to give completion. Hence, it requires a few of the materials definite; for incidence, when implementing organism series gathering, it maybe has a big constraint toward memory. And, to achieve the peak result in the execution of all job, we arrange tasks possessions to recognize a dissimilar state judgment variable where it is according to the material of task constraint set judgment variable. An agent collects associated information of all machine linked to this cloud machine, for example, central processing unit capability, existing memory, and bandwidth rate.

The cloud arena is compiled from different machines, where the assets of each node may differ. Also, the computing potential suggested by the existing central processing unit, the suitable size of memory, as well as bandwidth rate is different. Cloud computing offers the materials of all machine; the presented material of each machine probably will be different in a full of an active state. From task completion of the existing central processing unit, existing memory, and bandwidth rate is the necessary things for the execution [14].

Accordingly, from the study, the accessible central processing unit ability, the obtainable memory size, and bandwidth acquired as the threshold to estimate manager VM. A graphic instance of precise as given below:

(I)   Existing CPU $\geq$774 MB/s
(II) Existing memory $\geq$251,468 KB/s
(III) Bandwidth $\geq$9.21 MB/s

To distribute some works to distinct computing machines, in a way, the whole assignment price is to be the minimum amount that is known as the assignment problem.

If the numeral works are not the same as a numeral of the computing machines, then it is recognized to be an unequal assignment problem.

Suppose a problem that consists of a set of 'i' machines $M = \{M_1, M_2, ..., M_i\}$. A set of '$j$' works $\{W = W_1, W_2, ..., W_j\}$ is considered to be allocated for finishing on 'm' existing computing machines, the finishing cost of each work on all the computing machines is recognized and mentioned in the matrix of the order of n. The goal is to come to a decision the minimum cost. This problem is solved by an established method called Hungarian method.

4 Proposed Work

Whenever the matrix with a problem is a non-square matrix, that is, when the amount of sources machine is unequal to the quantity of destinations machine, the assignment matrix problem is described an unbalanced assignment matrix problem.

4.1 Algorithm

To in attendance an algorithmic image of the technique, consider a problem consisting of a set of 'i' machines $M = \{M_1, M_2, ..., M_i\}$. A set of '$j$' works $W = \{W_1, W_2, ..., W_j\}$ is considered to be allocated for finishing on 'm' existing computing machines with the finishing value C_{mn}, where $m = 1, 2, ..., i$ and $n = 1, 2, ..., i$, where $j > i$, i.e., the numeral of works is more than a numeral of machines.

The development of the proposed algorithm is offered as follows:

Step 1: At first, do determine the average completion time of all tasks for all machines, in that order.

Step 2: Next, do determine the task having the highest average completion time.

Step 3: Now, do determine the unassigned machine having the minimum completion time for the task certain in Step 2. After that, this task is transmitted to the chosen machine for working out.

Step 4: Reiterate Step 2 to Step 3, in anticipation of each task have been assigned to the related machine entirely.

Step 5: Stop.

Example: Assume a corporation has six machines used for five works. Each work is able to be appointed to one and only one machine at the same time.

Experiment: Assignment Problem

Works/machines	M_{11}	M_{12}	M_{13}	M_{14}	M_{15}	M_{16}
W_{11}	12	21	10	14	20	19
W_{12}	11	20	17	19	32	21
W_{13}	15	25	21	17	18	21
W_{14}	23	21	17	19	37	21
W_{15}	17	18	11	20	23	13

Result:

12	21	10	14	*20*	19
11	20	17	19	32	21
15	25	21	*17*	18	21
23	21	*17*	19	37	21
17	18	11	20	23	*13*

5 Experiment

Figure 1 expresses the finishing time for all tasks at dissimilar computing machines. To calculate the operation of the proposed algorithm, our work is contrasted with

Fig. 1 The evaluation of the finishing time of each task at diverse machines

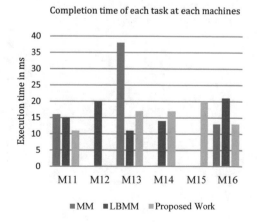

Completion time of each task at each machines

MM and LBMM showed in Fig. 1. Figure 1 expresses the evaluation finishing time of all executing machine with our approach, LBMM, and MM. Our approach completes the smallest finishing time and improved load balancing than other algorithms.

6 Result Analysis

Our current load balancing algorithm can get improved load balancing and act better than other available algorithms, for example, LBMM and MM from the following figures.

We will compare our proposed work with MM and LBMM algorithm with various machines and tasks from Figs. 2, 3, 4, 5, and 6. And, we will try to show how our proposed algorithm is better than others.

The cost of all works on all machines is specified in the following Table.

Fig. 2 Execution time in ns of six tasks in seven machines

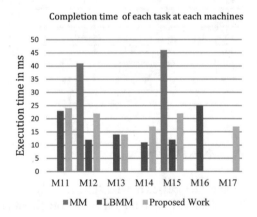

Fig. 3 Execution time in ns of five tasks in six machines

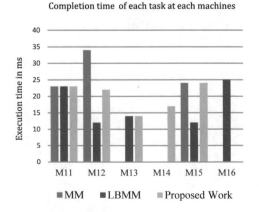

Fig. 4 Execution time in ns
of five tasks in six machines

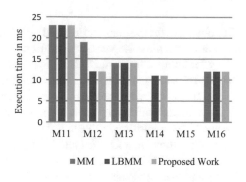

Fig. 5 Execution time in ns
of five tasks in four machines

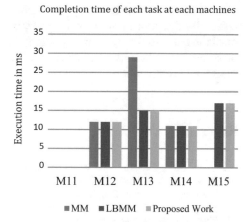

Fig. 6 Execution time in ns
of three tasks in four
machines

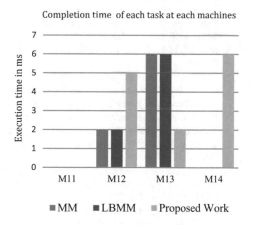

7 Conclusion

Our objective for this paper is to build up a successful load balancing algorithm using a method to reduce the various operation parameters, for example, network loads or delay for cloud computing. In this paper, a Max–Min has been presented to start the service of load balancing under cloud architecture.

In this paper, we have planned an algorithm for cloud networking to distribute the total loads to all computer machines along with their material ability in an equal manner. This algorithm is able to achieve superior load balancing and operation than other algorithms, like LBMM and MM from the experiment.

This effort intends to obtain load balancing by our proposed method that arranges all machines in an execution state. At the other side of, in our work, our proposed method is to develop to make the smallest execution time on the machine of all tasks. Besides, in a comprehensive case, the cloud technology is not only the static machine but also the dynamic machine. Additionally, our proposed scheme would be extending to preserve and execute while the machine is basic cloud computing machine in future work.

References

1. Höfer, C.N., Karagiannis, G.: Cloud computing services: taxonomy and comparison. J. Internet Serv. Appl. **2**(2), 81–94 (2011)
2. Pasquale, F., Ragone, T.A. Protecting health privacy in an era of big data processing and cloud computing. Stan. Tech. L. Rev. **17**, 595 (2013)
3. Tran, X.T., et al.: A new data layout scheme for energy-efficient MapReduce processing tasks. J. Grid Comput. **16**(2), 285–298 (2018)
4. Fernandes, D.A.B., et al.: Security issues in cloud environments: a survey. Int. J. Info. Secur. **13**(2), 113–170 (2014)
5. Mondal, R.K., et al.: Load balancing. Int. J. Res. Comput. Appl. Inf. Technol. **4**(1), 01–21 (2016)
6. Hu, J., et al.: A scheduling strategy on load balancing of virtual machine resources in the cloud computing environment. In: 2010 3rd International Symposium on Parallel Architectures, Algorithms and Programming, pp. 89–96 (2010)
7. Hung, C.L., et al.: Efficient load balancing algorithm for a cloud computing network. In: International Conference on Information Science and Technology, pp. 28–30 (2012)
8. Singh, R.M., et al.: Task scheduling in cloud computing. Int. J. Comput. Sci. Info. Technol. **5**(6), 7940–7944 (2014)
9. Wu, M.Y., et al.: Segmented min-min: a static mapping algorithm for meta-tasks on heterogeneous computing systems. In: Proceedings 9th Heterogeneous Computing Workshop (HCW 2000), IEEE, p. 375 (2000)
10. Kokilavani, T., Amalarethina, D.D.G.: Load balanced min-min algorithm for static meta-task scheduling in grid computing. Int. J. Comput. Appl. **20**(2), 43–49
11. Mondal, R.K., et al.: Load balancing with modify approach. Int. J. Comput. Tech. 2394–2231 (2015)
12. Mondal, R.K., et al.: Load balancing of unbalanced matrix with hungarian method. In: International Conference on Computational Intelligence, Communications, and Business Analytics, pp. 256–270 (2017)

13. Wang, S.-C., et al.: Towards a load balancing in a three-level cloud computing network. In: 2010 3rd International Conference on Computer Science and Information Technology, vol. 1, pp. 108–113. IEEE (2010)
14. Hung, C.-L., et al.: Efficient load balancing algorithm for a cloud computing network. In: International Conference on Information Science and Technology, pp. 28–30 (2012)

Load Balancing with Minimum Makespan in Cloud Computing

Ranjan Kumar Mondal, Payel Ray, Enakshmi Nandi and Debabrata Sarddar

Abstract Cloud computing is an innovative Internet technology which serves numerous accommodations to different location users. This technology grants a range of computing sources for the proper execution of voluminous-scale tasks with the appropriate size. There is a huge number of servers attached to web supply of several accommodations to users. By overcoming all constraints, it is required that all servers perform a huge amount of tasks in the same period. So it is not admitted to executing each task at the same time. So proper balancing of a huge amount of load is necessary for this area which cuts the makespan with runs each task as a specific way. It is almost impossible to maintain the same number of systems to run equivalent quantity jobs. Often tasks to be run in the cloud environment would be more than attached systems, and sometime, it will happen that some system has to allot more number of tasks than their available servers. To overcome the problem, we will propose an algorithmic method to get optimization load balancing with least makespan.

Keywords Cloud computing · Load balancing · Hungarian method

1 Introduction

A cloud-based system is a universal term made use of to explain a novel group of network-oriented computing taking position across the web over utility computing,

R. K. Mondal (✉) · P. Ray · E. Nandi · D. Sarddar
Department of Computer Science & Engineering, University of Kalyani, Kalyani, India
e-mail: ranjan@klyuniv.ac.in

P. Ray
e-mail: payelray009@gmail.com

E. Nandi
e-mail: nandienakshmi@gmail.com

D. Sarddar
e-mail: dsarddar1@gmail.com

© Springer Nature Singapore Pte Ltd. 2020
J. K. Mandal and S. Mukhopadhyay (eds.), *Proceedings of the Global AI Congress 2019*, Advances in Intelligent Systems and Computing 1112, https://doi.org/10.1007/978-981-15-2188-1_27

a grouping of integrated and network hardware, software and web infrastructure [1]. Using the web to communicate and transportation provides hardware, software and networking services to end users [2]. These platforms conceal the complexity of the underlying infrastructure from clients and applications giving an elementary application programming interface (API) [3]. Also, the platform gives on-demand services, which are always on, anytime, anyplace and anywhere. You should pay according to your utilization. The hardware and software services exist for the public, corporations and markets. Cloud system is a term made use to refer to web-based services [4].

The cloud system is mainly made use of because it is distributed in nature. It is designed for remote sharing and heterogeneous storage. It also gives some services like computation capabilities and applications, etc. [5]. Cloud system comes into today's picture because it provides high reliability over a vast network. However, the demand comes are dynamic, so resource allocation called dynamic allocation method required here [6]. Because of dynamic resource allocation in the cloud-based system requires proper load balancing method. Balancing the load means the dispersion of a huge amount of tasks from resources among the clients so that neither machine is overloaded nor inactive [7]. If the load balancing is not correct, then the effectiveness of some overloaded devices is able to degrade their performance and hence lead to SLA violation [8]. The load balancing algorithms are of three categories according to their functions, such as static scheduling, dynamic scheduling algorithm and the third one is mixed scheduling algorithms [9].

(1) Static load balancing algorithm is appropriate for small-scale distributed systems with maximum speed in communication and less delay.
(2) Dynamic load balancing algorithm mainly made use of for leveraging duration period and communication delays and time needed for completion of the operation; therefore, it is primarily made use of in the large distributed system.
(3) Mixed load balancing algorithm is used to proportional delivery of computing task and scaling down distribution cost for respective machines. As stated above, it is seen that the cloud system is into second types. In conventional distributed system, migration process is not costly, whereas the process migration rate is high due to a granularity of data. Thus, cloud system background needs a load balancing schedule which could serve to change the client demands while providing optimized load balancing [10]. To measure the effectiveness of an algorithm in cloud computing background, reliability, adaptability, fault tolerance, throughput and waiting time [11] are necessary.

2 Related Works

The intent of load balancing making bigger the resources utilization with light or idle load is thereby minimizing loads. The method tries to deal out the total load among all the obtainable resources. At a time, the method desires to decrease in size of the makespan [12].

In distributed systems consist of standardized and devoted servers. Though, these techniques will not act glowing in cloud architecture for a reason that scalability, heterogeneity and self-sufficiency [13]. This method formulates the load balancing plan for cloud computing extra problematical and an exciting subject for a number of a scientist.

The traditional that is not traditional differs from the conservative, predictable techniques that make the most positive results in a miniature period. There is no excellent process for cloud systems. An alternative is to choose an apposite scheduling practice to occupy in known cloud environs for the reason that the independence of the tasks, servers and system heterogeneity [14].

LBMM [15] is a static balancing algorithm. This algorithm employs load balancing among machines taking into account it. The Min–Min algorithm decides the lowest completion time for all jobs. Then, it favors the jobs with the least completion time among all the jobs. The algorithm proceeds by assigning the jobs to the resource creating the smallest lifespan. Min–Min states similar procedures in anticipation of all jobs are scheduled. The algorithm completes the Min–Min schedule and makes a choice of the node with the top makespan. Consequent to the machine desires the job with the smallest execution time. The execution time of the preferred job is considered for each resource. The peak execution time of the preferred job is evaluated with the makespan. If the execution time is lowest, then the preferred job is distributed to the device, having the maximum makespan. Else, the afterward maximum makespan of the task has chosen and the steps are persistent. The procedure stops if each machine with each job is allocated. In situations where the figure of undersized jobs is greater than the quantity of abundant jobs in meta-job, these algorithms get improved presentation. This algorithm does not assume small and elevated machines heterogeneity and jobs.

Two-Phase Scheduling [16] is the arrangement of OLB and LBMM algorithms to run superior completion competency and goes on with the system load balancing. OLB scheduling keeps on each machine in functioning status to get hold of the purpose of balancing the load with LBMM algorithm employed to diminish the completion of the period of all jobs on the device by minimizing the entire execution time. This algorithm is working to get recovered the process of resources and picked up the effort competency.

Min–Min [17] assumes MM with load balancing plan. In this technique, all tasks are separated into subtasks. So the completion times of each subtask in every server, as well as the threshold, are estimated. All subtasks along with the server to present the smallest amount of execution time for the subtask go into Min time. As a result, the subtask having least amount of execution time among all subtasks is preferred and dispensed to the associated server.

LB3M [18] scheme estimates the average completion time of all subtasks on all servers. The subtask is having highest average completion time, and the server is giving the least completion time chosen. At present, the chosen subtask will be executed by the related server. And if the server is before consigned, then the procedure appraises the completion time of the server. For the allocated server, makespan is the summation of the makespan of allotted tasks and completion time for the running subtask. For the unallocated server, makespan is the execution time of the existing task. This progress is continual in the subsequently unallocated subtasks.

3 Background

Task scheduling and distribution of machines are the essential attributes of cloud computing which have an effect on the activity of the server. To attain maximum throughput, different task scheduling has been proposed by the researchers for scheduling and scaling of resources. The adjustment of the suitable algorithm settles on the function of the server [19].

3.1 Task Scheduling

In cloud systems, appropriate task scheduling in cloud computing is mandatory to develop competence and to reduce the execution time. The objective of task scheduling is to identify a specific resource of all the presented machines, so that execute the competence of the computing background enhances [20].

3.2 Makespan of Tasks

The total time used for all the jobs in meta-task to get performed makespan. Meta-task is a category of the queue whereas jobs, which are set to perform, are stored. It is to evaluate the throughput of the various systems.

Cloud computing is a modern web-based computing technology. Cloud computing gives us a convenient way to use large data centers instead of desktop and portable PCs. This innovative technology changed the life of the IT industry. Besides that, there are many pros and cons of cloud computing. Such as [21]:

- Promising effective access control remains a question mark.
- For data evolution stage delivers proper security to the data which is a big problem.
- Data unavailability.
- In the case of data swamping and transitive expectation matters, there are many faults occurred.
- Complexity arises in the situation of load balancing in cloud computing.

4 Proposed Work

The problem of an assignment is not a square matrix, namely the number of sources is not equivalent to the number of destinations, and the assignment matrix is called an unbalanced assignment. We have discussed it with the following example.

Algorithm

1. *To compute the average completion time for each task of all nodes, correspondingly.*
2. *To find a task having the highest average completion time.*
3. *To find the node having the least completion time. Then, this task is to be sent out to the selected node for execution.*
4. *A node selects at most two jobs with least completion time based on the highest average completion time where all nodes assign at least one job.*
5. *Do again above Step 2 to Step 4, until all tasks have been dispatched to all nodes.*
6. *Then it is to add all completion time of all nodes, respectively.*
7. *End.*

4.1 *Example*

Suppose there is a business group having five machines used for five jobs. Each job can be distributed to one machine at a time. The time of each job of the corresponding machine is specified in the subsequent Table.

Jobs/machines	M_{11}	M_{12}	M_{13}	M_{14}	Avg
J_{11}	10	19	8	15	13
J_{12}	12	18	7	16	13.25
J_{13}	13	15	9	14	12.75
J_{14}	12	19	8	18	14.25
J_{15}	14	17	10	19	15

5   Comparison

In our proposed algorithm, we can get enhanced load balancing and better results compared with other algorithms, for instance, LBMM and MM from the following figure. We compare our proposed work with MM and LBMM algorithm with various machines and tasks from Figs. 1, 2, 3, 4, 5 and 6. And we will try to show how our proposed algorithm is quietly better than other existing algorithms.

The figure demonstrates the makespan for each task at different computing nodes. To analyze the performance of our proposed work is evaluated with other techniques by the case shown in figures that displays the comparison of the makespan of each computing node among our approach. Our technique attains the least makespan and improved load balancing than other existing algorithms.

Fig. 1 Evaluation of the makespan of five tasks for four machines

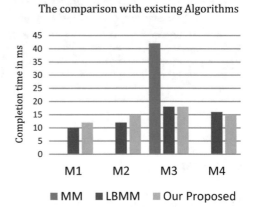

Fig. 2 Evaluation of the makespan of six tasks for five machines

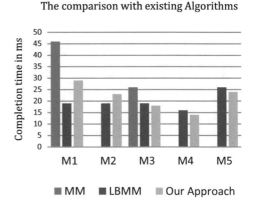

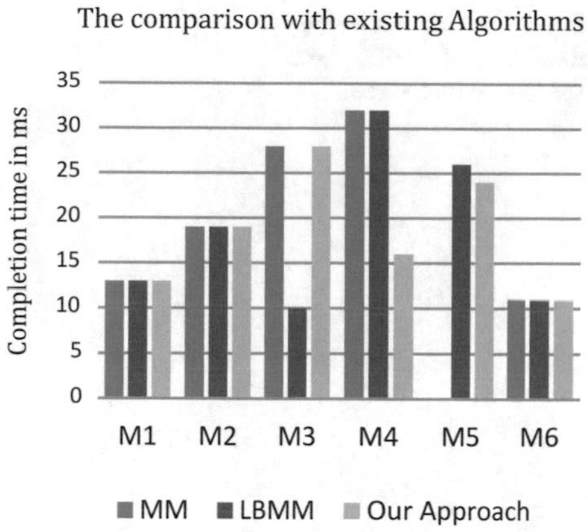

Fig. 3 Evaluation of the makespan of seven tasks for six machines

The comparison with existing Algorithms

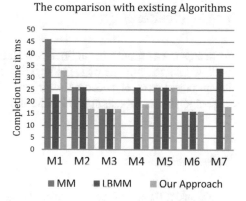

Fig. 4 Evaluation of the makespan of eight tasks for seven machines

The comparison with existing Algorithms

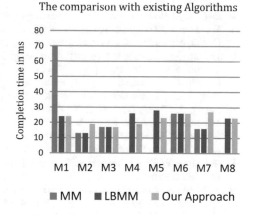

Fig. 5 Evaluation of the makespan of eight tasks for eight machines

The comparison with existing Algorithms

Fig. 6 Evaluation of the makespan of thirteen tasks for twelve machines

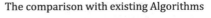

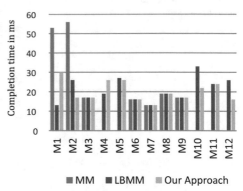

6 Conclusion

In our paper, we proposed a qualified new algorithm for the cloud computing networks to allow jobs to compute nodes along with their resource facility. Correspondingly, this proposed work can accomplish improved load balancing and performance than other algorithms, for example, LBMM and MM from the case study.

References

1. Kavitha, R.N., et al.: Cloud computing integrated with testing to ensure quality. J. Sci. Indus. Res. **75**, 77–81 (2016)
2. Berl, A., et al.: Energy-efficient cloud computing. Comput. J. **53**(7), 1045–1051 (2010)
3. Wu, J., et al.: Cloud storage as the infrastructure of cloud computing. In: 2010 International Conference on Intelligent Computing and Cognitive Informatics, pp. 380–383. IEEE (2010)
4. Iyanda, O.: Big data and current cloud computing issues and challenges. Int. J. Adv. Res. Comput. Sci. Softw. Eng. **4**(6), 1192–1197 (2014)
5. Subashini, S., Kavitha, V.: A survey on security issues in service delivery models of cloud computing. J. Netw. Comput. Appl. **34**(1), 1–11 (2011)
6. Zhang, Q., Cheng, L., Boutaba, R.: Cloud computing: state-of-the-art and research challenges. J. Internet Serv. Appl. **1**(1), 7–18 (2010)
7. Rathore, N., Chana, I.: Load balancing and job migration techniques in grid: a survey of recent trends. Wireless Pers. Commun. **79**(3), 2089–2125 (2014)
8. Voorsluys, W., et al.: Cost of virtual machine live migration in clouds: a performance evaluation. In: IEEE International Conference on Cloud Computing, pp. 254–265 (2009)
9. Kanmani, A., Sukanesh, R.: Adequate algorithm for effectual multi service load balancing in cloud-based data storage. J. Sci. Ind. Res. **74**, 614–617 (2015)
10. Buyya, R., et al.: Intercloud: utility-oriented federation of cloud computing environments for scaling of application services. In: International Conference on Algorithms and Architectures for Parallel Processing, pp. 13–31 (2010)
11. Hameed, A., et al.: A survey and taxonomy on energy-efficient resource allocation techniques for cloud computing systems. Computing **98**(7), 751–774 (2016)

12. Kaur, K., Dhindsa, K.S.: Energy-efficient resource scheduling framework for cloud computing. J. Cloud Comput. **2**(4), 1 (2015)
13. Kaur, T., Chana, I.: Energy efficiency techniques in cloud computing: a survey and taxonomy. ACM Comput. Surv. **48**(2), 22 (2015)
14. Bowman-Amuah, M.K.: Method for providing communication services over a computer network system. US Patent 6,332,163, issued 18 Dec 2001
15. Kokilavani, T., Amalarethinam, D.D.G.: Load balanced min-min algorithm for static meta-task scheduling in grid computing. Int. J. Comput. Appl. **20**(2), 43–49 (2011)
16. Wang, S.-C., et al.: Towards a load balancing in a three-level cloud computing network. In: 2010 3rd IEEE International Conference on Computer Science and Information Technology, vol. 1, pp. 108–113. IEEE (2010)
17. Wu, M.-Y., et al.: Segmented min-min: a static mapping algorithm for meta-tasks on heterogeneous computing systems. In: Proceedings 9th Heterogeneous Computing Workshop (HCW 2000), p. 375, IEEE (2000)
18. Hung, C.-L., et al.: Efficient load balancing algorithm for a cloud computing network. In: International Conference on Information Science and Technology, pp. 28–30 (2012)
19. Kumar, E.S., et al.: A hybrid ant colony optimization algorithm for job scheduling in computational grids. J. Sci. Ind. Res. **74**, 377–380 (2015)
20. Hamscher, V., et al.: Evaluation of job-scheduling strategies for grid computing. In: International Workshop on Grid Computing, pp. 191–202. Springer, Berlin, Heidelberg (2000)
21. Kavitha, R., et al.: Cloud computing integrated with testing to ensure quality. J. Sci. Ind. Res. **75**, 77–81 (2016)

Circulation Desk-less Library Management System Using Radio-Frequency Identification(RFID) Technology and Computer Vision

Abhinaba Basu ⓘ and Arnab Biswas ⓘ

Abstract Library circulation involves lending materials to patrons and checking the material. There were several efforts to simplify the circulation process using bar codes, and then using RFID-based self-check-in checkout system. Despite that, the circulation counters are still a necessary component of a library. Our objective is to provide brief insight into circulation desk-less system in the library using radio-frequency identification (RFID) and computer vision. For that, the system will detect user, track their activities and let the user check-in or check out books and freely move out. It will save time for patrons and increase the efficiency of the circulation process. The system can handle impersonation, restrict borrowing books more than allotted quota and misplacing materials by alerting library staffs instantly. The system can also provide smart inventory management which will instantly detect the location of any material and detect any misplaced item. Authorized library staff can get complete physical stock verification from the system instantly. The system can completely revolutionize the way library works and increase its efficiency manifolds. A small prototype system has been developed and provided satisfactory results.

Keywords Radio-frequency identification (RFID) · Face recognition · Computer vision · Smart library

1 Introduction

A library is a collection of books and various academic materials which acts as a storehouse of knowledge. Library members (patrons) can borrow books for a specific time. The library has mainly three processes, i.e., Acquisition (adding new books), Circulation and Inventory Management. Here we will discuss improving efficiency

A. Basu (✉)
IIIT Allahabad, Prayagraj, Uttar Pradesh 211015, India
e-mail: mail@abhinaba.com

A. Biswas
Camellia School of Engineering & Technology, Barasat, West Bengal 700124, India

© Springer Nature Singapore Pte Ltd. 2020
J. K. Mandal and S. Mukhopadhyay (eds.), *Proceedings of the Global AI Congress 2019*, Advances in Intelligent Systems and Computing 1112, https://doi.org/10.1007/978-981-15-2188-1_28

349

on circulation process, inventory management (i.e., stock verification, place materials properly in designated shelf) using RFID technology and computer vision techniques.

Circulation process involves issuing items to the user and then checking in the returned items. It involves maintaining a proper catalog of the available books as well as logging all issues and returns, to check availability of any item.

Inventory Management process is used to properly keep all items in the properly designated shelf so that the user can find it easily.

"Radio-Frequency Identification (RFID) is the use of radio waves to read, capture, and interact with information stored on a tag. The reader and tag don't have to be in a direct line of sight for interaction." [2]. Based on the tag type, the band used, reader, the read range can vary from 2 cm to 200 m. The system is mainly dependent on RFID technology to detect all activities by a user (picking up a book, or putting it in a shelf). It is also used to detect misplaced book and stock verification.

Computer vision is used for face detection and identification. An array of cameras is used to capture all activities in the library and the video feed is analyzed using computer vision techniques.

2 Preliminary

2.1 Circulation Process

Circulation Process of Library involves lending books and academic materials like Charts, Maps, electronic storage device containing academic matters to the patrons. It also handles the process of check-in (returning) of those books by the patrons.

Libraries normally use any of the following techniques for circulation:

Circulation by Library Staff In this process, library staff issue or return items using the item accession number. Item records and circulation information can be stored in a physical register or electronically. For reading accession number, bar codes can be used. With the advent of RFID technology, RFID-based circulation method was adopted. Here all books will have RFID tags and the system has the information of all tags and can associate tag information with a particular book. Now using an RFID reader, the circulation process can be done.

Self Circulation In this process, patrons issue or return items themselves using the item accession number. Book records and circulation information must be stored electronically. Here the user will first have to authenticate (using user name/password or biometric information or Smart Card) themselves in the circulation desk. A book can be issued using bar code or RFID reader/writer based on the security system implementation. In case of manual verification at exit gates, a print receipt will be generated and it will be used for checking. RFID gates can be used instead of manual physical verification, which will simply check all items passing through the gates are issued or not, based on the associated tag.

Circulation Desk-Less System (Proposed) In this system, patrons issue or return items themselves without going to the circulation desk or self-checkout kiosks. The system automatically tracks all items a patron is taking away with him and issue them to that patron automatically. It will also track when the user carries issued book back to the library and automatically make those book available to other patrons and consider those items as returned.

2.2 Inventory Management

One of the primary aims of the library is to let users find the item they are looking for, as fast as possible using minimum effort. For this, all books are divided into specific categories and subcategories by following standard library item classification method (example Dewey Decimal Classification). Shelves were arranged based on the classification and books are stored accordingly. However, patrons often misplaced books while reading them. Inventory management consists of proper placement of items in the designated shelf. It also involves physical stock verification to find out any missing books and update the library catalog accordingly.

The new system will detect all misplaced items, provide instant stock verification status of the entire library in seconds.

2.3 RFID

Radio-frequency identification (RFID) technology is used to detect books and identify patrons. Patrons must carry their RFID-enabled library card. All books in the library must be tagged before placing it in circulation. The system has the information to identify patron and books based on the tags. UHF band RFID is used for the system, which conforms to ISO/IEC 18000-6:2013 standard and has read range of up to 12 m.

2.4 Computer Vision Techniques

While RFID technology can completely provide all necessary features required for the system, computer vision technology is used for mainly security purposes. Face detection of every patron is achieved by using OpenCV pre-trained Haar cascade classifier [1]. At the time of entry, it only matches the face with the patron photo on record (using the RFID enabled library card). Every time the patron enters the library it captures face data in the dataset, which will be used to track his action. Also, a PTZ camera module detects the number of faces in front of a shelf and if required detect those faces.

3   Problem

3.1   Circulation

Problem in library circulation Currently, patrons must visit the circulation counter (staff operated or self circulation kiosk) to issue or return books. The circulation counter may get crowded soon. The system saves valuable time of library staffs, hence increasing efficiency and decreasing cost.

Security Issues It is very easy to conceal the book in dress, in case of manual security check. However, with the advent of RFID, RFID scanner/gate can detect books concealed within the dress.

3.2   Inventory Management

Patron often misplaces books from one shelf to another. It is very challenging for library staff to find a misplaced book or to find if any books are misplaced on a particular shelf.

4   Requirement of Infrastructure

The following infrastructure is the prerequisite to implement the system.

The entrance of the library will have half-height revolving gates to put some delay in entry. Staffs will be placed near the entrance and exits to handle the situation if the alarm situated on both entry and exit gate triggered. The whole library including the reading arena has RFID-enabled floor mat. All shelves have a total of four RFID antennas [two right-hand circular polarization (RHCP) and two right-hand circular polarization (LHCP)] to cover the whole shelf. All shelves have their boundary shielded with aluminum foils so that RFID antennas of a particular shelf does not track other shelf items. All reading table will have an also set of RFID readers (based on dimension). All shelf and reading area have set of the camera which can detect the frontal as well as a top view of all patrons passing by. All books will have RFID tags attached to them. The ID card of the patron is also an RFID tag, and the patron must carry his/her ID card always, inside the library.

The system will have the location of each RFID readers and cameras deployed. The system will also have details of every book and patron (tag data, face and top-view datasets), library policy, issuance quota of each user, circulation history of the patron, catalog of the library, restricted books, etc.

All RFID readers will always read all tags available in its range in every $100\,\mu s$ and pass the information to the system.

UHF RFID technology (conforming to ISO/IEC 18000-6:2013 standard, 865–867 MHz frequency band for India) is used for both book tags and patron cards, due to their range (2 cm–12 m) falls on the optimum range in the library. Also, the tags are the cheapest among all other RFID variants. All tags are passive by nature.

The brief description of each device and its usage is given below:

4.1 RFID Floor Mat Antenna

RFID Floor Mat Antenna is commercially available RFID UHF antenna, housed in molded polyurethane or rubber or EVA enclosure. It can withstand multiple body weight. It is connected to the RFID reader using an RP-TNC connector (Fig. 1).

4.2 RFID Reader

RFID Reader is a commercially available UHF RFID reader. It can be connected to single or multiple antenna using RP-TNC connector. The output interface contains RJ45 connector, RJ232 connector, Wi-Fi Antenna Connector (SMA Female), USB Connector. RJ45 connector is ideally suited for the implementation. All RFID readers will be connected to layer 3 network switch and have separate static IP address. RFID reader scan all RFID tags within the range of the connected antennas and pass the data to server using network (Fig. 2).

Fig. 1 Commercial UHF RFID Mat antenna [3]

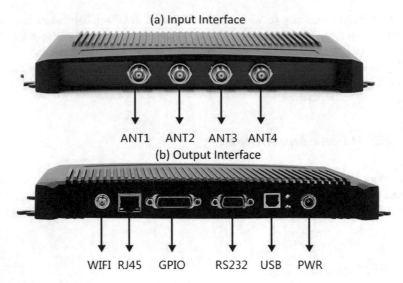

Fig. 2 Commercial UHF RFID reader [4]

4.3 RFID Antenna

"RFID Antennas are responsible for emitting and receiving waves that allow us to detect RFID chips. When an RFID chip crosses the antenna field, it is activated and emits a signal" [5]. RFID Reader converts that signal to data and passes it to the server for further processing.

4.4 PTZ Camera Module

The camera module comprises of Raspberry Pi 3B+, Pi supported Camera, Pan Tilt Bracket, 2 servos. It will detect faces using HAAR cascade classifier and then it will pan and tilt the camera to get a picture of all faces in the range of the camera. The face detection and camera movement logic reside in Raspberry PI. Raspberry PI is connected to the server using the local Ethernet network. It sends the number of faces detected by the camera to the server periodically (100 ms). If the module is installed at the entry gate, then it sends the face image of the patron to the server for processing. The server can also send an instruction to the camera module for images in certain cases. In such a case, the Raspberry PI will send the images to the server.

Table 1 RFID read range

Device type	Range	Other specification
RFID Floor Mat antenna	2 cm–2.5 m	Linear polarization, Input Impedence 50 Ω, Dimension 500 * 350 * 18 mm
RFID reader	2 cm–5 m	4 port reader, Max read rate: Up to 700 tag reads per second
RFID antenna (Reading Area, Shelf)	2 cm–1.5 m	RHCP/LHCP (combination of both)

4.5 Network Switch

"A network switch is a computer networking device that connects devices on a computer network by using packet switching to receive, process, and forward data to the destination device" [6]. All RFID readers and the server should be connected to the switch and each of the devices will be treated as a node. Based upon the number of nodes, multiple network switches may be used.

4.6 Server

The server control all RFID readers and process data. All system logic is implemented in the server. It provides an interactive system for the user to get a report (inventory, device performance-related, etc.) or query misplaced book. Based on the system settings, it will communicate to library staff as per requirement.

4.7 Device Specifications

The read range and device specifications are mentioned in Table 1.

5 Proposed Method

The system works in the way described in the diagram (Fig. 3).

Fig. 3 System flow diagram

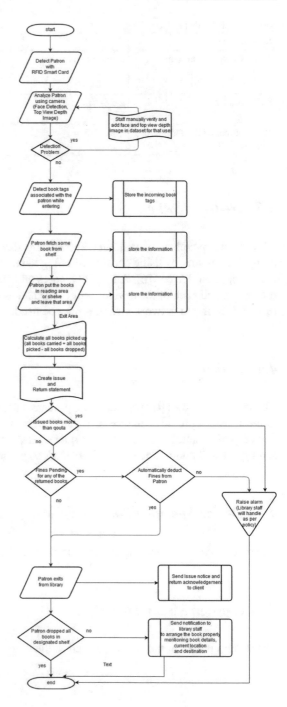

5.1 Detect Patron with Smart Card

When a patron enters the library, the system will detect the patron using the RFID tags. UHF RFID floor mats are used to detect the RFID tag of the patrons. Based on the first digit of the tag hex code, the system will detect if the tag is for a book or patron. The scanner will omit all book tags. Patrons will be strictly instructed to carry his ID Card always with him, within the library.

5.2 Analyze Patron Using Camera

Analyze Patron using the camera using both 3D face detection technique (PTZ camera), top-view depth image (top-view camera). All detection is one-to-one match, i.e., match with only the patron image dataset identified by the RFID patron tag.

Face detection is achieved using OpenCV pre-trained Haar cascade classifier [1] by using publicly available open-source code written in Python-based Keras library. PTZ Camera will auto track the patron's face. The system also checks the liveness of the face by testing if the patron is blinking or not based on publicly available technique [7].

In case the system fails to recognize the patron, it will raise an alarm to move the patron to library staff. Library staff will verify and then let the patron enter into the library. New images of the patrons are added to the dataset to increase the accuracy of the system.

5.3 Detect Incoming Books

An RFID Reader will detect incoming books by RFID tags along with the user. If the book is not issued to the user, the system will assume that the book is being returned by the patron whom the book is originally issued and send a return acknowledgment. If the book is issued to the patron carrying it, store the information.

5.4 Track Books Picked by Patron

The system will track all books picked by the user. It can be done using the fusion of RFID readers of the shelf, mats and the camera. Once a book is removed from a shelf, RFID reader of that shelf will automatically trigger a pick book event. Each shelf will have a PTZ camera module installed, covering total 180° view in front of the shelf. The camera will detect all patrons in front of the shelf. When the book pickup event occurs, The system will search all patrons near the shelf using RFID floor mat antenna and matches the count using the data retrieved from the PTZ camera. If there is a match, and there is only a single patron, the system assign the book to that patron. If there are multiple patrons and the system is unable to decide which patron have picked it conclusively, it will start tracking the book and also mark all possible patrons using the RFID system installed in floor mats. After some time, there will be only one patron from all the possible patron with that book and then the system will assign that patron with that book. If there is a mismatch between the number of faces detected using a PTZ camera in front of the shelf and from RFID Mat Antenna, system will try to identify using the face detection technique. It will then detect the patron who is missing from the RFID scanner but visible in camera. A notification will be sent to library staff to attend there and contact the patron roaming without an RFID ID card. Instead of this method, a borrowed book can be easily tracked at the time of exit, but there are certain security challenges. If a patron conceals the book using aluminum foil, then the system won't be able to detect it.

5.5 Track Patron Book Drops

A patron may drop any book he picked from the shelves or have carried with him, in the reading area or on another shelf. If the patron adds a book in the shelf, RFID reader designated for the shelf instantly identifies it. If he put it in the reading table, RFID reader of reading tables will identify it. Also if for some reason, the book was kept on the floor, the system can detect it using the RFID enabled floor mat. Maintain the drop list in real-time.

5.6 Validate Exit of a Patron

In case a patron wants to leave, the library system will calculate total book currently should be with him (books carried by the patron in the library + books picked up by the patron in the library—books carried by the patrons in the library, but not issued to him—dropped book list). The system will also match it with the book currently in possession with the patron. It should match in normal cases, but if it does not match, then the system will trigger an alarm so that library staff can handle the issue.

If the patron is taking away more books than permitted as per library policy or restricted books, a similar alarm will be raised.

If the patron returned books which have due, the first system will try to deduct the amount from the patron deposit (if configured so) or alarm may or may not raise based on library policy (most of the library does not allow checkouts if fees are outstanding).

Other validation may be configured as per library policy.

After the patron exits, send a notification to the patron (issue notice or return acknowledgment).

5.7 Inventory Management

The system periodically sends a notification to the library staff to place all dropped books in the proper place (in case the dropped books are not kept properly) specifying book details, current location, and destination.

5.8 Physical Stock Verification

Physical stock verification can be done instantly, by reading all book tags from all RFID readers. It can also provide the current location of every book instantly.

5.9 Finding Any Misplaced Book

Finding any misplaced book is very easy. Based upon the RFID tag of that book, the system will identify the nearest reader which can read the tag. The system can then just provide the location of that reader to the staff.

6 Implementation and Result

6.1 Implementation Technique

The system will consist of three parts

Data Update Workers For each RFID readers, a separate process will work on the background. They will periodically (100 ms) fetch the data from the RFID reader and update the data in the database. No data processing is done here. All data from the camera module will also get updated here.

Data Monitor and Controller This will just check the database and based on the system logic, act accordingly. If a book is being moved from a shelf or from the reading table, book tracking module will be called. At the time of entry or exit, it will contain all the logic mentioned in the system flow diagram, except tracking of the book. It will also actively communicate with the library staff using SMS or another method, based on the logic.

Book Tracking Module Once a book movement occurs, the system will start tracking the book based on input provided by both the camera module and the RFID module. This module will start tracking by querying the database periodically with the current location of the book and all patrons near it. Once it determines the patron carrying the book, it will pass the value to the Data Monitor and Controller module. In case there is a mismatch of headcount between the camera module and the RFID module, it requests an image from the relevant camera module and then does face identification of all faces detected by the camera module. It then identifies the patron who is roaming without RFID card, by tallying it with the RFID data.

6.2 Infrastructure Requirement Analysis of a Library

A study on a central library of a reputed institute was done. It consists of about 60,000 books housed in around 370 shelves. The total area is around 470 m^2. To cover the whole library, we require around 1000 mat antennas (each covering 0.5 m^2, 5% extra to cover non-rectangular spaces), 620 RFID reader (1 RFID reader for each 4 mat antenna and 1 RFID antenna for each shelf), 400 PTZ camera module. We also require around 20 numbers of 52 port switch. Apart from that around 62,500 passive RFID tags are required for all books and patrons (including library staff).

6.3 Prototype System

A small prototype of the system (except camera module, mat antenna) with a very small section of patrons has been developed and implemented as a prototype, which provides a satisfactory result. Only two shelves were made RFID enabled, and two separate RFID antenna was used instead of mat antenna. The prototype system can detect incoming patrons using RFID, detect all books on a particular shelf, detect the removal of any book of RFID enabled shelf, locating all nearby patrons in the RFID shelf. Most of the functions described here worked flawlessly, while few functionalities cannot be tested due to unavailability of the hardware.

7 Conclusion

In this paper, we present an experimental study on the implementation of a circulation desk-less library management system. The proposed system worked flawlessly as a small prototype, but it is yet to be tested on a real library and a large number of patrons. There are some studies and implementation of such a system in the retail sector, but those were mainly based on computer vision. The proposed system is based on mainly RFID technology, while computer vision technology is used just for security purpose.

While the system makes the life of the library staff as well as patrons much easier, but the implementation requires a lot of technical infrastructures, which can be quite costly for a library. While it can save cost for a library as the system require less number of staff to operate but it requires one-time capital investment. The proper cost-benefit analysis must be done before the implementation of the system.

References

1. Viola, P., Jones, M.: Rapid object detection using a boosted cascade of simple features. In Proceedings of the 2001 IEEE Computer Society Conference on Computer Vision and Pattern Recognition, pp. I-I. 10.1109/CVPR.2001.990517 (2001)
2. How Does RFID Technology Work?, https://www.makeuseof.com/tag/technology-explained-how-do-rfid-tags-work/. Last accessed 21 July 2019
3. UHF RFID Floor Mat Antenna, https://www.indiamart.com/proddetail/rfid-uhf-floor-mat-antenna-17233088473.html. Last accessed 21 July 2019
4. UHF RFID Reader, https://www.aliexpress.com/item/32847095804.html. Last accessed 21 July 2019
5. How does a UHF RFID System Work? https://www.dipolerfid.com/cn/blog/How-UHF-RFID-System-Works. Last accessed 21 July 2019
6. Network switch, https://en.wikipedia.org/wiki/Network_switch. Last accessed 21 July 2019
7. Real-time face liveness detection with Python, Keras and OpenCV, https://towardsdatascience.com/real-time-face-liveness-detection-with-python-keras-and-opencv-c35dc70dafd3. Last accessed 21 July 2019

Intelligent Water Drops-Based Image Steganography

Pinaki Prasad Guha Neogi, Saptarsi Goswami and Joy Mustafi

Abstract The level of impersonation from nature is astonishingly high in the domain of computational intelligence, particularly in the field of evolutionary computation and swarm-based frameworks. Attempts are being made to develop more number of algorithms, which imitate nature and the activities that take place in a specific natural phenomenon. One of the most recent optimization algorithms developed being inspired by the nature is the intelligent water drops (IWD) algorithm, which is based on the phenomenon taking place between the drops of water that flow in a river and the soil present in the riverbed. This paper puts forward the proposal of a novel intelligent water drop-based complex region determination (IWD-CRD) technique for proficiently concealing information within images. Superiority in the performance of the proposed algorithm over the existing ones has been demonstrated in terms of hiding capacity (HC), mean squared error (MSE) and peak signal-to-noise ratio (PSNR) of the stego-images.

Keywords Steganography · Intelligent water drops · Least significant bit · Hiding capacity · Mean squared error · Peak signal-to-noise ratio

1 Introduction

Ever since the commencement of the era of Internet, a standout amongst the most crucial components of information technology and correspondence has been the security of data. It is very much important to render protection to the data so that

P. P. Guha Neogi (✉)
Department of CSE, Meghnad Saha Institute of Technology, Kolkata, India
e-mail: ratul.ng@gmail.com

S. Goswami
A.K. Choudhury School of Information Technology, University of Calcutta, Kolkata, India
e-mail: saptarsi007@gmail.com

J. Mustafi
MUST Research, Hyderabad, India
e-mail: joy@must.co.in

© Springer Nature Singapore Pte Ltd. 2020
J. K. Mandal and S. Mukhopadhyay (eds.), *Proceedings of the Global AI Congress 2019*, Advances in Intelligent Systems and Computing 1112,
https://doi.org/10.1007/978-981-15-2188-1_29

they are accessible only to the authorized people and anyone who is unauthorised can not have any access to that information. Cryptography emerged as a technique for safeguarding the method of communication and various strategies have been developed for the purpose of encryption and decryption of the information involved in the communication. While the ultimate goal of cryptography is to keep the data protected, it often fails to maintain the secrecy and people can easily say that a message has been encrypted, though they might not be able to decode it without the key. Hence, in order to keep the presence of a message secret, a new concept, called steganography, came into existence.

Steganography is the technique and art of imperceptible correspondence and is achieved by concealing data in other data, therefore, concealing the presence of the imparted data. The term "steganography" is an amalgamation of two different Greek words—"stegos" and "grafia". The literary meaning of the word "stegos" is "cover" and that of "grafia" is "writing" [1], thus defining "steganography" as "covered writing". Though the secret information can be hidden in a wide variety of mediums including texts, digital audios and even videos, images act as the best medium for this concealing purpose because of its high redundancy. In case of image steganography, images act as the medium for covering up the data or information to be hidden. It is one of the efficacious techniques for information concealing that shields information from unapproved or undesirable exposure, and hence finds a wide range of applications in a number of fields like defence, confidential communication, surreptitious information storing, preventing information modification, research, medicines, etc.

In case of steganography, the original file can be alluded to as a cover-medium (cover-picture, cover-text, cover-audio, etc.), and in the wake of embedding the secret message, it is alluded to as a stego-medium. A stego-key is utilized for the concealing or encoding procedure so as to limit the discovery or abstraction of the embedded information. The resulting stego-picture, thus, formed is having a message encoded in it that is undetectable to the human eye, which implies that a person can not discover the disparity between the cover image and the stego-image. The secret message is embedded into the original image by utilizing an algorithm and then this message is acquired from the stego-picture by utilizing a reverse algorithm.

The overall layout of the paper is as follows: Sects. 2 and 3 give an overview of the concepts of steganography and LSB technique, respectively; Sect. 4 elaborates the method being proposed in this paper; experimental set-up and the result and analysis of the experiment have been provided in Sect. 5 and ultimately, the paper is concluded in Sect. 6.

2 Steganography

The term "steganography" can be defined as a technique for confidential communication in which a snippet of data, called secret message, is covered up into another piece of seemingly innocuous data, known as a cover. The secret message is covered up inside the cover-medium in such a manner that it fails to elicit any kind of suspicion in people's mind [2], thus proving it to be an efficient approach for maintaining the secrecy in the communication and as a matter of fact, plays a crucial role in case of data security as well. It can thus be called as an art of imperceptible correspondence by hiding data inside other data.

The first noted utilization of the word "steganography" dates back to the year 1499 in the book Steganographia, written by German author Johannes Trithemius, which is basically a treatise on steganography and cryptography, camouflaged as a book on enchantment. For the most part, the concealed messages seem, by all accounts, to be something different—pictures, shopping records, articles, or some different cover-text. For instance, the concealed message might be written in invisible ink between a book's noticeable lines. A few executions of steganography that come up short on a common secret is types of security via obscurity, and key-oriented steganographic plans cling to Kerckhoffs' principle [3].

When compared to cryptography, the supremacy that steganography enjoys is that it is a more secure method of communication as the information being hidden does not pull in attention towards itself as an item of inspection. Regardless of how unbreakable they are, evidently noticeable encoded messages initiate curiosity among people and may in themselves be implicating in nations where encryption is illegitimate [4]. While cryptography is the act of hiding and safeguarding the contents of a message only, steganography also camouflages the fact that a message is hidden in the cover-medium besides hiding the contents of the message.

Image steganography is an important and widely used type of steganography, which alludes to the methodology of concealing information inside an image file. The picture utilized for this purpose is termed as the cover image and the picture that is resulted after the intended information is hidden inside it is called the stego-image. Digitally, a picture is depicted as an $N * M * 3$ (in the event of colour images) or $N * M$ (in the event of greyscale images) matrix in the machine's memory, with every single entry embodying a pixel's intensity value. In case of image steganography, the secret data is embedded into the cover image by modifying the intensity values of certain pixels that are selected according to the encryption algorithm that is utilized for this purpose. At the same time, the recipient of the stego-image should also be apprised of the same encryption technique so as to discover the image pixels where the message is hidden and decrypt it. Figure 1 shows the basic steps in image steganography. The technique of locating the hidden message inside the cover image is termed as steganalysis.

Fig. 1 Basic steps in image steganography

3 Least Significant Bit (LSB) Technique

Several techniques are prevalent for the purpose of concealing the secret message within a normal file in steganography, but amongst them, the technique of least significant bit (LSB) is considered to be the most popular and widely utilized one.

This strategy alters a byte's last few bits for the purpose of encoding a message in it, which is particularly beneficial in case of an image file, where green, blue and red estimations of every pixel are denoted by 8 bits (1 byte) stretching from 0 to 255 in case of decimal or from 00000000 to 11111111 when considered in binary. LSB strategies ordinarily accomplish high capacity [5]. This is because, for example, changing the last two significant bits of a fully red pixel from 11111111 to 11111100 alters its red value from 255 to 252, and this change is too meagre to be perceptible to human eyes and as a consequence, both the cover and the stego-images almost look alike.

The idea of LSB embedding technique is pretty straightforward. It makes utilization of the fact that in plenty of image formats, the extent of precision exceeds the limit recognizable by normal human vision by a great deal. As a matter of fact, a modified picture with slight varieties in its hues will appear almost identical to the actual image for a person, just by taking a gander at it. In LSB algorithm, at the very beginning, both the cover photo and the secret message are transformed into their corresponding binary form from their pixel organization. Following which, the least significant bits (LSBs) of the binary represented pixel values of the cover images are substituted with the bits of the message to be hidden into this cover image, thus resulting in the stego-image. Figure 2 shows an example of least significant bits (LSBs) substitution by substituting the two least significant bits of the cover-medium pixel values with the hidden message bits.

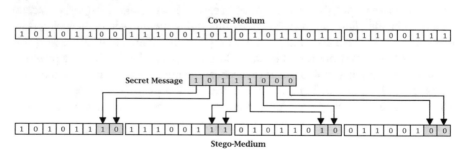

Fig. 2 LSB substitution by substituting the two least significant bits of the cover-medium pixel values with the hidden message bits

4 Proposed Method

The most fundamental and critical step in concealing data in the complex region of an image is the discovery of the complex regions in any given cover image. A number of prevalent techniques are available for the purpose of locating the complex regions in a photo, like Roberts cross edge detection technique, Deriche edge detector, Canny edge detection approach, the Sobel operator, differential edge detector and the Prewitt operator. But, these are basically edge detection techniques which incorporates an assortment of numerical techniques that go for recognizing points in an image at which the picture brightness shows a sharp variation or, all the more formally, has discontinuities. Though they are very efficient in detecting the edges in any given picture, these strategies do not totally comply with the definition of complex regions. Most of the prevalent techniques discover feeble and disjointed edge pixels and regard them as the image's true complex regions and also sometimes, take into account pixels that are not a part of the edges and are also susceptible to noise.

In this paper, an intelligent water drop (IWD) algorithm-based novel approach has been followed for the purpose of discovering the complex regions in a given cover image, followed by the concealment of the hidden message in these complex regions using LSB steganography. In IWD-based complex region recognition approach, a soil matrix is constructed based on the movement of the numerous intelligent water drops (IWD), whose motion is steered by the local variations of the intensity values of the image pixels. At first, the IWD-based approach is initialized and then iterated for N times in order to construct the soil matrix. The procedure performs both the development and the updated steps iteratively. Finally, at the end, the pixels belonging to the cover image's complex region are determined utilizing the decision process. The entire procedure is elaborated as follows.

4.1 Initialization

A digital image can be seen as composed of an array of pixels whose intensity level can be considered as I. In the proposed method, a grayscale image of dimension $r \times c$ is considered as the cover image ($\mathbb{C}$). Every single pixel that constitutes the cover image ($\mathbb{C}$) is thought of an individual node in the system and an aggregate of K IWD particles are arbitrarily assigned on these nodes. To start off with the process of detecting the complex region, each component of the soil matrix is initialized to a constant InitSoil and at the same time, the initial velocity with which the IWD moves is set to another constant InitVel.

4.2 Construction

The construction stage comprises several steps of computation and iterations. At the nth step, one of the K intelligent water drops is arbitrarily picked up and the chosen IWD is allowed to move for a total of N_{mov} steps over the cover image. If it is considered that an IWD (say, i) is currently located at Node(x, y), then the next node for the ith, say Node(p, q), is chosen based on a probability function given as:

$$\mathbb{P}_{(x,y),(p,q)}^{\text{IWD}(i)} = \frac{\frac{1}{0.01+s_{(x,y),(p,q)}}}{\sum_{(i,j)\in\Omega_{(x,y)}}\frac{1}{0.01+s_{(x,y),(i,j)}}} \tag{1}$$

such that

$$s_{(x,y),(p,q)} = \begin{cases} (\text{soil}(p,q)), if\left(\min_{(w,z)\in\Omega_{(x,y)}}(\text{soil}(w,z))\right) \geq 0 \\ (\text{soil}(p,q)) - \left(\min_{(w,z)\in\Omega_{(x,y)}}(\text{soil}(w,z))\right), \text{else} \end{cases} \tag{2}$$

and, $\Omega_{(x,y)}$: The neighbour pixels (4-connected or 8-connected) of the pixel (x, y).

4.3 Updating Stage

As the IWD travels from the Node(x, y) to Node(p, q), its velocity $\left(\vartheta^{\text{IWD}(i)}\right)$ would increase less if the amount of soil in the Node(p, q) is more and would increase more if amount of soil in the Node(p, q) is less, i.e. the increase in velocity of the IWD particle is inversely proportional to the amount of soil that it comes across in its path while travelling. The velocity of the IWD particle i, as it flows from Node(x, y) to Node(p, q), is updated according to (3).

$$\vartheta^{\text{IWD}(i)}(t+1) = \vartheta^{\text{IWD}(i)}(t) + \left(\frac{\alpha_v}{\beta_v + (\gamma_v \cdot (\text{soil}(p,q)))}\right) \tag{3}$$

where α_v, β_v, γ_v are constants whose values are tabulated in Table 1.

Besides velocity, the soil content of the IWD particle $(s^{IWD(i)})$ also gets updated as it moves from Node(x, y) to Node(p, q), which is represented with the help of (4).

$$s^{IWD(i)} = s^{IWD(i)} + \xi((x,y),(p.q); \vartheta^{IWD(i)}) \tag{4}$$

Table 1 Control parameters for the proposed approach

Control parameters	Values
InitSoil	1000
InitVel	100
α_s	1000
β_s	0.01
γ_s	1
α_v	1000
β_v	0.01
γ_v	1
δ	0.9

Such that

$$\xi\left((x,y),(p.q);\, \vartheta^{IWD(i)}\right) = \left(\frac{\alpha_s}{\beta_s + \left(\gamma_s \cdot \left(\mathsf{t}\left((x,y),(p.q);\, \vartheta^{IWD(i)}\right)\right)\right)}\right) \tag{5}$$

where α_s, β_s, γ_s are constants whose values are tabulated in Table 1 and $\mathsf{t}\left((x,y),(p.q);\, \vartheta^{IWD(i)}\right)$ represents the time that the IWD particle takes to reach from Node(x, y) to Node(p, q) with a velocity of $\left(\vartheta^{IWD(i)}\right)$ and is computed as

$$\mathsf{t}\left((x,y),(p.q);\, \vartheta^{IWD(i)}\right) = \frac{||\sigma(x,y) - \sigma(p,q)||}{\max(0.0001, \vartheta^{IWD(i)})} \tag{6}$$

where the function $\sigma(.)$ corresponds to the two-dimensional positional vector of the pixel nodes and in the denominator the function max is used to choose the maximum value between 0.0001 and the velocity $\vartheta^{IWD(i)}$ so as not to count in the negative or zero values of the velocity.

At the same time, the amount of soil present in the path followed by the IWD particle also needs to be updated. As the IWD moves across the Node(x, y) to reach the $Node(p, q)$, the soil content of the path also diminishes and this reduction in the path-soil is directly proportional to the increment in the soil content of the IWD, i.e. $S^{IWD(i)}$. This can be illustrated with the help of (7).

$$\text{soil}(x, y) = (1 - \delta) \cdot (\text{soil}(x, y)) - \delta \cdot \xi\left((x, y),\, (p.q);\, \vartheta^{IWD(i)}\right) \tag{7}$$

where δ is a small fractional value lesser than 1 and in this experiment, the value of δ has been taken as 0.9.

4.4 Decision Stage

The decision technique is the final and most important stage where a decision is taken whether a particular pixel belongs to the smooth region or the complex region of the cover image. The decision is taken based on the final soil contents of the pixel nodes in the soil matrix and for this purpose, a threshold soil value (soil^t) is reckoned.

The average of the values of the soil matrix is taken as the initial threshold (soil^{ti}), following which, each of the soil matrix entries is categorized into two classes—one which comprises the values exceeding the initial threshold (soil^{ti}) and the other containing values lesser than the initial threshold (soil^{ti}). Then again, the average value is computed for each of the two classes and the mean of the averages of the two classes is defined as the new threshold. This process for reckoning the threshold value is continued till a stable value is achieved, which is then considered as the final threshold value (soil^{tf}).

The ultimate decision for a pixel (x, y), as to whether it belongs to the complex region or not, is taken based on the final threshold value (soil^{tf}) and the soil content soil(x, y) at position (x, y) as illustrated in (8).

$$\text{Bin}_{(x,y)} = \begin{cases} 1, & \text{soil}(x, y) \le \text{soil}^{tf} \\ 0, & \text{otherwise} \end{cases} \tag{8}$$

where Bin: The binary texture image.

If the soil value for a particular pixel (x, y) is lesser than the threshold value, then the pixel is considered to be in the complex region, else it belongs to the smooth region of the image.

4.5 Data Hiding Process

For the purpose of information concealing, the least significant bit (LSB) substitution technique has been taken into account. This stage performs the task of concealing the secret message in the LSBs of the pixels belonging to the complex regions of the cover image $\mathbb{C}$. For this purpose, the entire image is explored and each pixel $\mathbb{C}_{(x,y)}$ is examined whether it falls into the smooth region or the complex region. If the pixel belongs to the smooth region, then it is left as it is and the next pixel is taken into account, but if the pixel $\mathbb{C}_{(x,y)}$ falls into the complex region then the least significant bits of this pixel are replaced with the secret message.

From (8), it can be seen that a pixel $\mathbb{C}_{(x,y)}$ is considered to be in the complex region if its corresponding $\text{Bin}_{(x,y)} = 1$ and in the smooth region if its corresponding $\text{Bin}_{(x,y)} = 0$. Now, let a secret message SecMsg to be concealed in the cover image $\mathbb{C}$ and $\mathbb{S}$ be the final stego-image acquired as a result of hiding this information. Then, the stego-image can be represented as shown in (9).

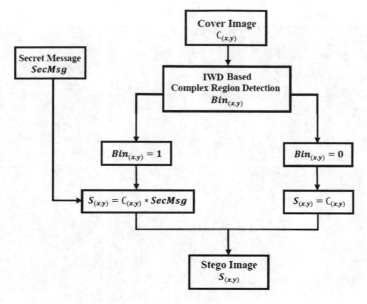

Fig. 3 Diagram showing the overall idea of the proposed method

$$\mathbb{S}_{(x,y)} = \begin{cases} \mathbb{C}_{(x,y)} * \text{SecMsg}, & \text{Bin}_{(x,y)} = 1 \\ \mathbb{C}_{(x,y)}, & \text{Bin}_{(x,y)} = 0 \end{cases} \tag{9}$$

The overall idea that is being put forward in this paper can be pictorially represented as shown in Fig. 3.

5 Experimental Result and Analysis

The performance of the proposed IWD-based approach of steganography has been evaluated on a number of standard and popular images used for steganography experiments like Lena, Mandrill, Cameraman, Pepper, Jelly Beans, House, Tree and Tiffany. Each of these images is of the same dimensions of 128 × 128. Each of these images has been utilized as cover images and the IWD-based proposed method has been utilized for finding out the complex regions within them, where the secret messages are embedded, thus resulting into the stego-images. Figure 4 shows some sample cover images and their respective stego-images. The base framework prerequisites for the proposed method are Ram 2 GB, CPU 1.5 GHz Dual Core, operating system Windows 7, programming language MATLAB 2015a.

To analyse the quality of the stego images, thus formed, a number of evaluation metrics like hiding capacity (HC), mean squared error (MSE) and peak signal-to-noise ratio (PSNR) have been taken into account.

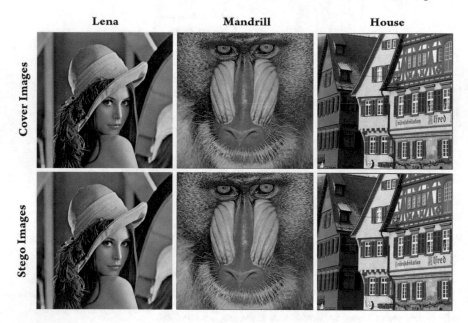

Fig. 4 Sample cover images and their respective stego-images

The term hiding capacity (HC) for an image can be defined as the amount of data that can be concealed within any given cover image, without debasing its quality. For a steganographic approach to be efficient in concealing data within a cover image, the hiding capacity (HC) should be high so that without degrading the image quality, the maximum amount of information can be hidden within it. It can be computed numerically as shown in (10).

$$HC = \frac{\text{Total number of hidden bits of information}}{\text{Total number of bits in the Cover image}} \times 100\% \qquad (10)$$

$$\Rightarrow HC = \frac{m}{r \times c \times 8} \times 100\% \qquad (11)$$

where m is the number of bits that the secret message consists of and r and c are the number of rows and columns of pixels that the cover and the stego-images have.

The mean squared error (MSE) is defined as the average of the squared variations in the intensities of the pixels between the pixels of the cover photo and their corresponding pixels in the stego-image. The higher the estimation of MSE, the higher is the error. For any image with a dimension of $r \times c$, the mean squared error (MSE) is reckoned as shown in (12).

$$MSE = \frac{\sum_{x=1}^{r} \sum_{y=1}^{c} [\text{Cover}(x, y) - \text{Stego}(x, y)]^2}{r \times c} \qquad (12)$$

where r and c are the quantities of rows and columns of pixels that the images have.

Peak signal-to-noise ratio (PSNR) is another vital evaluation metric, which is defined as the ratio of a signal's most extreme conceivable intensity to the intensity of adulterating noise that influences the precision of its delineation. The higher the value of PSNR, the more superior the quality of the picture is. Generally, PSNR is measured in the logarithmic decibel calibration due to the reason that numerous signals possess a very wide and extensive dynamic range. For a given image, the peak signal-to-noise ratio (PSNR) can be calculated as shown in (13).

$$PSNR = 10 \log_{10}\left(\frac{R^2}{MSE}\right) \tag{13}$$

where the term R corresponds to the highest intensity value of the pixels comprising an image and for scenarios where the pixels are organized in 8-bit format, the value of R is 255. Hence, for our case, (13) can be reframed as (14).

$$PSNR = 10 \log_{10}\left(\frac{255^2}{MSE}\right) \tag{14}$$

For each of the stego-image, thus, obtained by following the proposed approach on the considered set of cover images, the hiding capacity (HC), mean squared error (MSE) and peak signal-to-noise ratio (PSNR) have been computed and the acquired values for each of these images have been tabulated in Table 2.

Table 3 shows the comparison of the output obtained by using the proposed method, with other recent state-of-art techniques like the ones put forward by Honsinger et al. [6], Fridrich et al. [7], Goljan et al. [8], Khan et al. [9], Vleeschouwer et al. [10], Sahib et al. [11] and Macq and Dewey [12]. All these approaches have been compared on the basis of hiding capacity (HC) and peak signal-to-noise ratio (PSNR) of the resulting stego-images, and for this purpose, only the Lena and the Mandrill images have been taken into consideration.

Table 2 Hiding capacity (HC), mean squared error (MSE) and peak signal-to-noise ratio (PSNR) of various images using the proposed method

Cover images	Hiding capacity (%)	MSE	PSNR (dB)
Lena	4.690920	0.42707	51.82581296
Mandrill	5.804968	1.15500	47.50498377
Cameraman	4.287816	1.21352	47.29033423
Pepper	4.760704	0.87934	48.68923532
Jelly Beans	4.218032	0.96915	48.26689361
House	4.195776	1.03488	47.98190367
Tree	5.449496	1.20428	47.32352887
Tiffany	4.767048	1.18538	47.39222765

Table 3 Comparison of the proposed method with other recent state-of-art techniques

Approaches	Lena		Mandrill	
	Hiding capacity (bpp)	PSNR (dB)	Hiding capacity (bpp)	PSNR (dB)
Honsinger et al. [6]	<0.0156	–	<0.0156	–
Fridrich et al. [7]	0.0156	–	0.0156	–
Vleeschouwer et al. [10]	0.0156	30.00	0.0156	29.00
Macq and Deweyand [12]	<0.03125	–	<0.03125	–
Goljan et al. [8]	0.36	39.00	0.44	39.00
Khan et al. [9]	0.33	46.23	0.669	44.12
Sahib et al. [11]	0.36	50.78	0.44	45.67
Proposed approach	0.38	51.83	0.49	47.50

6 Conclusion

The proposed method of intelligent water drop-based complex region determination (IWD-CRD) proves to be an efficient approach for hiding the secret information within the images and this commendably exploits the shortcomings of less sensitivity to the variations in the cover image's complex regions that were associated with HVS. For a steganographic approach to be more stable and efficient, it should have a greater hiding capacity (HC) and higher PSNR value, which the proposed method exhibits. This procedure results in remarkably superior quality of stego-images with high data concealing limit and comparison with other recent state-of-art techniques reveals the supremacy of the proposed method.

References

1. Devi, K.J.: A secure image steganography using LSB technique and pseudo random technique technique. Department of Computer Science and Engineering National Institute of Technology, Rourkela Odisha, Bachelor Thesis (2013)
2. Mishra, M., Adhikary M.C.: An easy yet effective method for detecting spatial domain LSB steganography. Int. J. Comput. Sci. Bus. Inform. 8(1) (2013)
3. Fridrich, J., Goljan, M., Soukal, D.: Searching for the stego-key. In: Proceedings of Security, Steganography and Watermarking of Multimedia Contents, vol. 5306, pp. 70–82 (2004). https://doi.org/10.1117/12.521353
4. Pahati, O.J.: Confounding Carnivore: How to Protect Your Online Privacy. AlterNet. Archived from the original on 2007-07-16 (2001)
5. Chan, C.: Hiding data in images by simple LSB substitution. J. Pattern Recogn. Soc. (2003)
6. Honsinger, C.W., Jones, P.W., Rabbani, M., Stoffel, J.C.: Lossless recovery of an original image containing embedded data. U.S. Patent No. 6,278,791 (2001)
7. Fridrich, J., Goljan, M., Du, R.: Invertible authentication. In: Photonics West 2001-Electronic Imaging. International Society for Optics and Photonics (2001)

8. Macq, B., Dewey, F.: Trusted headers for medical images. In: DFG VIII-D II Watermarking Workshop, Erlangen, Germany, vol. 10 (1999)
9. Goljan, M., Fridrich, J.J., Du, R.: Distortion-free data embedding for images. In: Information Hiding. Springer, Berlin (2001)
10. Khan, S., Ahmad, N., Ismail, M., Minallah, N., Khan, T.: A secure true edge based 4 least significant bits steganography. In: International Conference on Emerging Technologies (ICET), Islamabad, Pakistan, pp. 1–4 (2015)
11. Khan, S., Bianchi, T.: Ant Colony Optimization (ACO) based data hiding in image complex region. Int. J. Electr. Comput. Eng. **8**, 379–389 (2018)
12. Vleeschouwer, C.D., Delaigle, J.F., Macq, B.: Circular interpretation of histogram for reversible watermarking. In: IEEE Fourth Workshop on Multimedia Signal Processing (2001)

Hierarchical Multi-objective Route Optimization for Solving Carpooling Problem

Romit S. Beed, Sunita Sarkar, Arindam Roy and Durba Bhattacharya

Abstract Rapid urbanization has resulted in traffic congestion leading to air and noise pollution. An easy and effective solution to this problem is carpooling. The objectives of the carpooling system are generally conflicting in nature and hence obtaining an optimal route falls under the domain of multi-objective optimization. Most of the literature, available in this domain, treats these objectives at a horizontal level. This work proposes a hierarchical approach of classifying objectives; first at the micro level to choose a particular passenger based on passenger location characteristics and then optimizing at a macro level to obtain the most profitable route. The most optimum routes are generated which maximizes the profit for the service provider by minimizing the travel cost and the passenger pickup-drop cost and maximizing the capacity utilization of the car. The proposed algorithm generates Pareto optimal solutions.

Keywords Route · Multi-objective · Optimization · Carpooling · Genetic algorithm · Hierarchical

R. S. Beed (✉)
Department of Computer Science, St. Xavier's College (Autonomous), Kolkata, India
e-mail: rbeed@yahoo.com

S. Sarkar
Department of Computer Science & Engineering, Assam University, Silchar, India
e-mail: sunitasarkar@rediffmail.com

A. Roy
Department of Computer Science, Assam University, Silchar, India
e-mail: arindam_roy74@rediffmail.com

D. Bhattacharya
Department of Statistics, St. Xavier's College (Autonomous), Kolkata, India
e-mail: durba0904@gmail.com

© Springer Nature Singapore Pte Ltd. 2020
J. K. Mandal and S. Mukhopadhyay (eds.), *Proceedings of the Global AI Congress 2019*, Advances in Intelligent Systems and Computing 1112,
https://doi.org/10.1007/978-981-15-2188-1_30

1 Introduction

In the last few decades, the car ownership has risen significantly. As cars have become an absolute necessity in daily commute, it also has worsened the air quality leading to air pollution and increase in lung diseases; traffic jams leading to wastage of time and noise pollution due to honking; need for more parking space leading to increase in expenses and space crunch. In the present scenario, the effects of rising population are quite vivid and clear especially in developing and under-developed countries [1]. It leads to reduction in greenery to make way for buildings and roads. Rapid urbanization is accelerating global warming forcing people to opt for comfort; traveling in air-conditioned cars is no more a luxury but a necessity. Public transport may not be always freely available, and time has become precious leading to the increase in the individual ownership of cars. Owning a car may no more be a problem due to financial schemes being offered by banks, but this leads to various other problems—the need for parking space, the need for road space, and finally emission of poisonous gases. Carpooling is an effective solution to this grave problem [2].

According to the Collin's English Dictionary, Carpool is '… an arrangement to take turns in driving fellow commuters to and from work or friends' children to school and back, so as to avoid the unnecessary use of several under-occupied vehicles.' The main benefit of carpooling is definitely improving the quality of life in metropolitan cities through the reduction of carbon emission [3]. Carpooling ensures reduction in the number of cars on the road and proper utilization of the car's seating capacity. Carpooling also helps in bridging the social divide and harnessing better bonding among citizens through social interactions during rides. Although many carpooling algorithms do exist, not much research has taken place to support passengers as well as service providers in choosing most optimal routes based on various conflicting objectives. Most of the research has focused on finding the shortest journey distance and route ignoring various other parameters which significantly contribute toward such decision making. This paper aims to help both the individual passenger and the service provider/driver and maximize profit by choosing the most optimal route. The system takes into consideration various constraints and suggests a handful of optimal routes which would maximize returns. A hierarchical structure is proposed to overcome the weakness of a flat structure. At the upper level, the focus is on route optimization, leading to maximization of profit for the company. The conflicting objectives here are minimization of travel distance, maximization of car utilization, and minimization of individual passenger's cost. At the lower level, the optimization involves minimizing the passenger's cost which comprises of the pickup and drop cost as well as availability of more customers in the neighborhood.

Genetic algorithms (GAs) are adaptive, exploration algorithm which work on historical data and model the evolutionary concepts based on heredities. In general, these algorithms, although randomized, tend to use heuristic data to exploit promising regions within the search space [4]. The fundamental concept of GAs focuses on simulating natural processes which are modeled on the principle of evolution as illustrated by Charles Darwin in his 'survival of the fittest' philosophy. As seen in nature, there exists a strong struggle among entities due to a limited supply of resource. These result in stronger entities commanding over the weaker entities, so the tougher entities have higher chances to transmit their genes to future offspring over the weaker ones. Pareto optimality is defined to be an efficient allocation of assets from which it is difficult to reallocate assets in order to create any single preference criterion better off without creating at least a single preference criterion worse off [5]. Multi-objective optimization problems may be solved by a posteriori techniques, wherein all Pareto optimal solutions are generated. It may be mathematical program based or evolutionary algorithm based. In the former type, the algorithm is repeated and each execution of the process produces one Pareto optimal solution, whereas in the latter, a set of Pareto optimal solutions are produced by executing the algorithm. The biggest advantage of using the latter is that it generates a set of results, permitting calculation of an estimate of the complete Pareto front. However, one of the disadvantages of this technique is its slow processing speed and there exists no surety of generating the Pareto front.

In spite of the drawbacks, GAs are appropriate for solving MOOPs because it is a population-based approach. This paper proposes to generate results from the Pareto front and improve these results further across generations using genetic algorithms. Section 2 focuses on literature survey on the various techniques used so far in solving the carpooling problem. Section 3 proposes the new hierarchical model of multi-objective optimization in carpooling. Section 4 comprises of the experimental details and results. Section 5 concludes the paper and discusses the achievements and scope for further work.

The work derives motivation from the philosophy of greener and cleaner earth. The work aims at providing solutions which would not only reduce traffic congestion but in turn help in reducing pollution, both air and noise. It will also increase social communication among the users and help in cooperative growth. Another major contribution would be financial savings as the cost to travel will reduce substantially whereby people would be saving on daily expenses. This would lead to a lower cost of living and a stronger economy. This work will also contribute in increasing business returns for the service provider and in enhancing business performance. It will help the driver take decisions based on scientific grounds rather than intuitions.

2 Literature Survey

Several research works have been conducted in this domain of carpooling focusing on the routing process, allocating riders, matching customer requirements, etc. Various techniques like weighted sum approach and genetic algorithms have been used to solve these problems. He et al. [6] focused on routing in carpooling system to increase the desirability of the ride. Various GPS trajectories were mined to obtain most frequent routes using route splitting, grid mapping, and route grouping. Mined frequent routes were stored in a database and used during route generation process afterward for a particular set of riders. Schreieck et al. [7] focused on matching ridesharing offers with ride requests and also storing and retrieving routes in inverted index data structures. Google API was used for geocoding the source and destination address. Based on the results, it was seen that this approach performed well for real-time applications than existing optimization-based techniques. The model proposed by Boukhater [8] applied genetic algorithm for solving the carpooling problem. The input was taken as a map which provided ample number of cars sufficient to provide rides for all passengers, considering their personal preferences. Initial populations were generated and their fitness values were checked based on various constraints. This algorithm was applied on a set of 119 students in Lebanon wishing to travel between 4 different universities. The proposed algorithm involving GA showed better results than the standard algorithm.

Masum et al. [9] proposed another system based on GA involving selection (select a certain section of the population to work on) and reproduction (through crossover and mutation obtain next generation of solutions). Each chromosome represented a set of points for a particular route. The process included a reparation phase which audits the child for any duplicate genes within it as a result of crossover and muta-tion. A heuristic approach was used to re-insert excluded passengers. Penalty concept was also included for selecting the chromosomes for further generations. Although genetic approach did find some good results, it was seen that it took many genera-tions to find valid results. It was also noticed that applying moderate level of heuristic methods for inserting missing customers that were lost did play a significant role in obtaining a better solution. Zhang et al. [10] proposed a carpooling system based on data mining techniques. The approach involved incorporating the non-carpooling taxicab services to develop an optimal taxi-sharing technique as most of these taxi-cabs at present are equipped with GPS and communication devices. These cars would periodically update their location and status to data centers where data mining tech-niques would be used to identify passenger requirements and occupancy ratios of these taxicabs. The system was then used for recommending share rides whenever a new request was found mainly by calculating the shortest detour distance. This model was then compared to a heuristic model to test its effectiveness. It is seen that the proposed system based on data mined from historic taxicab rides increased the effectiveness of the system, much more than a heuristic model. When applied to a real-life data model comprising of 14,000 instances, it showed 60% reduction in total mileage, 41% in passenger's waiting time, and 28% in total travel time.

Mallus et al. [11] suggested a model named CLACSOON, which is based on a partial ridesharing approach. A restricted ridesharing occurs if the source and destination of the on-route passengers fall on the vehicle's original route, or close to it. In this method, the client may have to take a short walk to a point where a drive has to pick them up or drop them off close to a point of destination. This ensures that the detour is avoided. The system was designed for the city traffic with aim to offer dynamism as well as real-time services. Google API provided information regarding the route. The matching algorithm also included a search radius for the driver and a timeout system for the user which decided how much time they would agree to wait for a particular ride. The system also gave freedom to the rider in terms of choosing between one of many matched rides, hence giving a better chance for it to find a near perfect ride. CLACSOON is popular among the android and IOS users, and it managed to attract around 3000 users up until the end of 2016 in the Italian city of Cagliari. Bruglieri et al. [12] introduced the 'PoliUniPool Project' which included time window, favorable and unfavorable list of co-passengers among the student community. The optimization system was modeled on a linear objective function.

3 Proposed Carpooling Model

The proposed model is a real-world example of multi-objective optimization comprising of conflicting objectives. The model proposes to obtain an optimized route which would generate the maximum profit for the service provider. Hierarchical decision structures [13] have widespread applications in various competitive economic organizations, which operate under several conflicting objectives. This paper formulates the carpooling problem as a multi-level, multi-objective optimization model. In the proposed model, the conflicting objectives being local to a particular passenger are considered at the lower level, while the passenger's pickup and drop cost is decided at the higher level. This can be justified as the upper level parameters do influence but does not totally regulate the objectives of the subordinate level. If decentralized components be given choice making abilities, and if the system can be properly synchronized and coordinated, it is bound to improve the performance of the overall system [14] In this regard, dedicated subalgorithms may lead to noteworthy improvements in solving such hierarchical MOOPs.

Let us assume that Passenger A wants to make a journey between points 'A' and 'a.' Figure 1 shows there are three possible routes between the source and destination of Passenger A, marked in orange green and blue. Route Green is the shortest distance route, but it has only three passengers in this route. Route Blue has the maximum number of passengers in the route, but it relatively longer than the Route Green. Route Orange may not have the maximum number of passengers on the route, but all

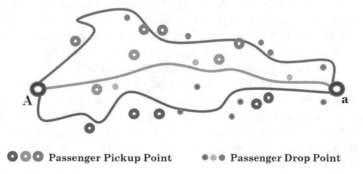

●○◐ Passenger Pickup Point   ●●● Passenger Drop Point

Fig. 1 Higher-level parameters

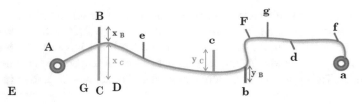

Fig. 2 Lower-level parameters

the customers are located quite close to the main route and there is very little detour distance to be covered by the car.

The service provider would always want to maximize profit driving the vehicle for the least possible distance. More the car travels, more are the expenses incurred in terms of fuel consumption. Maximization of trip profit and minimization of traveling distance are thus conflicting in nature. Coming to the next set of conflicting objectives, maximization of car utilization and minimization of traveling distance, if one tries to maximize the car utilization, there is a high chance that the car needs to travel greater distance. Finally, the last set of conflicting objectives, trying to minimize the pickup + drop cost may lead to improper car utilization. This summarizes the multi-objective nature of the problem.

In Fig. 2, considering $(X_B + Y_B)$ to be the pickup and drop distance associated with Passenger B, it much smaller than the pickup and drop cost associated with Passenger C $(X_C + Y_C)$. However, it is worth noting that there are two other passengers in the vicinity of C in contrast to none in the vicinity of B. It would definitely be more profitable for the service provider to opt for C ahead of B as the car capacity will be better utilized. The detour distance for picking a particular passenger and dropping the passenger at his desired location affects the decision associated with this passenger selection. Similarly, the likelihood of finding other passengers in the neighborhood helps in the overall objective of maximizing the profit. These two are again conflicting in nature as the passenger with the minimum detour distance may not be living in a densely populated area and finding other passengers in that area may be difficult. These two objectives only contribute toward a particular passenger's selection and not as important as the other objectives which affect the route selection. There is a marked difference in the influence of the previous objectives in comparison with these objectives. Thus, it is best to design a hierarchical model for this problem as this problem displays two categories of conflicting objectives.

In Fig. 3, at the upper level, the focus is on route optimization, leading to maximization of profit for the company. The conflicting objectives are minimization of travel distance, maximization of car utilization, and minimization of individual passenger's pickup + drop cost. At the lower level, the optimization involves minimizing the passenger specific detour distance and availability of more customers in the neighborhood of this passenger.

At the **higher level**, the three conflicting objectives are as follows:
Conflicting objectives:
f_1: Minimize the total distance travelled:

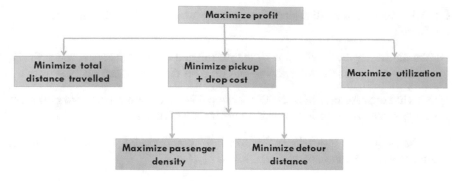

Fig. 3 Hierarchical optimization structure

Distance is defined as the total distance between the source and the destination of the first passenger among the n possible routes. It is assumed that the end location of the trip is the end location of the first passenger.

For the sth possible route,

$$D_s = \sum d_s(i, j) \tag{1}$$

is the distance. Lesser the distance lesser is the cost on fuel and greater is the overall profit.

f_2: Maximize utilization:

The company would want to maximize the utilization of the car, x; i.e., there are minimum number of vacant seats.

f_3: Minimize individual passenger cost:

The company would want to spend the minimum to pick up a passenger. This would again help in maximizing the overall profit. This is achieved by optimizing at a lower level as discussed below.

Combining the above conflicting objectives, we formulate the profit maximization as follows:

Main objective:

$$\text{maximize } f \equiv (f_1; f_2; f_3) \tag{2}$$

Constraints:

$d_s(i, j) \geq 0 \, \forall \, s$
$0 \leq x \leq 4$, as the maximum carrying capacity of the vehicle
It is assumed to be 4 for general vehicles.

At the **lower level,** the conflicting objectives are as follows:

f₄: Minimize detour distance: The pickup and drop off distance should be minimized.

For the ith passenger, let x_i and y_i be the pickup and drop off distance, respectively.
    Hence, $(x_i + y_i)$ is the total pickup and drop off distance of the ith passenger.

f₅: Maximize passenger density: Denser populated areas are expected to give more passengers, thereby ensuring maximum utilization of the car.

Let n_i be the passenger density within a radius r corresponding to the ith passenger.
Main objective: Minimize individual's cost

$$f_3 \equiv (f_4; f_5) \tag{3}$$

Constraints:

$$x_i, y_i \geq 0 \,\forall i$$

$$n_i \geq 0 \,\forall i$$

If a person is made to wait for a long time, he may cancel the ride and opt for another vehicle. So, for the ith passenger, the time for pickup t_i should be less than or equal to a prefixed positive quantity $t(>0)$.
    This algorithm starts off by generating 'm' random routes between the source and the destination of the first passenger, taking into consideration all possible routes available between these two points. Once the routes are generated, the concept of Pareto optimality is used to determine 'n' non-dominated, optimal routes. Routes from the Pareto front are shortlisted and sent for further improvement in the next rounds along with 'm-n' newly generated random routes. This process continues for 'p' generations. During each generation, they undergo selection, crossover, and mutation for refinement and to add randomness. Every route is assessed based on the three conflicting objectives—minimum distance, maximum utilization, and minimum individual passenger cost, and only the non-dominated solutions are shortlisted for further improvement. A request matrix is maintained which keeps track of all requests generated, the latitude, longitude, request time, and nearest junction for each passenger's request for availing the cab.
    The algorithm proposed is as follows:

Algorithm Best Route Selection

Step 1: Read source and the destination of Passenger 1.
Step 2: Generate 'm' random routes between source and destination of first passenger.
Step 3: Apply genetic algorithm (selection, crossover, and mutation) on the 'm' random routes.

Step 4:   Generate '*n*' Pareto optimal routes from among '*m*' random routes and forward it to the next generation of GA.

Step 5:   Add '*m-n*' random routes to the set of routes generated in Step 4 and proceed with Step 3. Repeat this for '*p*' generations.

Step 6:   Display the '*n*' Pareto optimal routes after '*p*' generations.

Any one of these routes can be considered to be a solution as all the '*n*' routes are optimal in nature. Coming to the uniqueness of this work, the hierarchical structure— the conflicting objectives considered so far have only focused on choosing the most optimal route at a higher level without considering the cost associated of choosing one passenger over the other. The question that comes to one's mind is that choosing a particular passenger at the cost of another should definitely make, though small, but some influence on the total cost.

The hierarchical model performs this optimization at the lower level using the following algorithm:

The algorithm starts by considering any drop request for that particular geographic location. If there exists a drop request, then the passenger is dropped and the vacancy counter is incremented by one. The request matrix is then checked for any new pickup request from that specific location. If there exists a request, then the passenger is picked up and the vacancy counter is decremented by one. If there is further vacancy in the car, the car then proceeds to pick up the next optimal customer generated based on the conflicting objectives. This algorithm provides the selection of individual customers at the lower level of the hierarchical structure using multi-objective optimization.

Algorithm Passenger Selection

Step 1:   Drop passenger at current location.

Step 2:   Increase vacancy count by 1.

Step 3:   Read request matrix for passenger request at current location.

Step 4:   If passenger request exists and there is vacancy in car, add passenger to car and store drop location of passenger.

Step 5:   Decrease vacant count by 1.

Step 6:   If vacancy exists, select non-dominant Pareto optimal chromosomes satisfying minimum detour distance and maximum neighbor density.

Step 7:   Repeat Steps 1–6 till final destination is reached.

This process needs to be repeated for each route that exists. The request matrix needs to be refreshed dynamically every time it is accessed.

4   Experiment and Results

The proposed algorithm was implemented on the actual route map of the Salt Lake (Bidhannagar) area, Kolkata. There were 77 junction points in the map. Most of the roads in this area were straight and the junctions at right angles. Google APIs were used to obtain the actual distance between the junction points. The request matrix was dynamically updated every time a route search was performed.

A sample run of the program yields the following results:

Enter the source point (range [0, 77]): 24
Enter the destination point (range [0, 77]): 46

Route No.	Detour_dist	Total_Ocpncy	Dist_Trvld
1	2400	4	7549
2	4200	2	9245
3	4592	24	10,299
4	3996	9	9245
5	1198	3	12,295
6	9598	11	7549
7	5484	26	10,299
8	4790	9	10,299
9	1200	5	10,299
10	2000	8	10,745

Route no. 1 is a non-dominated route.
Route details: 24I39I42I56I55I54I27I33I46I
Route no. 3 is a non-dominated route.
Route details: 24I37I41I42I40I35I17I34I54I14I28I32I46I
Route no. 4 is a non-dominated route.
Route details: 24I37I26I18I26I17I5I14I13I28I45I46I
Route no. 5 is a non-dominated route.
Route details: 24I25I20I25I19I18I26I17I15I14I13I12I28I29I32I46I
Route no. 6 is a non-dominated route.
Route details: 24I37I43I55I54I14I28I33I46I
Route no. 7 is a non-dominated route.
Route details: 24I25I36I40I44I54I27I45I51I63I51I50I46I
Route no. 8 is a non-dominated route.
Route details: 24I25I26I18I25I37I43I55I54I14I27I33I46I
Route no. 9 is a non-dominated route.
Route details: 24I39I25I26I17I5I14I15I14I28I29I33I46I
Route no. 10 is a non-dominated route.
Route details: 24I25I26I17I16I17I34I16I15I6I14I28I33I46I

All distances are measured in meters. It is to be noted that Route 2 does not qualify as a probable route because it is dominated with respect to all the three parameters by Route 1. However, if Routes 1 and 3 are compared, none of the routes dominate the other and hence both belong to the non-dominated set. In this way, all the remaining routes are checked for dominance and the set of non-dominated routes are generated. Another run of the same program keeps the start and end locations the same but at a different hour of the day yields a different set of optimal paths. This shows the effect of the varying request list though the journey start and end points are the same.

Route No.	Detour_dist	Total_Occpncy	Dist_Trvld
1	2200	3	7549
2	3098	11	11,145
3	996	10	9549
4	2000	3	10,745
5	1392	6	10,299
6	1392	6	10,299
7	3392	12	10,299
8	3996	9	9245
9	0	3	10,299
10	996	5	10,745

Route no. 1 is a non-dominated route.
Route details: 24|19|18|17|15|14|28|33|46|
Route no. 2 is a non-dominated route.
Route details: 24|25|36|37|36|26|18|4|17|15|14|28|29|31|46|
Route no. 3 is a non-dominated route.
Route details: 24|25|36|40|35|44|55|54|27|33|46|
Route no. 7 is a non-dominated route.
Route details: 24|25|37|42|36|35|16|5|15|14|28|33|46|
Route no. 8 is a non-dominated route.
Route details: 24|19|25|37|36|17|16|5|14|28|45|46|
Route no. 9 is a non-dominated route.
Route details: 24|19|18|4|16|5|14|13|12|29|30|31|46|
Route no. 10 is a non-dominated route.
Route details: 24|20|25|26|18|17|34|44|34|44|54|27|45|46|

Figure 4a displays the simulated model during pickup of Passenger A. Three possible routes marked green, red, and yellow are displayed along with location of requests. Figure 4b shows the movement of the car for picking up two passengers. Once the passengers are picked up, their drop locations are displayed in Fig. 4c. Finally in Fig. 4d, the car reaches its final destination, i.e., the destination of Passenger A. It is a predecided constraint that the destination of the first passenger would be the final destination of the route. The program has worked successfully for multiple runs considering varying starting and end locations as well as different request lists.

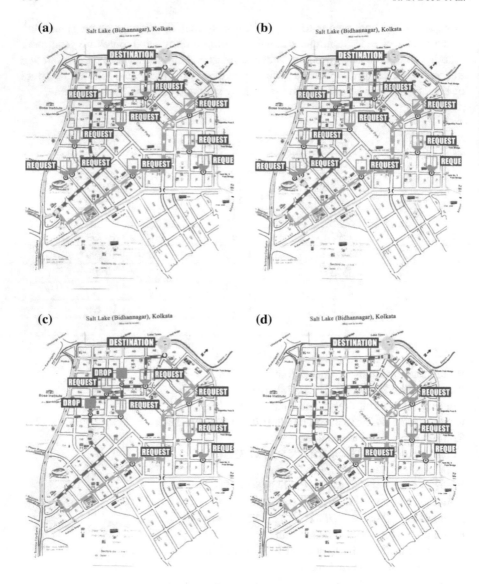

Fig. 4 **a** Route map start of Passenger 1. **b** Car detour to pick new passenger. **c** Drop location of passengers. **d** Car reaches final destination

5 Conclusion

Most of the carpool algorithms designed so far had concentrated on routing techniques and passenger selection by simply considering the shortest distance criteria. But, in reality it is not the sole criteria which decided whom to pick and whom not to pick. This work has aimed at adding a new dimension to this carpooling concept by introducing parameters which are conflicting in nature. Not only they are conflicting but are hierarchical in nature. The simulation over a specific area of the town of Kolkata has shown favorable results. The routes provided by the system are all realizable and when checked on Google APIs have shown similar performances. The hierarchical structure has lent a better reflection of the real-world scenario.

The use of Pareto optimal solutions has done away with the problems of the weighted sum technique, the most popular technique for scalarizing the multiple conflicting objectives and expressing them in terms of their fitness values. The use of genetic algorithms has ensured refinement of the results over the generations. Not only does it provide refinement, it also avoids getting trapped in local optima due to the inclusion of randomness among chromosomes within the algorithm.

This work can be further expanded to include further conflicting parameters like co-passenger compatibility, traffic congestion, etc. Secondly, the system limits itself to only one vehicle, but in future it can consider multiple cars at the same instance of time. This algorithm does not guarantee that all passengers on the route will be serviced. When multiple cars are available, there exists a chance that this passenger may be serviced in the meantime by another vehicle. In Fig. 2, Passenger B may not be serviced by the car after picking up Passengers C, D, and G as the optimality is re-calculated after picking the new passengers. This person may become an optimal passenger for another car at this point of time. Thirdly, the first passenger's start and end locations are considered to be source and destination of the route. The system may be conveniently generalized to continue beyond the destination of the first passenger if needed. The look-ahead clustering that has been proposed considered only a fixed radius from the intended passenger's location. There is scope for improving this clustering technique as well as taking it to another look-ahead level ahead using heuristics. This work presently does not encompass dynamic cancelation both from passenger and driver's perspectives. It does not include the waiting time delays due to a passenger or a technical snag or traffic signal. In spite of these limitations, the model has given considerably satisfactory results for different trial runs and promises to give even better results after the improvements.

References

1. Mulders, C.: Carpooling, a vehicle routing approach. Universit_e catholique de Louvain, Thesis submitted for the Master in Computer Science and Engineering, option Artificial Intelligence, 2012–13
2. Martino, S.D., Galiero, R., Giorio, C., Ferrucci, F., Sarro, F.: A matching-algorithm based on the cloud and positioning systems to improve carpooling. In: DMS, Knowledge Systems Institute, 2011, pp. 90–95
3. Riccardo Manzini, A.P.: A decision-support system for the car pooling problem (2012)
4. Deb, K.: Multi-objective Optimization Using Evolutionary Algorithms. Wiley, New York (2001)
5. Konak, A., Coitb, D.W., Smith, A.E.: Multi-objective optimization using genetic algorithms: a tutorial. Reliab. Eng. Syst. Saf. **91**, 992–1007 (2006)

6. He, W., Hwang, K., Li, D.: Carpool routing for urban ridesharing by mining GPS trajectories. IEEE Trans. Intell. Transp. Syst. **15**(5), 2286–2296 (2014)
7. Schreieck, M., Safetli, H., Siddiqui, S.A., Pflügler, C., Wiesche, M., Krcmar, H.: A matching algorithm for dynamic ridesharing. Transp. Res. Procedia **19**, 272–285 (2016)
8. Boukhater, C.M., Dakroub, O., Lahoud, F., Awad, M., Artail, H.: An intelligent and fair GA carpooling scheduler as a social solution for greener transportation. In: MELECON, 2014-2014 17th IEEE Mediterranean Electrotechnical Conference, Beirut, 2014, pp. 182–186
9. Masum, A.K.M., Shahjalal, M., Faisal Faruque, M., Iqbal Hasan Sarker, M.: Solving the vehicle routing problem using genetic algorithm. Int. J. Adv. Comput. Sci. Appl. **2**(7) (2011)
10. Zhang, D., He, T., Liu, Y., Lin, S., Stankovic, J.A.: A carpooling recommendation system for taxicab services. IEEE Trans. Emerg. Top. Comput. **2**(3) (2014)
11. Mallus, M., Colistra, G., Atzori, L., Murroni, M., Pilloni, V.: Dynamic carpooling in urban areas: design and experimentation with a multi-objective route matching algorithm (2017)
12. Bruglieri, M., Davidovic, T., Roksandic, S.: Optimization of trips to the university: a new algorithm for a carpooling service based on the variable neighborhood search. In: Proceedings of REACT 2011 Shaping Climate Friendly Transport in Europe: Key Findings and Future Directions, Belgrade, Serbia, 16–17 May 2011, pp. 191—199
13. Baky, I.A.: Solving multi-level multi-objective linear programming problems through fuzzy goal programming approach. Appl. Math. Model. **34**(9), 2377–2387 (2010)
14. Takama, N., Loucks, D.P.: Multi-level optimization for multi-objective problems. Appt. Math. Model. **5**, 173–178 (1981)

Bio-molecular Event Trigger Extraction by Word Sense Disambiguation Based on Supervised Machine Learning Using Wordnet-Based Data Decomposition and Feature Selection

Amit Majumder, Asif Ekbal and Sudip Kumar Naskar

Abstract Event extraction is a task of extracting detailed biological phenomenon from biomedical literature. This task needs to extract the words which represent the biological phenomena in terms of natural language in the form of text data. Trigger words are often ambiguous and can convey different meanings in different contexts in biomedical data. We use a supervised approach to disambiguate senses of such ambiguous trigger words. In this paper, we propose a supervised machine learning approach for Word Sense Disambiguation using Genetic Algorithm (GA) for feature selection. We take help of *Wordnet* dictionary to disambiguate words. Our experiments are applied on BioNLP-2011 datasets and we find recall, precision and F-score of 70.71%, 83.70% and 76.66%, respectively, in bio-molecular event trigger extraction.

Keywords Event extraction · Trigger detection · Genetic Algorithm · Feature selection · Wordnet

1 Introduction

Internet is a huge repository for digital data. Electronic documents in biomedical domain are generated daily and volume of these data is being increased day by day. These huge data need to be organized in such a way so that information can be extracted easily and efficiently in less amount of time. This can be immensely ben-

---

A. Majumder (✉)
JIS College of Engineering, Kalyani, West Bengal 741235, India
e-mail: cseamit49@gmail.com

A. Ekbal
IIT Patna, Patna 801106, India
e-mail: asif@iitp.ac.in

S. K. Naskar
Jadavpur University, Kolkata 700032, India
e-mail: sudip.naskar@gmail.com

© Springer Nature Singapore Pte Ltd. 2020   391
J. K. Mandal and S. Mukhopadhyay (eds.), *Proceedings of the Global
AI Congress 2019*, Advances in Intelligent Systems and Computing 1112,
https://doi.org/10.1007/978-981-15-2188-1_31

eficial to the practitioners and researchers who are working in the field of medicine, biology and other allied disciplines. The most recent focus of NLP researches is on extracting fine-grained information from biomedical text [1]. This was addressed in some text mining challenges like BioNLP-11, BioNLP-13 and BioNLP-16 [2–4]. Biomedical event triggers are classified into nine potential types. Among these, five are simple which correspond to *gene expression, transcription, protein catabolism, phosphorylation* and localization. The rest four events, namely *binding, regulation, positive regulation* and *negative regulation* are relatively complex. We use supervised WSD approach to disambiguate the ambiguous words. Our system for trigger word extraction is evaluated using the dataset of BioNLP-2011 event extraction task.

In the field of computational linguistics, word sense disambiguation (WSD) [5] is a problem of natural language processing. WSD is used to identify in which sense (i.e., meaning) a word is used in a particular context, when the word has multiple meanings. The conventional approaches for WSD are *dictionary and knowledge-based methods, semi-supervised or minimally supervised methods, supervised methods and unsupervised methods.*

Different methods have been applied to extract trigger words by researchers. Some of the common methods which have been applied in bio-molecular event trigger extraction are SVM-based machine learning [6, 7], deep learning [8, 9], stacking, model combination [10], classifier ensemble approach, knowledge-based approach [11], etc. For our experiment in event trigger extraction, we used supervised method for applying WSD using *Wordnet* as dictionary of words to disambiguate the ambiguous trigger words. It has been observed that the context information can provide enough evidence on its own to disambiguate the ambiguous words in machine learning using supervised methods.

Table 1 presents a snapshot of the BioNLP-2011 dataset, from which it can be observed that the same word is used in different event types, which indicates the need for WSD. In Table 1, the first two columns present a word and its sense (i.e., trigger type) and the third column shows the number of occurrences of that word sense pair in the dataset. The table shows that same word is used in different sense. For example, the word *accumulation* is used three times as *Localization* type trigger and 33 times as *Positive_regulation* type trigger.

2   WordNet

WordNet[1] is a lexical database for the English language. It is included in Natural Language Toolkit (*NLTK*) which is a module for building Python programs that can work with natural language data. We can use it as a reference for getting the meaning of words and definition. A collection of similar words is called lemmas. The words in *WordNet* are organized with nodes and edges where the nodes represent the word

---

[1]http://www.nltk.org/howto/wordnet.html.

Table 1 Statistics of some example words along with their senses

Word	Sense of word	Number of occurrence
abnormal	Negative_regulation	2
abnormal	Regulation	2
absence	Binding	1
absence	Gene_expression	1
absence	Negative_regulation	23
absence	Positive_regulation	1
absence	Regulation	1
absence	Transcription	1
absent	Gene_expression	1
absent	Negative_regulation	3
absent	Transcription	1
accumulation	Localization	3
accumulation	Positive_regulation	33
activating	Positive_regulation	6
activation	Gene_expression	1
activation	Positive_regulation	286
...	–	–
...	–	–

text and the edges represent the relations between the words. Below we will see how we use the *WordNet* module for applying WSD in trigger detection.

3 Proposed Approach

We applied WSD method to disambiguate biomedical trigger words using *Wordnet* for getting the synonyms of words. In this experiment, the whole dataset is divided into some subsets of data (i.e., *decomposition of data*) wherein each subset we apply the optimization technique based on NSGA-II for feature selection. The whole process consists of the following steps.

1. We make two dictionaries of words. Among these, one is dictionary of ambiguous word (say, U) and the other is dictionary of non-ambiguous (say, NA) trigger words from training dataset.

2. For each trigger word, say t, (belonging to ambiguous or non-ambiguous) we create an empty dataset E_t. That is, create $m+n$ number of empty datasets identified by trigger words, where m and n are sizes of ambiguous and non-ambiguous dictionaries, respectively. One additional empty dataset (say, default dataset iden-

tified as *D*) is created. These can be expressed symbolically in the following way.

$E\_t \leftarrow \{\}$ for $t \in trigger\_words$ (ambiguous or non_ambiguous)
$D \leftarrow \{\}$

3. For each example word (say *w*) from the dataset perform the following actions.

   (a) if the word belongs to dictionary of ambiguous words or dictionary of non-ambiguous words, then, the word along with its feature value (say, F_w) is inserted to the corresponding dataset.
   i.e.,
   $$E\_w \leftarrow E\_w \cup F\_w$$

   (b) Else if one synonym (say, *s*) of the word *w* (using *Wordnet*) belongs to one dictionary, then, the word along with its feature value (say, F_w) is inserted to the corresponding dataset where match occurs.
   i.e.,
   $$E\_s \leftarrow E\_s \cup F\_w$$

   (c) Else insert the word along with its feature value (say, F_w) into the default dataset.
   i.e.,
   $$D \leftarrow D \cup F\_w$$

4. We apply feature selection by NSGA-II for each dataset and create classifier using Support Vector Machine (SVM)[2] as classification algorithm. As the feature selection method is applied on different datasets, the task is performed parallely.

4 Architecture

In the proposed approach, we create different dataset based on the method as explained in Sect. 3. Several types of features have been considered to prepare dataset in feature-format for trigger word extraction. Among these features content, syntax and contextual features are very effective. We used surface wordforms, bag-of-words (BOW), chunk, named entity, stem, PoS, bi-gram, tri-gram features along with some linear features and dependency path features [12]. For every dataset, we find out optimized feature-set by applying feature selection using NSGA-II [13]. The whole process is visualized in Fig. 1.

---

[2]https://www.cs.cornell.edu/people/tj/svm_light/svm_multiclass.html.

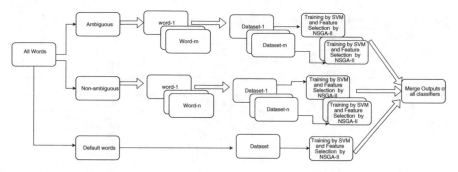

Fig. 1 WSD approach of proposed method

Table 2 Trigger word extraction using *Method1*

Phosphorylation	93.06	89.33	91.16
Positive_regulation	63.30	77.88	69.84
Gene_expression	79.85	88.61	84.00
Regulation	50.72	71.14	59.22
Binding	67.70	87.44	76.32
Localization	74.42	76.19	75.29
Entity	41.86	66.06	51.25
Negative_regulation	55.65	76.95	64.59
Transcription	49.57	90.63	64.09
Protein_catabolism	100.00	91.30	95.45
Overall	63.99	80.33	71.24

5 Experimental Result

We apply WSD on the BioNLP-2011 genia event dataset[3] for trigger detection and our WSD model is tested on the development data. Experimental results have been tabulated in Tables 2 and 3. Table 2 displays the result of trigger word extraction using normal method based on feature selection on whole data at a time (say *Method1*) and Table 3 shows the result of trigger word extraction using our proposed method which performs feature selection on different partial dataset generated from whole dataset based on decomposition using *Wordnet* (say *Method2*). Overall result is shown in Table 4, where we find that trigger word detection using *Method2* has provided better result (5% increase in f-score value) than the *Method1*.

---

[3]http://2011.bionlp-st.org/home.

Table 3 Trigger word extraction using *Method2*

Phosphorylation	90.27	92.85	91.54
Positive_regulation	70.28	83.56	76.35
Gene_expression	81.93	90.16	85.85
Regulation	60.28	76.36	67.37
Binding	76.65	88.73	82.25
Localization	76.74	82.50	79.51
Entity	53.48	66.66	59.35
Negative_regulation	62.50	82.03	70.94
Transcription	59.82	93.33	72.91
Protein_catabolism	100.00	87.50	93.33
Overall	70.71	83.70	76.66

Table 4 Comparison between *Method1* and *Method2* for Trigger detection

Method	Recall	Precision	F-score
Without WSD (i.e., de)	63.99	80.33	71.24
With WSD	70.71	83.70	76.66

5.1   Analysis of Experimental Results

From Tables 2 and 3, it is clear that in most of the cases, *Method2* (i.e., our proposed method) has provided better result than the *Method1* (i.e., without WSD). For example, in case of *Positive_regulation* type trigger, f-score values are 71.24% and 76.66% in *Method1* and *Method2*, respectively. The *Gene_expression* type trigger words were extracted with f-score values of 84.00% and 85.85% in *Method1* and *Method2*, respectively. For the *Regulation* type trigger words, f-score values are 59.22% and 67.37% in *Method1* and *Method2*, respectively. Therefore, the method with WSD (i.e., *Method2*) approach shows 7, 2 and 8% increase in f-score values in case of *Positive_regulation*, *Gene_expression* and *Regulation* type trigger words. From the overall result, we see that the method with WSD (i.e., *Method2*) shows increment of 5% f-score in comparison with *Method1*. This improvement is due to mainly two reasons. One reason is, we applied feature selection for each ambiguous word separately. Another reason is, we relied on the *Wordnet* dictionary which provides us relevant information to disambiguate the ambiguous words.

5.2   Comparison with Existing Systems

Trigger detection is the first step in extracting genia event expressions. Event extraction systems generate results for event expression, but we have done experiments on trigger detection. Therefore, it is not possible to compare our experimental results with the results of event extraction systems. We compare our results with the existing results on trigger detection. The state-of-the-art result for trigger detection on BioNLP-2011 dataset is 67.3% [14] f-score, whereas our trigger detection system results in 76.66% f-score (+9%). Therefore, our trigger detection system performs better than the state-of-the-art trigger detection systems.

6 Conclusion

In this paper, we reported an approach based on supervised WSD using *Wordnet* dictionary to disambiguate word sense in biomedical data for event trigger detection. We have relied on the *Wordnet* dictionary, though it is not completely adapted with biomedical data. Because it has been seen that there are some words which are present in biomedical literature but not available in *Wordnet*. In the future, we would like use one dictionary which is adapted with biomedical data. The present work is focused to disambiguate the ambiguous words and extract the trigger words. In the future, we would like extend this work to extract the arguments of the extracted trigger words.

References

1. Kim, J.D., Ohta, T., Pyysalo, S., Kano, Y., Tsujii, J.I.: Overview of BioNLP'09 shared task on event extraction. In: BioNLP '09: Proceedings of the Workshop on BioNLP, pp. 1–9
2. Nédellec, C., Bossy, R., Kim, J.D., Kim, J.J., Ohta, T., Pyysalo, S., Zweigenbaum, P.: Overview of BioNLP shared task 2011. In: Proceedings of BioNLP Shared Task 2011 Workshop, pp. 1–6
3. Li, L., Wang, Y., Huang, D.: Improving feature-based biomedical event extraction system by integrating argument information. In: Proceedings of the BioNLP Shared Task 2013 Workshop, pp. 109–115
4. Kim, J.-D., Wang, Y., Colic, N., Beak, S.H., Kim, Y.H., Song, M.: Refactoring the genia event extraction shared task toward a general framework for IE-driven KB development. In: Proceedings of the 4th BioNLP Shared Task Workshop, pp. 23–31
5. Kågebäck, M., Salomonsson, H.: Word sense disambiguation using a bidirectional LSTM. In: 5th Workshop on Cognitive Aspects of the Lexicon (CogALex). Association for Computational Linguistics
6. Joachims, T.: Multi-class support vector machine. Cornell University, Department of Computer Science, USA
7. Cortes, C., Vapnik, V.: Support-vector networks. Mach. Learn. **20**, 273–297 (1995)
8. B. örne, J., Salakosk, T.: Biomedical event extraction using convolutional neural networks and dependency parsing. In: Proceedings of the BioNLP 2018 workshop, pp. 98–108 (2018)
9. Björne, J., Salakoski, T.: Generalizing biomedical event extraction. In: Proceedings of BioNLP Shared Task 2011 Workshop
10. Riedel, S., McClosky, D., Surdeanu, M., McCallum, A., Manning, C.D.: Model combination for event extraction in BioNLP 2011. In: Proceedings of BioNLP Shared Task 2011 Workshop, pp. 51–55
11. Henry, S., Cuffy, C., McInnes, B.: Evaluating feature extraction methods for knowledge-based biomedical word sense disambiguation. In: BioNLP 2017, pp. 272–281. Association for Computational Linguistics, Vancouver, Canada (2017)

12. Majumder, A., Ekbal, A., Naskar, S.K.: Feature selection and class-weight tuning using genetic algorithm for bio-molecular event extraction. In: NLDB
13. Deb, K., Pratap, A., Agarwal, S., Meyarivan, T.A.M.T.: A fast and elitist multiobjective genetic algorithm: NSGA-II. IEEE Trans. Evol. Comput. **6**, 181–197 (2002)
14. Wang, J., Wu, Y., Lin, H.F., Yang, Z.H.: Biological event trigger word extraction based on deep syntactic parsing. Comput. Eng. **40**, 25–30 (2014)

Multimodal System for Emotion Recognition Using EEG and Customer Review

Debadrita Panda, Debashis Das Chakladar and Tanmoy Dasgupta

Abstract Emotion is a basic expression of a human being using which he/she can communicate with the external world. Emotion can be recognized using various media such as movie, image, facial expression, and audio. Emotion recognition can be performed by the brain signal (e.g., Electroencephalogram). A multimodal system of emotion recognition using Electroencephalogram (EEG) and sentiment analysis of customer has been proposed. Four types of emotion, namely: Happy, sad, relaxed and anger have been recognized. The proposed multimodal framework accepts the combination of temporal (EEG signal) and spatial (customer reviews/comments) information as inputs and generates the emotion of user during watching the product on computer screen. The proposed system learns temporal and spatial discriminative features using EEG encoder and text encoder. Both of the encoders transform the features of EEG and text into common feature space. The methodology is being tested on a dataset of 30 users, consisting of EEG and customers' review data. An accuracy of 98.27% has been recorded.

Keywords Natural language processing · Machine learning · EEG · Sentiment analysis

1 Introduction

For every marketer, the main challenge is to impose something new into their products that can pull customers toward their brand instead of another. Every organization aims to develop products and services that increase profit as well as market share. Therefore, studying brain responses of consumers is growing with unparal-

D. Panda (✉) · T. Dasgupta
Department of Business Administration, University of Burdwan, Bardhaman, India
e-mail: DebadritaPanda1@gmail.com

D. D. Chakladar
Department of Computer Science and Engineering, Indian Institute of Technology Roorkee, Roorkee, India

© Springer Nature Singapore Pte Ltd. 2020
J. K. Mandal and S. Mukhopadhyay (eds.), *Proceedings of the Global AI Congress 2019*, Advances in Intelligent Systems and Computing 1112, https://doi.org/10.1007/978-981-15-2188-1_32

lel pace. Observation of brain activities for marketing field is the main foundation logic behind neuromarketing. This constantly developing field focuses on consumer behavior to understand the cognitive process as well as their influence on decision-making. Research [1] has highlighted how enough money is being invested by firms for customer engagements, still not able to make them motivate for the brand and ultimately causing a great waste of money moreover ideas. As a preventive step, the primary objective of neuromarketing is to make clear why consumers are selecting or why they are not selecting a brand or product at the time of decision-making [2]. In a study conducted by Vecchiato et al. [3] revealed that more than 70% of new products in different product lines like cars, shoes, etc have seen extreme failure within first six months even though the concept was tested through traditional marketing techniques with questionnaires and psychological interviews. The underlying logic is that people generally cannot express true views about any product or brand when interviewed. Plassmann et al. [4] identified five main ways that neuromarketing can help marketing professionals by identifying, validating, redefining, extending existing marketing theories, by measuring implicit processes, the understanding difference for individual purchase, improvement of prediction and dissociation between many psychological processes. With the help of neuroscience and other research technique related to the physiological field, neuromarketing is getting very popular in many marketing-related domains like consumers' behavior, preferences, decision-making [5]. Some popular tools used for emotion detection in the neuromarketing field are functional magnetic resonance imaging (fMRI), magnetoencephalography (MEG), and electroencephalography (EEG). Besides these, eye tracking [6], heart rate, and facial images [7] are some of the widely used physiological aspects which also can be measured to gain insight about customers' experience. But the costs associated with these experiments are very expensive in nature. Moreover, additional costs like maintenance, insurance, etc limit the use of fMRI. To overcome this, EEG is being preferred to examine brain activities due to inexpensive setup along with high temporal resolutions. Advantages like easy handling, lower maintenance costs, wireless connectivity have make EEG very popular for emotion detection. Many studies have discussed how emotion detection through EEG helps in prediction for consumer behavior [8, 9]. Along with emotions, sentiments do have an important role to play. Sentiment classifiers [10], recommender systems [11], opinion extraction [12], are some of the common forms of sentiment analysis. This type of analysis extracts overall sentiment exists in a form of the written document as a form of feedback, recommendation, experience sharing, etc. The outcome is expected to understand the overall expression, either positive or negative, or somewhere in between. Focused areas of this paper are listed below:

i. We propose a new architecture to combine two different sources, namely EEG and customer's review used for emotion recognition.
ii. The proposed model implements a combined approach of spatio-temporal adaptation using machine learning and natural language processing (NLP).

Section 2 provides some glimpses of related work, Sect. 3 elaborates different phases of the proposed model. Result and analysis part is mentioned in the next section (Sect. 4). At the end of the paper, Sect. 5 provides conclusion.

2 Literature Review

This review has been done based on three subsections: emotion recognition using EEG (Sect. 2.1), sentiment analysis using NLP (Sect. 2.2), and multimodal framework for emotion recognition (Sect. 2.3).

2.1 Emotion Recognition Using EEG

This subsection is related to those studies which have combined EEG signal activity along with self-reported ratings for predicting consumer behavior. Ambler et al. [13] had shown how brand choice can be predicted through brain imaging. In this study, subjects were seen to watch the virtual video about supermarket visit and finally asked to select one of the three brands. This study provides a guideline that customer decision-making can be predicted using brain imaging. A strong correlation had been noticed between brain activation and brand familiarity. "Correlation-Based Subset Selection" and LDA classifier have been implemented for EEG-based emotion recognition technique [14]. They have achieved 82% accuracy. In paper [15], both eye tracker and EEG signal have been simultaneously used for 18 subjects to predict two sets of emotions (like and dislike) while making choice from a set of three images. Mohammadi et al. [16] proposed wavelet-based emotion recognition using KNN classifier.

2.2 Sentiment Analysis Using NLP

In paper [17], authors tried to understand the brand preference of consumers through Facebook data by using map reducing paradigm with the iterative process of data preprocessing. Day et al. [18] used deep-learning methods to predict stock price by using financial news as a source. A convolutional neural network (CNN)-based visual sentiment analysis has been implemented in paper [19]. Different classifiers like SVM, Naïve Bayes, Maximum entropy had been used for classifying the emotions from twitter comments [20].

2.3 Multimodal Framework for Emotion Recognition

An increasing number of researches is being conducted by combining brain activity tracking using EEG and sentiment analysis for emotion detection, video tagging, etc. For analyzing user comments' from Youtube Web site, a multimodal framework had been established with the combination of both EEG signal and sentiment analysis [21]. With the help of four modes like audio/video and still photographs, a multimodal emotion recognition test consisting of two variants each for five emotion families (total 10 emotions) had been presented by Banziger et al. [22]. Zheng et al. [23] suggested a multimodal emotion recognition model by combining EEG signal along with eye tracking of five participants with the help of fifteen emotional film clips. The decision and feature fusion scheme of the multimodal system has improved the performance up to 74% and 73%, respectively.

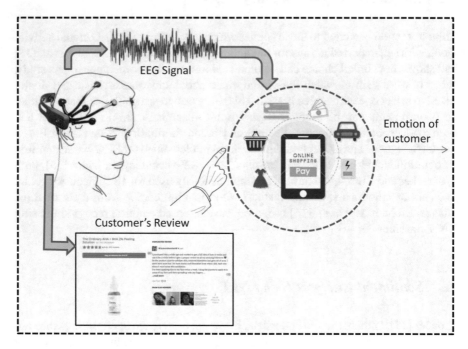

Fig. 1 Pictorial representation of multimodal framework for recognizing emotions using EEG signals and analysis of customers' review

3 Proposed Methodology

In this section, we discuss emotion recognition using multimodal framework, namely EEG and sentiment analysis. Moreover, we also discuss the system setup, prepossessing of EEG and text, detailed analysis of the proposed framework.

The proposed multimodal emotion recognition system comprised of two different unimodal sources, namely: EEG signal and customers' review. The review of a customer can be examined using sentiment analysis which will be beneficial to identify the emotion of the customer during watching a product on an e-commerce platform. The proposed multimodal system has been represented in Fig. 1. EEG signals are being captured while the participant is watching the product during an online shopping Web site (e.g., amazon, Flipkart, etc.). The comments of a participant are also being considered during the experiment for sentiment analysis. The EEG signal acquisition is done using the Emotiv EPOC+ device. The device has 14 channels and the electrode channels are, namely: AF3, AF4, F7, F3, F4, F8, FC5, FC6, T7, T8, P7, P8, O1, and O2.

3.1 Prepossessing and Feature Extraction of EEG

In this subsection, we discuss the prepossessing of the EEG signals. We also discuss the feature extraction technique after signal smoothing. The details are mentioned below.

3.1.1 Discrete Wavelet Transform (DWT)

By nature EEG signal is noisy, so it needs to be prepossessed before further processing. Therefore, we have used a bandpass filter (1–32Hz) to remove the high-frequency components or noise elements from EEG signal.

Discrete wavelet transform (DWT) has been used for extracting the relevant features from smoothed EEG signal. In DWT, input signal is decomposed into two coefficients: approximate and detail coefficient. Low pass and high pass filters have been used for extracting the approximate and detail coefficients, respectively. Coefficients in the ith level are downsampled by two to generate the coefficients in the next $(i + 1$th) level. DWT segregated the signal x(t) into wavelet ($\Psi_{j,k}$) and scaling ($\omega_{j,k}$) functions which are represented in the (1) and (2), respectively, where k represents the samples of data records, j is the small natural number and $j, k \epsilon Z$.

$$\Psi_{j,k}(t) = 2^{-j/2}\Psi(2^{-j}t - k) \tag{1}$$

$$\omega_{j,k}(t) = 2^{-j/2}\omega(2^{-j}t - k) \tag{2}$$

The approximation coefficient (A_j) and detail coefficient (D_j) at jth levels are calculated using (3) and (4) respectively.

$$A_j(t) = \frac{1}{\sqrt{T}} \sum_t (x(t)\Psi_{j,k}(t)) \tag{3}$$

$$D_j(t) = \frac{1}{\sqrt{T}} \sum_t (x(t)\omega_{j,k}(t)) \tag{4}$$

Here, we use Daubechies-4 (DB4) wavelet transform as DB4 decomposes the brain signal into five small signal components equivalent to the frequency of different EEG bands (alpha, beta, delta, theta, and gamma). So using DWT decomposition, we extract the features of specific bands from the smoothed EEG signal.

3.2 Prepocessing of Customers' Review

Generally, review consists of many noisy data which do not contribute much for analysis of sentiment. Hence, preprocessing is mandatory. This process has been initiated with elimination of stop words, punctuation marks, and those words which repeatedly occur. Description of activities involved in preprocessing is listed below.

1. Tokenization: This is the first step in text analytics. In tokenization, the text paragraph is broken down into smaller chunks such as words or sentences. In the next stage, each word mapped to meaningful part of speech.
2. Part of speech (POS) tagging: Here, individual word has been assigned with a tag, for example, noun (NN), verb (VB), and adjective (JJ), etc. The POS lexicon maintains the mapping between lexeme of words and corresponding POS.
3. Stopwords: Stopwords are considered as noise in the text. Some common forms of Stopwords in review, like "a," "an," "the," "after," "am," etc.
4. Lemmatization: With the help of vocabulary and morphological analysis, transformation of some words which are represented in different ways has been replaced with the root word.
5. Filtering of repeating words: While writing the review, sometimes customers uses repetitive characters like "Helloooo," "Noooooo," "Hiiiiiii." So these words need to be suppressed during the prepossessing step.
6. Special characters: Removal of special Characters like "!, @, #, $, %", from the review is done at this stage.

3.3 Multimodal Framework of Emotion Recognition

The input of our proposed model is a pair containing EEG signal and the review of a customer. The neural activity (recorded EEG signal) and review text have different

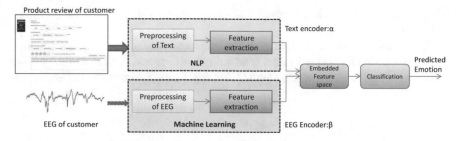

Fig. 2 Proposed multimodal emotion recognition framework using EEG and customer's review

structures. So, we convert both heterogeneous inputs into a common feature space before further processing. The proposed framework has been plotted in Fig. 2. The preprocessing and feature extraction process of EEG signal has been performed by a machine learning algorithm, whereas preprocessing and feature extraction of review text has been done using natural language processing (NLP). EEG band-related features from DWT have been used for emotion recognition. Unigram, Bigram, and n-gram features are used by the NLP system. Once the features of both unimodal systems have been extracted using EEG encoder and text encoder, then those features are combined into common/embedded feature space.

3.3.1 Embedded Feature Space

Let $S = \{e_j, t_j\}$, where $j = 1, 2, 3, \ldots, N$, be a dataset consists of EEG signals (e_j) and review text (t_j). The EEG signals are captured while the customers are watching the product. Assume γ is the space of EEG signal and θ is the space of the text. Now, our objective is to train two encoders (EEG and text) such that they convert the neural response and text data into an embedded space. The transformation/encoding of the input EEG signal and text data into a common space is performed by the EEG encoder (β) and text encoder (α). The EEG encoder converts the temporal features into a 2D feature vector, whereas the text encoder extracts the relevant features using bag-of-words model (BoW) which converts the features into a 2D vector. Once the feature vector of embedded space is ready, then we classify the emotion based on the common feature vector. Here, we use random forest classifier to classify the emotion of customer.

4   Results and Analysis

Here, we have discussed results and performance of our proposed framework. The dataset is consisting of 30 users of different age group from Indian Institute of Technology Roorkee, India. The detailed description of this dataset has been presented in Table 1.

Customers provide their review on the computer while watching the product. Different words like "good," "fine," "amazing," etc. words related to the positive sentiment of a customer, whereas "worst," "disappointing," "not effective," etc. words

Table 1 Age group-wise subjects distribution

Group	Distribution of age	Frequency of participants
Group 1 (G1)	18–30	12
Group 2 (G2)	25–45	10
Group 3 (G3)	40–50	8

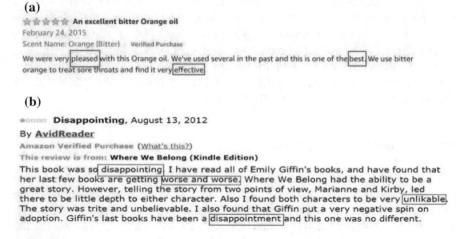

(a)

★ ★ ★ ★ ★ **An excellent bitter Orange oil**
February 24, 2015
Scent Name: Orange (Bitter)   Verified Purchase

We were very pleased with this Orange oil. We've used several in the past and this is one of the best. We use bitter orange to treat sore throats and find it very effective

(b)

★ **Disappointing**, August 13, 2012
By **AvidReader**
Amazon Verified Purchase (What's this?)
This review is from: **Where We Belong (Kindle Edition)**
This book was so disappointing I have read all of Emily Giffin's books, and have found that her last few books are getting worse and worse. Where We Belong had the ability to be a great story. However, telling the story from two points of view, Marianne and Kirby, led there to be little depth to either character. Also I found both characters to be very unlikable. The story was trite and unbelievable. I also found that Giffin put a very negative spin on adoption. Giffin's last books have been a disappointment and this one was no different.

Fig. 3 Different reviews of customer: **a** positive review, **b** negative review. The red colored word signifies the different type of sentiments of customer

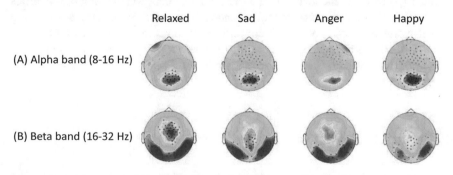

Fig. 4 Topographical view of brain for different emotions: **a** spectrum power of alpha band, **b** spectrum power of beta band

related to the negative sentiments. Our text dictionary stores several popular words of different sentiments that have been used in online e-commerce Web site and compare those words with customer review to identify the proper emotion. A sample of positive and negative review of a product has been shown in Fig. 3. The specific words related to different sentiments have been marked with red color.

The excitation of different lobes within the brain during watching a product has been measured by topographical image. The topographical view of each emotion has been displayed in Fig. 4. Excitation of alpha-band power in parietal lobe and excitation of beta-band power in temporal and occipital lobe has been observed for all emotional states.

4.1 Performance Analysis

This section revealed the effectiveness of the proposed multimodal system. The classification of the emotional state of a user has been done by random forest classifier. It is an ensemble classifier which performs best for imbalanced class distribution. Here, the text length and number of matching words within the text vary from one review to another, then the "n-gram"-based class distribution for each emotion is largely varied. The classification accuracy reaches up to 98.27%. Here, we started the threshold value from 0.1 and gradually increase it for computing the true positive rate (TPR) and false positive rate (FPR) for each threshold. For evaluating the performance of the model, we increase the TPR and FPR of the model and build

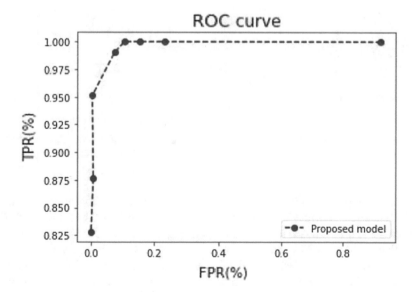

Fig. 5 ROC curve of the proposed multimodal system

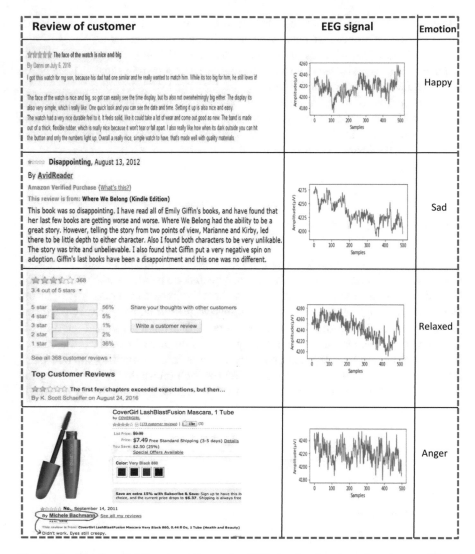

Fig. 6 Identification of different emotions using EEG signal and customer's review on different products. Note: EEG data corresponding to electrode F7 is plotted for visualization purpose

the Receiver Operating Characteristics (ROC) curve. The threshold value should be carefully chosen such that TPR will be maximized and FPR will be minimized. The ROC curve of the proposed model is shown in Fig. 5. From the ROC curve, it can be noted that at 0.2 the TPR reaches to 100%, so we should set the optimum threshold nearly at 0.2. Variation in EEG signal with the customer review of different emotions has been presented in Fig. 6. EEG signal has been captured when participants were watching images of the different products on the computer screen.

5 Conclusion

In this paper, we have combined two different sources to identify the emotion of customer during online shopping. Two powerful sources have been merged, namely: EEG and NLP to correctly identify the emotion. The features of two different modalities have been fused into a common platform, and then the classification is done using random forest classifier. The review of the customer is being analyzed using natural language toolkit (NLTK) whereas machine learning algorithms are implemented for analyzing the EEG signal. The proposed model achieves 98.27% classification accuracy. Text Analytics has lots of applications in today's online world. Companies like Amazon, Flipkart, etc. can understand user feedback or review on the specific product, and the marketing team of those companies can change their business policy based on customer feedback/emotions. Corporate and business need to analyze textual data to understand customer opinion, and feedback to successfully derive their business. So this multimodal emotion recognition will impact largely in neuromarketing as well as corporate and business sectors.

In the near future, we will extend this work with the deep neural network model using different modalities.

References

1. Morin, C.: Neuromarketing the new science of consumer behavior. Society **48**(2), 131–135 (2011)
2. Plassmann, H., Kenning, P., Ahlert, D.: Why companies should make their customers happy: the neural correlates of customer loyalty. ACR North American Advances (2007)
3. Vecchiato, G., Astolfi, L., De Vico, F., Fallani, J.T., Aloise, F., Bez, F., Wei, D., Kong, W., Dai, J., Cincotti, F., et al.: On the use of EEG or MEG brain imaging tools in neuromarketing research. Comput. Intell. Neurosci. **2011**, 3 (2011)
4. Plassmann, H., Weber, B.: Individual differences in marketing placebo effects: evidence from brain imaging and behavioral experiments. J. Mark. Res. **52**(4), 493–510 (2015)
5. Stanton, S.J., Sinnott-Armstrong, W., Huettel, S.A.: Neuromarketing: ethical implications of its use and potential misuse. J. Bus. Ethics **144**(4), 799–811 (2017)
6. Huddleston, P., Behe, B.K., Minahan, S., Thomas Fernandez, R.: Seeking attention: an eye tracking study of in-store merchandise displays. Int. J. Retail Distrib. Manage. **43**(6), 561–574 (2015)
7. Savran, A., Ciftci, K., Chanel, G., Mota, J., Hong Viet, L., Sankur, B., Akarun, L., Caplier, A., Rombaut, M.: Emotion detection in the loop from brain signals and facial images (2006)
8. Boksem, M.A.S., Smidts, A.: Brain responses to movie trailers predict individual preferences for movies and their population-wide commercial success. J. Mark. Res. **52**(4), 482–492 (2015)
9. Liu, Y., Sourina, O., Hafiyyandi, M.R.: EEG-based emotion-adaptive advertising. In: 2013 Humaine Association Conference on Affective Computing and Intelligent Interaction, pp. 843–848. IEEE (2013)
10. Hearst, M.A.: Direction-based text interpretation as an information access refinement. In: Text-Based Intelligent Systems: Current Research and Practice in Information Extraction and Retrieval, pp. 257–274 (1992)
11. Terveen, L., Hill, W., Amento, B., McDonald, D., Creter, J.: Phoaks: a system for sharing recommendations. Commun. ACM **40**(3), 59–63 (1997)

12. Morinaga, S., Yamanishi, K., Tateishi, K., Fukushima, T.: Mining product reputations on the web. In: Proceedings of the Eighth ACM SIGKDD International Conference on Knowledge Discovery and Data Mining, pp. 341–349. ACM (2002)
13. Ambler, T., Braeutigam, S., Stins, J., Rose, S., Swithenby, S.: Salience and choice: neural correlates of shopping decisions. Psychol. Mark. **21**(4), 247–261 (2004)
14. Chakladar, D.D., Chakraborty, S.: EEG based emotion classification using "correlation based subset selection". Biol. Inspired Cogn. Archit. **24**, 98–106 (2018)
15. Khushaba, R.N., Greenacre, L., Kodagoda, S., Louviere, J., Burke, S., Dissanayake, G.: Choice modeling and the brain: a study on the electroencephalogram (EEG) of preferences. Expert Syst. Appl. **39**(16), 12378–12388 (2012)
16. Mohammadi, Z., Frounchi, J., Amiri, M.: Wavelet-based emotion recognition system using EEG signal. Neural Comput. Appl. **28**(8), 1985–1990 (2017)
17. Dasgupta, S.S., Natarajan, S., Kaipa, K.K., Bhattacherjee, S.K., Viswanathan, A.: Sentiment analysis of Facebook data using Hadoop based open source technologies. In: 2015 IEEE International Conference on Data Science and Advanced Analytics (DSAA), pp. 1–3. IEEE (2015)
18. Day, M.-Y., Lee, C.-C.: Deep learning for financial sentiment analysis on finance news providers. In: 2016 IEEE/ACM International Conference on Advances in Social Networks Analysis and Mining (ASONAM), pp. 1127–1134. IEEE (2016)
19. Campos, V., Jou, B., Giro-i-Nieto, X.: From pixels to sentiment: fine-tuning CNNs for visual sentiment prediction. Image Vis. Comput. **65**, 15–22 (2017)
20. Anto, M.P., Antony, M., Muhsina, K.M., Johny, N., James, V., Wilson, A. (2016). Product rating using sentiment analysis. In: 2016 International Conference on Electrical, Electronics, and Optimization Techniques (ICEEOT), pp. 3458–3462. IEEE (2016)
21. Gauba, H., Kumar, P., Roy, P.P., Singh, P., Dogra, D.P., Raman, B.: Prediction of advertisement preference by fusing EEG response and sentiment analysis. Neural Netw. **92**, 77–88 (2017)
22. Bänziger, T., Grandjean, D., Scherer, K.R.: Emotion recognition from expressions in face, voice, and body: the multimodal emotion recognition test (MERT). Emotion **9**(5), 691 (2009)
23. Zheng, W.-L., Dong, B.-N., Lu, B.-L.: Multimodal emotion recognition using EEG and eye tracking data. In: 2014 36th Annual International Conference of the IEEE Engineering in Medicine and Biology Society, pp. 5040–5043. IEEE (2014)

A Modified Dragonfly Algorithm for Real Parameter Function Optimization

Sabari Pramanik and S. K. Setua

Abstract In this article, our aim is to suggest an algorithm which optimizes the real parameter function. We improve the performance of normal dragonfly algorithm, which is a swarm intelligence algorithm, by incorporating opposition-based learning with some alternation. We have applied our proposed algorithm over a set of 22 functions for real parameter function optimization. From the output, it can be found that the proposed algorithm gives a better result than the normal dragonfly algorithm for these function sets. The Wilcoxon rank-sum test, which is a statistical test, is also performed to confirm the superiority of our proposed algorithm.

Keywords Swarm intelligence · Dragonfly algorithm · Opposition-based learning · Real parameter function optimization · Metaheuristics

1 Introduction

In this day and age, optimization techniques are seeking the interest of the scientist community for its vast significance in the field of research. All of the state-of-art problems need a good optimization technique for predicting the result in case of maximization and minimization. For almost all territory, a good optimization algorithm is required, ranging from computer science to mathematics, from economics to finance, from bioinformatics to remote sensing. Nature with all its treasures reveals a wide range of opportunity to us by providing a full range of swarms with optimizing social behaviour. Various swarm intelligence algorithms have been presented for optimization day by day. Grey wolf optimization [1] algorithm is established

S. Pramanik (✉)
Department of Computer Science, Vidyasagar University, Midnapore, West Bengal, India
e-mail: sabari.pramanik@mail.vidyasagar.ac.in

S. K. Setua
Department of Computer Science and Engineering,
University of Calcutta, Kolkata, West Bengal, India
e-mail: sksetua@gmail.com

© Springer Nature Singapore Pte Ltd. 2020
J. K. Mandal and S. Mukhopadhyay (eds.), *Proceedings of the Global AI Congress 2019*, Advances in Intelligent Systems and Computing 1112,
https://doi.org/10.1007/978-981-15-2188-1_33

on the hunting and social nature of the grey wolf community. Particle swarm optimization [2] technique is suggested build on the social behaviour of fish schooling or bird flocking. Ant colony optimization technique is a well-known technique for optimization [3]. Though it is an old algorithm, its importance is increasing day by day as it mimics the behaviour of ants. Several modifications of ant colony optimization has been proposed [4–6], and these modified algorithms are aiming for a better research in different optimization fields. In recent time, a new proposed algorithm, salp swarm algorithm [7], is proposed. It imitates the social behaviour of sapls, which are basically a marine planktonic tunicate. Grasshopper optimization algorithm [8] simulates the social nature of grasshoppers. Grasshoppers movement against enemy and towards food source is all mathematically modelled for optimization. All of these algorithms are based on some behaviour of swarms. These type of algorithms are categorized as swarm intelligence algorithm.

Opposition-based learning is a machine learning strategy [9] which is motivated from the opposite relationship among various units. When we searched among a set of opposite swarms, the chances to find a near optimum value are increased. The grasshopper optimization algorithm with induced opposition-based learning is proposed [10], and it can be found that it is an improved version of normal grasshopper optimization algorithm in all respect. An efficient approach of using oppositional learning strategy over krill herd algorithm [11] is proposed [12], and this algorithm is a good move towards function optimization [12] and load despatch problem [13].

In this paper, we have used a modified opposition-based learning approach with dragonfly algorithm. Dragonfly algorithm [14] is a metaheuristic search algorithm which imitates the social characteristics of dragonflies. We have applied our proposed algorithm over a set of 22 real parameter functions of type unimodal, multimodal and composite [15]. It can be found that the obtained results got a better performance than normal dragonfly algorithm for all functions.

The remaining part of the paper is organized as follows: the design logic is discussed in Sect. 2. The dragonfly algorithm and opposition-based learning are discussed in the sub-sections of Sect. 2. Section 3 introduces the concept of our newly proposed algorithm, the modified opposition-based learning algorithm (MODA). Section 4 justifies the results of different experiments. The paper concludes in Sect. 5.

2   Design Logic

2.1   Dragonfly Algorithm

Dragonfly algorithm (*DA*) is a quite new metaheuristic, swarm intelligence algorithm, proposed by Mirjalili [14]. It mimics the characteristics of dragonflies. The dragonfly algorithm begins with a set of random artificial dragonflies (solution) for optimization process. In each iteration, the dragonfly which have the best fitness value (or objective function value) is termed as the food source (X^{best}) and the worst fitness value is termed as enemy (X^{worst}). The solution space is known as neighbourhood. For exploration and exploitation, all dragonflies (N) follow three basic norms

- Separation, which is used to avoid collision among the dragonflies in the neighbourhood. It is calculated as,

$$S_j = -\sum_{k=1}^{N} X - X_k \tag{1}$$

where X is the position of the updating dragonfly and X_k are the positions of the neighbouring dragonflies.
- Alignment, which is used to equate the velocity of neighbouring individuals. It is calculated as,

$$A_j = \frac{\sum_{k=1}^{N} V_k}{N} \tag{2}$$

where V_k is the velocity of kth neighbouring grasshopper.
- Cohesion, which is used to attract a grasshopper towards the centre of the neighbourhood area. It is calculated as,

$$C_j = \frac{\sum_{k=1}^{N} X_k}{N} - X \tag{3}$$

where X is the position of updating grasshopper and N_k are the positions of neighbouring grasshopper.

Attraction towards the food source and repulsion from the enemy is calculated as,

$$F_j = X^{\text{best}} - X \tag{4}$$

and

$$E_j = X^{\text{worst}} - X \tag{5}$$

where X is the position of updating grasshopper.

The flowchart of grasshopper algorithm is shown in Fig. 1. The algorithmic structure of grasshopper algorithm is shown in Algorithm 1.

Algorithm 1: Dragonfly Algorithm

---

1 Randomly initialize the dragonflies X_j, j=1 to n ;
2 Randomly initialize step vectors ΔX_j, j=1 to n ;
3 **while** *terminating condition is not reached, i.e., t < maxT* **do**
4   Calculate the fitness values of all dragonflies ;
5   Update the best dragonfly as food source (X^{best}) and the worst dragonfly as enemy (X^{worst}) ;
6   **foreach** *dragonfly in the population* **do**
7   Calculate S, A, C, F and E using eq.1 - 5 ;
8   Select randomly the scaling parameters w,s,a,c,f and e ;
9   Update neighbouring radius r ;
10   **if** $r > 1$ **then**
11   update velocity vector
               $\Delta X_{t+1} = (sS_j + aA_j + cC_j + fF_j + eE_j) + w\Delta X(t)$;
12             $X_{t+1} = X_t + \Delta X_t$;
13   **end**
14   **else**
15   update X using Levy flight $X_{t+1} = X_t + Levy(D) * X_t$;
16   **end**
17   t=t+1 ;
18   **end**
19 **end**

2.2 Opposition-Based Learning

Opposition-based learning is a strategy that uses opposite number concept [16] to explore the search region. We can define opposite number as,

$$\hat{X} = a + b - X \tag{6}$$

where X is a number in the range $[a, b]$ and $\hat{X}$ is the opposite number of X.

When we are looking for an optimum solution of a real parameter function, our aim is to increase the search region by modifying the search points for betterment. By using opposite numbers we can observe new solution in the opposite direction along with the original points. In this way, we can discover a new set of more effective solutions.

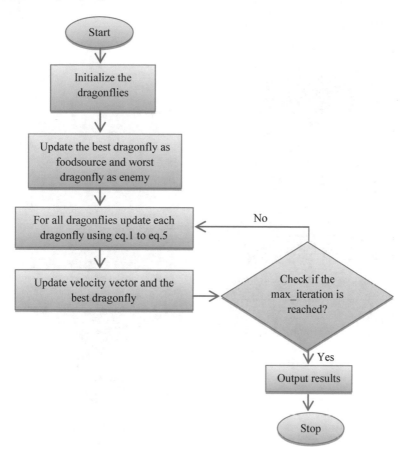

Fig. 1 Flowchart of Dragonfly algorithm

3 The Modified Opposition Inspired Dragonfly Algorithm (MODA)

Our proposed algorithm, the modified opposition inspired dragonfly algorithm (MODA)is a modification of normal dragonfly algorithm. We have incorporated the oppositional learning strategy with a twist to the dragonfly algorithm. Figure 2 shows the flowchart of our proposed algorithm. The algorithm is shown in Algorithm 2. Here, we include the oppositional learning logic into our algorithm in two phases, at the initial phase and at the updating phase. At the initialization phase, we have randomly selected the dragonfly seed value, and then use oppositional learning theory over the seed value for a better seed value. Then, we execute the dragonfly algorithm in the normal way. In each iteration, after finding the set of new dragonflies, we again use the opposition learning-based strategy. From the updating dragonflies and their

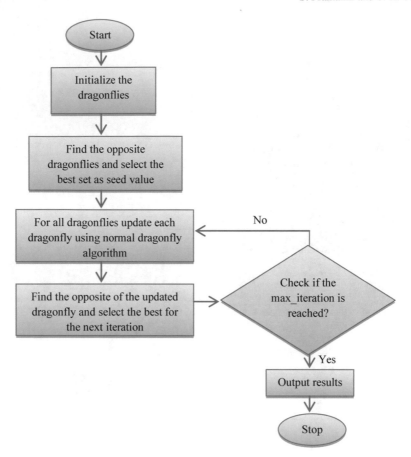

Fig. 2 Flowchart of modified opposition inspired dragonfly algorithm

opposite dragonflies, the best solutions are selected for the next iteration. This small twist shows a big improvement in the results.

After executing this proposed algorithm on a set of real parameter functions several times, it can be shown that the proposed algorithm gives striking results for all functions.

4 Results and Discussion

We run our proposed algorithm MODA over a set of real parameter functions [15]. Our test function sets consist of 22 functions, among which five are unimodal functions (F1–F5), nine are multimodal functions (F6–F14), and eight are composite functions (F15–F22). Unimodal functions are those functions which have only one solution peak. Multimodal functions on the other hand have several optimum solution peaks. Composite functions are the functions which are generated using more than one functions and the result lies in the area of all functions that are combined to form the composite function. Table 1 elaborates the detailing of each function with upper and lower bounds.

Algorithm 2: Modified Dragonfly Algorithm
1 Initialize the dragonflies population (n) randomly;
2 Find the oppositional dragonflies using opposite number strategy;
3 Find the best n no. of dragonflies from the random initial population and its opposite according to the Objective function value;
4 **while** *iteration < maxiteration* **do**
5   **foreach** *dragonfly in the population* **do**
6   calculate the objective function values of the dragonflies;
7   update the food source (best) and enemy (worst);
8   update all parameters;
9   update velocity and position of grasshopper according to the dragonfly algorithm;
10   find the opposite of the velocity vector;
11   select the best one from theis two dragonflies for next iteration;
12   **end**
13 **end**

Figure 3 shows the graph of unimodal ($F1$), multimodal ($F8$) and composite ($F18$) functions.

We run our algorithm MODA over the set of functions 20 times with random seed value. The same seed value is used for DA, ODA and MODA for an appropriate comparison. The number of dragonflies for each population is 50. For each run, the number of iterations is 700. We collect the results in form of best value (best), average value (mean) and standard deviation(s.d.) of all runs. Table 2 shows a comparison of our proposed algorithm (MODA) with normal dragonfly algorithm (DA) and opposition-based learning in initialization phase only (ODA).

After careful observation of the results, it is found that for all functions MODA gives a better results than other two algorithms in best value, mean value and standard derivative value. We have taken the best value from the 20 runs. The mean value shows the average best value among from 20 runs. The standard deviation depicts the range of the results of 20 runs. Figure 4 shows the convergence graphs of all 22 functions. For each function picture, the convergence curve of DA, ODA and our proposed algorithm MODA is given. From the graphs, it can be seen the convergence speed of MODA is better than other two algorithms for all graphs.

Table 3 shows a non parametric statistical test, the Wilcoxon rank-sum test results. This test is used for checking whether there is any significant differences between two independent sets.

Table 1 Illustrations of the Benchmark Test Functions

F. Num.	Func.	Range
F1	Shifted Sphere Function $F_1(x) = \sum_{i=1}^{D} x_i^2$	$[-100, 100]^D$
F2	Shifted Schwefel's Problem $F_2(x) = \sum_{i=1}^{D}(\sum_{j=1}^{i} x_j)^2$	$[-100, 100]^D$
F3	Shifted Rotated High Conditional Elliptic Function $F_3(x) = \sum_{i=1}^{D}(10^6)^{\frac{i-1}{D-1}} x_i^2$	$[-100, 100]^D$
F4	Shifted Schwefel's Problem with Noise in Fitness $F_4(x) = (\sum_{i=1}^{D}(\sum_{j=1}^{i} x_i)^2) * (1 + 0.4\|N(0, 1)\|)$	$[-100, 100]^D$
F5	Schwefel's Problem with Global optimum on Bounds $F_5(x) = \max \|\mathbf{A}_i\mathbf{x} - \mathbf{B}_i\|, i = 1, 2, \ldots, D,$ $x = x_1, x_2, \ldots, x_D$ $f(x) = max(\|x_1 + 2x_2 - 7\|, \|2x_1 + x_2 - 5\|),$ $i = 1, 2, \ldots, n, x^* = [1.3], f(x^*) = 0$ **A** is a $D * D$ matrix	$[-100, 100]^D$
F6	Shifted Rosenbrock's Function $F_6(x) = \sum_{i=1}^{D-1}(100 * (x_i^2 - x_{i+1})^2 + (x_i - 1)^2)$	$[-100, 100]^D$
F7	Shifted Rotated Griewank's Function without Bounds $F_7(x) = \sum_{i=1}^{D} \frac{x_i^2}{4000} - \prod_{i=1}^{D} \cos(\frac{x_i}{\sqrt{i}}) + 1$	$[0, 600]^D$
F8	Shifted Rotated Ackley's Function with Global Optimum on Bounds $F_8(x) = -20\exp(-0.2\sqrt{\frac{1}{D}\sum_{i=1}^{D} x_i^2})-$ $\exp(\frac{1}{D}\sum_{i=1}^{D}\cos(2\Pi x_i) + 20 + \exp(1)$	$[-32, 32]^D$
F9	Shifted Rastrigin's Function $F_9(x) = \sum_{i=1}^{D}(x_i^2 - 10\cos(2\Pi x_i) + 10)$	$[-5, 5]^D$
F10	Shifted Rotated Rastrigin's Function $F_{10}(x) = \sum_{i=1}^{D}(z_i^2 - 10\cos(2\Pi z_i) + 10), \mathbf{z} = \mathbf{x} * \mathbf{M}$ **M**: Linear Transformation matrix	$[-5, 5]^D$
F11	Shifted Rotated Weierstrass Function $F_{11}(x) = \sum_{i=1}^{D}(\sum_{k=0}^{kmax}[a^k \cos(2\Pi b^k(z_i + 0.5))]-$ $D\sum_{k=0}^{kmax}[a^k \cos(2\Pi b^k(z_i * 0.5)])$ $a = 0.5, b = 3, \mathbf{z} = \mathbf{x} * \mathbf{M}$ **M**: Linear Transformation matrix	$[-0.5, 0.5]^D$
F12	Schwefel's Problem $F_{12}(x) = \sum_{i=1}^{D}(\mathbf{A}_i - \mathbf{B}_i(x))^2$ $\mathbf{A}_i = \sum_{j=1}^{D}(a_{ij}\sin\alpha_j + b_{ij}\cos\alpha_j),$ $\mathbf{B}_i(x) = \sum_{j=1}^{D}(a_{ij}\sin x_j + b_{ij}\cos x_j);$ **A**, **B** are two $D*D$ matrix, a_{ij}, b_{ij} are integer random numbers in the range $[-100, 100]$ $\alpha = [\alpha_1, \alpha_2, \ldots, \alpha_D],$ α_j are random numbers in the range $[-\Pi, \Pi]$	$[-\Pi, \Pi]^D$

(continued)

Table 1 (continued)

F. Num.	Func.	Range
F13	*Shifted Expanded Griewank's plus Rosenbrock's Function* $F_{13}(x) = F8(F2(x_1, x_2)) + F8(F2(x_2, x_3)) + \cdots + F8(F2(x_{D-1}, x_D)) + F8(F2(x_D, x_1))$	$[-3, 1]^D$
F14	*Shifted Rotated Expanded Scaffer's Function* $F_{14}(x) = F(z_1, z_2) + F(z_2, z_3) + \cdots +$ $(F(z_{D-1}, z_D) + F(z_D, z_1)$ $F(x, y) = 0.5 + \frac{(\sin^2(\sqrt{x^2+y^2})-0.5}{(1+0.001(x^2+y^2))^2}$ $\mathbf{z} = \mathbf{x} * \mathbf{M}$	$[-100, 100]^D$

For the following composition function (F15−F25)

F(x): new composition function

$f_i(x) = i^{th}$ basic function used to construct the composition function

n: number of basic functions

$$F(x) = \sum_{i=1}^{n} w_i * [f_i']$$

F15	*Hybrid Composition Function* $F_{1-2}(x)$: Rastrigin's Function $F_{3-4}(x)$: Weierstrass Function $F_{5-6}(x)$: Griewank's Function $F_{7-8}(x)$: Ackley's Function $F_{9-10}(x)$: Sphere function **M** are identity matrices	$[-5, 5]^D$
F16	*Rotated Version of Hybrid Composition Functions F15* $F_{1-2}(x)$: Rastrigin's Function $F_{3-4}(x)$: Weierstrass Function $F_{5-6}(x)$: Griewank's Function $F_{7-8}(x)$: Ackley's Function $F_{9-10}(x)$: Sphere function **M** are linear transformation matrices	$[-5, 5]^D$
F17	*F16 with Noise in Fitness* $F_{17}(x) = G(x) * (1 + 0.2|N(0, 1)|)$ where G(x) = F16	$[-5, 5]^D$
F18	*Rotated Hybrid Composition Function* $F_{1-2}(x)$: Ackley's Function $F_{3-4}(x)$: Rastrigin's Function $F_{5-6}(x)$: Sphere Function $F_{7-8}(x)$: Weierstrass Function $F_{9-10}(x)$: Griewank's function **M** are rotation matrices	$[-5, 5]^D$
F19	*Rotated Hybrid Composition Function with* *narrow basin global optimum* All settings are the same as F_{18} except σ and λ	$[-5, 5]^D$

(continued)

Table 1 (continued)

F.Num.	Func.	Range
F20	*Rotated Hybrid Composition Function* *with Global Optimum on the bounds* All settings are the same as F_{18} except after load the data file, set $o_{1(2j)} = 5$ for $j = 1, 2, \ldots, \lfloor D/2 \rfloor$	$[-5, 5]^D$
F21	*Rotated Hybrid Composition Function* $F_{1-2}(x)$: Rotated Expanded Scaffer's F6 Function $F_{3-4}(x)$: Rastrigin's Function $F_{5-6}(x)$: F8F2 Function $F_{7-8}(x)$: Weierstrass Function $F_{9-10}(x)$: Griewank's function **M** are orthogonal matrices	$[-5, 5]^D$
F22	*Rotated Hybrid Composition Function with* *High Condition number matrix* All settings are the same as F_{21} except **M** 's condition numbers are [10 20 50 100 200 1000 2000 3000 4000 5000]	$[-5, 5]^D$

Table 2 The results of DA, ODA and MODA over 22 functions

Function	DA			ODA			MODA		
	best	mean	s.d.	best	mean	s.d.	best	mean	s.d.
F1	6.10e−07	2.142−00	5.86e−00	2.60e−10	3.78e+01	1.19e+02	6.73e−11	1.41e−01	4.44e−01
F2	4.61e−00	2.09e+02	1.13e+02	1.46e+02	2.11e+02	6.42e+01	3.85e−05	5.41e+01	6.08e+01
F3	1.01e+06	2.59e+06	1.88e+06	7.17e+05	2.03e+06	1.11e+06	1.78e+05	9.89e+05	8.65e+05
F4	7.92e−04	7.65e−01	1.00e−00	3.43e−06	4.65e−01	1.06e−00	0.00e−00	0.00e−00	0.00e−00
F5	1.94e−00	2.26e+02	2.63e+02	4.98e−00	2.60e+02	5.39e+02	0.00e−00	1.47e+02	2.09e+02
F6	4.78e−01	7.27e+04	1.72e+05	1.83e−00	2.83e+05	6.87e+04	1.87e−01	1.09e+04	3.28e+04
F7	4.28e+01	7.80e+01	3.59e+01	3.42e+01	7.75e+01	2.83e+01	1.12e+01	5.93e+01	2.47e+01
F8	2.00e+01	2.01e+01	5.76e−02	2.00e+01	2.01e+01	5.28e−02	2.00e+01	2.01e+01	4.04e−02
F9	4.97e−00	2.18e+01	1.31e+01	3.11e−00	1.51e+01	7.88e−00	9.94e−01	1.11e+01	5.68e−00
F10	7.53e−00	3.63e+01	2.36e+01	1.17e+01	3.03e+01	9.65e−00	6.12e−02	1.49e+01	9.22e−00
F11	3.39e−00	3.98e−00	4.71e−01	3.58e−00	4.02e−00	6.40e−01	2.52e−00	3.96e−00	3.12e−01
F12	1.31e+03	3.72e+03	1.64e+03	2.08e+03	4.11e+03	2.59e+03	1.98e+02	3.08e+03	1.55e+03
F13	4.03e−01	4.11e−00	5.40e−00	8.6e−01	2.79e−00	2.76e−00	1.9e−01	1.00e−00	7.04e−01
F14	1.56e−01	8.90e−01	5.40e−01	1.13e−01	1.02e−00	4.99e−01	1.12e−01	4.57e−01	2.91e−01
F15	6.25e+02	7.81e+02	1.60e+02	6.20e+02	8.07e+02	1.45e+02	6.07e+02	7.66e+02	1.10e+02
F16	5.30e+02	7.00e+02	1.50e+02	4.43e+02	6.43e+02	1.39e+02	4.28e+02	4.81e+02	4.26e+01
F17	3.74e+02	5.35e+02	1.53e+02	3.93e+02	5.00e+02	1.13e+02	3.18e+02	4.14e+02	5.67e+01
F18	8.62e+02	9.51e+02	6.53e+01	8.32e+02	9.13e+02	5.59e+01	7.67e+02	8.52e+02	4.66e+01
F19	7.90e+02	9.24e+02	9.34e+01	7.96e+02	9.27e+02	8.86e+02	7.07e+02	8.11e+02	6.63e+03
F20	8.41e+02	9.17e+02	8.67e+01	8.21e+02	9.04e+02	7.13e+02	6.70e+02	8.19e+02	5.88e+01
F21	6.11e+02	9.53e+02	1.90e+02	6.82e+02	9.65e+02	1.70e+02	5.00e+02	8.24e+02	1.11e+02
F22	7.47e+02	9.51e+02	2.00e+02	7.95e+02	8.77e+02	9.54e+01	6.45e+02	7.87e+02	8.14e+01

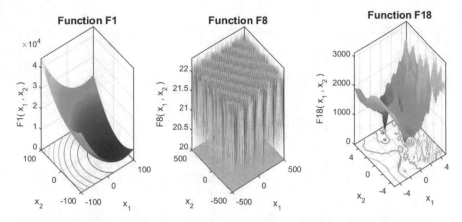

Fig. 3 Example of unimodal test function, multimodal test function and composite test function

If the results (*p*-value) are smaller than 0.05 then some significant differences occur in the sets. We have applied Wilcoxon rank-sum test on DA and MODA. After observing the results, we find that only 3 out of 22 function does not show a significant difference. So from the results, the superiority of our proposed algorithm MODA is statistically proved.

5 Conclusion

We have designed a modified opposition-based dragonfly algorithm, which is normal dragonfly algorithm with infinitesimal modification. Though it is a small change in the original algorithm, it gives a huge improvement in the results. We have incorporated opposition-based learning strategy in two phases, the initialization phase and in the updating phase. A set of 22 real parameter functions are used as experimental media. We have applied our proposed algorithm to the set of 22 functions and compare the results with normal dragonfly algorithm and opposition-based dragonfly algorithm (at initialization phase). From the results, we hypothesize that MODA gives a better result in all respect. The converge speed of MODA is also good over other two algorithms. These circumstances are true for all 22 functions. We prove the supremacy of our algorithm using Wilcoxon rank-sum test. The result of the Wilcoxon rank-sum test depicts a significant difference exists between the algorithm DA and MODA. Though there are three functions out of 22 functions does not show a significant difference, still we can say that our proposed algorithm is performed better. By seeing all artifacts we can say that the proposed algorithm MODA undoubtedly be a good algorithm for real parameter optimization.

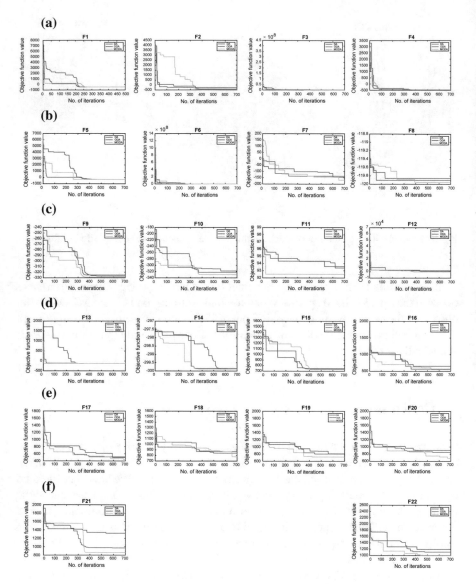

Fig. 4 Convergence graph of all 22 functions for DA, ODA and MODA

Table 3 Result of Wilcoxon Rank-sum Test of MODA with DA

Function	p-value	Function	p-value	Function	p-value
F1	0.0376	F9	0.0357	F17	0.0211
F2	0.0028	F10	0.0211	F18	0.0058
F3	0.0211	F11	0.7913	F19	0.0091
F4	0.0000	F12	0.6232	F20	0.0452
F5	0.0032	F13	0.0140	F21	0.0452
F6	0.0232	F14	0.0451	F22	0.0312
F7	0.0097	F15	0.6232		
F8	0.0474	F16	0.0000		

References

1. Mirjalili, S., Mirjalili, S.M., Lewis, A.: Grey wolf optimizer. Adv. Eng. Softw. **69**, 46–61 (2014)
2. Kennedy, J.: Particle swarm optimization. Encycl. Mach. Learn. 760–766 (2010)
3. Dorigo, M., Di Caro, G.: Ant colony optimization: a new meta-heuristic. In: Proceedings of the 1999 Congress on Evolutionary Computation-CEC99 (Cat. No. 99TH8406), vol. 2. IEEE (1999)
4. Wang, J., et al.: An improved ant colony optimization-based approach with mobile sink for wireless sensor networks. J. Supercomput. **74**(12), 6633–6645 (2018)
5. Engin, O., Güçlü, A.: A new hybrid ant colony optimization algorithm for solving the no-wait flow shop scheduling problems. Appl. Soft Comput. **72**, 166–176 (2018)
6. Mahi, M., Baykan, Ö.K., Kodaz, H.: A new hybrid method based on particle swarm optimization, ant colony optimization and 3-opt algorithms for traveling salesman problem. Appl. Soft Comput. **30**, 484–490 (2015)
7. Mirjalili, S., et al.: Salp Swarm Algorithm: a bio-inspired optimizer for engineering design problems. Adv. Eng. Softw. **114**, 163–191 (2017)
8. Saremi, S., Mirjalili, S., Lewis, A.: Grasshopper optimisation algorithm: theory and application. Adv. Eng. Softw. **105**, 30–47 (2017)
9. Mahdavi, S., Rahnamayan, S., Deb, K.: Opposition based learning: a literature review. Swarm Evol. Comput. **39**, 1–23 (2018)
10. Ewees, A.A., Elaziz, M.A., Houssein, E.H.: Improved grasshopper optimization algorithm using opposition-based learning. Expert Syst. Appl. **112**, 156–172 (2018)
11. Gandomi, A.H., Alavi, A.H.: Krill herd: a new bio-inspired optimization algorithm. Commun. Nonlinear Sci. Numer. Simul. **17**(12), 4831–4845 (2012)
12. Wang, G.-G., et al.: Opposition-based krill herd algorithm with Cauchy mutation and position clamping. Neurocomputing **177**, 147–157 (2016)
13. Bulbul, S.M.A., et al.: Opposition-based krill herd algorithm applied to economic load dispatch problem. Ain Shams Eng. J. **9**(3), 423–440 (2018)
14. Mirjalili, S.: Dragonfly algorithm: a new meta-heuristic optimization technique for solving single-objective, discrete, and multi-objective problems. Neural Comput. Appl. **27**(4), 1053–1073 (2016)
15. Suganthan, P.N., et al.: Problem definitions and evaluation criteria for the CEC special session on real-parameter optimization. KanGAL report 2005005 (2005), 2005

16. Tizhoosh, H.R.: Opposition-based learning: a new scheme for machine intelligence. In: International Conference on Computational Intelligence for Modelling, Control and Automation and International Conference on Intelligent Agents, Web Technologies and Internet Commerce (CIMCA-IAWTIC'06), vol. 1. IEEE (2005)

Design Considerations of a Medical Expert System for Differential Diagnosis of Low Back Pain (ES$_{LBP}$)

Debarpita Santra, J. K. Mandal, S. K. Basu and Subrata Goswami

Abstract Low back pain is a communal musculoskeletal ailment that deprives many individuals worldwide of doing their daily and normal activities. With the absence of external biomarkers, most of the symptoms of low back pain diseases seem similar, making the diagnosis process quite difficult. Application of artificial intelligence is beneficial in this regard. The paper deals with the design of an efficient knowledge base and a reliable inference engine for a medical expert system for treatment of low back pain. As many low back pain diseases have common clinical signs, consideration of only the dissimilar patterns of the diseases in the design of knowledge base would surely overcome the problem of processing the same symptoms over and over. The acquired knowledge is represented with a discernibility matrix that captures only the disparities among low back pain diseases. An inference mechanism has also been proposed, which uses the discernibility matrix for offering the diagnostic conclusions in a timely manner. The designed system has been tested with patient records empirically selected from the repository of ESI Hospital Sealdah, Kolkata. The test results show that the diagnostic inference generated by the proposed inference engine conforms to the conclusions made by the expert physicians.

Keywords Medical expert system · Low back pain · Knowledge representation · Discernibility matrix · Inference engine

D. Santra (✉) · J. K. Mandal
Department of Computer Science & Engineering, University of Kalyani, Kalyani, Nadia, West Bengal 741235, India
e-mail: debarpita.cs@gmail.com

J. K. Mandal
e-mail: jkm.cse@gmail.com

S. K. Basu
Department of Computer Science, Banaras Hindu University, Varanasi 221005, India
e-mail: swapankb@gmail.com

S. Goswami
ESI Institute of Pain Management, Kolkata 700009, India
e-mail: drsgoswami@gmail.com

© Springer Nature Singapore Pte Ltd. 2020
J. K. Mandal and S. Mukhopadhyay (eds.), *Proceedings of the Global AI Congress 2019*, Advances in Intelligent Systems and Computing 1112, https://doi.org/10.1007/978-981-15-2188-1_34

1 Introduction

Medical expert systems [1] have been developed and in use since 1970 for several diseases like meningitis and bacteremia infections, heart diseases, diabetes, asthma, glaucoma, oncology, etc. A medical expert system can provide quality assurance of decision by reproducing doctor's good judgment through rapid analysis of huge clinical data and reducing the diverse possibilities of a diagnosis to critical decision points. Effects of shortage of expert physicians in the health care delivery model can be mitigated to a certain extent through use of medical expert systems.

Medical expert systems are beneficial at the preliminary screening phases of diseases like low back pain (LBP), where most of the causes are unspecific and unspecified. Unfortunately, many who experience LBP undergo inappropriate or unnecessary diagnostics and treatments in India. According to researches, LBP is liable for developing more disability among individuals compared to other diseases [2]. The cases of LBP in India have been estimated to be about 6.2% in general population and 92% in worker population [3]. A huge no. of inhabitants in India with LBP take no medical consultation, and majority prefers traditional treatment. In spite of availability of numerous upgraded techniques for assessing and diagnosing LBP, the precise reason behind occurrence of pain cannot be well recognized at times unless the patient undergoes intricate pathological investigations.

To offer fast, quality, and reliable medical advice in India, the aim is to develop a dependable and proficient medical expert system for improving the conventional process of assessing and managing LBP. Its diagnosis is full of uncertainties, as the exact cause is not well identified due to absence of external biomarkers [4]. X-ray, MRI, blood tests, and other investigations do not always lead to the right direction. Also, with many LBP diseases having apparently similar signs and symptoms, the diagnosis of LBP becomes quite difficult. As there are various common clinical factors among LBP diseases, the traditional data-driven inference techniques that take care of the fact that how much match is there among the input information and the knowledge stored in knowledge base may involve redundant computations. This paper proposes the construction of an optimal knowledge base and an efficient inference mechanism for offering differential diagnoses for LBP, which overcomes the problem of redundancy in computation. Differential diagnosis refers to a list of probable diseases found against a set of clinical signs and symptoms.

The organization of the article is given here. Section 2 provides an outline of related works. In Sect. 3, knowledge representation issues as well as the inference technique for the intended expert system have been discussed. Section 4 gives the experimental results of the proposed methods. Lastly, Sect. 5 concludes the paper.

2 Related Works

Since 1970s, some notable medical expert systems are MYCIN [5], CASNET [6], INTERNIST [7], ONCOCIN [8], PUFF [9], and so on. The expert systems use different types of knowledge representation techniques [10, 11]. While MYCIN used the production rules for representing the acquired knowledge, CASNET used a network structure for capturing the diagnostic, prognostic, and therapeutic knowledge about glaucoma. In some other expert system, knowledge has been represented with frame data structures [12]. Case-based reasoning has been used for representing knowledge in many applications [13] in the form of cases.

The expert systems use a variety of inference mechanisms for providing evidence-based conclusions. Forward or backward chaining inference mechanism [14] is generally used by the rule-based expert systems. Some expert systems use Bayesian network-based inference technique for reaching to non-conflicting diagnostic conclusions. Artificial neural network (ANN) [15] is also a popular approach that uses large training dataset for achieving more accurate inference results compared to the traditional approaches.

In recent years, an ontology-based expert system was proposed for diabetic patients [16]. The benefit of ontology-driven approach is that the medical knowledge can be captured in a formal way ensuring its reuse in later stages. Also, there is a web-based development for gestational diabetes, where finite automata have been used for assessing patient's metabolic conditions [17]. The system also constructed rule-based knowledge base for generating therapeutic decisions. An ANN-based medical expert system has been developed for prediction of heart-failure risks [18]. The ANN-based approach, combined with a fuzzy analytic hierarchy process, offers dependable diagnostic inference as compared to the traditional ANN-based approaches. Another clinical decision support tool has been designed for reducing the duration of hospital stay and the in-hospital mortality rate, and the developed system has been successfully tested with 64,512 patients [19]. A fuzzy clinical decision support application has been constructed very recently for screening of diabetic retinopathy [20]. The application uses hierarchical fuzzy decision tree data structure for classifying the patients into many classes depending on the risk factors the patients are exposed to.

The approaches reported in the literature for knowledge representation and inference engine design were developed based on the application scenarios. As the nature of medical knowledge varies from one disease to another, the process of reasoning also changes. The field of LBP, being a domain full of unspecified and unclear symptoms, lacking clear object-oriented directives, should follow different approaches, induced from the existing methodologies, for both the knowledge representation and clinical resolution.

3 Design

The paper aims to develop a computational model for medical expert system for LBP management (ES_{LBP}). ES_{LBP} has been designed using four building blocks: user interface (UI_{LBP}), knowledge base (KB_{LBP}), working memory (WM_{LBP}), and inference engine (IE_{LBP}). UI_{LBP} acts as the gateway for the pertinent users (medical professionals) to interact with the system with relevant clinical information of LBP patients. WM_{LBP} preserves the inputted patient information for use in later stages. The KB_{LBP} stores the domain knowledge regarding the LBP diseases gathered from existing literature (journals, books, etc.) and the expert physicians having several years of experience of clinical practice in this medical field. IE_{LBP} takes help of KB_{LBP} and WM_{LBP} to infer reliable diagnostic and therapeutic decisions.

UI_{LBP} presents to the users a set of questions, which are set by the expert physicians after identifying an optimal set of clinical parameters / attributes associated with LBP. Relevant clinical information of LBP patients involving their demographic details, clinical history, local / general examination reports, pathological test results, etc., are fed into ES_{LBP} in appropriate format through UI_{LBP}. The interface transmits all the entered information to WM_{LBP}. WM_{LBP} stores the clinical information / records patient-wise in its database, allowing easy retrieval of the stored information whenever required. IE_{LBP} interacts with WM_{LBP} and UI_{LBP} for accepting patients' medical information as input. Also it interacts with KB_{LBP} for retrieving related medical knowledge. IE_{LBP} primarily performs a matching process between the input information and the retrieved knowledge in an efficient manner, and offers differential diagnosis of LBP for particular inputs. The inference outcomes, which basically form a list of probable diseases along with a numeric measure for each disease indicating its chance of occurrence, are then stored in WM_{LBP} against particular individuals for future reference, and are reflected in UI_{LBP} for assisting the users in providing healthcare to the sufferers. The block diagram of ES_{LBP} is presented in Fig. 1. As

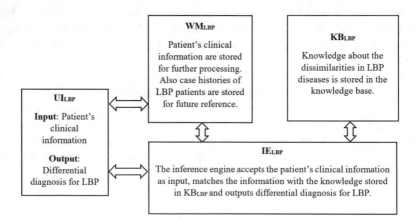

Fig. 1 Block diagram showing the components of ES_{LBP}

construction of user interface and working memory is mostly straightforward, the paper focuses on design of KB_{LBP} and IE_{LBP} as these two components pose as pillars of development of ES_{LBP}. The details of the design issues are provided subsequently.

3.1 Design of Knowledge Base

Diagnosis of LBP is quite challenging because most of the symptoms for different LBP diseases are alike. It would be an advantageous approach if knowledge can be extracted from the existing literature or domain experts regarding how an LBP disease is different symptomatically from other LBP diseases. With little difference between any two LBP diseases, consideration of only the dissimilar symptoms leads to efficient inferencing. Also, less space is required for storing only the dissimilarities in diseases in knowledge base.

The design considers a finite set A of n (>0) clinical attributes such that $A = \{a_i | a_i$ $(1 \leq i \leq n)$ is a clinical attribute from LBP domain}. Assuming that there are x (>0) LBP diseases in the literature, a finite set D is constructed as $D = \{d_j | d_j$ $(1 \leq j \leq x)$ is an LBP disease}. Every disease d_j is characterized by all the attributes in A, where any particular attribute $a_i \in A$ can accept Boolean values as 1 or 0 to represent its presence or absence for d_j, respectively. As an attribute represents a clinical sign / symptom or carries other information regarding the clinical examination / investigation, absence of the attribute from a particular disease may have significant clinical impact. The implicit assumption, in this paper, is that all the attributes have same clinical significance while diagnosing LBP.

A disease d_j can be formally represented as 3-tuple $\langle d_j, C_j, f_j \rangle$, where d_j $(1 \leq j \leq x)$ is an LBP disease, C_j is defined as a set represented as: $\cup_{i=1 \text{ to } n} \{(a_i, t_i)\}$, with a_i $(1 \leq i \leq n)$ being a clinical attribute belonging to A, and t_i being the Boolean truth value of a_i for disease d_j, and f_j is a mapping function between d_j and C_j defined as $f_j: C_j \rightarrow d_j$. If there are two different diseases d_{j1} and d_{j2} belonging to D with C_{j1} $\cap C_{j2} = C (\neq \emptyset$, and $(n/2 < |C| < n))$, then it would be better to consider only those attributes in A which are not in C for obtaining differential diagnosis. From clinical records and the knowledge of expert physicians, it has been seen that most of the LBP diseases share common attribute-value pairs which are more than half of the total no. of attributes in A. So, processing of only the dissimilar attribute-value pairs would require less time, compared to that of alike pairs during inferencing. That is why information about each disease would be stored in the knowledge base (KB_{LBP}) only in terms of the dissimilar attribute-value pairs with respect to the rest of the LBP diseases. The KB_{LBP} is represented as a lower triangular discernibility matrix of size $(x \times x)$ with the (k, l)th entry denoted as $KB_{LBP}[k, l]$ is shown in Eq. 1. Thus, each entry $KB_{LBP}[k, l]$ holds the set of attribute-value pairs upon which the disease d_k differs from the disease d_l. In Eq. 1, the value t_i of attribute a_i corresponding to disease d_k has been denoted as $t_i[d_k]$. The entries above the diagonal in the matrix have been kept empty.

	d_1	d_2	d_3	d_4	d_5
d_1	$\emptyset$				
d_2	$<a_1, 0>, <a_2, 0>,$ $<a_4, 1>, <a_5, 1>,$ $<a_6, 0>$	$\emptyset$			
d_3	$<a_1, 0>, <a_2, 0>,$ $<a_3, 0>, <a_6, 0>$	$<a_3, 0>, <a_4, 0>,$ $<a_5, 0>$	$\emptyset$		
d_4	$<a_1, 0>, <a_2, 0>,$ $<a_3, 0>, <a_5, 1>$	$<a_3, 0>, <a_4, 0>,$ $<a6, 1>$	$<a_5, 1>,$ $<a_6, 1>$	$\emptyset$	
d_5	$<a_1, 0>, <a_3, 0>,$ $<a_4, 1>, <a_6, 0>$	$<a_2, 1>, <a_3, 0>,$ $<a_5, 0>$	$<a_2, 1>,$ $<a_4, 1>$	$<a_2, 1>,$ $<a_4, 1>,$ $<a_5, 0>,$ $<a_6, 0>$	$\emptyset$

Fig. 2 Design of KB_{LBP} for the considered scenario

$$KB_{LBP}[k, l] == \begin{cases} \emptyset, & \text{if } k == l \\ \{(a_i, t_i[d_k])|t_i[d_k] \neq t_i[d_l] \text{ with } 1 \leq i \leq n\}, & \text{if } k > l \\ \text{nil}, & \text{if } k < l \end{cases} \quad (1)$$

The proposed scheme is illustrated with six clinical attributes from LBP domain, namely 'pain at rest' (a_1), 'pain at forward bending' (a_2), 'pain at backward bending' (a_3), 'tenderness at midline' (a_4), 'tenderness at low back' (a_5), and 'positive FABER test' (a_6). Five major LBP diseases such as *Sacroiliac Joint Arthropathy* (SIJA) (d_1), *Facet Joint Arthropathy* (FJA) (d_2), *Discogenic Pain* (DP) (d_3), *Myofascial Pain Syndrome* (MFPS) (d_4), and *Prolapsed Intervertebral Disc Disease* (d_5) (PIVD) have been considered here. The KB_{LBP} is constructed as shown in Fig. 2.

3.2 Design of Inference Engine

Inference engine of ES_{LBP} accepts patients' clinical information as input, matches the inputted patient information against the knowledge stored in KB_{LBP}, processes the matched knowledge, and offers differential diagnosis for LBP as output. For each disease that is concluded by the inference engine, a metric called 'chance of occurrence' (ch) is associated. The chance of occurrence lies between 0 and 1, both inclusive. If the inputted patient information completely matches with the information for a particular disease d, then it is concluded that the ch for d against the particular input is 1; that is, there is 100% match between the input and the stored knowledge. On the other hand, if the entered information does not at all match with the information for another disease d' ($\in D$), then it is said that the ch for d' against the particular input is 0. If p ($<n$) attribute-value pairs in the input vector match with the information for another disease d'' ($\in D$), then the chance of occurrence for d'' is p/n.

Input to the inference engine basically acts as a vector of n tuples, where each tuple is represented as $\langle a_i, I_i \rangle$ with $a_i \in A$ ($1 \leq i \leq n$) and I_i being the inputted Boolean truth value for a_i. The inference engine performs the inference process taking into account only those attributes from the input vector which have been used in the knowledge base KB_{LBP}. The inference process proceeds disease-wise. When a patient X comes with LBP, the system initially thinks that patient X is suffering from all the x diseases in D, with the chance of occurrence for each disease being 1. Then, for every disease $d_i \in D$ ($1 \leq i \leq x$), a set S_i is constructed using the corresponding attribute-value pairs as noted in KB_{LBP}. From the corresponding entries in KB_{LBP}, it is now observed how d_j differs from the other diseases in D' ($= D - \{d_j\}$). For each element $e \in S_i$, it is examined from the entries in KB_{LBP} those diseases in D' that differ from d_j upon the attribute-value pair e. Suppose the diseases that are different upon e constitute a set D'' ($\subseteq D'$). Then, it is checked whether e matches with the corresponding entry in the input vector. If not, then the chance of occurrence of every disease belonging to D'' is reduced by a factor of $1/n$. Otherwise, the chance of occurrence of disease d_j is decreased by $1/n$. To keep track of the information regarding which attribute is responsible for reducing the chance of occurrence of a disease, a 2-D matrix M of size ($x \times |S|$) is constructed, with S being the set of all the attributes in KB_{LBP}. Initially, M is empty. The (q, r)th entry of M, denoted by $M[q, r]$, holds 1 if the chance of occurrence of disease $d_q \in D$ has reduced due to attribute $a_r \in S$; otherwise $M[q, r]$ would hold 0 if the chance of occurrence for d_q is not reduced due to a_r. This process continues until the matrix M is *full*, that is, no column in M is empty, and every column holds at least one 1. The algorithm followed by IE_{LBP} is given in *Algorithm* 1. For the sake of clarity and easy understanding, the algorithm uses different symbols to denote the attributes, sets, and other necessary variables.

Procedure Inference_Differential_Diagnosis: Algorithm 1

Input: The knowledge base KB_{LBP}; A patient input vector I_p; A matrix M, which is initially empty

Output: A set of diseases L_D

Method: The algorithm retrieves relevant information from the knowledge base disease-wise, observes how a disease differs from the other LBP diseases based upon various clinical parameters under consideration, fills up the entries in matrix M accordingly, and finally produces differential diagnosis for LBP.

Begin
  For $i = 1$ to x do
    $S_i = $ **Extract**(KB_{LBP}, d_i)   // The procedure **Extract**() extracts the attribute-
                        value pairs from the ith column corresponding to disease d_i in KB_{LBP}
    $ch_i = 1$
    If $(S_i \neq empty)$ then
        $A'_i = $ **Find_Attribute**(S_i) // The procedure **Find_Attribute**() collects only
                        the attributes from S_i and the attributes are kept in a set A'_i
        $Ip^i = $ **Collect_Input**(A'_i)   // The procedure **Collect_Input**() collects only
                        the relevant attribute-value pairs from I_p based on the
                        attributes in A'_i
      If $(M != full)$ then
        For $j = 1$ to $|A'_i|$ do
           $a_j = A'_i[j]$ // $A'_i[j]$ denotes the jth attribute in A'_i
           $^iD_j = $ **Discernibility**(KB_{LBP}, a_j, d_i)   // The procedure
                      Discernibility() returns a set of LBP diseases from
                      which disease d_i differs upon a_j
           $t_j = $ **Fetch_Value**(S_i, a_j) // The procedure **Fetch_Value**() fetches
                      the value of attribute a_j as mentioned in S_i
           $col = $ **Find_Column**(M, a_j) // The procedure **Find_Column**()
                      finds the column no. in M corresponding to attribute a_j
           If $(M[i, col] == empty)$ then
              If $(Find(Ip^i, <a_j, t_j>) != true)$ then //The procedure **Find**() finds
                        whether the attribute-value pair $<a_j, t_j>$ matches with
                        the input information as mentioned in Ip^i
                $M[i, col] = 0$
                For $k = 1$ to $|^iD_j|$ do
                  $d_k = {}^iD_j[k]$ // $^iD_j[k]$ denotes the kth item of set ^iD_j
                  $ch_k = ch_k - (1/n)$
                  $row = $ **Find_Row**(M, d_k) // The procedure **Find_Row**()
                      finds the row no. in M corresponding to disease d_k
                  $M[row, col] = 1$
                End for
              Else
                $M[i, col] = 1$
                $ch_i = ch_i - (1/n)$
                For $k = 1$ to $|^iD_j|$ do
                  $d_k = {}^iD_j[k]$
                  $row = $ **Find_Row**(M, d_k)
                  $M[row, col] = 0$
                End for
              End if
           End if
        End for
      End if
    End if
  End for
  $L_D = \varnothing$
  For $i = 1$ to x do
    $L_D = L_D \cup \{<d_i, ch_i>\}$
  End for
  Return L_D
End **Inference_Differential_Diagnosis**

The proposed mechanisms for knowledge base construction and design of inference engine are advantageous from two major aspects, i.e., time efficiency and space

efficiency. In traditional inferencing approaches, each attribute-value pair for every disease needs to be compared with the inputted patient information for finding the similarity between them. As an LBP disease is characterized by n clinical attributes, then for each attribute-value pair in the input vector, the corresponding attribute-value pairs for all the x diseases would be checked. This process would continue for all the n attributes, and the whole process is time-consuming. Rather, in ES_{LBP}, only the dissimilarities among the diseases are being taken into account, without having explicit attribute-by-attribute comparisons. Also, as the designed knowledge base holds only the limited attribute-value pairs instead of maintaining the detailed knowledge about each disease, the space requirement is optimized.

The inference mechanism followed by IE_{LBP} is illustrated using the KB_{LBP} constructed as shown in Fig. 2, and an input vector I_p is taken as $[\langle a_1, 0\rangle, \langle a_2, 1\rangle, \langle a_3, 0\rangle, \langle a_4, 1\rangle, \langle a_5, 0\rangle, \langle a_6, 1\rangle]$. Using *Algorithm* 1, S_1 is constructed as $\{\langle a_1, 0\rangle, \langle a_2, 0\rangle, \langle a_3, 0\rangle, \langle a_4, 1\rangle, \langle a_5, 1\rangle, \langle a_6, 0\rangle\}$ for disease d_1, S_2 as $\{\langle a_2, 1\rangle, \langle a_3, 0\rangle, \langle a_4, 0\rangle, \langle a_5, 0\rangle, \langle a_6, 1\rangle\}$ for disease d_2, S_3 as $\{\langle a_2, 1\rangle, \langle a_4, 1\rangle, \langle a_5, 1\rangle, \langle a_6, 1\rangle\}$ for disease d_3, S_4 as $\{\langle a_2, 1\rangle, \langle a_4, 1\rangle, \langle a_5, 0\rangle, \langle a_6, 0\rangle\}$ for disease d_4, and S_5 as $\emptyset$ for disease d_5. Initially, $ch_1 = ch_2 = ch_3 = ch_4 = ch_5 = 1$, and the matrix M is empty.

The set S_1 is first considered to determine the ch for disease d_1. It is observed how d_1 differs from other diseases against each attribute in S_1. KB_{LBP} shows that for attribute a_1, d_1 differs from the set of other diseases $^1D_1 = \{d_2, d_3, d_4, d_5\}$; for attribute a_2, d_1 differs from the set of other diseases $^1D_2 = \{d_2, d_3, d_4\}$; for attribute a_3, the set of other diseases $^1D_3 = \{d_3, d_4, d_5\}$; for attribute a_4, the set of other diseases $^1D_4 = \{d_2, d_5\}$; for attribute a_5, the set of diseases $^1D_5 = \{d_2, d_4\}$; and for attribute a_6, d_1 differs from the set of other diseases $^1D_6 = \{d_2, d_3, d_5\}$.

Starting with $^1D_1$ obtained against a_1, the Boolean truth value for a_1 for any element in $^1D_1$ is compared against the value of a_1 in I_p. As there is a match ($a_1 = 0$), then ch_1 is updated as $(ch_1 - 1/6) = 0.83$, and the matrix M is also updated with $M[1, 1] = 1, M[1, 2] = 0, M[1, 3] = 0, M[1, 4] = 0$, and $M[1, 5] = 0$. Next, the set $^1D_2$ obtained against a_2 is taken into account. In this case also, the truth value for a_2 for any element in $^1D_2$ is compared with the corresponding truth value in I_p. As there is a mismatch, the following updations are made: $ch_2 = ch_3 = ch_4 = 0.83$; $M[1, 2] = 0, M[1, 2] = 1, M[1, 3] = 1$, and $M[1, 4] = 1$. After completing the processing with sets $^1D_3$, $^1D_4$, $^1D_5$, and $^1D_6$ in the similar way, ch_1 is finally obtained as 0.50. So, it can be concluded that, against the patient input I_p, the chance of occurrence for disease d_1 is 0.50. Also, the non-final chances of occurrences for diseases d_2, d_3, d_4, and d_5 are 0.50, 0.67, 0.67, and 0.83, respectively. After the computation for disease d_1 is over, the matrix M is filled up as shown in Fig. 3.

From the matrix M, it can be easily found that, out of thirty entries, only seven entries $M[2, 5]$, $M[2, 3]$, $M[3, 4]$, $M[4, 4]$, $M[3, 5]$, $M[5, 5]$, and $M[4, 6]$ remained empty. So, the next disease d_2 is considered for further execution.

Similar to the process as followed for d_1, the set S_2 is taken into account for determining the chance of occurrence for disease d_2. For this, only the attribute a_3 would be used as the effect of other attributes in S_2 on disease d_2 has already been examined (as shown in Fig. 3). From KB_{LBP}, it is observed that d_2 differs from the set of LBP diseases $^2D_3 = \{d_3, d_4, d_5\}$ for attribute a_3. With a match found for the

	a_1	a_2	a_3	a_4	a_5	a_6
d_1	1	0	1	1	0	0
d_2	0	1	-	0	1	1
d_3	0	1	0	-	-	1
d_4	0	1	0	-	1	-
d_5	0	-	0	0	-	1

Fig. 3 Instance of matrix M after the execution for disease d_1 is complete

corresponding attribute-value pair for both the I_p and an element of set S_2, ch_2 for disease d_2 is updated as $(ch_2 - 1/6) = 0.33$. As the corresponding entries in M under attribute a_3 for diseases d_3, d_4, and d_5 have already been filled with 0, no changes are made for the chances of occurrences for these three diseases. $M[2, 3]$ is now updated as 1. After the execution for d_2 is over, the final chance of occurrence for d_2 is 0.33, and some fields in matrix M still remain empty.

The same process is followed for the other diseases until the matrix M is filled with 0 or 1s, with each column having at least one entry filled with 1. When the matrix M is full, the process stops. Against the patient information in I_p, the chances of occurrence for diseases d_1, d_2, d_3, d_4, and d_5 are 0.5, 0.33, 0.5, 0.5, and 0.83, respectively.

4   Results and Discussion

The ES_{LBP} has been validated with ten LBP patients' cases which are empirically selected from the patient records at ESI Hospital Sealdah, Kolkata. Validation has been done with eleven important clinical attributes for diagnosis of three LBP diseases SIJA, FJA, and PIVD. Among ten patients, six patients have been diagnosed with only SIJA, and rest have been diagnosed by expert physicians with more than one LBP diseases. There are two patients who have arthropathy in both the sacroiliac and facet joints. Two patients are there with disc prolapse along with pain in facet or sacroiliac joints.

More specifically, a set is constructed constituting ten patients as {p1, p2, p3, p4, ..., p10}. According to expert physicians, six patients p1, p3, p4, p5, p6, and p9 are suffering from only SIJA, patients p2 and p7 are suffering from both SIJA and FJA, patient p8 is diagnosed with both PIVD and SIJA, and p10 is treated with both FJA and PIVD. The disease-wise patient distribution scenario has been depicted in Fig. 4.

As ES_{LBP} has been implemented with limited parameters, if the chance of occurrence for a particular disease d for an LBP patient is equal or greater than 0.50, then

Fig. 4 Disease-wise
distribution of ten LBP
patients for validation of
ES_{LBP}

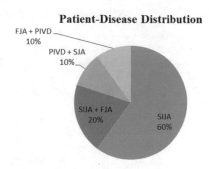

Patient-Disease Distribution

FJA + PIVD 10%

PIVD + SJA 10%

SIJA + FJA 20%

SIJA 60%

the inference engine concludes that the patient is suffering from d. If, for any patient, the chances of occurrence for more than one disease are greater or equal to 0.50, then the inference engine concludes that the patient is suffering from those diseases. This scenario is basically described as differential diagnosis for the patient based on his/her clinical symptoms. The patient-wise execution result is provided in Fig. 5. From the bar chart shown in Fig. 5, it is clearly visible that all the patients p1, p3, p4, p5, p6, and p9 have been accurately diagnosed with SIJA by ES_{LBP}; patients p2 and p7 have been treated with SIJA and FJA as differential diagnosis; patient p8 suffers from diseases SIJA and PIVD; and patient p10 has been diagnosed by ES_{LBP} with two differential diagnosis FJA and PIVD.

Based on comparisons between the obtained results (as shown in Fig. 5) with the expected outcomes (expert conclusions), it is seen that the diagnostic conclusions of all ten patients inferred by IE_{LBP} have matched with the conclusions made by the expert physicians at the ESI Hospital, Scaldah. Considering the expected chance of occurrence as 0.50 (as few clinical parameters have been used during this study), Fig. 6 shows dispersion of attained results with respect to the expected ones, where all the expected outcomes are at the level of 0.50 according to the vertical axis, implying that expert physicians opine on disease outcomes with at least 50% of certainty. As there are some patients who have been diagnosed with more than one diseases, there

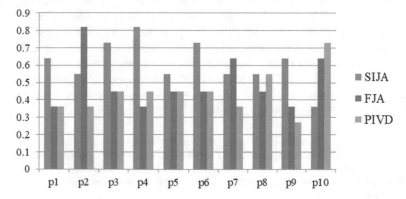

Fig. 5 Patient-wise inference outcomes for ES_{LBP}

Observed results with respect to the expected outcomes

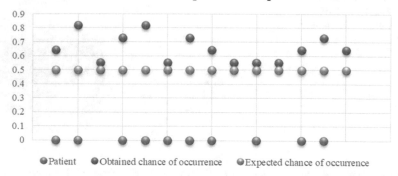

Fig. 6 Dispersion of observed results from the expected outcomes in ES_{LBP}

are basically a total of 14 outcomes obtained both by expert physicians and ES_{LBP} for ten patients. The graph has considered only those outcomes of the expert system that are associated with equal of greater than the chances of occurrence of 0.50.

From Fig. 6, it can be definitely concluded that the attained outcomes conform to the expected results.

5   Conclusions

Development of a medical expert system is a notable solution for assessment of LBP, as it would reduce the chance of misdiagnosis in case of unclear symptoms or other kinds of uncertainties, with the help of a carefully designed knowledge base and a reliable inference engine. The proposed technique of knowledge representation using discernibility matrix leads to a knowledge base design that ensures completeness, non-redundancy, and optimality. Having this kind of knowledge base design, a quick and reliable inference engine has been proposed for offering differential diagnosis of low back pain. With limited number of LBP diseases in literature, ES_{LBP} shows computational efficiency with respect to both time and space. The limitation of the approach is that it has been tested using limited numbers of patient records, with restrictions in choosing clinical parameters. Only three LBP diseases have been considered as test cases. ES_{LBP} needs to be exhaustively validated with large no. of patient records spanning over a no. of geographical areas, for making the intended medical expert system widely acceptable.

Acknowledgements The authors are sincerely thankful to the director and other faculty members at the ESI Institute of Pain Management, ESI Hospital Sealdah, West Bengal, India, for providing exhaustive domain knowledge. Also, the authors are very grateful to the hospital authority (ESI Hospital) and the members of ethics committee for supporting this research by allowing to access sufficient patient records.

References

1. Shortliffe, E.H.: Medical expert systems—knowledge tools for physicians. West. J. Med. **145**(6), 830 (1986)
2. Hoy, D.: The global burden of low back pain: estimates from the global burden of disease 2010 study. In: Ann Rheum Dis. doi **10** (2014)
3. Bindra, S., Sinha, A.G.K., Benjamin, A.I.: Epidemiology of low back pain in Indian population: a review. Int J Basic Appl. Med. Sci. **5**(1), 166–179 (2015)
4. Duthey, B.: Background paper 6.24 low back pain. In: Priority Medicines for Europe and the World. Global Burden of Disease (2010), pp. 1–29 (2013)
5. Shortliffe, E.H.: MYCIN: a rule-based computer program for advising physicians regarding antimicrobial therapy selection. In: Stanford univ calif dept of computer science. No. AIM-251 (1974)
6. Kulikowski, C.A., Weiss, S.M.: Representation of expert knowledge for consultation: the CASNET and EXPERT projects. Artif. Intell. Med. **51** (1982)
7. Miller, R.A., Pople Jr., H.E., Myers, J.D.: Internist-I, an experimental computer-based diagnostic consultant for general internal medicine. N. Engl. J. Med. **307**(8), 468–476 (1982)
8. Shortliffe, E.H., Scott, A.C., Bischoff, M.B., Campbell, A.B., Van Melle, W., Jacobs, C.D.: An expert system for oncology protocol management. In: Buchanan, B.G., Shortiffe, E.H. (eds.) Rule-Based Expert Systems, pp. 653–65 (1984)
9. Aikins, J.S., Kunz, J.C., Shortliffe, E.H., Fallat, R.J.: PUFF: an expert system for interpretation of pulmonary function data. Comput. Biomed. Res. **16**(3), 199–208 (1983)
10. Chakraborty, R.C.: Artificial intelligence–knowledge representation–issues, predicate logic, rules (2008)
11. Sittig, D.F., Wright, A., Simonaitis, L., Carpenter, J.D., Allen, G.O., Doebbeling, B.N., Sirajuddin A.M., Ash J.A., Middleton, B.: The state of the art in clinical knowledge management: an inventory of tools and techniques. Int. J. Medi. Info. **79**(1), 44–57 (2010)
12. Arsene, O., Dumitrache, I., Mihu, I.: Expert system for medicine diagnosis using software agents. Expert Syst. Appl. **42**(4), 1825–1834 (2015)
13. Bichindaritz, I., Marling, C.: Case-based reasoning in the health sciences: what's next? Artif. Intell. Med. **36**(2), 127–135 (2006)
14. Al-Ajlan, A.: The comparison between forward and backward chaining. Int. J. Mach. Learn. Comput. **5**(2), 106 (2015)
15. Zurada, J.M.: Introduction to artificial neural systems, vol. 8, St. Paul, West Publishing Company (1992)
16. Sherimon, P.C., Krishnan, R.: OntoDiabetic: an ontology-based clinical decision support system for diabetic patients. Arab. J. Sci. Eng. **41**(3), 1145–1160 (2016)
17. Caballero-Ruiz, E., García-Sáez, G., Rigla, M., Villaplana, M., Pons, B., Hernando, M.E.: A web-based clinical decision support system for gestational diabetes: automatic diet prescription and detection of insulin needs. Int. J. Med. Informatics **102**, 35–49 (2017)
18. Samuel, O.W., Asogbon, G.M., Sangaiah, A.K., Fang, P., Li, G.: An integrated decision support system based on ANN and Fuzzy_AHP for heart failure risk prediction. Expert Syst. Appl. **68**, 163–172 (2017)
19. Al-Jaghbeer, M., Dealmeida, D., Bilderback, A., Ambrosino, R., Kellum, J.A.: Clinical decision support for in-hospital AKI. J. Am. Soc. Nephrol. **29**(2), 654–660 (2018)
20. Romero-Aroca, P., Valls, A., Moreno, A., Sagarra-Alamo, R., Basora-Gallisa, J., Saleh, E., Marc, B.B, Puig, D.: A clinical decision support system for diabetic retinopathy screening: creating a clinical support application. Telemed. e-Health. **25**(1), 31–40 (2019)

Survey: Classification and Reconstruction of Archaeological Artefacts

R. Priyadarshini and A. Soumya

Abstract Cultural heritage is the benefaction of our bygone civilization, including physical and intangible properties that express the ways of living developed by a community. Those properties available convey the culture they followed and pass on the information from generations to generations, help the common people to understand, preserve and educate them to inherit the culture that is feasible and makes life easier. The physical properties might be ruined for various reasons. The deliberate act for preserving those matters from damage is by reconstruction. As the digital era is employing everywhere, technology can be used to document the fragments obtained in the excavation sites to be maintained for research purpose, educating and rebuilding without affecting the original fragments. This paper describes about different techniques used to classify the collected fragments into similar sets based on texture, colour and rearrange them to obtain its original structure and also in identifying the missing parts and reconstruct it accordingly.

Keywords Artefacts · Fragments · 3D surface registration · Photogrammetry · Terrestrial scanning · 3D virtual models

R. Priyadarshini (✉)
Department of Computer Science Engineering, REVA University, Bengaluru, India
e-mail: priyadarshinir@reva.edu.in

A. Soumya
Department of Computer Science Engineering, R V College of Engineering, Bengaluru, India
e-mail: soumyaa@rvce.edu.in

© Springer Nature Singapore Pte Ltd. 2020
J. K. Mandal and S. Mukhopadhyay (eds.), *Proceedings of the Global AI Congress 2019*, Advances in Intelligent Systems and Computing 1112,
https://doi.org/10.1007/978-981-15-2188-1_35

1 Introduction

Historical monuments and artefacts symbolize our culture. Cultural heritage which explains the livelihood of the past generations is a knowledge to be traversed with the future generations. It gives enough knowledge about the ways that the present and future generations live. Cultural heritage is included in all the livelihood of every individual for the purpose of identification and integration of people. It provides an automatic sense of unity and belonging within a group and allows to better understand the previous generations and the history to where the modern generation has reached. The artefacts not only helps in educating the people about the past disseminating culture but also helps the general public to view to communicate with heritage documents and in virtual museums.

The tangible properties of the ancient society can be obtained accidentally during the road or building construction or by visually observing the objects on the surface of the earth [1]. The archaeological fragments could be gathered more by using the instruments like radars, drones to capture the images of excavation sites and also ariels or satellite imagery to identify and locate the sites. The remains may be pottery pieces, ancient temples, writings on stones, monuments, buildings, townscapes, sculptures, paintings, etc. The places where the past generations survived, after the site has been abandoned, due to various reasons like natural disaster, wars, developments, vandalism etc., buildings, materials used by those people are deteriorated. The sites where those matters exist are called as archaeological sites.

The obtained artefacts need to be preserved to gather the information related to their existence and usage of those materials by the people and also help the archaeologists to conduct various research activities. Thus, it can be preserved by documenting through books, artefacts, objects, pictures, photographs, art and oral tradition.

In the area of reconstruction of the archaeological fragments, in ancient days, the fragments were manually classified by a researcher or an archaeologist based on the survey of the past information and documents in the archaeological site, colour of the obtained fragments, texture of it. Later, the classified fragments were joined together to build it accordingly as it existed. But, due to this traditional method, fragments obtained were more damaged by this wear and tear method; the result of reconstruction towards reaching to its original structure was very low. As the information and communication technologies (ICT) occupied to capture the world, the technologies like pattern recognition, computer graphics, image processing, etc., helped the researcher to capture the images of the obtained pieces of archaeological objects and move towards the reconstruction without touching the original materials (Table 1).

Table 1 Summarizes the different phases included in archaeological sites

Archaeological phases	Description
Establishment of sites	Places developed by the prehistoric civilization
Deterioration of sites	After the site is deserted, the natural forces control the area and instigate to weaken the structures built
Identification of sites	Sites may be recognized by common people and directed to archaeologists to excavate for detecting the remains buried, using different methods
Data acquisition	By excavating the place, gathering information from the fragments using different means and surveying the people living around the site
Conservation of the fragments	To preserve the artefacts, various measures have been implemented by the government and archaeological departments

2 Issues in Classifying and Reconstructing the Artefacts

The archaeological sites are identified either accidentally by common people when the materials are found on the surface of the earth, on the banks of river, during construction of roads, buildings, etc. Archaeologists intentionally inspect areas using group of people to walk overland and place the flags, when the materials are found and map the areas that contain more flags to term as archaeological site. Modern equipments like drones, ariel/satellite imagery, air balloons are used to capture sites in different elevations. Shovel testing, sensors, multi-spectral scanners present in satellites and geographical information system also help to recognize the materials present under the earth. Pieces procured during excavation might be related to common object or different objects [1].

The major issues faced by archaeologists are to identify the fragments, concerned to the similar object or not. If similar object, whether it belongs to different civilizations, classifying the fragments into similar group, documenting the physical properties, knowledge gathered by the people persisting in the same site, identifying the missing parts, its reconstruction to meet the original structure. We will briefly introduce the issues:

i. **Classification of fragments**. In early generations, archaeologists acquire images of the fragments through various gadgets; the features of the snaps taken were restricted to 2D, categorizing those images of fragments were done using manual method, thus resulting in improper documentation. Complexity increases as the differences occur in measurements of the fragments such as 2D profile curve and 3D line.

ii. **Identifying the missing parts**. During excavation, few fragments of the objects are obtained; reconstructing such objects is a difficult task as geometric shape of the original objects cannot be determined. Archaeologists face difficulties during concur the fragments to reconstruct an object, especially when there exist gaps between the fragments.

iii. **Reconstruction of the fragments into artefacts**. Archaeologists try to restore the unstructured pieces of the object, based on conditions like colour and appearance, meet towards its origin. It helps to gather the lifestyle, document civilization of the ancient people in that era.

In the early 90s, archaeologists try to manually identify the fragments obtained in the archaeological sites. Based on the past knowledge gathered from various sources, they trial to assemble together the fragments and build to reach its original structure. But in vain, some archaeologists would rather damage the fragments during rejoin as the fragments may be weak due to age factor, etc. Today, we have resources that help capture the images of the fragments using cameras in 2D or 3D that would help store the fragments and act as a data sets to reconstruct or restore the artefacts according to the ancient civilization.

3   Technologies and Tools for Collecting the Similar Fragments and Reconstructing It

As the computing technology replaced the manpower, classification of the fragments also took place based on the images captured by the various tools. Many researchers are helping archaeologists in investigating and developing technologies to classify based on the local colour and texture of the fragments. Some try to collect the adjacent pieces of the fragments. There also arises a major challenge in reconstructing or restoring the fragments obtained during the excavation in the archaeological sites to retain its original shape, identifying the missing parts, filling the gaps when its appropriate piece is not found, during collection of similar fragments especially during chronic cases when filling the gaps, etc.; here are the works done by different researchers using different techniques applied for classification and reconstruction.

Rasheed. N. A, Nordin, M. J. used algorithm to classify fragments based on local and global features to identify the colour and texture properties of the archaeological fragments [2]. The images of the fragments are captured by a Nikon camera and stored as the data sets in the memory. Algorithm helps to identify the RGB properties of the images to check the highest value when intersected images with other. Fragments are also classified based on the texture properties recognized using Euclidean distance considering the local binary pattern (LBP) as a vector to identify the pixels into a histogram. They also explained about reconstruction process as, first step with all reconstruction activities are by converting the images captured from 2D to 3D using the 3D laser scanning device. The work done by them had various phases like acquisition of 3D model and applies the preprocessing to eliminate the noise and calculate geometric features after extracting each part and divided into subsets. Identify the sets that match the pair of fragments with respect to geometrical x using neural networks and aligning the fragment images to form an object. If not completed, then retain the fragments in the data set until the other parts of object are found.

Terrestrial laser scanning technology (TLS), an efficient tool, can be used for documenting and 3D modelling of the monuments. The tool generates laser beam that is reflected from object to capture the in-depth 3D models, 2D drawings to collect the images of the object in different angles. Based on the slope, distance and angles, data are converted from 3D to 2D objects to be stored in memory [3]. Mariana Calin et al., the data sets are converted from 2D to 3D through data acquisition resulting in 3D modelling to virtually reconstruct the object. 3D model was constructed using cyclone software, and data filtering process was used to remove unnecessary information like noise, redundant edges, etc. 3D modelling technique is depended on the cultural properties that are obtained during excavation [3] and also based on the historical documents gathered from the people around the archaeological sites and research. 3D modelling also concentrates on virtually reconstructing the missing and also damaged parts. TLS can be used to collect the fragments by capturing the images in 2D and convert them into 3D to perform reconstruction.

Traditional or laser scanning technology is very complex to implement. Architecture, engineering and construction (AEC) tools can be used integrated with parametric programming languages to classify the fragments that include computer graphics for identifying the fragments through the geometrical points and other parameters. AEC tools are helpful in creating a library that includes the fragments describing the parameters [4]. Helena Rua and Ana Gil used building information modelling (BIM) tools and geographic information system (GIS) that help to construct the virtual models and assign properties that replicate the 3D geo-referencing to the modelling components. It is a very efficient technique as it helps in maintaining the database that includes alphanumeric and graphical and other types of information [4].

Carlos Sanchez and E. V. Vidal used the laser scanning technology to capture the 3D digital models. Once the images are captured, using the B-rep model [5] are used to calculate the projection models. The search process includes geometric translations, rotations and projections of the image captured, visibility including self-occlusions in fragments, discretization and comparison between samples. Different search algorithms were used to identify the accuracy during the reconstruction of the objects. The exhaustive approach, GPU depth map, is the efficient technique used to calculate all the possible alignments between the obtained and also the missing fragments [5]. The authors also used other searching algorithms to find the pieces of artefacts like counter-matching techniques, surface matching techniques. Once the input was collected, searching process would be started to find the appropriate fragment to reconstruct the original fragment. The searching technique is based on the geometric features that include the translations, rotations and projection of the fragments. GPU-based calculations are used to determine the geometric features like visibility, discretization and finding similarities between samples.

Documented and stored the 2D and 3D captured images of ceramic and glass artefacts, in a multimedia database including the multilingual-polytonic text [6]. This was first done by a Cultural and Educational Technology Institute (CETI) for storing the detailed description of morphology, historical data and scientific measurements that help in reconstruction. Using this technology, they were able to store the information including chemical, mineralogical, physical and mechanical properties of the material, archaeological data information, etc. N. Tsirliganis et al., the collected images and its related documents were used in presentation to construct the virtual image of the artefacts through VRML, to the researcher or an archaeologist in finding its structure, history, to interact with its 3D representation by performing rotation, scaling, transformation, zooming, panning, etc. [6].

Rendering techniques and 2D photography were used to capture the images of the monuments. Conversion from 2D to 3D was done using the 123D-Catch, an inexpensive 3D photogrammetry. Multi-image photogrammetry also called as Structure from Motion (SFM) [7] software to calculate the geometric information from large datasets, to compare the large sets of images simultaneously, and identifies matching features in an accurate, cost-effective, user-friendly manner. It also uses the memento product, a high-resolution 3D point cloud or solid surface mesh, generated through a network of computers to analyse the data and animate them to reach the virtual image. Photogrammetry was used to determine the geometrical properties of the captured image and to reconstruct the monuments based on the obtained fragments using some animation tools like ambient occlusion (AO). The photogrammetry uses the protocols based on the texture and geometric resolution of the fragments. It includes the initial analyses of monuments, subsequent project memento digital processing methods, method of deliberate animation and manipulation of the digital components. The animation tools help to classify and match the suitable fragment and finally to restore or reconstruct the monument [7].

Digital photogrammetry technique was also used to create the 3D models and scaled orthophotos, pans, etc. The deteriorated objects must be reconstructed using the captured data. During excavation of the archaeological sites, the measurements taken by the experts help in 3D modelling from digital images, while modelling the surface and complex archaeological structures. The photogrammetric images describe the construction, spatial layout, in the form of handmade drawings, CAD drawings, 3D visualizations and animations [8]. The input photos must undergo the series of processing steps including interior, exterior, relative orientation resulting in generating a document or a manual. The manual generated to create a digital 2D or 3D plans containing metric information during the excavation and classification. It also generates a 3D model of surfaces, projection of real textures [8].

F. Menna et al. explored the importance of the underwater cultural heritage. It is called as the Heritage-at-(great)-Risk since documenting and recording the findings are a difficult task, and deterioration is at the highest rate. The author discussed the underwater excavation through reconnaissance, mapping or high-resolution monitoring for data acquisition [9]. F. Menna et al. used 3D digital recording to extract the geometric measurements of the captured images of the objects, mapping techniques to analyse the structure during its existence, 3D data accuracy and resolution provided by the sensors to illustrate the structure of missing parts based on distance between the fragments [9].

Multi-scale feature extraction technique and a keypoint selection strategy driven by saliency and a fast one-to-one registration algorithm based on a three-level hierarchical search were used to extract a set of formalized geometrical conditions of the fragments obtained excavation site. Carlos Sanchez and E. V. Vidal proposed execution cycle for the automatic reconstruction process. He used extension of the persistent feature histogram descriptor to reconstruction approaches and a graph-based global reconstruction algorithm that uses individual matches to perform the final reassembly of the original artefact [10].

According to virtual archaeologist technology [11], the images of the fragments obtained from the 3D or 2D scanner were digitized. Virtual manipulation and visual representation techniques were applied to acquire the physical and chemical properties of the fragments and were archived. Later, the archives were given as an input to the virtual archaeologist to reconstruct the monument accordingly where the first step included was to find the correct match between captured pieces of the fragments from geometric view point, using broken surface morphology technique. This technique determines the fractured faces of the buildings, assembles the similar fragments into complete or partially complete entities. Further, mesh segmentation was done to glue the potentially related fragments, and fragment matching was done to estimate the relative picture obtained and determine the structure of a monument during its existence.

Reference No.	Year, name of the journal/conference	Author	Technology used	Accuracy	Challenges	Limitations
[2]	2018, Journal of KingSaud University-Computer and Information Sciences	Rasheed, N. A, Nordin, M. J.	Neural networks	Classification: 96.1% Reconstruction: 100% precision **Object:** pottery	Neural networks Challenges in filling gaps and assembling the fragments efficiently	Classifying and reconstructing the fragments including angles as an additional feature
[3]	2015, International Conference on: Agriculture for life, Life for Agriculture	Mariana Calin, George Damian, Tiberiu Popescu, Raluca Manea, Bogdan Erghelegiu, Tudor Salagean	Geometry, graphics	Result: approximate 3D models can be generated based on knowledge of history of monuments **Object:** monuments	To generate high fidelity 3D models	To create the 3D database using terrestrial laser scanning technology
[4]	2014, manuscript: Elsevier publication	Helena Rua, Ana Gil	geometry	Result: approximate 3D models can be generated based on knowledge of history of monuments **Object:** monuments	BIM software does not satisfy the deformations in the sets obtained when compared to the information obtained during excavation Data sets are unstable, and the failure rate is comparatively high	Generating the models of the monuments that are inaccessible Include the interaction between the machine to give the inputs at various stages of classification and restoration of monuments

(continued)

(continued)

Reference No.	Year, name of the journal/conference	Author	Technology used	Accuracy	Challenges	Limitations
[5]	2012, 18th International Conference on virtual systems and Multimedia	Carlos Sánchez Belenguer, Eduardo Vendrell Vidal	Computer graphics	Results: 99.6% faster and achieving towards optimal solution **Object:** pottery	Exhaustive approach uses more memory space requirement and performance-related issues that include loss of data Exhaustive approach requires lot of memory for the characterization of fragments obtained. It also leads to less optimization, thus decrease in the resolution	Include Optimization techniques to increase the efficiency in searching algorithms, reducing the data storage, etc. Searching techniques need improvisation to refer to the complex scenarios where certain amount of irregular fragments occur Storing the images of fragments using surface point database information system

(continued)

(continued)

Reference No.	Year, name of the journal/conference	Author	Technology used	Accuracy	Challenges	Limitations
[6]	2002, IEEE Proceedings of the First International Symposium on 3D Data Processing Visualization and Transmission	N. Tsirliganis, G. Pavlidis, A. Koutsoudis, D. Papadopoulou, A. Tsompanopoulos, K. Stavroglou, Z. Loukou, C. Chamzas	JAVA VRML, Multimedia database	Results: integrated and complete documentation of cultural objects **Object:** ceramic and glass fragments	VRML standard does not support for the multi-user environments	The client–server interface can still be enhanced by including the 2D charts containing technological and functional information controlled by the VRML applet 3D data and image compression Increase the percentage of efficiency by reducing the data transmission
[7]	2016, Journal of Archaeological Science: Reports	Robin Ann Johnson, Ariel Solis	Digital processing, animation	Results: obtained objects after animation are fragile and finely scaled objects **Object:** monument	Images may lack in the resolution due to megapixel limits Difficulty in providing the reference objects around the monuments for software stitch software	Enhancement can be made in modelling and in finding the contact surfaces Protocol to match the relevant fragments and thus improve in the field of pattern matching

(continued)

(continued)

Reference No.	Year, name of the journal/conference	Author	Technology used	Accuracy	Challenges	Limitations
[8]	2009, Journal of Cultural Heritage	Gungor Karauguz, Ozsen Corumluoglu, Ibrahim Kalayc, Ibrahim Asri	geometry	Result: analogy using photogrammetry is consequentially lower costs **Object**: buildings	The surface on the obtained fragments is not determined and complex Speed in generating the restored monument is comparatively low	Increasing the speed to generate the information by including several other tools
[9]	2018, Journal of Cultural Heritage	Fabio Menna, Panagiotis Agrafiotis, Andreas Georgopoulos	Sensors, geometry	Result: document generation at a faster rate **Object:** Underwater objects	Low contrast images are obtained. In sediment covered wreck sites	Using the photogrammetry devices to increase the contrast, with an optimistic resolution images

(continued)

(continued)

Reference No.	Year, name of the journal/conference	Author	Technology used	Accuracy	Challenges	Limitations
[10]	2014, International Conference on Virtual systems and Multimedia	Carlos Sánchez Belenguer, Eduardo Vendrell Vidal	Computer graphics and vision	Result: fragments are reconstructed into artefacts without huge memory requirement **Object:** monuments	Accuracy rate for large datasets	To improve robustness, by enriching a penetration detection algorithm to detect the incorrect alignments between fragments/clusters. Extract information to globally to register the fragments for future usage
[11]	2011, IEEE Computer Graphics and Applications	Papaioannou, G., Karabassi, E.-A., & Theoharis, T.	Computer graphics and vision	Results: since the genetic algorithm is used, execution time is less. Global reconstruction error is reduced **Object:** monuments	Fragment matching is computationally expensive System incrementally invokes updates to file of matching errors when a new piece is added to the collection	Depth first resolution can be adjusted to trade accuracy for computational speed. Typical dimensions for the depth buffer is limited to 100×100 and 256×256 pixels

4 Conclusion and Future Enhancement

Cultural heritage is the assets of our ancestors that should be conserved restored to provide as the precious asset, evidence about human activities and cultural values. The assets are needed for research by the archaeologists, for the general people to understand the culture, living style of the past civilization. This paper discusses the different methods to classify the fragments of a similar object captured using various methods like checking the texture, colour of the fragments directly, cameras to capture the images of fragments, scanning technology, that determines all the properties of the images captured including geometric features, texture, etc. The authors also discussed about the reconstruction of the collected fragments of the similar objects obtained in the excavation sites. Reconstruction of an archaeological artefact requires the historical knowledge about the ancient civilization, techniques and tools like graphics, neural networks, to determine the original sculpture.

In this paper, we have presented various technologies to classify and join the fragments to reconstruct the archaeological artefact. But, the authors referred had used the techniques of pattern recognition, computer graphics and neural networks to find the fragments of the same object. In these concepts, human intervention is high and no automation. Further, research will be made on automating the machines to identify the similar fragments whenever the input is given and reconstruct it using the artificial intelligence algorithms.

References

1. http://anthropology.msu.edu/anp203h-ss16/2016/02/10/how-do-archaeologist-find-sites
2. Rasheed, N.A., Nordin, M.J.: Classification and reconstruction for the archaeological fragments (2018)
3. Calin, M., Damian, G., Popescu, T., Manea, R., Erghelegiu, B., Salagean, T.: 3D modeling for digital preservation of Romanian heritage monuments (2015)
4. Rua, H., Gil, A.: Automation in heritage—parametric and associative design strategies to model inaccessible monuments: the case-study of eighteenth century Lisbon Aguas Livers. Aqueduct, Digital Applications on Archaeology and Cultural Heritage (2014)
5. Belenguer, C.S., Vidal, E.V.: Archaeological fragment characterization and 3D reconstruction based on projective GPU depth maps (2012)
6. Tsirliganis, N., Pavlidis, G., Koutsoudis, A., Papadopoulou, D., Tsompanopoulos, A., Stavroglou, K., Loukou, Z., Chamzas, C.: Archiving 3D cultural objects with surface point-wise database information. In: Proceedings of the First International Symposium on 3D Data Processing Visualization and Transmission (2002)
7. Johnson, R.A., Solis, A.: Using photogrammetry to interpret human action on Neolithic monument boulders in Ireland's Cavan Burren (2016)
8. Karauguz, G., Corumluoglu, O., Kalayc, I., Asri, I: 3D photogrammetric model of Eflatunpinar monument at the age of Hittite empire in Anatolia (2009)
9. Menna, F., Agrafiotis, P., Georgopoulos, A.: State of the art and applications in archaeological underwater3D recording and mapping (2018)

10. Belenguer, C.S., Vendrell Vidal, E.: An efficient technique to recompose archaeological artifacts from fragments (2014)
11. Papaioannou, G., Karabassi, E.-A., Theoharis, T.: Virtual archaeologist: assembling the past. IEEE Comput. Graph. Appl. **21**(2), 53–59 (2001). https://doi.org/10.1109/38.909015

Assessment of Black Tea Using Low-Level Image Feature Extraction Technique

Amitava Akuli, Abhra Pal, Tamal Dey, Gopinath Bej, Amit Santra, Sabyasachi Majumdar and Nabarun Bhattacharyya

Abstract This paper proposes a low-level image feature extraction and data analysis technique for rapid assessment of quality of tea to overcome the drawback of human perception-based organoleptic method of sensory panel during 'tea tasting'. An image capturing system has been developed to capture the image of tea liquor samples under controlled illumination. The experiment has been performed using 36 CTC (cut-tear-curl) tea sample collected from the Tocklai Tea Research Institute, Jorhat, Assam. Firstly, 27 low-level image features have been extracted using three different colour models and are then analysed using principal component analysis (PCA) and linear discriminant analysis (LDA) for visualization of underlying information. Finally, different classification models based on statistical regression techniques, e.g. multiple linear regression (MLR), principal component regression (PCR) and partial least square regression (PLSR), are investigated to find out a correlation between extracted image features with the tea tasters' score.

Keywords Tea quality · Tea taster · Image processing · Colour features · PCA · LDA · MLR · PCR · PLSR

A. Akuli (✉) · A. Pal · T. Dey · G. Bej · A. Santra · S. Majumdar · N. Bhattacharyya
Centre for Development of Advanced Computing (C-DAC), Salt Lake, Kolkata, India
e-mail: amitava.akuli@cdac.in

A. Pal
e-mail: abhra.pal@cdac.in

T. Dey
e-mail: tamal.dey@cdac.in

G. Bej
e-mail: gopinath.bej@cadc.in

A. Santra
e-mail: mailme.amitsantra@gmail.com

S. Majumdar
e-mail: sabyasachi.majumdar@cdac.in

N. Bhattacharyya
e-mail: nabarun.bhattacharya@cdac.in

© Springer Nature Singapore Pte Ltd. 2020
J. K. Mandal and S. Mukhopadhyay (eds.), *Proceedings of the Global AI Congress 2019*, Advances in Intelligent Systems and Computing 1112,
https://doi.org/10.1007/978-981-15-2188-1_36

1 Introduction

Tea, the widest spread drink of choice everywhere in the world, is being consumed as health drink with a growing market. The assessment of tea quality is a multi-faceted problem because of the presence of hundreds of chemical compounds and their presence in different concentrations to determine the final quality of tea [1]. The quality of tea varies with the geographical locations, seasonal variations, clone type, weather conditions, soil type, plucking quality and finally the art of tea processing. Once tea leaf is plucked from a garden, it passes through a chain of processing stages like withering, preconditioning, cutting, tearing and curling (CTC), fermentation, drying and packaging [2]. It is well known that the quality of tea cannot be enhanced after plucking, but can be destroyed due to improper processing methods. The colour of 'black' tea is formed due to the presence of multiple chemical constituents like theaflavins (yellowish brown), thearubigins (reddish brown), flavonol glycosides (light yellow), pheophorbide (brownish), pheophytin (blackish) at different concentrations. Theaflavins (TF) and thearubigins (TR), two important pigments, produced during tea processing are responsible in the formation of the colour and brightness of tea liquor [3]. Presently, the quality of tea is assessed by the human senses of vision, nose and taste of 'tea tasters'. The human perception may be influenced by psychological factors like mental state, age, sex, individual biasness to a particular tea brand, etc., which cause the variation in quality score of a same tea [2]. Thus, the quality index assigned by the tea tasters sometimes produces inconsistent result which is subjective.

The tea scientists are using the analytical methods to measure the above bio-chemical compounds [4] in tea. The analytical instrumental analysis methods [5, 6] like spectrophotometry [7] and high-performance liquid chromatography (HPLC) are accurate, repeatable and reproducible. However, the instrumental methods are restricted to high procurement and maintenance cost of the instruments, costly reagents, complicated and time-consuming sample preparation method, longer analysis time, requirement of skilled operators [8]. Hence, the analytical methods may not be feasible for assessment of tea quality in tea auction centres and tea testing laboratories where a large number of tea samples are tested daily. So, there is a dare need to develop a low-cost, rapid measuring instrument for objective and quantifiable assessment of black tea quality. The application of 'electronic nose (E-nose)' and 'electronic tongue (E-tongue)' is well-known technology, nowadays, in the field of electronic tea tasting. The application of electronic eye (E-eye) for quality analysis of agricultural produces has been reported in a number of research publications. The application of E-tongue using voltammetric technique was reported for estimating the percentage of TF [5, 6] and TR [6]. The measurement of tea quality index (TQI) during fermentation [9] was attempted using computer vision. The photo-micrographic image analysis technique was proposed by capturing the image of bulk tea sample using a microscope, and an artificial neural network technique was applied to identify the different tea grades [10]. A study on changing in colour during fermentation using red, green, blue (RGB) colour model was reported by Sharma and Thomas

[11]. The use of computer vision to detect the end point of fermentation using RGB and hue, saturation and intensity (HSI) colour model was proposed by Borah and Bhuyan [12, 13]. Laddi et al. [14] studied the requirement of optimum illumination intensity during image analysis to discriminate the black tea varieties. Gill et al. [15] proposed image analysis algorithms for colour and texture analysis for grading of made tea. A computer vision system for analysis of tea liquor using histogram colour matching technique was proposed by Kumar et al. [16]. Lekamge and Ratnaweera [17] proposed an image processing technique to identify the stalk particles using fuzzy logic-based colour segmentation.

In the present work, an electronic vision (E-vision) system is proposed to assess the tea quality by extracting low-level image features from the tea liquor image and a suitable computational model is developed to correlate with tea taster's score. This paper explores the performance of different computational models (MLR, PCR and PLSR) to predict the tea taster's appraisal.

2 Materials and Methods

The materials and the methodology to assess tea quality and correlation with tea tasters' appraisal are described below. The CTC-made tea sample was used to prepare the tea liquor for quality assessment using E-vision system, and the same tea liquor was presented to tea taster for quality appraisal.

2.1 Electronic Vision (E-vision) Set-Up

The instrument for appearance-based tea quality consists of a digital colour camera (Logitech®), a controlled illumination arrangement, a sliding sample holding arrangement, constant power supply source and a computer loaded with the application software. The block diagram of the E-vision system is shown in Fig. 1. The camera is mounted over the tea liquor sample at a distance of thirty centimetre [8]. The camera is connected with the computer using USB 2.0 (Universal Serial Bus). The above components are placed inside a cabinet whose internal surface is coated with black colour. The configuration of camera during image capturing is manual mode, fixed focus, no zoom, no flush, auto white balance, image resolution of 640 × 480 (N × M) pixels and 24-bit BMP (Microsoft© Bitmap) image storage format. The tea sample is illuminated using four high-intensity DOME light-emitting diodes (LEDs) [3], placed at four corners inside a closed cabinet. A fixed luminance on the sample is maintained using a constant power source. The tea liquor in white porcelain bowl is placed manually in the predefined slot on the sliding tray. Sliding tray allows the user to push the tea sample smoothly inside the cabinet and place the sample below the camera at a predefined position.

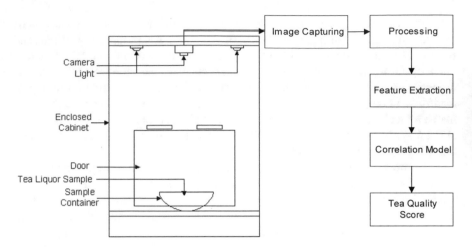

Fig. 1 Block diagram of E-vision system

Image analysis and feature extraction: Image processing software is developed using Microsoft ®Visual C++, loaded in a computer. The tea liquor preparation steps are described in Sect. 2.2. Image processing and analysis steps are described below. Image of the tea liquor sample is captured by the digital camera. The imaging set-up is shown in Fig. 2(a), and a representative image of tea liquor sample is shown in Fig. 2(b).

Image Processing: The images are taken under a controlled environment with a fixed light intensity about 110 lx. However, some noise in the image is introduced due to the presence of suspended tea particles in the liquor and the water bubbles formed during preparation of the tea liquor. A median filter with 5×5 window size is used for noise elimination. This filter is applied in each colour plane (R, G, B) of the image separately.

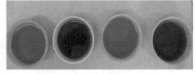

(b). Image of tea Liquor sample

(a). Electronic Vision System

Fig. 2 (a) Electronic vision system. (b) Image of tea liquor sample

Image Segmentation: Experimental result shows that the value of R (red) is prevailing over B (blue) and G (green) values in the liquor images. Hence, the image segmentation of the liquor image from the background image is done using R plane. A global threshold value is considered for segmentation by observing the histogram of the R plane. Binary image is formed after segmentation. The localization of the pixel in the binary image is used to extract the liquor image from the processed colour image.

Extraction of Features: The bowl containing tea liquor is concave shape with maximum depth at middle and diminishing the depth of liquor while approaching towards the edge of the bowl. As a result, the colour value at different locations of liquor imager is not same. The different locations in the liquor image (top view) are marked as C, N, S, E and W in Fig. 3 [3]. It is observed that the colour values (R, G) are found to be decreasing from the middle region (C) to any side (N, S, E or W) with minor change in blue (B) value.

Experimentally, it is observed that colour value changes with the depth of the tea liquor at different positions in the testing bowl. Hence, the colour features at five representative locations are identified in our experiment, e.g. at location C (centre), location N (north), location E (east), location S (south) and location W (west). A small image of size 40×40 pixels from each location (C, N, S, E, W) (Fig. 3) is considered for analysis. It is observed that colour value does not change much within this region (here, $d = 40$ pixels). Based on our observation and assumption, twenty-seven image features are selected for colour analysis for each image. The mean of different colour components (mean 'L', mean 'a', mean 'b'), at different locations (C, N, E, S, E), is calculated, forming fifteen image features. Rest, twelve features are calculated by subtracting the mean colour values at different locations (N, E, S, W) from the centre (C) for three colour components. List of extracted features using CIELAB colour model is shown in Table 1. Also, similar features are extracted using RGB and HSI colour model. Here, we are using raw values of the image and low-level

Fig. 3 Different marked regions in tea liquor image

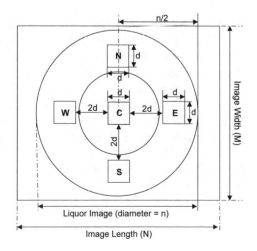

Table 1 List of extracted features using L*a*b colour space

Features	Description
LC, aC, bC	Mean L, a, b value of extracted image at location C
LN, aN, bN	Mean L, a, b value of extracted image at location N
LW, aW, bW	Mean L, a, b value of extracted image at location W
LS, aS, bS	Mean L, a, b value of extracted image at location S
LE, aE, bE	Mean L, a, b value of extracted image at location E
LC–N	Difference of mean L in extracted images at location C and N
LC–W	Difference of mean L in extracted images at location C and W
LC–S	Difference of mean L in extracted images at location C and S
LC–E	Difference of mean L in extracted images at location C and E
aC–N	Difference of mean a in extracted images at location C and N
aC–W	Difference of mean a in extracted images at location C and W
aC–S	Difference of mean a in extracted images at location C and S
aC–E	Difference of mean a in extracted images at location C and E
bC–N	Difference of mean b in extracted images at location C and N
bC–W	Difference of mean b in extracted images at location C and W
bC–S	Difference of mean b in extracted images at location C and S
bC–E	Difference of mean b in extracted images at location C and E

image features are extracted by simply calculating mean and applying subtraction operations, which do not require any complex computation.

2.2 Sample Collection and Preparation

The tea sample was collected from the Tocklai Tea Research Institute, Jorhat, Assam. Experiment had been carried out with thirty-six CTC tea samples (black tea). The standard procedure for the preparation of the tea sample was followed. The 2.5 gm of dry tea was taken into a brewing pot, and infuse it for 5 min in 100 ml distilled water. The liquor is then strained into the tasting bowl, and the residual leaves remain at bottom. The liquor was allowed to cool till it would reach at 50–55 °C. Five images were captured for each of thirty-six tea samples using E-vision system. Same tea sample was presented to a tea taster to assess the quality and provide a score between 1 and 10. Tea taster's score obtained among 36 tea samples varied from 5 to 10 with a mean of 7.05 and standard deviation (SD) of 0.984. The data analysis was performed using MATLAB®.

3 Data Analysis

The choice of a suitable data analysis method is dependent on the nature of the application and the data set. However, the steps are usually similar regardless of the method chosen. In our experiment, the image data set is assembled at first. Secondly, a suitable colour model is selected appropriately for the application. Thirdly, different features are extracted from the image. Thereafter, the features are exposed to the different classification methods. Finally, a performance criterion is set to select the best method.

3.1 *Data Pre-processing*

The images are captured by the E-vision system. The raw image data are then used for computation and finally for classification. Sometimes, the classification problem causes difficulty as the raw image data may be induced with noise due to the characteristics and limitations of the image sensors. In our experiment, the extracted features from images are huge variation in their ranges. Prior to applying classification technique, the raw data are normalized. An appropriate normalization technique may improve the classification accuracy. Several standard normalization techniques are available in the literature [18]. Different normalization techniques bear different meanings; e.g. relative scale gives a global compression of values with a maximum value of 1. Autoscale technique is helpful when responses are of different magnitudes. It sets the mean at the origin and sets the variance to 1. Range scale 1 and range scale 2 set the limits at [0, 1] and [−1, 1], respectively. Baseline subtraction method removes the offset. In this project work, range scale 1 technique [Eq. (1)] is used for normalization.

$$A_{ij} = \left(A_{ij} - \min\left(A_j\right)\right)/\left(\max\left(A_j\right) - \min\left(A_j\right)\right) \tag{1}$$

From the above equation,

A feature matrix,
A_{ij} ith sample of the jth features,
A_j all responses for features j.

A_i contains all responses for the features at the ith sample.

3.2 Principal Component Analysis (PCA) and Linear Discriminant Analysis (LDA)

PCA and LDA have been applied on the normalized data set to recognize the classes of tea sample having different tasters' scores [18]. The application of PCA on the image features transforms the data into two or more coordinates, which may be used to represent the information about whole data set. The transform projects the multidimensional data into two or more coordinates that maximize the variance and minimize the correlation in the data set. The 2D PCA plots using different colour models (RGB, HIS and Lab) with 27 features show the similar classes (tea sample with similar tea tasters' score) group together and dimensionality reduction process helps the user to spot trends, patterns and outliers in the data.

Like PCA, LDA also considers the information related to both the within-class and the between-class distributions. It finds the projection directions such that for the projected data, the between-class variance is maximized relative to the within-class variance. The difference between PCA and LDA is that PCA forms the cluster from the data set without knowing the number of classes present, while in LDA, the class information is required as input during computation.

3.3 Choice of Colour Model for Image Feature Extraction

The class separability criterion has been considered to compare the performance of different colour models using 27 features to form the cluster. The separability index is represented by the ratio of the 'between-class scatter matrix' (SB) to that of the 'within-class scatter matrix' (SW). It is calculated using the following expressions (2, 3, 4).

$$S_B = \sum_{i=1}^{c}(m_i - m)(m_i - m)' \tag{2}$$

$$S_W = \sum_{i=1}^{c}\left(\sum_{j=1}^{n_i}(x_{i,j} - m_i)(x_{i,j} - m_i)' \right) \tag{3}$$

$$m = \frac{1}{c}\sum_{i=1}^{c} m_i \tag{4}$$

where c represents the number of classes,

m is the mean of the entire data set
m_i is the mean vector of class i, $i = 1, 2, ..., c$
n_i is the number of samples within class i, $i = 1, 2, ..., c$
$x_{i,j}$ is the jth sample of ith class.

3.4 Mathematical Modelling Technique

Three regression models have been used in this study, namely multiple linear regression (MLR), principal component regression (PCR) and partial least square regression (PLSR).

Multiple Linear Regression (MLR)—Regression analysis is a mathematical modelling technique for finding out the relationship between a real-world continuous dependent variable and one or more independent variables. The goal is to formulate a function that describes, as closely as possible, the relationship between dependent and independent variables. The aim is to predict the value of the dependent variables using a range of values for independent variables.

Principal Component Regression (PCR)—PCR analysis technique works on the output of principal component analysis (PCA). The basic idea behind PCR is to calculate the principal components and then use some of these components as predictors in a linear regression model fitted using the typical least square procedure. A significant benefit of PCR is that by using the principal components, it can avoid the problem of multi-collinearity between the variables in the data set. The application of PCA on the raw data produces linear combinations of the predictors that are uncorrelated. Instead of taking all the components of PCA, only few components are considered during PCR analysis. It excludes some of the low-variance components in the regression step, reducing the cost of computation.

Partial Least Square Regression (PLSR)—PLS regression technique combines features from PCA and multiple regression. PLSR is a useful method of analysis because of lower sample size, the minimal demands on measurement scales and residual distributions. PLSR method is used when the goal of the analysis is to predict a set of variables from a set of predictors (features). PLSR is used to predict a whole table of data, and it can also handle the case of multi-collinear predictors (i.e. when the predictors/features are not linearly independent). PLSR can be used with very large data sets for which standard regression methods fail.

4 Results and Discussion

4.1 Data Clustering Using PCA and LDA

Exploratory data analysis on 36 samples has been performed using PCA and LDA to identify inherent clusters within the images captured using E-vision system. The plots are shown in Figs. 4 and 5. The PCA plots indicate the tendency to form clusters corresponding to the tea samples with closer tea taster's score.

The PCA plot in Fig. 4(a) shows that tea taster's score indicated by magenta (✿) and black (✿) markers is overlapping with each other. The PCA plot in Fig. 4(b) represents the PCA plot of image analysis results of tea taster's score using HSI colour model. It explains that tea taster's score indicated by green (*), cyan (□),

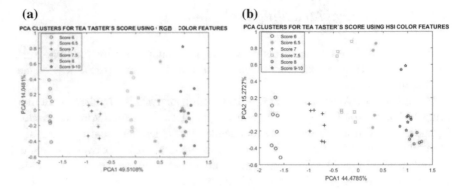

Fig. 4 PCA plot with 27 features using (**a**) RGB and (**b**) HSI colour models

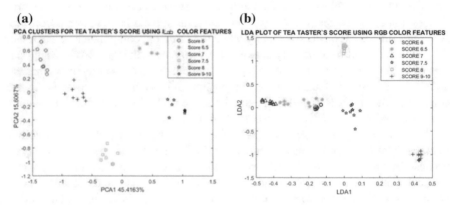

Fig. 5 (**a**) PCA plot with 27 features using Lab colour model and (**b**) LDA plot with 27 features using RGB colour model

magenta (✿) and black (✿) markers is overlapping with each other. The PCA plot in Fig. 5(a) presents the PCA plot of image analysis results for tea taster's score using Lab colour model. The plot shows the formation of good clusters produced by the tea sample with tea taster's score, indicated by red (o), green (*), blue (+), cyan (□), magenta (✿) and black (✿) markers. It may be concluded from the PCA plot that there is a complex and nonlinear interaction between the colour features extracted by three colour models and tea taster's score. The PCA plot using CIELAB colour model also provides the best cluster formation.

On other representation, the LDA plots also indicate the tendency to form clusters corresponding to the tea samples with closer tea taster's score. The LDA plot in Fig. 5(b) presents the image analysis results for tea taster's score using RGB colour model. The plot explains the formation of good clusters produced by the tea sample with tea taster's score, indicated by blue (Δ), magenta (+), cyan (□) and black (✿) markers. But other groups represented by red (o)–green (*) markers are indicating overlapping clusters of tea taster's score. Figure 6(a) presents the LDA plot of elec-

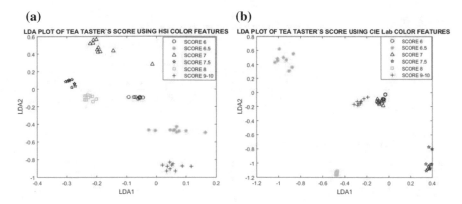

Fig. 6 LDA plot with 27 features (**a**) using HSI and (**b**) using CIELAB colour model

tronic vision image analysis results for tea taster's score using HSI colour model. The tea taster's score indicated by red (o), blue (Δ), magenta (+), cyan (□), black (✿) and green () markers forms notable clusters. But the markers green (*) and blue (Δ) are indicating tea taster's score minutely scattered cluster. Figure 6(b) represents LDA clusters of tea taster's score using CIELAB colour model. The tea taster's score indicated by red (o), blue (Δ), magenta (+), cyan (□), black (✿) and green (*) markers forms notable cluster. But the tea taster's score indicated by red (o) and blue (Δ) markers represents overlapping cluster with each other.

PCA and LDA plots show the formation of distinct clusters corresponding to each of the six grades of the tea. It is found that the separability measure for CIELAB colour model is maximum with 27 features.

4.2 Performance of Regression Models for Classifications

Data Set Preparation: 180 CTC tea samples (36 tea samples with 5 repetitions each) were used for development of MLR, PCR and PLSR models. The total data set obtained for analysis using 27 features was obsessed 180 × 27, respectively. The data set was divided into training set of size 150 × 27, respectively, and test set of size 30 × 27, respectively.

Selection of Performance Parameters: Three types of regression models are studied in this experiment. In order to compare the performance among three models, the following performance parameters are considered as shown in Table 2. The testing results using E-vision system are compared with the manual analysis results.

These performance parameters were calculated during both training and testing phases. However, only the testing results are reported. The most important parameters are RMSE and PA.

Prediction of Tea Taster's Score Using Multiple Linear Regression (MLR): A multiple linear regression model was developed using 27 components of response

Table 2 Performance parameters

Performance parameters	Description
PA	Prediction accuracy
SSE	Sum square prediction error
MSE	Mean square prediction error
RMSE	Root-mean-square prediction error

Table 3 Prediction of tea taster's score using MLR

Colour model	PA	SSE	MSE	RMSE
RGB	81.78	10.74	1.79	1.34
CIELAB	83.02	14.68	2.45	1.56
HSI	81.79	15.58	2.59	1.61

matrix. The accuracy prediction for correlation of tea taster's score was 83.07% for 27 features with CIELAB colour model which is best among others. Also, RMSE obtained 1.56 with 27 features. The multiple linear regression performance statistics are summarized in Table 3.

Prediction Tea Taster's Score Using PCR: A principal component regression model was developed using 27 components of response matrix. The accuracy prediction for correlation of tea taster's score was 92.07% for 27 features with CIELAB colour model which is best among others. Also, RMSE obtained 0.63 for 27 features, respectively. We may conclude that CIELAB model with 27 features can be used for prediction of correlation of tea taster's score with reasonable accuracy. The principal component regression performance statistics are summarized in Table 4.

Prediction of Tea Taster's Score Using PLSR: A partial least square regression model has been developed using 27 components of response matrix. The accuracy prediction for correlation of tea taster's score was 81.15 for CIELAB colour model, and RMSE is low for CIELAB colour model. So, it determines that CIELAB model can be used for prediction of correlation of tea taster's score with equitable accuracy. The partial least square regression (PLSR) performance statistics are summarized in Table 5.

Table 4 Prediction of tea taster's score using PCR

Colour model	PA	SSE	MSE	RMSE
RGB	89.21	3.41	0.57	0.75
CIELAB	92.07	2.38	0.39	0.63
HSI	90.90	4.18	0.70	0.83

Table 5 Prediction of tea taster's score using PLSR

Colour model	PA	SSE	MSE	RMSE
RGB	78.18	194.15	5.39	2.32
CIELAB	81.15	103.71	2.88	1.69
HSI	80.49	217.35	6.04	2.45

5 Conclusion

A low-cost imaging set-up using a digital camera under a controlled illumination has been described in this paper. A solution has been developed for assessment of quality of tea by capturing the image of the tea liquor sample. Image analysis techniques have been employed for extraction of different low-level colour features using three colour models. A panel of very renowned tea taster from North East India is employed for building up the organoleptic database using a large number of CTC tea samples. PCA and LDA clustering techniques have been explored to see the formation of clusters and 27 extracted features using three colour models. The PCA and LDA plots show that the developed solution can clearly discriminate among different grades of tea. Statistical regression models are used for correlation of electronic vision data with tea taster's score. The accuracy prediction for correlation of tea taster's score was 92.07% for 27 features with CIELAB colour model using PCR which is best among others. Also, RMSE obtained 0.63. Also, high R^2 value (0.86) using 27 Lab features with PCR model indicates reasonably better correlation compared to other models. A comparison of responses of predicted and measured tea taster's score using PCR model is shown in Fig. 7. We may conclude that PCR with 26 features using Lab colour model can be used for prediction of tea taster's score with

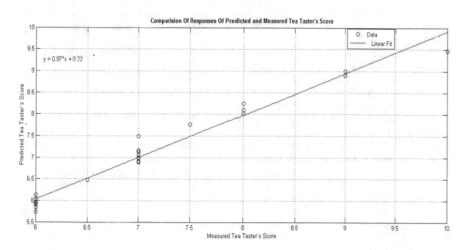

Fig. 7 Comparison of responses of predicted and measured tea taster's score using PCR model

reasonable accuracy. The reason may be the CIELAB colour model closely follows the human perception. However, the data used in this experiment are collected from a single tea garden, which does not possess wide variability. The system can be made versatile by incorporating knowledge of tea samples from multiple gardens across various agro-climatic zones in India and other tea-producing countries. Nevertheless, the electronic vision solution described in this paper, being low cost and portable, is affordable by the tea industries for daily use and has potential to rapid assessment of quality of tea.

References

1. Roberts, E.A.H.: The Phenolic substances of manufactured tea, their origin as enzymic oxidation products in fermentation. J. Sci. Food Agric. **9**, 212–216 (1950)
2. Obanda, M., Owuor, P., Mangoka, R.: Changes in the chemical and sensory quality parameters of black tea due to variations of fermentation time and temperature. Food Chem. **75**, 395–404 (2001)
3. Akuli, A., Pal, A., Bej, G., Dey, T., Ghosh, A., Tudu, B., Bhattacharyya, N., Bandyopadhyay, R.: A machine vision system for estimation of theaflavins and thearubigins in orthodox black tea. Int. J. Smart Sens. Intell. Syst. **9**(2), 709–731 (2016)
4. Robertson, A., Bendall, D.S.: Production and HPLC analysis of black tea theaflavins and thearubigins during in vitro oxidation. Phytochemistry **22**, 883–887 (1983)
5. Ghosh, A., Tamuly, P., Bhattacharyya, N., Tudu, B., Gogoi, N., Bandyopadhyay, R.: Estimation of Theaflavin content in black tea using electronic tongue. J. Food Eng. **110**, 71–79 (2012)
6. Ghosh, A., Tudu, B., Tamuly, P., Bhattacharyya, N., Bandyopadhyay, R.: Prediction of theaflavin and thearubigin content in black tea using a voltammetric electronic tongue. Chemometr. Intell. Lab. Syst. **116**, 57–66 (2012)
7. Roberts, E.A.H., Smith, R.: Spectrophotometric measurements of theaflavins and thearubigins in black tea liquors in assessment of quality of teas. Analyst (London) **86**, 94–98 (1961)
8. Akuli, A., Pal, A., Joshi, R., Dey, T., Bhattacharyya, N.: A new method for rapid detection of Total Colour (TC), Theaflavins (TF), Thearubigins (TR) and Brightness (TB) in orthodox tea using electronic vision system. In: IEEE International Conference on Sensing Technology, pp. 23–28 (2012)
9. Singh, G., Kamal, N.: Machine Vision system for tea quality determination—tea quality index (TQI). IOSR J. Eng. (IOSRJEN) **3**(7), 46–50 (2013)
10. Suprijanto, Rakhmawati, A., Yuliastuti, E.: Compact computer vision/or black tea qualify evaluation based on the black tea particles. In: IEEE 2nd International Conference on Instrumentation Control and Automation, pp. 87–91 (2011)
11. Sharma, M., Thomas, E.V.: Electronic vision study of tea grains color during Infusion. Int. J. Eng. Sci. Invention **2**(4), 52–58 (2013)
12. Borah, S., Bhuyan, M.: Computer based system for matching colours during the monitoring of tea fermentation. Int. J. Food Sci. Technol. **40**, 675–682 (2005)
13. Borah, S., Bhuyan, M.: Quality indexing by machine vision during fermentation in black tea manufacturing. In: Sixth International Conference on Quality Control by Artificial Vision, SPIE, 5132, pp. 468–475 (2003)
14. Laddi, A., Prakash, N., Sharma, S., Kumar, A.: Discrimination analysis of Indian tea varieties based upon color under optimum illumination. J. Food Meas. Charact. **7**(2), 60–65 (2013)
15. Gill, G.S., Kumar, A., Agarwal, R.: Monitoring and grading of tea by computer vision—A review. J. Food Eng. **106**(1), 13–19 (2011)
16. Kumar, A., Singh, H., Sharma, S.: Color analysis of black tea liquor using image processing techniques. IJECT **2**(3), 292–296 (2011)

17. Lekamge, B.M.T., Ratnaweera, D.A.A.C.: A hybrid approach for online tea color separation. In: IEEE international conference on Industrial and Information Systems (ICIIS), pp. 70–75 (2011)
18. Palit, M., Tudu, B., Dutta, P.K., Dutta, A., Jana, A., Roy, J.K., Bhattacharyya, N., Bandyopadhyay, R., Chatterjee, A.: Classification of black tea taste and correlation with tea taster's mark using voltammetric electronic tongue. IEEE Trans. Instrum. Meas. **59**(8), 2230–2239 (2010)

A Comparison Among Three Neural Network Models for Silk Content Estimation from X-Ray Image of Cocoons

Gopinath Bej, Tamal Dey, Abhra Pal, Sabyasachi Majumdar, Amitava Akuli and Nabarun Bhattacharyya

Abstract Silk cocoon is the one and only raw material for silk industry. In cocoon trading, price of the cocoon is determined by guessing the silk content within it. Human expert visually inspects the cocoon by its shape, size, color, etc., and feels the toughness of the cocoon by pressing with thumbs. This process is subjective and varies from person to person. Invasive techniques are there to estimate the silk content in cocoons, but those techniques are time-consuming, expensive, laborious, laboratory-oriented, and difficult to implement on large scale. So, it is essential to develop a methodology which will be able to estimate the silk content in cocoons using noninvasive manner. This paper proposes a nondestructive X-ray imaging technique to estimate the silk content in cocoon. The technology applies advanced image processing followed by appropriate data analysis techniques. Different significant features from the X-ray image were extracted at first. The aim of the project was to develop a suitable mathematical model to estimate the silk content in the cocoon. Three neural network models, namely general regression neural network (GRNN), radial basis function neural network (RBFNN), and feed-forward back-propagation neural network (FFBPNN), were studied under this research. A comparative study shows that the general regression neural network (GRNN) provides the best performance with a reasonable accuracy of 85%.

G. Bej (✉) · T. Dey · A. Pal · S. Majumdar · A. Akuli · N. Bhattacharyya
Centre for Development of Advanced Computing (C-DAC), Salt Lake, Kolkata, India
e-mail: gopinath.bej@cdac.in

T. Dey
e-mail: tamal.dey@cdac.in

A. Pal
e-mail: abhra.pal@cdac.in

S. Majumdar
e-mail: sabyasachi.majumdar@cdac.in

A. Akuli
e-mail: amitava.akuli@cdac.in

N. Bhattacharyya
e-mail: nabarun.bhattacharya@cdac.in

© Springer Nature Singapore Pte Ltd. 2020
J. K. Mandal and S. Mukhopadhyay (eds.), *Proceedings of the Global AI Congress 2019*, Advances in Intelligent Systems and Computing 1112,
https://doi.org/10.1007/978-981-15-2188-1_37

Keywords Silk content · X-ray image · Soft X-ray · Digital image analysis · Shell thickness · Gum content · GRNN · RBFNN · FFBPNN · Morphological opening · Median filter · Remove boundary objects · K-fold cross-validation

1 Introduction

Within agricultural commodities, silk is premium priced and the sheer volume is less than one percent of the market for natural textile fibers [1]. Unit price of silk is approximately twenty times of cotton. Thus, silk is an important source of revenue in developing countries [2]. In advanced countries, the growth of sericulture has declined steadily due to the high cost of labor, fast industrialization, and also climatic hinders. As a developing country, India has a better opportunity in sericulture field [3]. Silk has been interlaced with the life, culture, and tradition of the Indians. Out of many silk commodities, silk sarees are cordially accepted by the Indians, and also, it is one of the living examples of the craftsmanship of the weaver [4]. Raw material of silk is cocoon which is produced by the caterpillars belonging to the genus Bombyx [5, 6]. Cocoon is made of a continuous, strong, very fine, and lustrous thread of raw silk of 500–1500 m in length [1]. Three basic elements of cocoon are cocoon shell, pupa, and some fecal matters. Cocoon shell contains raw silk and the gum [7]. Silk extraction is a manual process. A chemical treatment is used to remove gum content form cocoon, and a long, continuous silk thread is extracted from cocoon [8]. Amount of silk content in a cocoon depends on some factors like environmental condition, eco race, proper feeding of silkworms. Two whole cocoons may have the same weight; however, the amount of silk content may be different. Many farmers are easily duped by the cunning dealer as there is no such standard tool or methodology to estimate the silk content in cocoons and determine the price accordingly. Hence, it is a challenge to develop a nondestructive, rapid measurement technique to estimate the amount of silk content in a cocoon.

2 Present Methodology

Presently, the estimation of silk content in cocoons is carried out in two processes by the traders. One of the processes is visual inspection, i.e., observing the size, shape, color, texture, etc., and feeling the toughness of cocoon by pressing it slightly using thumbs. This method is subjective and depends on the honesty and experience of the sorter. Another process is destructive in nature. In this method, some sample cocoons are collected from each lot and pupa are removed from the cocoon by cutting the shell. Then the cocoon shells are dried out, and weight of the dried shells is measured using a standard weighing machine. It is assumed that amount of silk content of the lot is the average shell weight, and accordingly, price is determined. This method is scientific but not accepted by the traders as it destroys cocoons and also adds some financial cost and time delay.

In this research, a nondestructive approach of silk content estimation of tasar cocoons using X-ray imaging technique was explored. Advanced digital image processing techniques followed by a suitable data analysis technique were proposed to estimate the silk content in a cocoon. In our previous study, general regression neural network (GRNN) and self-organized map (SOM) were explored for silk content estimation and quality grading of cocoons [9, 10]. The present study explores the performance of three neural network models, e.g., feed-forward back-propagation neural network (FFBPNN), radial basis function neural network (RBFNN), and general regression neural network (GRNN) for the estimation of silk content. Above mathematical models were cross-validated using tenfold cross-validation method. Experimental result shows that the GRNN model provides the best performance among three.

3 Materials and Methods

Figure 1 shows the proposed experimental process of silk content estimation from raw cocoons. Image processing and analysis algorithms [11, 12] were applied on X-ray image of cocoons for extracting features. Also, manually silk content was extracted by chemical treatment and weight was recorded. These features and the weight of the extracted silk content were fed to neural network models. Appropriate data analysis model was selected for silk content estimation best on the performance of the network [13].

Detail description of the process is described below step-by-step starting with sample collection.

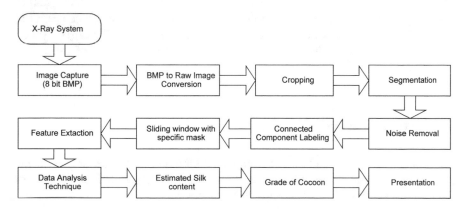

Fig. 1 Steps performed in this experiment

Fig. 2 Collected sample cocoons

3.1 Sample Collection

A total of 594 cocoons of different size, shape, and weight were collected from the field of Jharkhand. Collected cocoons were kept in a ventilated paper box. Cocoons were marked using a permanent marker pen serially from 1 to 594 for identification and tracking purposes during experiment. Figure 2 shows an image of collected cocoons.

3.2 Image Acquisition

X-ray images of cocoon were captured in a medical X-ray clinic by applying soft X-ray with controlled current as 8 mA and voltage as 45 kVp. A block diagram of the X-ray imaging setup is shown in Fig. 3. The resolution of the image was 1722 × 1430 pixels. The images were stored in 8-bit BMP (Microsoft® bitmap) grayscale format and subjected to further processing. A sample X-ray image of cocoons is shown in Fig. 4.

3.3 Image Processing

Silk content estimation starts with reading the 8-bit grayscale image. Several image processing and analysis techniques were applied on the image, and finally, a set of six features were extracted. The steps are discussed below.

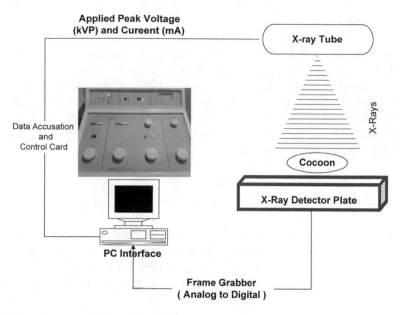

Fig. 3 X-ray image acquisition diagram

Fig. 4 Sample X-ray image

3.3.1 Image Cropping

Rectangular cropping was performed on the image to discard the unwanted region at four sides. Sixty-five (65) number of pixels were discarded from top, bottom, left, and right border of the original image.

3.3.2   Image Segmentation

The segmentation technique separates the background from the cocoons. Steps involved in segmentation are discussed below.

Morphological Opening. Opening of an image (IM) means dilation of the eroded image using specific structuring element (SE) [11, 12]. The process of opening smooths the contour of cocoon shells and pupa. It breaks narrow block connectors and eliminates small, thin protrusions from the reference image [11, 12]. Opening operation was performed using Eq. (1).

$$
\begin{aligned}
&\texttt{eroded\_image = Erosion(IM, SE)} \\
&\texttt{opened\_image = Dilation (eroded\_image, SE)}
\end{aligned} \tag{1}
$$

Applying grayscale morphological opening on the cropped image, it was observed that each cocoon shell from the image was vanished. So subtracting this opened image from the cropped image retained only the cocoon shell components.

Image Subtract. This is a process of pixelwise subtraction of two images. Image subtraction algorithm has been shown in (2).

$$
\begin{aligned}
&\texttt{Result(i, j) = IM\_1(i, j) - IM\_2(i, j)} \\
&\texttt{for each pixel} \\
&\texttt{if Result(i, j) > 255} \\
&\quad \texttt{Result(i, j) = 255} \\
&\texttt{else if Result(i, j) < 0} \\
&\quad \texttt{Result(i, j) = 0} \\
&\texttt{End}
\end{aligned} \tag{2}
$$

Image subtraction operation was performed between the cropped image and the resultant image of the opening operation. Subtracted image contains the cocoon shell component and some noises.

Global Thresholding. As the foreground pixels were pretty much different from the background pixels, a global thresholding [11] was performed on the subtracted image. Thresholding was performed using (3).

$$
\begin{aligned}
&\texttt{for each pixel(i, j) in image} \\
&\quad \texttt{if pixel(i, j) > lower\_cutoff and pixel(i, j) < upper\_cutoff} \\
&\quad \texttt{pixel(i, j) = 255} \\
&\quad \texttt{else} \\
&\quad \texttt{pixel(i, j) = 0} \\
&\quad \texttt{end} \\
&\texttt{end}
\end{aligned} \tag{3}
$$

Applying above thresholding operation, grayscale image was transformed to the binary image.

3.3.3 Noise Removal

After segmentation operation, it was found that a lot of noise was there in the image. So a noise removal technique was applied to remove the noise from the image. Below mentioned techniques were applied step by step to perform noise removal.
Median Filter. Median filter [11] is a widely used nonlinear noise removal technique. It was used to remove the noise from the image. Here, a window of fixed size was slide over the image and replaces the current pixel value by the median of the neighbors of window. Median filter was applied to the thresholded image using (4).

```
for each pixel(i,j) in the image
    1. find the window/neighbours
    2. find the median of the neighbours
    3. pixel(i,j) = meadian value
end                                                    (4)
```

Border Objects Removal. Objects linked to the border of the image were removed using morphological reconstruction [11, 12]. Algorithm shown in (5) was used to extract the image with border objects. This image was subtracted from input image to remove border objects.

```
Performed morphological reconstruction with
    1. Mask image = input image
    2. Marker image = zero everywhere except along the
    border, where it equals the mask image
    3. Connectivity = 8                                (5)
```

Figure 5 is the image after segmentation and noise removal. It represents only the cocoon shell component.

3.3.4 Connected Component Labeling

Detecting and labeling of connected components in a binary image are a fundamental step in digital image processing and analysis [11]. Here, each connected objects or blobs were uniquely labeled to separate from other blobs. Each pixel in a connected blob had the same label. Algorithm is shown in (6).

```
for all pixel in the image
```

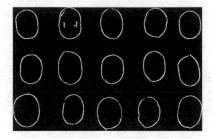

Fig. 5 Segmented image

```
if the current pixel is foreground pixel and it
is unlabeled then,
  assign a new label to the pixel and recursively
  assign same label to all the neighbour
  foreground pixels
else
  continue the raster scan
end
end                                                                    (6)
```

As this was a recursive algorithm, stack was used to reduce the time complexity. After the process, all the objects or the components in the image were labeled uniquely.

3.3.5 Masking Window

As the shell thickness of the cocoon was not uniform, after the segmentation sometime it was found that a cocoon shell has more than one labeled component. Those disconnected components were of same cocoon. To combine all those different labeled components to a single component, a sliding window of mask size 300 × 250 was propagated through the image. Masking width and height were set in such a way that it cannot be inferior to the cocoon's width and height. Algorithm is shown in (7).

```
for each rectangular window of specific mask
  if all the window edges doesn't cut a foreground
  component and the foreground pixel count is greater
  than a cutoff value then
    combine those foreground components to single
    component
end                                                                    (7)
```

Resultant image had all cocoons marked with unique label, so the features could be extracted from each cocoon for further data analysis.

3.4 Feature Extraction

Feature extraction is a process of dimensionality reduction where specific attributes of the image have been selected to represent the whole of the images. This selected feature will help on further data analysis techniques. In this step, six features were identified and extracted for the silk content estimation of individual cocoon. Features were given below.

- **Shell Area**. Shell area is the number of foreground pixel within the labeled component.
- **Average Shell Thickness**. Thickness is the linear distance between the inner and outer edges. Thickness of the shell wall was measured at 10 different locations. Then the average was calculated from those 10 thickness values.
- **Inner Height**. Inner height is the maximum Euclidean distance between the inner edges of the cocoon shell.
- **Inner Width**. Inner width is the minimum Euclidean distance between the inner edges.
- **Outer Height**. Outer height is the maximum Euclidean distance between the outer edges.
- **Outer Width**. Outer width is the minimum Euclidean distance between the outer edges.

All the height and width were calculated as (8).

$$\text{Euclidean}_{\text{Distance}} = \sqrt{(X_1 - X_2)^2 + (Y_1 - Y_2)^2},\tag{8}$$

where (X_1, Y_1) and (X_2, Y_2) are pixel co-ordinates.

An image with features explanation has been given in Fig. 6.

3.5 Manual Silk Extraction from Cocoon

After capturing image of the cocoon, same cocoon was chemically treated to extract the silk content. This extracted silk was dried enough in sunlight before measuring the weight [9]. This weight and the extracted features from the image were used to train and validate the neural network models.

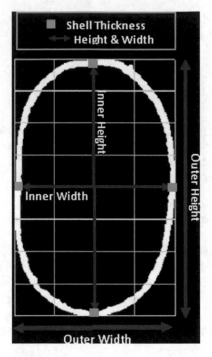

Fig. 6 Explanation of different features

3.6 Data Preprocessing

Extracted features were having different ranges of data. Hence, it is required to perform a preprocessing of the raw data set to have a uniform range of data preferably within the range of zero to one. Input data set was prepared by normalizing the input parameters using Eq. (9). This normalized data set was the input to the data analysis models.

$$X_{\text{Normalized}} = \frac{(X - X_{\text{Mean}})}{X_{\text{Standard Deviation}}} \tag{9}$$

3.7 Data Analysis Techniques

This part of the experiment was the decision-making part where the best suitable technique was chosen to fulfill the requirement. Extracted features and the known silk weight (in gram) were the inputs to the data analysis techniques. Each neural network was trained and validated with known data. Further, each neural network was tested with the unknown data set. Accuracy of each neural network calculated and

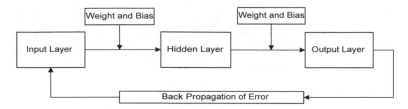

Fig. 7 Architecture of FFBPNN

compared with each other's for finding the appropriate one. Here, 594 cocoons were used for this experiment. Data of 414 cocoons were used for training and validation of the system, and 180 cocoons were used for testing of the system. Each network has six input nodes and one output node. Six input nodes can accept six image features, and one output node gives us the result. In this paper, three neural network models, i.e., feed-forward back-propagation neural network (FFBPNN), radial basis function neural network (RBFNN), and general regression neural network (GRNN), were used and a comparison result of the models has been shown in a later section.

Feed-Forward Back-Propagation Neural Network (FFBPNN) [14–17] is a supervised network, i.e., input and target data are known. It consists of an input layer, a hidden layer and an output layer as shown in Fig. 7. Hidden layer consists of a nonlinear activation function (tan-sigmoid), and output layer consists of a linear activation function. FFBPNN uses back-propagation technique for learning purposes. In the training time, calculations are performed with input data and feed-forwarded to the output layer from the initial layer. The error value is then propagated back to the previous layers. In the learning process, weight and bias vectors are being updated. Learning process stops on iteration limit reached or minimum error cutoff value achieved.

Radial Basis Function Neural Network (RBFNN) [18–20] consists of an input layer, a hidden layer, and an output layer. Hidden nodes, which are basically known as radial centers, implement the radial basis function, i.e., Gaussian function and the output layer implements the linear summation function. Radial centers can be chosen earlier or in the training time. During training, the weights of the hidden layer to the output layer are being updated and finalized. Output of the RBFNN is represented as (10).

$$y(x) = \sum_{i=1}^{n} w_i \emptyset_i(x), \quad \emptyset_i(x) = \emptyset(||x - c_i||), \tag{10}$$

where w_i = weight of ith center and $\emptyset$ = Radial function.

General Regression Neural Network (GRNN) [21, 22] is a kind of neural network architecture which can perform estimation on any arbitrary function having linear or nonlinear relationship between the dependent and independent variables, and the estimated result can meet the optimal regression surface even on the samples of sparse data. Basically, GRNN does not try to draw any relationship between

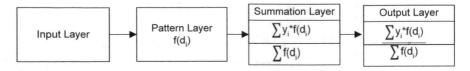

Fig. 8 Architecture of GRNN

the independent variables and dependent variables; rather, it uses a joint probability density function based on normal distribution to calculate the estimation. GRNN is a feed-forward network architecture implemented on supervised learning process. It has four layers input: pattern (Gaussian function), summation, and the output layer as shown in Fig. 8.

- **Input Layer**: It receives the input signal and passes it to the pattern layer for further processing.
- **Pattern Layer**: It calculates the Euclidean distance between the input vector and training vector and passes this value to the next layer, i.e., summation layer as an activation function (Gaussian function).
- **Summation Layer**: This layer has two neurons. One is numerator and another one is denominator. Numerator neuron calculates the weighted sum where a weight corresponds to a training sample output. Denominator neuron calculates unweighted sum of outputs from pattern layer.
- **Output Layer**: This layer calculates the output by dividing numerator by denominator. Basically, the output is the weighted average of the entire training sample.

Output of GRNN is calculated as follows:

$$Y(x) = \frac{\sum Y_i e^{-\left(\frac{d_i^2}{2\sigma^2}\right)}}{\sum e^{-\left(\frac{d_i^2}{2\sigma^2}\right)}} \tag{11}$$

where $d_i^2 = (x - x_i)^T (x - x_i)$, x is the test sample, and x_i is the training sample. Y_i is the training sample output. d_i^2 is the Euclidean distance between test and training sample. $Y(x)$ is the output of test sample (x). σ is called the spread constant or smoothing parameter. During training of GRNN, σ is optimized.

Above data analysis, techniques were applied in this experiment for estimating the silk content in cocoons and the outcome of the experiment is shown in the next section.

4 Results and Discussion

As mentioned earlier, total 594 cocoons were used here for the proposed technique of silk content estimation. Three models were validated for the whole set of data with tenfold cross-validation technique [23, 24]. K-fold cross-validation technique is a process where all data are divided into equal K parts, and for each of K data set, it uses $(K - 1)$ parts as the training data and the remaining part as testing data. Result of tenfold cross-validation technique has been shown below for FFBPNN, RBFNN, and GRNN. To test the repeatability of the data set, training and testing data sets were chosen randomly at each run. For each data analysis techniques, the validation has been performed eight times for repeatability testing [23, 24] and results have been shown in Tables 1, 2, and 3. Here selected indicators for performance evaluation were sum of the squared error (SSE), mean squared error (MSE), and the accuracy [13]. The results of all three models were validated by the manually extracted silk content of the cocoon. The accuracy of estimating the silk content of individual cocoon and average accuracy of the model were calculated using (12),

$$\text{Accuracy}\% = \left(1 - \frac{\left|\text{manul}_{\text{result}} - \text{system}_{\text{result}}\right|}{\text{manul}_{\text{result}}}\right) \times 100$$

Table 1 FFBPNN—cross-validation result

Sl. No.	Average SSE	Average MSE	Average accuracy
1	22.0554	0.3723	59.8861
2	21.6321	0.3635	56.0643
3	13.2896	0.2232	69.2245
4	17.5168	0.2953	62.9941
5	22.9633	0.3876	56.0402
6	27.5481	0.4631	54.0471
7	32.9141	0.5546	50.1866
8	23.7748	0.3996	55.7164

Table 2 RBFNN—cross-validation result

Sl. No.	Average SSE	Average MSE	Average accuracy
1	4.7091	0.0792	81.2668
2	4.7241	0.0796	81.3985
3	4.6890	0.0789	81.4363
4	4.6829	0.0788	81.2486
5	4.7772	0.0804	81.2312
6	4.7151	0.0794	81.3044
7	4.7764	0.0804	81.1516
8	4.6410	0.0781	81.3816

Table 3 GRNN—cross-validation result

Sl. No.	Average SSE	Average MSE	Average accuracy
1	3.0886	0.0520	85.2288
2	3.1163	0.0525	85.1209
3	3.0949	0.0521	85.1979
4	3.1066	0.0524	85.1630
5	3.0941	0.0520	85.2087
6	3.0922	0.0521	85.1310
7	3.1105	0.0524	85.1732
8	3.1074	0.0523	85.1820

$$\text{Average\_Accuracy} = \frac{1}{n} \sum_{i=1}^{n} \text{Accuracy}(i) \qquad (12)$$

Comparison of the above results is summarized in Table 4. In this table, average data of SSE, MSE, and accuracy have been shown. For GRNN, prediction errors (SSE and MSE) are minimum and the accuracy is maximum with respect to FFBPNN and RBFNN. As far as the best accuracy is concerned, GRNN has outperformed others.

Ten-fold cross-validation technique was applied to estimate the performance of the neural network models with all 594 data set, and it represents satisfactory result. After performing above cross-validation, finally GRNN, FFBPNN, and RBFNN were trained with 414 data sets and tested with rest 180 data sets. Accuracy% of testing result of GRNN, FFBPNN, and RBFNN is shown in graphical representation in Fig. 9.

Table 4 Comparison of the cross-validation results

Algorithm	SSE	MSE	Accuracy%
GRNN	3.1013	0.0522	85.1756
RBFNN	4.7143	0.0793	81.3023
FFBPNN	22.7117	0.3824	58.0199

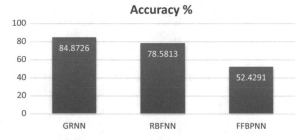

Fig. 9 Accuracy comparison of testing result

5 Conclusion

Above result and comparative study among three models GRNN, RBFNN, FFBPNN show that GRNN had exhibited a better performance in this experiment. Overall, accuracy of the silk content estimation using GRNN was in the tune of 85% which was the maximum accuracy in compare to RBFNN and FFBPNN. Also, the prediction error was higher in both cases, e.g., RBFNN and FFBPNN, in comparison with GRNN. GRNN model provides maximum accuracy and less time for training. Moreover, the X-ray imaging methodology is nondestructive. As there is no wastage of cocoons, the solution increases the turnover in cocoon trading. This automated process of cocoon counting and grading reduces the dependency on expert manpower. The developed solution is scientific and authentic, increases the trust between the farmers and the buyers during cocoon trading. There is no such system/tool available in the market that can estimate the silk content of cocoon in noninvasive manner. Thus, the proposed methodology may open up a new era of digital transactions on cocoon trading and beneficial to the sericulture/textile industries.

References

1. Lee, W.Y.: Silk Reeling and Testing Manual
2. The Major Industry Sectors: Fiber, Fabric, Finished Products, and Machinery Manufacturing (Chapter 3). https://www.princeton.edu/~ota/disk2/1987/8733/873305.PDF
3. Ahamad, S., et al.: Studies on the haemocytes of mulberry silkworm *Bombyx mori* L. in the region of District Amethi, Uttar Pradesh, India. Int. J. Entomol. Res. **1**(5), 29–32 (2016). ISSN: 2455-4758
4. Bowonder, B., Sailesh, J.V.: ICT for the Renewal of a Traditional Industry: A Case Study of Kancheepuram Silk Saree. http://planningcommission.nic.in/reports/sereport/ser/stdy_ict/7_kanchee.pdf
5. Goldsmith, M.: *Bombyx mori*. In: Encyclopedia of Genetics (2001)
6. Rockwood, D.N., et al.: Materials fabrication from *Bombyx mori* silk fibroin. Nat. Protoc. **6**(10), 1612–1631 (2011). https://doi.org/10.1038/nprot.2011.379
7. Gupta, Deepti, Agrawal, Anjali, Rangi, Abhilasha: Extraction and characterization of silk sericin. Indian J. Fibre Text. Res. **39**, 364–372 (2014)
8. Luong, T.-H., Ngoc Dang, T.-N., Pham Thi Ngoc, O., Dinh-Thuy, T.-H., Nguyen, T.H. Vo, T., Hoang, D., Son, H.: Investigation of the silk fiber extraction process from the Vietnam natural *Bombyx mori* silkworm cocoon. IFMBE Proc. **46** (2014). https://doi.org/10.1007/978-3-319-11776-8_79
9. Bej, G., Akuli, A., Pal, A., Dey, T., Chaudhuri, A., Alam, S., Khandai, R., Bhattacharyya, N.: X-ray imaging and general regression neural network (GRNN) for estimation of silk content in cocoons. In: Permin'15, pp. 71–76 (2015). ISBN 978-1-1-4503-2002-3
10. Bej, G., Akuli, A., Pal, A., Dey, T., Bhattacharyya, N.: Quality inspection of cocoons using X-ray imaging technique. In: 2014 International Conference on Control, Instrumentation, Energy and Communication (CIEC'14), 31 Jan–2 Feb 2014, pp. 106–110. https://doi.org/10.1109/ciec.2014.6959059
11. Gonzalez, R.C., Woods, R.E.: Digital Image Processing. Addition-Wesley Publishing Company, Reading, MA (1992)
12. Gonzalez, R.C., Woods, R.E., Eddins, S.L.: Morphological Reconstruction From Digital Image Processing Using MATLAB

13. Panneerselvam, R.: Research Methodology, 1st edn. PHI Learning, New Delhi (2009)
14. Vora, K., Yagnik, S.: A survey on backpropagation algorithms for feedforward neural networks. Int. J. Eng. Dev. Res. **1**, 193–197 (2014). ISSN: 2321-9939
15. Sazli, Murat: A brief review of feed-forward neural networks. Commun. Fac. Sci. Univ. Ankara **50**, 11–17 (2006). https://doi.org/10.1501/0003168
16. Fayaz, M., Shah, H., Mohammad Aseere, A., Mashwani, W., Shah, A.S.: A framework for prediction of household energy consumption using feed forward back propagation neural network. Technologies **7**, 30 (2019). https://doi.org/10.3390/technologies7020030
17. Svozil, D., Kvasnicka, V., Pospichal, J.: Introduction to multi-layer feed-forward neural networks. Chemometr. Intell. Lab. Syst. **39**, 43–62 (1997)
18. Bullinaria, J.A.: Radial basis function networks: algorithms. Neural Comput.: Lect. **14** (2015). http://www.cs.bham.ac.uk/~jxb/INC/l14.pdf
19. Radial Basis Function Neural Network Tutorial. http://www.csc.kth.se/utbildning/kth/kurser/DD2432/ann12/forelasningsanteckningar/RBFNNtutorial.pdf
20. Bonanno, F., Capizzi, G., Napoli, C., Graditi, G., Tina, G.M.: A radial basis function neural network based approach for the electrical characteristics estimation of a photovoltaic module. Appl. Energy **97**, 956–961 (2012)
21. Specht, D.F.: A general regression neural network. IEEE Trans. Neural Netw. **2**(6) (1991)
22. Bauer, M.M.: General Regression Neural Network, GRNN—A Neural Network for Technical Use. A thesis submitted in partial fulfillment of the requirements of the degree of Master of Science (Chemical Engineering) at the University of Wisconsin-Madison
23. Stone, M.: Cross-validatory choice and assessment of statistical predictions. J. R. Stat. Soc. Ser. B **36**(1), 111–147 (1974)
24. Refaeilzadeh, P., Tang, L., Liu, H.: Cross-Validation. Arizona State University. http://leitang.net/papers/ency-cross-validation.pdf

A Comparative Study on Disaster Detection from Social Media Images Using Deep Learning

Arif, Abdullah Omar, Sabah Ashraf, A. K. M. Mahbubur Rahman, M. Ashraful Amin and Amin Ahsan Ali

Abstract The availability of images of events almost in real time on social media has a prospect in many application developments. A humanitarian technology for disaster type and level assessment can be developed using the images and video available on social media. In this paper, we investigate the potential use of various available deep learning techniques to develop such an application. For our research, based on the use of publicly available image data, we have started collecting disaster images from various sources from South Asia. We created the South Asia Disaster (SAD) image dataset containing 493 images from various online news portals. Using the Keras as our framework to run our models: Visual Geometry Group (VGG-16 and VGG-19), Inception-V3, and Inception-ResNet-V2 (ResNet: Residual Network). However, to boost up the training speed, we dropped the fully connected layer and added a small, fully connected model. To identify the five different disasters: fire disaster, flood disaster, human disaster, infrastructure disaster, and natural disaster; our proposed method with VGG-16 model's recognition accuracy was 83.37%, which is the highest accuracy on the SAD dataset.

Arif · A. Omar · S. Ashraf · A. K. M. M. Rahman · M. A. Amin (✉) · A. A. Ali
Computer Vision and Cybernetics Group, Department of Computer Science and Engineering,
Independent University, Bangladesh (IUB), Dhaka, Bangladesh
e-mail: aminmdashraful@iub.edu.bd

Arif
e-mail: arif1881996@gmail.com

A. Omar
e-mail: mdabdullahomar97@gmail.com

S. Ashraf
e-mail: sabah.ashraf97@gmail.com

A. K. M. M. Rahman
e-mail: akmmrahman@iub.edu.bd

A. A. Ali
e-mail: aminali@iub.edu.bd

© Springer Nature Singapore Pte Ltd. 2020
J. K. Mandal and S. Mukhopadhyay (eds.), *Proceedings of the Global
AI Congress 2019*, Advances in Intelligent Systems and Computing 1112,
https://doi.org/10.1007/978-981-15-2188-1_38

Keywords Disaster image · Humanitarian technology · Standard disaster dataset ·
Convolutional neural network · VGG-16 · VGG-19 · Inception-V3 ·
Inception-ResNet-V2 · Keras model

1 Introduction

Bangladesh is one of the most vulnerable countries that suffer from huge climate
change, as well as disasters. Bangladesh's flat topography, low-lying, and climatic
features are responsible for natural disasters, i.e., floods, cyclones, droughts, and
earthquakes. Moreover, its population density and socioeconomic environments
make it highly susceptible to many human-made hazards that include fire, building
collapse, infrastructural damage, road accident, etc. Also, there are other disasters
happening like infrastructure or non-natural disasters. Day after day, people are suffering from different kinds of disasters (e.g., fire incidents in Chawk Bazar, Banani
fire incident). These incidents which happened in March 2019 were featured by some
newspapers such as the famous online news portal Dhaka Tribune [1], The Daily Star
[2], etc. It is crucial in times of crisis that how emergency response workers reach
all those affected promptly. It would be great to have a system that would raise an
alert and determine the degree of damage of any disaster and inform the appropriate
authorities based on the automated analysis of the image data that are almost all the
time available in real time on various social media. There has been some effort along
this direction.

Rizk et al. [3] proposed a multimodal approach to automate crisis data analysis
using machine learning. The proposed multimodal two-stage framework relies on
computationally inexpensive visual and semantic features to analyze Twitter data.
Level I (one) classification consists of training classifiers separately of semantic
descriptors and combinations of visual features. These classifiers' decisions are
aggregated to form a new feature vector to train the second set of classifiers in
Level II (two) classification. In this second classification phase, two approaches are
compared: score learning and majority vote.

Avgerinakis et al. [4] presents the algorithms that CERTH team deployed to
tackle disaster recognition tasks. Deep convolutional neural network (DCNN), DBpe-
dia Spotlight, and combMAX were implemented to tackle DIRSM. The model
GoogleNet was used to train on 5055 ImageNet concepts.

Giannakeris et al. [5] presented a novel warning system framework for detecting
people and vehicles in danger. The proposed framework provides a near real-time
localization solution for detecting and scoring severity and safety levels of people and
vehicles in flood and fire images. They chose to fine-tune the pretrained parameters
of the VGG-16 on Places 365 dataset to leverage useful distinctions between various
visual clues that relate to generic scenery images. Their Initial 3F-emergency dataset
is composed of 6 K images from Flickr.

Alam et al. [6] proposed a work where their image filtering module employs deep
neural networks and perceptual hashing techniques to determine whether a newly

arrived image is relevant for a given disaster response context. To train the relevancy filter, 3518 images were randomly selected from the severe and mild categories. They adopted a transfer learning approach where they have used the VGG-16 network pretrained on ImageNet data.

Mouzannar et al. [7] proposed a multimodal deep learning framework to identify damage-related information from social media posts. This framework combines multiple pretrained unimodal convolutional neural networks that extract features from raw texts and images separately. The inception convolutional neural network (CNN) model was adopted, which was pretrained on ImageNet to process images, and for words, they used a pretrained word embedding model to process the texts. The framework was evaluated on a homegrown labeled dataset of multimodal social media posts.

In the paper [8], images posted on social media platforms during natural disasters to determine the level of damage caused by the disasters are analyzed. In this study, Imran et al. used the VGG-16 network trained on the ImageNet dataset. They collected data from the Web (such as Typhoon Ruby, Hurricane Matthew, and Nepal Earthquake) and made their dataset. They ran tests on their training sets, and the highest accuracy they achieved was when they combined their Google, Ruby, and Matthew datasets.

In this research paper, we propose a deep learning-based method to automate the effective extraction of information from social media posts to direct relief resources efficiently. Since posts on social media contain text, photographs, and videos, we are performing a deep learning framework for multimodal identification of damage-related information. Also, our objective is to collect and work with South Asian disaster images, which include disaster images of Bangladesh and similar countries.

2   Methodology

Our proposed system uses Keras as a framework to implement our models and modified deep learning algorithms to classify five different types of disaster images. And for all deep learning-based methods, data is the most important component. The reason for using Keras is that Keras is easy to use and very flexible. It supports the models we used for our paper. Compared to other neural network API Keras is light-weighted and very user-friendly. Keras is consistent and simple yet very powerful thanks to its model and sequential APIs, and hence, it helps create very complex neural networks in a matter of minutes. Another main reason for us to use Keras is that it is very popular, and Keras development is backed by key companies in deep learning ecosystem [9].

2.1 Dataset Description

Standard database is a prerequisite for better model creation [10]. The main issue using a deep learning system for classification is a well-defined and diversified dataset. For this experiment, we collected the images from [11]. There are six different damage categories; they are: fire damage, flood damage, damage infrastructure, damage nature, human damage, and non-damage. There is a total of 5885 images, and the total number of images for each category is given in Table 1.

The fire disaster images contain fire elements, smoke, and burning objects and burned objects. Flood disaster images contain a huge volume of water, objects underwater, and objects submerged in water. Infrastructure disaster images contain collapsed buildings, rusted, and damaged objects. Nature disasters contain broken trees, buildings, and roads caused by earthquake or cyclone. Human disaster image contains bleeding, burned face, torturing people, and human damaging the environment. Non-disaster images contain products, cosmetics, books, human models, and people eating food. Some sample images from the [11] dataset are given in Fig. 1. From Table 1, we can observe that non-damage has the highest number of images, 2972 than

Table 1 Distribution of the number of disaster images for each class from [11] dataset

Categories	Number
Fire damage	349
Flood damage	385
Damage infrastructure	515
Damage nature	1418
Human damage	240
Non-damage	**2972**
Total	5885

Fig. 1 Sample images from [11] Dataset (left-right): fire disaster, flood disaster, human disaster, infrastructure disaster, natural disaster, and non-disaster [3]

Table 2 Distribution of the number of disaster images for each class from SAD dataset

Categories	Number
Fire damage	82
Flood damage	82
Damage infrastructure	65
Damage nature	73
Human damage	81
Non-damage	**110**
Total	493

https://archive.ics.uci.edu/ml/datasets/Multimodal+Damage+ Identification+for+Humanitarian+Computing

other classes; the lowest number of images is human damage 240. Damage nature has the second-highest number of images 1418. Other classes have an adjacent number of images.

In the sample images note that the images are from mostly from outside South Asia. We intend to implement our system for Bangladesh and to check the performance on the native data, and we collected a small dataset for this project. We are calling it, **SAD (South Asian Disaster)** image dataset. We have collected 493 data, and class-wise distribution can be found in Table 2.

In Fig. 2, we gave samples of the SAD image dataset. Images were collected from online news portal, social news network, independent news organization, and online benefactor for people. We got 32 images from each online news portal such as The Times of India, The Hindu, and Indian express. 32 images are collected from the UN news (news portal), 32 images from World Bank (online benefactor for people), 32 images from Rappler (social news network), 100 images from BBC

Fig. 2 Sample images from SAD image dataset (left-right): fire disaster, flood disaster, human disaster, infrastructure disaster, natural disaster, and non-disaster

(online news portal), 100 images from Fox News (online news portal) and rest from Al Jazeera (independent news organization). From visual comparison among images from Figs. 1 and 2, it is clear that there are differences in perception of similar classes.

2.2 Convolutional Neural Networks (CNNs)

CNNs have shown state-of-the-art efficiency in a multitude of computer vision challenges, including image classification and retrieval, object detection, and segmentation of images [12]. A typical CNN comprises of various processing layers, including convolution, max pooling, and fully connected layers. These layers are cascaded in a way that one layer's output becomes the next layer's input. A set of kernels are introduced to the input information to produce characteristic maps at each convolution layer [12].

Our study is mainly based on performing different types of experiments with different models on our SAD dataset and [11] dataset and compares and analyzes their results. To perform our experiments, we have decided to use four models from three CNN families: Visual Geometry Group (VGG) [13], Inception [14] by GoogLeNet, and Inception Residual Network (ResNet) [15] by Microsoft. Our objective is to compare results; hence, we have decided to compare the results of different CNN families' model as their architectures are different from one another. The VGG architecture is deep, using only 3×3 convolutional layers stacked on top of each other, which increase depth. Volume size is reduced by max pool layer. It has cascaded convolution max pool layers [13]. The Inception architecture, on the other hand, is "wider" rather than "deep" like the VGG architecture. The architecture typically consists of filters of three different sizes: 1×1, 3×3, and 5×5. After max pooling, the outputs are concatenated and sent to the next inception module. The newer version of the architecture has an extra added 1×1 convolution before the 3×3 and 5×5 convolutions for cost-effectiveness [14]. One of the most popular CNN family in the present is the ResNet family by Microsoft. This architecture goes deeper than the VGG architecture with nodes having stems of different inception modules [15]. In [5] and [8] VGG-16 CNN model was implemented and in [7] Inception-V3, Inception-V4, VGG-16, and Inception-ResNet-V2 was implemented. For our problem, we used VGG-16, VGG-19, Inception-V3, and Inception-ResNet-V2.

Visual Geometry Group (VGG) [13]: The VGG nets were originally designed to detect and recognize objects in images. VGG nets proved to be efficient in the identification of objects in still pictures. VGG is a very profound model that significantly enhanced a wide variety of visual recognition functions, including detecting objects, semantic segmentation, picture captioning, and recognition of the action on video. VGG-16 has three fully connected layers following a stack of convolutional layers, and the final layer is the softmax layer. It has a total of 16 layers; all hidden layers are equipped with rectification. VGG-19 model is very similar to the VGG-16 model. The main difference between VGG-16 and VGG-19 is that the VGG-19 has 19 layers instead of 16.

Inception [14]: Though VGG net has architectural simplicity and convincing features, this comes at a high price of relatively large computation time to evaluate the network. Inception architectures of GoogLeNet were intended to perform well even under rigorous memory and computer budget limitations. Inception's computational cost is also much smaller than the VGG net. This made it possible to use Inception networks in large-data situations where large amounts of data required to be processed at a reasonable price or where memory or computing capability are inherently restricted. Inception-V1 has nine inception modules stacked linearly, and 27 layers deep, including pooling layers. In Inception-V2 the 5×5 convolution was factorized to two 3×3-convolution operations to improve computational speed and reduce computational expense. The Inception-V3 (INv3) is 42 layers deep; however, the computation is only about 2.5 times more than Inception-V1 and much more efficient than VGG net.

Inception-ResNet-V2 [15]: Inception-ResNet-V2 (INRv2) is a 164 profound layer deep neural network that can classify pictures into 1000 categories of objects. Inception-ResNet-V2 has introduced residual connections that add the output of the convolution operation of the inception module to the input. For residual addition to work, the input and output after convolution must have the same dimensions. The pooling operation inside the main inception modules was removed to be replaced by the residual connections. To increase stability, the residual activations were scaled by a value within 0.1–0.3 range. It was found that Inception-ResNet models were able to achieve higher accuracies at a lower number of epochs.

The models that we are using for our experiment are: VGG-16, VGG-19, Inception-V3, and Inception-ResNet-V2. In Table 3, a comparative view of the number of parameters of all four above-mentioned CNN models is given. However, we also implemented a small model and added that at the end of each of the models mentioned above; this is done by dropping the convolutional fully connected layer at the end. The reason for adding this small model is to speed up the training process.

In Fig. 3, we gave the schematic view of the models used. We have used ReLU activation on the 512 dense layers and also on the 256 dense layers so that all the layers are connected to each other. And then we used dropout, and it helps reduce over-fitting, by preventing a layer from seeing twice the same pattern. In the end, we used six dense layers because we have six classes. As optimizer we used Adam, and as loss function, sparse categorical cross-entropy is used.

Table 3 Parameters of the models

Models	Parameters
VGG 16	138 Million
VGG 19	144 Million
Inception-V3	23 Million
Inception-ResNet-V2	54 Million

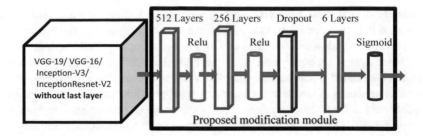

Fig. 3 Schematic view of the proposed deep neural model

3   Experimental Setup

Images are usually of different sizes as they are collected from non-restricted sources. To feed them into the deep neural networks, they were resized to 150 by 150 pixels. For training the networks, 50 epochs have been used, and epoch means training the whole dataset in one cycle; however, it takes a lot of space in the memory. Therefore, the dataset is divided into batches to train. The batch size 11 and 12 are used for [11] training and testing dataset, respectively. For SAD dataset, the batch size is 29 for the whole dataset. To run this experiment, we are using Intel core i5 processor and 8 GB DDR4 RAM. The experiments that we are performing are:

(1) **Experiment 01**: Cross-validation performance measure while training the CNNs with [11] dataset testing them on [11] dataset. Table 4 shows the results of this experiment.
(2) **Experiment 02**: Performance measure while training with 100% [11] dataset and test with 100% SAD dataset. Table 5 shows the results of this experiment
(3) **Experiment 03**: Cross-validation performance measure while training with the 100% [11] dataset plus 60% of the SAD dataset and test with the remaining 40% of the SAD dataset. Table 6 shows the results of this experiment.

In Table 4, note that all four classifiers perform in the same range on average. However, different classifiers show slightly higher classification accuracy class of disaster.

In Table 5, the first issue to note is that, as we test the classifiers trained with [11] dataset and test with our SAD dataset, the accuracy reduces by about 7%. This means that the [11] dataset does not contain well-distributed and diverse data from all demography, but the SAD dataset is different from [11] dataset (Fig. 4). Something very noticeable in this experiment is that the precision for human damage is significantly low for all classifiers (Figs. 1 and 2). This would mean that identifying human damage by a deep neural net trained with [11] dataset only is very difficult. VGG 16 is an older version of the deep neural network; however, the overall accuracy of this classifier is slightly higher than other models for experiment 2.

In experiment 3, we added 60% of the SAD data with the [11] dataset, and it shows the improvement in performance. In Table 6, note that the precision of human damage

Table 4 Cross-validation average performance of only [11] dataset

	Precision (%)				Recall (%)				Accuracy (%)			
	VGG 19	VGG 16	INv3	INRv2	VGG 19	VGG 16	INv3	INRv2	VGG 19	VGG16	INv3	INRv2
Fire damage	63.31	58.74	**64.44**	58.22	68.00	67.33	67.96	**75.66**	95.51	95.62	95.72	**96.08**
Flood damage	**59.40**	51.63	41.93	51.12	54.82	60.64	**63.32**	63.12	93.92	**94.37**	93.91	94.25
Human damage	57.63	**59.11**	49.92	56.00	**71.21**	61.32	69.00	69.23	**97.07**	96.54	96.82	96.79
Damage infrastructure	**70.92**	70.12	63.33	69.84	65.13	63.55	**67.66**	67.16	83.87	83.04	83.54	**84.33**
Damage Nature	37.75	47.00	48.10	**49.33**	**52.95**	51.00	47.91	50.94	**91.28**	91.13	90.40	91.06
Non-damage	83.52	82.71	85.55	**85.82**	**83.25**	83.13	79.84	81.83	**83.65**	82.91	81.79	82.60
Average	62.08	61.55	**67.20**	61.72	65.89	64.50	65.94	**67.99**	**90.88**	90.60	90.36	90.85

Table 5 Performance of training with 100% [11] dataset and testing with 100% SAD dataset

	Precision (%)				Recall (%)				Accuracy (%)			
	VGG 19	VGG 16	INv3	INRv2	VGG 19	VGG 16	INv3	INRv2	VGG 19	VGG 16	INv3	INRv2
Fire damage	46.33	59.73	**60.92**	37.83	97.44	83.11	81.92	100	90.87	**91.28**	**91.28**	89.66
Flood damage	45.12	**59.73**	46.33	43.93	74.02	62.83	80.52	75.00	88.24	87.42	**89.25**	88.24
Human damage	6.15	9.23	9.23	**10.73**	44.44	33.33	35.33	43.71	**86.61**	85.60	85.50	86.41
Damage infrastructure	**82.22**	75.33	79.52	68.52	**32.11**	36.63	26.92	30.11	71.60	77.08	64.91	**71.81**
Damage nature	30.83	32.12	17.31	**51.82**	56.81	54.23	51.82	51.22	**84.79**	84.38	83.77	83.98
Non-damage	**80.00**	71.82	61.83	70.00	53.71	**56.44**	54.44	51.34	80.12	**81.34**	79.92	78.51
Average	48.44	**51.32**	45.85	47.13	**59.75**	54.42	55.15	58.56	83.70	**84.51**	82.43	83.10

Table 6 Cross-validation performance of training with 100% [11] dataset plus 60% SAD dataset, and testing with 40% SAD dataset

	Precision (%)				Recall (%)				Accuracy (%)			
	VGG 19	VGG 16	INv3	INRv2	VGG 19	VGG 16	INv3	INRv2	VGG 19	VGG 16	INv3	INRv2
Fire damage	**72.22**	66.66	48.55	69.66	96.00	91.66	88.81	**100**	**94.95**	93.43	90.40	94.68
Flood damage	34.44	53.11	40.66	40.61	**84.63**	77.33	81.33	65.00	88.38	**89.90**	88.89	85.11
Human damage	17.43	**26.12**	8.69	21.72	36.44	35.32	**66.66**	38.55	86.87	85.86	**88.89**	84.04
Damage infrastructure	**87.81**	78.82	84.81	78.73	39.22	**44.12**	40.51	40.00	75.25	**79.80**	79.77	70.74
Damage nature	23.33	26.63%	**33.33**	23.33	**87.54**	50.00	45.55	50.00	**87.88**	84.85	83.84	86.17
Non-damage	80.81	78.71	**87.22**	76.52	56.73	**61.66**	58.62	57.12	80.81	**83.33**	82.32	76.06
Average	52.67	**55.01**	50.54	51.76	**66.76**	60.02	63.58	58.45	85.69	**86.20**	85.68	82.80

(a)　　　　　　　　　　　　　　　　(b)

(c)　　　　　　　　　　　　　　　　(d)

Fig. 4 (Left–Right; Top–Bottom) **a** SAD flood damage; **b** other countries flood damage; **c** SAD damage nature; **d** other countries damage nature

improves. Moreover, accuracy is also increased. It suggests that if the networks can be trained with significantly large and diverse data, performance will increase with the same network setup.

Comparing Tables 4 and 5, the accuracy dropped around 7% when performing the second experiment. The reason for this drop of accuracy maybe for some specific class, there is significant difference between [11] dataset and SAD dataset. In Fig. 4, we observe that SAD flooding disaster and nature disaster is different than other countries flooding disaster and natural disaster.

In Fig. 5, we provide some images associated with fire class; here, we observed that there are two types of images: during fire disaster and after a fire disaster. In

(a)　　　　　　　　　(b)　　　　　　　　　(c)

Fig. 5 (Left–Right) **a** during fire damage; **b** after fire damage; **c** damage infrastructure

(a)   (b)

(c)   (d)   (e)

Fig. 6 Wrong prediction by the proposed system; (left–right) **a** after fire damage classified as damaged infrastructure; **b** actual training damaged infrastructure image; **c** fire damage classified as flood damage; **d** and **e** actual flood damage training image

Fig. 5, note that during fire disaster and after fire disaster images are visually very distinctive. The difference is one has fire element, and smoke in it and the other one is all black, and there are ashes. Visually we ourselves are confused between the middle image (after fire disaster) and the right image (damaged infrastructure). Thus, the deep networks are also sometimes confused with after fire image and infrastructure disaster image.

In Fig. 6, we provide some more images that are misclassified by our system. (a) After fire damage classified as damaged infrastructure; (b) is an actual infrastructural damage training image; (c) another after fire damage classified as flood damage; in the image (d) the black and white floor and books floating resembles the state of the floor of (c), and in (e) the training flood image does not really look like a flood image, but there are pipes appearing in it which also appears in test image (c). We believe that some of the fire test images were classified as flood and infrastructure by Inception-V3 and Inception-ResNet-V2 models, because the test image does not contain any fire object in them.

4 Conclusion

Real-time support is the most significant challenge to minimize damages by any disaster; most of the time, the traditional method of assessing the occurrence of disaster and its characteristics takes too long, and in the meantime, the damage has been done. In this paper, we propose a method to determine the type of disaster automatically from social media images. To test the performance of the proposed idea publicly available dataset collected from [11] is used. Moreover, we extended the dataset by gathering 493 disaster images of six classes from this region, and we are calling it "South Asia disaster" (SAD). Our experiments revealed that the [11] dataset is not diverse and does not perform well recognizing disaster in SAD images. We implemented four different types of deep neural network models: VGG 16, VGG 19, Inception-V3, and Inception-ResNet-V2, for this research. However, to boost up the training speed, we removed the fully connected layer and added a small, fully connected model. To identify the five different disasters (fire disaster, flood disaster, human disaster, infrastructure disaster, and natural disaster;) our proposed method with VGG 16 models has recognition accuracy of 83.37% on the SAD dataset.

References

1. Sobhan, Z.: Dhaka Tribune, Kazi Anis Ahmed, viewed 20 June 2019 (2013). https://www.dhakatribune.com/
2. Anam, M.: The Daily Star, Mahfuz Anam, viewed 20 June 2019 (1991). https://www.thedailystar.net/
3. Rizk, Y., Awad, M., Castillo, C.: A computationally efficient multi-modal classification approach of disaster-related twitter images. In: 34th ACM/SIGAPP Symposium on Applied Computing, pp. 2050–2059. ACM, New York, USA (2019)
4. Avgerinakis, K., Moumtzidou, A., Andreadis, S., Michail, E., Gialampoukidis, I., Vrochidis, S., Kompatsiaris, I.: Visual and textual analysis of social media and satellite images for flood detection. In: MediaEval 2017, pp. 3–5. Multimedia Satellite Task, Dublin, Ireland (2017)
5. Giannakeris, P., Avgerinakis, K., Karakostas, A., Vrochidis, S., Kompatsiaris, I.: People and vehicles in danger—a fire and flood detection system in social media. In: IEEE Image, Video, and Multidimensional Signal Processing. Zagorochoria Greece (2018)
6. Alam, F., Imran, M., Ofli, F.: Image4Act : online social media image processing for disaster response. In: IEEE/ACM International Conference on Advances in Social Networks Analysis and Mining 2017, pp. 601–604. ACM, New York, USA (2017)
7. Mouzannar, H., Awad, M.: Damage identification in social media posts using multimodal deep learning. In: 15th International Conference on Information Systems for Crisis Response and Management. Rochester, New York, USA (2018)
8. Nguyen, D.T., Ofli, F., Imran, M., Mitra, P.: Damage assessment from social media imagery data during disasters. In: IEEE/ACM International Conference on Advances in Social Networks Analysis and Mining, pp. 569–576. ACM, New York, USA (2017)
9. Chollet, F.: Keras, GitHub, viewed on 15 Aug 2019 (2015). https://keras.io
10. Amin, M.A., Mohammed, M.K.: Overview of the ImageCLEF 2015 medical clustering task. In: CLEF (Working Notes) (2015)
11. Mouzannar, H., Rizk, Y., Awad, M.: Damage identification in social media posts using multimodal deep learning. In: 15th International Conference on Information Systems for Crisis Response and Management. Rochester, New York, USA (2018)

12. Muhammad, K., Ahmad, J., Baik, S.W.: Early fire detection using convolutional neural networks during surveillance for effective disaster management. Neurocomputing **288**, 30–42 (2018). Elsevier Science Publishers B.V., Amsterdam, Netherlands
13. Zhang, X., Zou, J., He, K., Sun, J.: Accelerating very deep convolutional networks for classification and detection. In: IEEE Transactions on Pattern Analysis and Machine Intelligence, pp. 1943–1955. IEEE Computer Society, Washington, DC, USA (2016)
14. Szegedy, C., Vanhoucke, V., Loffe, S., Shlens, J., Wojna, Z.: Rethinking the inception architecture for computer vision. In: Computer Vision and Pattern Recognitioin, Las Vegas, NV, USA (2016)
15. Szegedy, C., Ioffe, S., Vanhoucke, V., Alemi, A.: Inception-v4, Inception-ResNet and the Impact of residual connections on learning. In: Thirty-First AAAI Conference on Artificial Intelligence, pp. 4278–4284. AAAI Press, San Francisco, California, USA (2017)

Rank Based Pixel-Value-Differencing: A Secure Steganographic Approach

Debashis Das and Ratan Kumar Basak

Abstract Image steganography is a frequently used method in digital data security in modern communication system. The two prime objectives of steganography lie in hiding capacity and security of hidden data. In this paper, we propose a data hiding scheme, focusing on message security, employs pixel-value differencing to embed the secret bits where the bitstream is shuffled before embedding in each pixel block. The bit-shuffling process ensures the data security in a great extent. The original image is partitioned into non-overlapping blocks which are subsequently marked with specific ranks, estimated from pixel-difference value of each candidate block. A set of unique bins, same in number as the total number of ranks obtained, are constructed to store the secret bitstreams. In the embedding process, secret bits are considered from various bins that are selected based on the rank matrix. The proposed method is implemented and tested on standard gray-scale images. The experimental results establish the efficiency of the algorithm. A comparative analysis with the state-of-the-art method shows the strength of the proposed algorithm in the ground of security issues.

Keywords Image steganography · Rank bin · Pixel-value-differencing · Bit shuffling · Data security

D. Das
Department of Computer Science & Engineering,
Techno India University, Kolkata, India
e-mail: debashisitnsec@gmail.com

R. K. Basak (✉)
Department of Computer Science & Engineering,
University of Engineering & Management, Kolkata, India
e-mail: ratan.basak@uem.edu.in

© Springer Nature Singapore Pte Ltd. 2020   501
J. K. Mandal and S. Mukhopadhyay (eds.), *Proceedings of the Global AI Congress 2019*, Advances in Intelligent Systems and Computing 1112,
https://doi.org/10.1007/978-981-15-2188-1_39

1   Introduction

Secure digital data communication, nowadays, has become the most challenging task due to a number of online threats. Hence, it requires an obvious security before propagating the personal/secret information through the Web. Steganography is one alternative to meet the purpose. The core concept of steganography is to conceal secret data into a cover media without affecting the imperceptibility of the cover. Various mediums have been employed as the cover media like—image, video, audio signals, text, etc. In steganographic approaches, there are a large number of researches have been carried out in this domain for last few decades. Among various researches, LSB substitution is the easiest method of data hiding with high efficiency by maintaining good imperceptibility [1]. As this method performs simple replacement of few secret bits with original data bits, it is quite vulnerable from security attacks. Another method which gains massive popularity hides secret data based on the contiguous pixel differences of the cover media. The secret bits are managed to conceal by only adjusting the pixel values without any data replacement which is comparatively more secure against various threats. The pixel-value-differencing method [2] has been incorporated in many recent researches in several forms [3–6]. The hiding capacity in these techniques is generally measured on a special fixed and/or variable length range table. In recent works, most of the techniques are found to focus on increasing the hiding capacity [7, 8]. In contrast, we have put impact on secret message security in the present work. Here, we have shuffled the secret data before hiding into the cover media which are stored into several unique data bins. The data bins, on the other hand, are constructed based on the specified ranks of each pixel block of the cover. The data shuffling process, basically, enhances the security measure of the proposed algorithm.

Organization of this article is as follows. Subsequent to the brief introduction, a thorough literature survey is provided in Sect. 2. Section 3 describes the proposed method in detail. Experimental results are depicted in Sect. 4. Section 5 illustrates the security analysis, and the concluding remarks are drawn in Sect. 6.

2   Literature Survey

In this section, we have reviewed few steganographic methods that have been devised in recent past and specifically implemented on image files. LSB replacement approach is the earliest and easiest way to conceal secret data bit in the cover media by simply replacing secret bit(s) with the bits contained by the cover media. A number of variants of LSB replacement have also been proposed in the literature [1]. Pixel-value differencing is another popular technique [2] that has been used frequently in recent researches. This method hides secret data bits inside a pair of pixels by managing their corresponding difference value. Wang et al. proposed a high-quality steganographic scheme based on the original PVD method by incorporating an additional modulus function [6]. Tri-way pixel-value differencing (TPVD) algorithm was developed by Chang et al. [3] where the secret data were embedded in three different directions in each 2×2 pixel block which results significantly higher payload compared to original PVD. Few modified version of TPVD have also been devised in [4, 5]. Luo et al. designed another secure method where the image sub-blocks are rotated randomly before embedding data in the middle pixel of each three continuous pixel blocks [9]. On the other hand, a bi-directional approach was proposed by Himakshi et al. where two layers of data hiding were performed in each pixel block to enhance total embedding capacity [7]. An adaptive steganographic technique proposed in [8] by combining both LSB and PVD techniques. In contrast to the PVD and TPVD, Gulve et al. devised a pixel-value differencing approach by considering 2×3 non-overlapping blocks where five-pixel pairs were formed for embedding secret bits by keeping one common pixel for adjustment [10]. A number of algorithms designed in transformed domain where the secret data were hidden on the basis of image-specific properties [11, 12]. To the contrary, Jiang et al. [13] came up with a concept of quantum image steganography where simple LSB substitution is employed. A novel idea proposed in [14] where each pixel value of the cover image is divided by a certain value to obtain two separate parts, namely—quotient and remainder in which the secret bits are managed to conceal by employing PVD and LSB substitution. A completely different orientation of image steganography developed by Tao et al. [15], the prime objective of which is to enhance the robustness by embedding secret data into the transmission-channel-compressed image. Besides, a number of recent data hiding approaches discussed in the literature focused on the various security issues while transmitting through public channel [16–18].

3 Proposed Method

In this section, we have described the proposed algorithm in detail. The proposed secure message hiding scheme has been designed in such a way that it can be used with any steganographic algorithm. However, the core concept of data embedding in our method employs the pixel-value differencing (PVD) technique as an arbitrary choice. We have, therefore, discussed the original PVD method in a nutshell before elaborating the proposed idea.

3.1 Review of PVD Method

The original pixel-value-differencing method executes in the following steps. The cover image is partitioned into non-overlapping pixel pairs containing two pixels P1 and P2. The pixel-difference value, *diff* , is found out to estimate the hiding capacity in each candidate pair. A range table has been constructed to derive the number of hiding bits which entirely depends on the pixel difference. The difference range considered as [(0–7), (8–15), (16–31), (32–63), (64–127), (128–255)]. Mathematically,

$$B = \log_2 (H - L) \tag{1}$$

Here, L and H signify lower and higher limit of any range; B specifies the number of bits that can be hidden in a pair.

Now, B number of bits are collected from the secret bitstream and converted in its corresponding decimal value d. A new difference value is then estimated as:

$$new\_diff = L + d \tag{2}$$

This *new_diff* is now divided in two equal parts and embed into two candidate pixels as are expressed in Eqs. 3–6:

$$\left. \begin{array}{l} P1 = P1 + \lceil m/2 \rceil \\ P2 = P2 - \lfloor m/2 \rfloor \end{array} \right\} \text{ where } new\_diff \geq diff \text{ and } P1 \geq P2 \tag{3}$$

$$\left. \begin{array}{l} P1 = P1 - \lfloor m/2 \rfloor \\ P2 = P2 + \lceil m/2 \rceil \end{array} \right\} \text{ where } new\_diff \geq diff \text{ and } P1 < P2 \tag{4}$$

$$\left. \begin{array}{l} P1 = P1 - \lceil m/2 \rceil \\ P2 = P2 + \lfloor m/2 \rfloor \end{array} \right\} \text{ where } new\_diff < diff \text{ and } P1 \geq P2 \tag{5}$$

$$\left. \begin{array}{l} P1 = P1 + \lceil m/2 \rceil \\ P2 = P2 - \lfloor m/2 \rfloor \end{array} \right\} \text{ where } new\_diff < diff \text{ and } P1 < P2 \tag{6}$$

All the pixel pairs of the cover image follow the same procedure for embedding secret bits, collected from the secret bitstream sequentially, to obtain the stego-image.

In the extraction phase, the stego-image is partitioned similarly and hidden bits are extracted from each pixel pair. In a particular pair, difference value d_1 is calculated which is matched with range table to obtain the number of hidden bits embedded into the pair by following Eq. 1. Now, the decimal value V corresponds to the hidden bits is found as:

$$V = d_1 - L \quad \text{where } L \text{ is lower value of the lying range} \tag{7}$$

Convert V into B number of bits and store in the extracted stream list. The above procedure is performed on the entire stego-image and store all the extracted secret bits sequentially to obtain the hidden secret message.

The proposed method is now presented in the following sections.

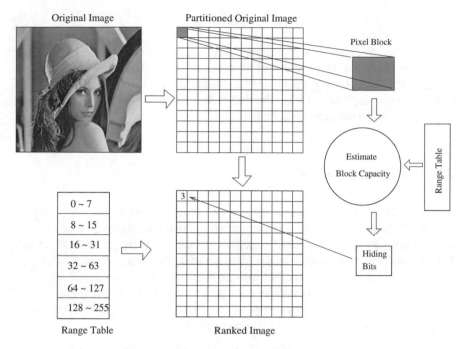

Fig. 1 Formation of a rank matrix against an input cover image

3.2 Rank Matrix Formation

The cover image is partitioned in several non-overlapping blocks in row-major order where each block contains two contiguous pixels. Pixel difference is then evaluated and hiding bit capacity is estimated, in similar fashion as PVD, with the help of a range table for each pair of pixels. Here, we have assumed the same range table as mentioned in the PVD method. Now, hiding capacity is considered as the rank of that particular pair which lies in the range $\{3, 4, 5, 6, 7\}$ and hence we obtain a Rank Matrix corresponds to an input cover image. Notably, for an input image having dimension $N \times N$, Rank Matrix will contain the size $N \times (N/2)$. The process of Rank Matrix formation is depicted in Fig. 1.

3.3 Secret Data Bin Construction

In the previous step, we have obtained a matrix marked with five different ranks. Now, five different data bins have been constructed, against five ranks, to store the secret message bits separately. The capacity or size of each bin is different which can be defined as:

$$C_{\text{bin}}^i = N^i \times i \quad \text{where} \quad i \in 3, 4, 5, 6, 7 \tag{8}$$

Here, C_{bin}^i and N^i specify the capacity of ith bin and counting of rank i, respectively.

Now, each bin is filled with the secret bitstream sequentially in order of 3-bit bin then 4-bit bin and so on. In this manner, the entire secret bitstream has been divided in five separate sequences and get shuffled in the order of embedding process which is discussed in the subsequent section.

3.4 Embedding Process

The cover image is now embedded with secret bits in each pixel pair by applying the original PVD embedding scheme by considering the Rank Matrix as marker image. In contrast to PVD method, the proposed algorithm selects the required secret bits, for any candidate pixel pair, from the corresponding data bins having similar rank. For instance, an arbitrary pixel pair marked with rank 4 in rank matrix should be embedded with 4 secret bits that are collected from data bin-4. Such approach of secret bit selection is the prime concept of data shuffling. The sequential steps of the proposed embedding algorithm are summarized next.

- Scan the entire image and find the blocks having same hiding capacity
- Rank each block by marking with the hidden bit capacity
- Construct five different data bins through a histogram between ranks and corresponding number of blocks
- Count the number of bits required for each bin as $Tot_{bit} = (\text{rank} \times \text{number of blocks in bin})$
- Construct five bitstreams by selecting continuous bits from the hiding bitstream as required by each of the bins
- Read each pixel block and apply PVD to hide data by following Eqs. 3–6 where secret bits should be selected from the specified bin.

The entire flow of the embedding process is depicted in Fig. 2.

3.5 Extraction Process

The proposed data extraction procedure executes the reverse process of the embedding algorithm. Bit extraction from each pixel pair of the stego-image follows the similar steps mentioned in the original PVD extraction. In contrast, the retrieved bits are stored in specified data bin based on the rank of that particular pixel pair. Now, the retrieved bits are combined by collecting all bits from bin-3, bin-4, bin-5, bin-6 and bin-7. This particular order should be maintained for combining the secret bits. The entire process of data extraction is summarized as follows:

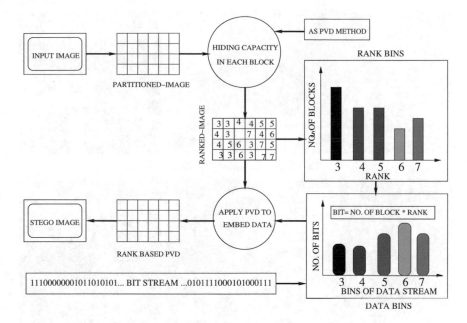

Fig. 2 Workflow model of the proposed embedding procedure

- Scan the stego-image and partition into pixel pairs in row-major order
- Construct five different bins as bin-3, bin-4, bin-5, bin-6 and bin-7 to store the extracted secret bits
- Apply PVD extraction to collect the bits and store into specific bin marked with same rank as of the pixel pair
- Combine the extracted bits from bin-3, bin-4 and so on in the specified sequence to obtain the entire secret message.

4   Experimental Results

The proposed algorithm has been implemented in C language on Linux platform and experimented on a set of standard gray-scale images [19]. To establish the algorithmic efficiency, we have selected a set of random standard gray-scale image in our experiment whereas the proposed algorithm can perfectly work on the same sequence of color images as well. The cover images of size 512×512 are selected and the same image is used as the secret information during the experiment. The experimental results on six different images have been depicted here in Figs. 3 and 4.

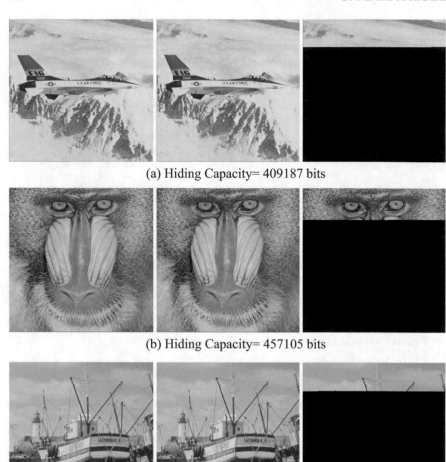

(a) Hiding Capacity= 409187 bits

(b) Hiding Capacity= 457105 bits

(c) Hiding Capacity= 418232 bits

Fig. 3 First column shows the original images; second column images represent corresponding stego-images after embedding secret information; third column images are the extracted hidden part of the secret image

5 Comparative Analysis

We have compared the proposed algorithm with the original PVD method in measure of (i) Payload, (ii) Mean Square Error (MSE), (iii) Peak Signal Noise Ratio (PSNR), (iv) Structural Similarity Index (SSIM) and (v) Histogram analysis. Payload defines the hiding capacity in a single pixel of cover media. MSE and PSNR are expressed as:

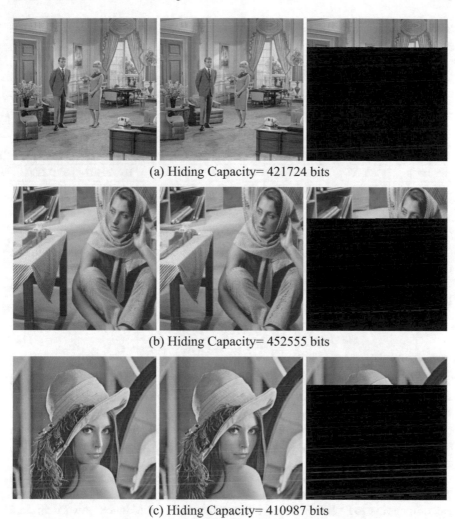

(a) Hiding Capacity= 421724 bits

(b) Hiding Capacity= 452555 bits

(c) Hiding Capacity= 410987 bits

Fig. 4 First column shows the original images; second column images represent corresponding stego-images after embedding secret information; third column images are the extracted hidden part of the secret image

$$\text{MSE} = \sum_{i=1}^{m} \sum_{j=1}^{n} \frac{\left[\text{Img}_{\text{org}}(i,j) - \text{Img}_{\text{stego}}(i,j)\right]^2}{m \times n} \tag{9}$$

$$\text{PSNR} = 10 * \log_{10} \left[\frac{255^2}{\text{MSE}} \right] \tag{10}$$

It is to be mentioned that lesser MSE and higher PSNR signify better image quality of the stego-image.

Table 1 Comparative measure between original PVD and proposed method against Payload, MSE and PSNR

Image	PVD			Proposed method		
	Payload (bpp)	MSE	PSNR (dB)	Payload (bpp)	MSE	PSNR (dB)
Figure 3a	1.560925	5.564037	40.676903	1.560925	5.397526	40.808857
Figure 3b	1.743717	12.330017	37.221169	1.743717	12.487408	37.166080
Figure 3c	1.595428	6.765411	39.827862	1.595428	6.481525	40.014030
Figure 4a	1.608749	6.934189	39.720848	1.608749	6.920078	39.729694
Figure 4b	1.726360	15.711540	36.168617	1.726360	16.629307	35.922062
Figure 4c	1.567791	5.097179	41.057503	1.567791	4.926098	41.205772

The comparative analysis between the proposed algorithm and original PVD in measure of Payload, MSE and PSNR is tabulated in Table 1.

We have also measured the algorithmic performance in terms of SSIM which can be expressed as:

$$SSIM\ (x, y) = \frac{(2\mu_x\mu_y + C_1)(2\sigma_{xy} + C_2)}{(\mu_x^2 + \mu_y^2 + C_1)(\sigma_x^2 + \sigma_y^2 + C_2)} \tag{11}$$

where x and y is cover and stego-media, respectively, μ_x is the mean of x, μ_y is the mean of y, σ_x^2 is the variance of x, σ_y^2 is the variance of y and σ_{xy} is the covariance of x and y. The constants are assigned as $C_1 = (k_1L)^2$, $C_2 = (k_2L)^2$ which are used to stabilize the division with weak denominator and parameters are fixed with specific pre-defined values as $L = 2^{\text{bits per pixel}} - 1 = 255$, $k_1 = 0.01$ and $k_2 = 0.03$, respectively.

The comparative analysis between two competitive methods has been provided in Table 2. The first data column provides SSIM measure between original image and stego-image due to PVD method, second column shows SSIM index between original image and stego-image obtained from proposed algorithm, and the last column is

Table 2 Comparative measure between original PVD and proposed method in measure of SSIM

Image	SSIM (PVD)	SSIM (Proposed)	SSIM (PVD vs. proposed)
Figure 3a	0.998479	0.998518	0.998650
Figure 3b	0.996722	0.996657	0.997087
Figure 3c	0.998723	0.998760	0.998899
Figure 4a	0.998360	0.998361	0.998542
Figure 4b	0.997466	0.997314	0.997686
Figure 4c	0.998980	0.998994	0.999117

(a) Stego image due to PVD method with corresponding histogram

(b) Stego image due to Proposed method with corresponding histogram

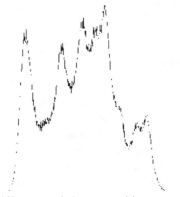

(c) Difference of above two histograms

Fig. 5 Histogram analysis between PVD and proposed method

Fig. 6 Comparison between retrieved hidden information due to the proposed and PVD extraction method

tabulated with the metric values between stego-images of PVD and proposed method. The metric values reflected in Table 1 reveal that the proposed method maintains the similar hiding capacity as it has used the same embedding technique as PVD while for most of the images PSNR values contain marginal higher value which indicates that the proposed method can produce more imperceptible stego-image. On the other hand, Table 2 shows that structural degradation of the stego-image due to the proposed algorithm is very minimal compared to the original PVD method but provides secret message security to a higher extent.

Besides, we have analyzed the experimental results with histograms to establish the efficacy of the proposed technique. Figure 5 shows the histogram-based comparison between PVD and proposed method which proves insignificant difference between both the stego-image.

5.1 Message Security

As the secret bitstream has been made shuffled before embedding into the cover image, normal extraction strategy of PVD method cannot extract those properly. Hence, it is claimed that the secret information is more secure compared to PVD based steganography. An experiment on lena image has been performed what if it is extracted by applying original PVD algorithm. The result is depicted in Fig. 6 which clearly reveals that the retrieved message is completely unreadable if extracted through PVD method.

6 Conclusion and Future Scope

In this paper, we have developed a secure steganography on pixel-value differencing scheme which is based on block ranking of cover media. The data security has been ensured through bit shuffling before the embedding process where the shuffling is carried out on the basis of block-based cover image ranking. Each pixel pair of the cover image is ranked with its potential hiding capacity and the secret bitstream is subsequently stored in appropriate data bins which basically provides the data shuffling. The secret bits have now been embedded by employing the original PVD method by selecting required hidden bits from specified data bin. The experimental results and analyses have established the proposed claim of data security while the proposed method maintains the same hiding capacity as the original PVD method.

The current research has been implemented and applied only on PVD method. In future research, the proposed algorithm may be extended in more generalized way so that it can be applied for any block-based image steganography in view of increasing the security of secret message.

References

1. Luo, W., Huang, F., Huang, J.: Edge adaptive image steganography based on lsb matching revisited. IEEE Trans. Inf. Forensics Secur. **5**(2), 201–214 (2010)
2. Wu, D.C., Tsai, W.H.: A steganographic method for images by pixel-value differencing. Pattern Recogn. Lett. **24**(9–10), 1613–1626 (2003)
3. Chang, K.C., Chang, C.P., Huang, P.S., Tu, T.M.: A novel image steganographic method using tri-way pixel-value differencing. J. Multimedia **3**(2) (2008)
4. Huang, H.S.: A combined image steganographic method using multi-way pixel-value differencing. In: Sixth International Conference on Graphic and Image Processing (ICGIP 2014), vol. 9443, p. 944319. International Society for Optics and Photonics (2015)
5. Lee, Y.P., Lee, J.C., Chen, W.K., Chang, K.C., Su, J., Chang, C.P.: High-payload image hiding with quality recovery using tri-way pixel-value differencing. Inf. Sci. **191**, 214–225 (2012)
6. Wang, C.M., Wu, N.I., Tsai, C.S., Hwang, M.S.: A high quality steganographic method with pixel-value differencing and modulus function. J. Syst. Softw. **81**(1), 150–158 (2008)
7. Himakshi, H.K.V., Singh, R.K., Singh, C.K.: Bi-directional pixel-value differencing approach for steganography. In: Proceedings of the Third International Conference on Soft Computing for Problem Solving: SocProS 2013, vol. 1, p. 109. Springer, Berlin (2014)
8. Khodaei, M., Faez, K.: New adaptive steganographic method using least-significant-bit substitution and pixel-value differencing. IET Image Process. **6**(6), 677–686 (2012)
9. Luo, W., Huang, F., Huang, J.: A more secure steganography based on adaptive pixel-value differencing scheme. Multimedia Tools Appl. **52**(2–3), 407–430 (2011)
10. Gulve, A.K., Joshi, M.S.: A high capacity secured image steganography method with five pixel pair differencing and LSB substitution. Int. J. Image Graph. Sig. Process. **7**(5), 66–74 (2015)
11. Al-Dmour, H., Al-Ani, A.: A steganography embedding method based on edge identification and XOR coding. Expert Syst. Appl. **46**, 293–306 (2016)
12. Feng, B., Lu, W., Sun, W.: Secure binary image steganography based on minimizing the distortion on the texture. IEEE Trans. Inf. Forensics Secur. **10**(2), 243–255 (2014)
13. Jiang, N., Zhao, N., Wang, L.: Lsb based quantum image steganography algorithm. Int. J. Theor. Phys. **55**(1), 107–123 (2016)

14. Swain, G.: Very high capacity image steganography technique using quotient value differencing and LSB substitution. Arab. J. Sci. Eng. **44**(4), 2995–3004 (2019)
15. Tao, J., Li, S., Zhang, X., Wang, Z.: Towards robust image steganography. IEEE Trans. Circuits Syst. Video Technol. **29**(2), 594–600 (2018)
16. Gupta, P., Bhagat, J.: Image steganography using LSB substitution facilitated by shared password. In: International Conference on Innovative Computing and Communications. pp. 369–376. Springer (2019)
17. Kini, N.G., Kini, V.G., et al.: A secured steganography algorithm for hiding an image in an image. In: Integrated Intelligent Computing, Communication and Security, pp. 539–546. Springer, Berlin (2019)
18. Yang, Z.L., Guo, X.Q., Chen, Z.M., Huang, Y.F., Zhang, Y.J.: RNN-stega: linguistic steganography based on recurrent neural networks. IEEE Trans. Inf. Forensics Secur. **14**(5), 1280–1295 (2018)
19. Standard grayscale dataset. http://decsai.ugr.es/cvg/CG/base.htm

Forest Covers Classification of Sundarban on the Basis of Fuzzy C-Means Algorithm Using Satellite Images

K. Kundu, P. Halder and J. K. Mandal

Abstract The present study deals with forest cover classification of Sundarban on the basis of fuzzy c-means algorithm using satellite images from 1975 to 2018. Four features are considered in this region such as dense forest, open forest, open land, and water bodies. The study reveals that during 1975–2018 dense forest and water bodies have gradually increased by 3.75% (113.65 km^2) and 4.38% (133.06 km^2), respectively, while other features like open forest and open land have progressively declined by 4.68% (142.22 km^2) and 3.44% (104.72 km^2) correspondingly. The net forest area (dense forest, open forest) has declined by 0.93% (28.57 km^2) during the study period. The increasing or decreasing trend is not uniform over the entire study period. Net precision and kappa coefficient are used to validate the classification correctness. The classification results are validate using the net precision (86.29, 82.14, 83.16%) and kappa co-efficient (0.817, 0.761, 0.775) for the year of 1975, 2000, and 2018 respectively. The study reveals that the forest declination rate may be increased over the upcoming years. From this study, policy makers may take appropriate decisions to take proper action to control the declination of forest of Sundarban.

Keywords Fuzzy c-means · Forest cover classification · Satellite image · Net precision · Kappa coefficient

K. Kundu (✉)
Department of Computer Science and Engineering, Government College of Engineering and Textile Technology, Serampore, Hooghly, India
e-mail: krishan_cse@rediffmail.com

P. Halder
Department of Computer Science and Engineering, Ramkrishna Mahato Government Engineering College, Purulia, West Bengal, India
e-mail: prasunhalder@gmail.com

J. K. Mandal
Department of Computer Science and Engineering, University of Kalyani, Kalyani, Nadia, West Bengal, India
e-mail: jkm.cse@gmail.com

© Springer Nature Singapore Pte Ltd. 2020
J. K. Mandal and S. Mukhopadhyay (eds.), *Proceedings of the Global AI Congress 2019*, Advances in Intelligent Systems and Computing 1112,
https://doi.org/10.1007/978-981-15-2188-1_40

1 Introduction

Sundarban is the biggest mangrove forest island in the planet and its total area is almost 10,000 km^2. The whole area about 40% resides in India and remaining is occupied in Bangladesh. It consists of mainly various islands that create the sediments put down by three main rivers, the Ganga, Brahmaputra, and the Meghna, and solid network of small rivers, creeks, and channels [1]. It is surrounded by west, east, south, and north, respectively, which are Hooghly River, Ichamati-Kalindi-Raimangal, Bay of Bengal, and Dampier Hodge line. In this region total island exists 102, among these 48 islands preserve by forest, and left over use by people for accommodation. Various biodiversities exist over the entire region [2].

Forest is the valuable important resource in the globe. It creates a natural barrier to protect soil along the river shoreline or sea surface area and mitigate the natural disaster like hurricanes, cyclones, and tsunamis. Its plays a significant part in biological balancing such as safeguard of soil, balanced biodiversity, and climate change evasion. It is widely used in state or national or international level for growth of economy through providing lumber industry, structure, source of medicine, etc. Reddy et al. [3] have studied at the state of Orissa, India, for the period of preceding 75 years (1935–2010) using remote sensing data and GIS. The study examined that forest cover area has gradually declined from the year 1935 to 2010. The study observes that almost 40.5% net forest areas were declined, and annual rate of forest degradation rate was 0.69% over the period. Moreover, study shows that patch size of forest cover area has gradually decreased, while the number of patches has increased over the period.

Schulz et al. [4] studied in Central Chile during 1975–2008 and observe that dry land forests were reduced and transformation of shrub land into farmland. The rate of annual deforestation was 1.7%, shrub land reduces to 0.7%, although other features like timber plantation, agricultural, and urban area were increased during the period. Redowan et al. [5] have reported that crowded forest and open land were turned down, whereas medium dense forests were increased in Khadimnagar National Park (KNP) which belongs to Bangladesh Sylhet Forest Division. Moreover, it examines that the main cause of dense forest degradation was illegally cutting trees, increased buildup area, increased population density, and encroachment, while medium dense forests were increased because new plantation project was initiated.

Kumar et al. [6] prescribed that distance parameter is the most important of forest edge which closely depends on the high chance of forest change. Three distance variables that mainly affect the forest areas are distance between the forest edge and road, distance between the forest edge and settlement, and distance between the forest edge and the slope position of forest due to anthropogenic pressure. Ghebrezgabher et al. [7] have observed that forest cover areas (including woodland forest) were declined during 1970–2014 and interpreted that cause of degradation of forest due to changes in climate and irregular rainfall in the environment reduces the biodiversity on the earth surface. Giri et al. [8] have reported that due to impact on sea level rise the water body areas were gradually increased while entire mangrove forest areas were progressively decreased. Recently, UNESCO has reported that sea level has increased by about 45 cm over the globe and other anthropogenic force also affects the Sundarban region, and its forest region may degrade about 75%.

Application areas of image clustering are extensively used in various fields such as medical imaging, computer vision, robotics, and information retrieval. The most popular image classification process is fuzzy c-means (FCM) algorithm [9, 10] which is based on the unsupervised methodology. In this technique, ambiguity of pixels is minimized to specific class, as a result overall clustering outcome is very well. FCM allow numerous features with different membership magnitude are constantly updated until reached to the specific iteration. Kaur et al. [11] have reported clustered satellite images into various classes using fuzzy c-means algorithm. In accordance with the distance between the sample point and cluster center, the membership magnitude value is assigned to the pixel.

Forest degradation is one of the main causes of climate changes in the world [12]. Various factors influence the changes in climate such as structure of wind, wind speed or pattern, ice and arctic temperature changes, rainfall quantity changes extensively, ocean salinity level, and changes of terrific weather characteristic over the entire globe like heavy rainfall, famine, intensity of waves, frequency of storm, seashore modification, and overflow of low-lying regions. In the last 100–150 years coastal erosion were happen in various places in the earth surface. It is observed that coastal erosion is mainly caused due to sea level rise. In the environment, mixing proportion of CO_2 concentration swiftly enlarged from about 315 ppm in 1960 to 390 ppm in 2010 [13]. The main intention of this study is to find out the changes in forest encircled areas. From this study, the decision makers or planers also take some decision or take some plans to sustain the forest cover region of Sundarban delta.

2 Martial and Methodology

2.1 Study Site

The present study area conducted in 21°40′20″–22°30′10″ N latitude and 88°32′10″–89°4′13″ E longitude in the district of 24 parganas (south) of West Bengal state in India (Fig. 1). The entire study area belongs to the reserve forest area and its area is almost 3038.68 km². The east, west, north, and south regions are encircled by edge of India and Bangladesh, River of Thakuran, Basanti block, and Bay of Bengal correspondingly. In this site also, there are various faunal diversities exist such as fish species, mammal's species, reptile's species, amphibian's species, and bird species. Royal Bengal Tiger, Indian python, crocodiles, and venomous snakes belong to the dense forest region. There are various tree species belong to this region like Heritiera fomes (locally known as Sundari), Goran and Gewa Xylocarpus, Sonneratia, Rhizophora and Nypa palm, Avicennia marina, Bruguiera sp., Phoenix paludosa,

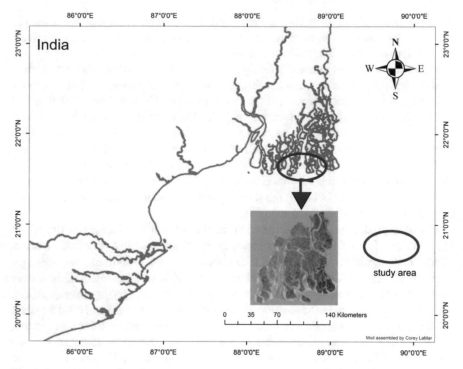

Fig. 1 Location map of study area

Excoecaria sp, and Ceriops sp. Among these tree species, Heritiera fomes (Sundari), Goran and Gewa are the most important and valuable species in this site.

2.2 Satellite Image Collection

In this study, we need three multispectral Landsat satellite images which are acquired from various web portals such as www.earthexplorer.usgs.gov/, www.scihub.copernicus.eu/ and. The satellite image information is illustrated in Table 1, and it is also undoubtedly observed that the entire satellite images are found in

Table 1 Satellite image descriptions

Satellite sensor	Spectral resolution	Image collection date	Path	Row	Pixel size (m)
Landsat 2 MSS	4 bands	05-12-1975	p-148	r-45	60
Landsat 7 ETM+	8 bands	17-11-2000	p-138	r-45	30
Landsat 8 OLI	11 bands	28-02-2018	p-138	r-45	30

the nearly similar time. All the Landsat images are almost blur-free and clear-cut. Landsat 2 satellite uses multispectral scanner (MSS) sensor including spectral bands which are green, red, blue, and near infrared (NIR), and its corresponding spectral reflectance value is (0.5–0.6 um), (0.6–0.7 um), (0.7–0.8 um) and (0.8–1.1 um), and spatial resolution is 60 m. The Landsat 7 enhanced thematic mapper plus (ETM+) sensor consists of eight spectral bands whose spatial resolutions is 30 m which excludes the thermal and panchromatic bands. Landsat 8 satellite uses operational land imager (OLI) sensor composed of eleven spectral bands. The topographic map 79C/9 was collected from the survey of India with scale of 1:50,000 which use for geo-referencing purpose.

2.3   Methodology

In this study, remote sensing data (1975, 2000, and 2018) are used to carry out the meaningful and significant information. Before working on the satellite images, image has to be pre-processed to rectify the error such as geometrically error and radiometric error. Satellite image contains any geometric distortion that can be removed or fixed. If images have some radiometric error, that can be eliminated by the radiometric calibration operation using some image processing softwares such as TNTmips professional 2019. Histogram equalization methods are used to enhance the interpretability and quality of remote sensed data. The correction of geometric distortion and radiometric calibration vectorized the satellite image and got the periphery line of study area. Next, combine the three-layer bands such as infrared, red, and green into a single layer band using layer stacking tool and produce the false color combination (FCC). For re-sampling, the dataset nearest-neighbor algorithm is used and picks the WGS84 datum, and UTM zone with 45 N has projected to the mapping purpose. Then, cut the study area to carry out the current work. Sundarban forest enclosed areas are considered as several features like dense forest, open forest, open land, and water bodies. The fuzzy c-means clustering algorithm is employed to categorize the features of forest enclosed images. In this classification technique initially set of model class has been constructed, and then obtained the membership grade for every feature for each cell. Using the membership grade to adjust the class allotted and estimate the new cluster center, this process is repeated until it converges or reaches the particular iteration. After classifying the satellite image, the specific class is assigned to the specific colors for the visual interpretation. Then, calculate the each feature areas in km^2 which is represented in Table 2. Precision judgment was found to validate the classification exactness through the net precision and the kappa coefficient. Image precision mainly depends on the relationship between the reference data and classified data. Figure 2 demonstrates the detailed representation of the flowchart of the methodology.

Table 2 Forest cover features areas (km^2) and % of area for the year of 1975, 2000, and 2018

Feature name	Year: 1975		Year: 2000		Year: 2018	
	Area (km^2)	%	Area (km^2)	%	Area (km^2)	%
Dense forest	784.36	25.81	817.24	26.91	898.01	29.56
Open forest	986.32	32.46	831.77	27.38	844.10	27.78
Open land	247.48	8.14	273.93	9.02	142.76	4.70
Water bodies	1020.31	33.58	1114.47	36.69	1153.37	37.96

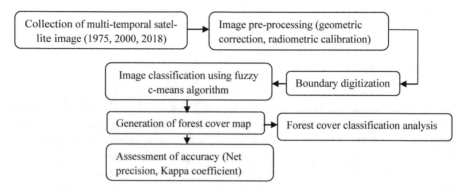

Fig. 2 Flowchart of the methodology

2.4 Fuzzy C-Means Algorithm

In this algorithm, allocate membership value to every data point subsequence to every cluster center according to the distance between the elements and cluster center. In a specific cluster center, more elements are closer to the cluster center, i.e., its membership value is high. For a particular cluster, every data point of membership aggregation should be equal to unit. After every step, each and every time fuzzy membership value and cluster centers change. Now, set of data points and cluster center are considered as $S = \{s_1, s_2, s_3, \ldots, s_n\}$ and $U = \{u_1, u_2, u_3, \ldots, u_p\}$, respectively. The FCM algorithm [9] is given below:

Step1. Arbitrarily choose 'p' cluster centers.
Step2. Estimate fuzzy membership 'μ_{ij}' using the following equation

$$\mu_{ij} = 1 / \sum_{k=1}^{p} \left(\frac{d_{ij}}{d_{ik}}\right)^{\left(\frac{2}{m}-1\right)} \tag{1}$$

where 'μ_{ij}' denotes the membership of ith data to jth cluster center, 'p' specifies the number of cluster center, 'm' is the fuzziness index m $\in$ [1, ∞], 'd_{ij}' indicates the Euclidean distance between ith data and jth cluster center.

Step3. Calculate fuzzy centers 'u_j' using the Eq. (2)

$$U_j = \frac{\sum_{i=1}^{n} (\mu_{ij})^m s_i}{\sum_{i=1}^{n} (\mu_{ij})^m} \quad where \; j = 1, 2, 3, \ldots \ldots p \tag{2}$$

where 'u_j' depicts the jth cluster center

Step4. Evaluate objective function using the Eq. (3)

$$Z(A, B) = \sum_{i=1}^{n} \sum_{j=1}^{p} (\mu_{ij})^m \| s_i - u_j \|^2 \tag{3}$$

where '$\| s_i - u_j \|$' is the Euclidean distance between ith data and jth cluster center

Step5. Repeat steps 2–4 until convergence or minimum 'Z' value is achieved

3 Results and Discussions

In the study, forest enclosed area features are considered as dense forest, open forest, open land, and water bodies. Dense forest area means where tree density canopy covers more than 50%, whereas less than 50% is considered as open forest area. Open land means where enclosed area includes wet land, degraded land, and cleared land. Water body includes lakes, small channels, and rivers. For this study, fuzzy c-means classification method is needed to classify the satellite image and it is the unsupervised classification technique. From Tables 2 and 3, it is noticeably depicted that overall forest area was turn down by about 0.93% (28.57 km²) during the period 1975–2018, although it was not equal over the period. Dense forest area gradually increased about 1.1% (32.88 km²) and 2.65% (80.77 km²) during the period 1975–2000 and 2000–2018, respectively. Therefore, overall dense forest areas were enlarged by approximately 3.75% (113.65 km²) during 1975–2018. During the interval of 1975–2000, open forest area was rapidly decreased by almost 5.08% (154.55 km²), while during the period 2000–2018 slightly increased by about 0.4% (12.33 km²), and net

Table 3 Forest cover features areas during the period 1975–2000, 2000–2018, and 1975–2018

Feature name	Year: 1975–2000		Year: 2000–2018		Year: 1975–2018	
	Area (km²)	%	Area (km²)	%	Area (km²)	%
Dense forest	32.88	1.1	80.77	2.65	113.65	3.75
Open forest	−154.55	−5.08	12.33	0.4	−142.22	−4.68
Open land	26.45	0.88	−131.17	−4.32	−104.72	−3.44
Water bodies	94.16	3.11	38.9	1.27	133.06	4.38

forest area lost by approximately 4.68% (142.22 km^2) during 1975–2018. During the 1975–2000, open land area marginally increased by about 0.88% (26.45 km^2), but its area has rapidly decreased by about 4.32% (131.17 km^2) in 2000–2018; therefore, overall area loss is about 3.44% (104.72 km^2) in 1975–2018 because some open land area may be flooded by the water bodies. Water body areas progressively increased by about 3.11% (94.16 km^2) and 1.27% (38.9 km^2) during the period 1975–2000 and 2000–2018 correspondingly. Therefore, overall water body areas are significantly increased by about 4.38% (133.06 km^2) during 1975–2018 due to rising sea level. Figures 3, 4, and 5 depict the visual representation of the forest cover classified results. Figure 6 shows the graphical representation of each feature area in year wise.

4   Precision Assessment

The image classification accuracy is obtained using puzzlement matrix. The image classification accuracy determine through the puzzlement matrix. It is the matrix correlation between reference data and classified data. The puzzlement matrix column represents the observation data that it may include ground inspection records or ground truth data, visual exposition records, and Google Earth records, and row represents the classified data which are obtained after classifying the image. The crossway elements of the puzzlement matrix are depicted by number of elements correctly classified. On the other hand, off-diagonal elements are represented that are not correctly classified elements. Net accuracy was estimated using the ratio of oblique positional whole number of elements into the entire number of elements used for the classification. Another statistical tool to measure the classification precision is the kappa coefficient, and it is the concurrence between the classified elements and the ground truth data. Generally, kappa coefficient value ranges from 0 to 1 and upper value indicates that categorization outcome is more proper. Now, to obtain the truthfulness of forest cover maps, puzzlement matrices or confusion matrix was created. Puzzlement matrices of 1975, 2000 and 2018 images are given in Tables 4, 5 and 6 correspondingly. Net precision, user precision, producer precision and kappa coefficient were calculated, and the net precision for 1975 MSS, 2000 ETM+ and 2018 OLI has been achieved 86.29%, 82.14% and 83.16%, respectively, and the kappa coefficient for these classification is 0.817, 0.761 and 0.755 correspondingly. The classification accuracy 82–86% indicates that 82–86% points are correctly classified.

5   Conclusion

The study exposed that forest encircle area was gradually declined during 1975–2018 over the last 43 years, although it was not uniform over the period. During the period of 1975–2018, open forest and open land area are declined by almost 4.68% (142.22 km^2) and 3.44% (104.72 km^2), respectively, while during the same

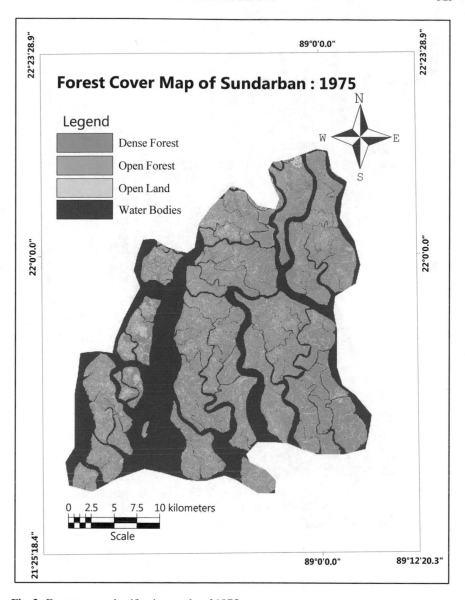

Fig. 3 Forest covers classification results of 1975

period dense forest and water bodies are increased by about 3.75% (113.65 km²) and 4.38% (133.06 km²) correspondingly. It is clearly observed that water bodies were progressively increased during the study period because of rising sea level. In addition, it is observably scrutinized that net forest areas such as dense forest and open forest were gradually declined by about 0.93% (28.57 km²) during the period 1975–2018. To validate the classification results, overall classification accuracy has

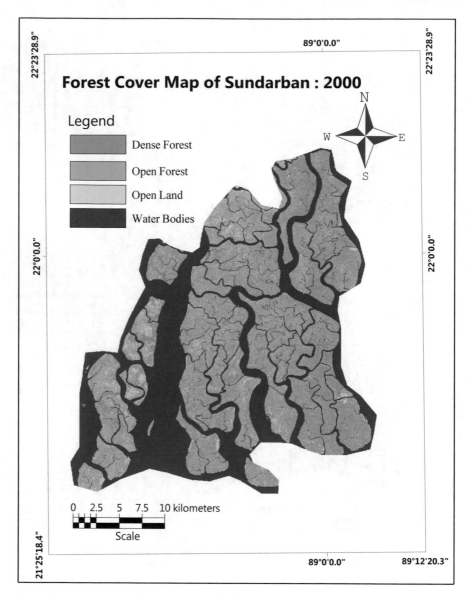

Fig. 4 Forest covers classification results of 2000

achieved between 82 and 86%. Therefore, from this study, it is undoubtedly observed that forest enclose areas were progressively turned down in future.

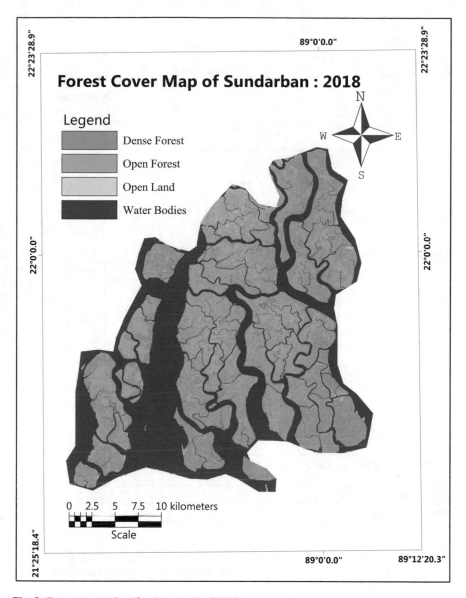

Fig. 5 Forest covers classification results of 2018

References

1. Ghosh, A., Schmidt, S., Fickert, T., Nüsser, M.: The Indian Sundarban mangrove forests: history, utilization, conservation strategies and local perception. Diversity **7**(2), 149–169 (2015)
2. Gopal, B., Chauhan, M.: Biodiversity and its conservation in the Sundarban Mangrove Ecosystem. Aquat. Sci. **68**(3), 338–354 (2006)

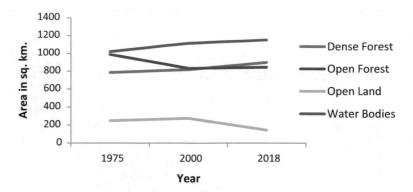

Fig. 6 Graphical representation of forest covers areas for the year of 1975, 2000, and 2018

Table 4 Puzzlement matrix for the year of 1975

Classified data	Observation data					Total	User's precision (%)
	Class name	Dense forest	Open forest	Open land	Water bodies		
	Dense forest	49	1	4	2	56	87.5
	Open forest	2	41	2	1	46	89.13
	Open land	5	2	34	3	44	77.27
	Water bodies	1	3	1	46	51	90.19
Total		57	47	41	52	197	
Producer's precision (%)		85.96	87.23	82.93	88.46		
Net precision = 86.29%							
Kappa coefficient = 0.817							

3. Sudhakar Reddy, C., Jha, C.S., Dadhwal, V.K.: Assessment and monitoring of long-term forest cover changes in Odisha, India using remote sensing and GIS. Environ. Monit. Assess. **185**, 4399–4415 (2013)
4. Schulz, J.J., Cayuela, L., Echeverria, C., Salas, J., Benayas, J.M.R.: Monitoring land cover change of the dryland forest landscape of Central Chile (1975–2008). Appl. Geogr. **30**, 436–447 (2010)
5. Redowan, M., Akter, S., Islam, N.: Analysis of forest cover change at Khadimnagar National Park, Sylhet, Bangladesh, using Landsat TM and GIS data. J. Forest. Res. **25**(2), 393–400 (2014)
6. Kumar, R., Nandy, S., Agarwal, R., Kushwah, S.P.S.: Forest cover dynamics analysis and prediction modeling using logistic regression model. Ecol. Ind. **45**, 444–455 (2014)
7. Ghebrezgabher, M.G., Yang, T., Yang, X., Wang, X., Khan, M.: Extracting and analyzing forest and woodland cover change in Eritrea based on landsat data using supervised classification. Egypt. J. Remote Sens. Space Sci. **19**(1), 37–47 (2016)

Table 5 Puzzlement matrix for the year of 2000

Classified data	Observation data					Total	User's precision (%)
	Class name	Dense forest	Open forest	Open land	Water bodies		
	Dense forest	35	1	5	2	43	81.39
	Open forest	2	41	2	1	46	89.13
	Open land	1	5	39	6	51	76.47
	Water bodies	4	3	3	46	56	82.14
Total		42	50	49	55	196	
Producer's precision (%)		83.33	82	79.59	83.63		
Net precision = 82.14%							
Kappa coefficient = 0.761							

Table 6 Puzzlement matrix for the year of 2018

Classified data	Observation data					Total	User's precision (%)
	Class name	Dense forest	Open forest	Open land	Water bodies		
	Dense forest	39	4	2	3	48	81.25
	Open forest	2	43	5	2	52	82.69
	Open land	3	2	32	4	41	78.05
	Water bodies	1	2	3	49	55	89.09
Total		45	51	42	58	195	
Producer's precision (%)		86.67	84.31	76.19	84.48		
Net precision = 83.16%							
Kappa coefficient = 0.775							

8. Giri, S., Mukhopadhyay, A., Hazra, S., Mukherjee, S., Roy, D., Ghosh, S., Ghosh, T., Mitra, D.: A study on abundance and distribution of mangrove species in Indian Sundarban using remote sensing technique. J. Coast. Conserv. **18**, 359–367 (2014)
9. Bezdex, J.C., Ehrlich, R., Full, W.: FCM: Fuzzy c-means clustering algorithm. Comput. Geosci. **10**(2–3), 191–203 (1984)
10. Bezdek, J.: A convergence theorem for the fuzzy ISODATA clustering algorithms. IEEE Trans. Pattern Anal. Mach. Intell. **2**, 1–8 (1980)

11. Kaur, R., Sharma, D., Verma, A.: Enhance satellite image classification based on fuzzy clustering and Marr-Hildreth algorithm. In: 4th IEEE International Conference on Signal Processing, Computing and Control (ISPCC 2017), Solan, India, pp. 13–137, 21–23 Sept 2017
12. Raha, A., Das, S., Banerjee, K., Mitra, A.: Climate change impacts on Indian Sunderbans: a time series analysis (1924–2008). Biodivers. Conserv. **21**(5), 1289–1307 (2012)
13. Ray, R., Jana, T.K.: Carbon sequestration by mangrove forest: one approach for managing carbon dioxide emission from coal-based power plant. Atmos. Environ. **171**, 149–154 (2017)

Artificial Neural Network: An Answer to Right Order Quantity

Saurav Dey and Debamalya Ghose

Abstract In recent years, the concept of artificial intelligence (AI) is being used in various functions across the globe. This is particularly because of the versatile range of utility of artificial intelligence. Artificial intelligence is also being used very much in logistics industry. As we all know, the six basic functional areas of logistics are as follows: (1) inventory planning and management, (2) warehousing, (3) procurement of goods and services, (4) packaging and storage, (5) transportation, and (6) customer service. Along with the six functional areas, there are also six key performance indicators. The following are the key performance indicators of logistics operations as mentioned and collected from various literature: (1) arrival precision, (2) pick up discrepancy alert, (3) number of incidents, (4) late delivery alert, (5) filling rate in transport equipment, (6) stock accuracy. The aim of this paper is to find the variables which play a vital role in the decision of the right quantity to be ordered and simultaneously to propose a model which can integrate inputs from all those variables and use the inputs through artificial neural network and finally propose a suitable model.

Keywords Forecasting · Demand · Artificial intelligence · Artificial neural network · Model

1 Introduction

AI is a branch of information technology that deals with the automation of intelligent behavior. AI is the attempt to program a computer so that it is able to process problems independently, similar to the way a human with the appropriate training would. Problem-solving means making decisions that constitute an appropriate response to the underlying problem within the specified time, based on data from various

S. Dey · D. Ghose (✉)
Department of Business Administration, Assam University, Silchar, India
e-mail: operationsdghosh@gmail.com

S. Dey
e-mail: sauravdey22121988@gmail.com

© Springer Nature Singapore Pte Ltd. 2020
J. K. Mandal and S. Mukhopadhyay (eds.), *Proceedings of the Global AI Congress 2019*, Advances in Intelligent Systems and Computing 1112, https://doi.org/10.1007/978-981-15-2188-1_41

sources (databases, sensors, video cameras, etc.). As we all know, the subject of logistics moves around its six basic functional areas, namely (1) inventory planning and management, (2) warehousing, (3) procurement of goods and services, (4) packaging and storage, (5) transportation, and (6) customer service. Successful logistics operation is nothing but proper implementation and usage of all these six functional areas. But, at the same point of time, some bottlenecks arise which hampers the smooth functioning of all these six functional areas, and hence, the entire logistical operation of the company is hampered which finally sometimes create dissatisfaction in the minds of the workers as well as customers, and hence, the overall profitability goal of the company is hampered. After much of research, the companies found a ray of hope in the form of artificial intelligence which can solve their problem, and hence, both these domains, i.e., logistics and artificial intelligence, need to be clubbed for achieving proper logistical operation and hence ensuring the smooth functioning of the company. One such bottleneck is the proper estimation of correct ordering quantity which hampers the companies performance and sometimes results in blockage of funds. In this paper, this particular bottleneck is dealt with.

2 Need for the Study

To find the right ordering quantity is an important task for many companies as excess quantity might result in blockage of funds, and sometimes for perishable items, it may be more hazardous for the organization. This study critically tries to find out the correct ordering quantity which can be ordered.

3 Literature Review

Ali et al. [1] in their paper entitled "Forecasting of optimum raw materials using Artificial Neural Network" which was published in the year 2011 in the International Journal of Operations and Quantitative Management have mathematically choked out and solved how demand forecasting is possible with the help of artificial neural network.

Bottani [2] in their paper entitled "Modelling Wholesale distribution operations: an artificial intelligence framework" which was published in the Journal of Industrial Management and Data Systems in the year 2019 proposed an artificial intelligence-based framework to support decision making in wholesale distribution with the aim to limit wholesale out-of-stocks by jointly formulating price policies and forecasting retailer's demand.

Willppu [3] of the Turku School of Economics and Business Administration, Turku, Finland, in the paper entitled "Neural Network and Logistics" very lucidly identified the relationship between logistic functions, their KPI, and how ANN can be used as an useful tool to solve various issues related to logistics.

Kochak and Sharma [4] in their research paper entitled "Demand forecasting using Neural Network" which was published in the International Journal of Mechanical Engineering and Robotics Research in the year 2015 very critically established the relationship between neural network and demand forecasting and have exhibited how proper utilization of the knowledge of neural network can actually help in demand forecasting.

Lo et al. [5] in their research paper entitled "Comparative Study on logical analysis of data (LAD), Artificial neural networks (ANN), and proportional hazards model (PHM) for maintenance prognostics" examined three CBM fault prognostics model, logical analysis of data (LAD) and artificial neural network (ANN).

Paul and Azeem [6] in their research paper entitled "An Artificial Neural Network Model for Optimization of Finished Goods Inventory" which was published in the International Journal of Industrial Engineering Computations in the year 2011 thrown light on the variables that are related to the maintenance of finished goods inventory.

Penpece and Elma [7] in their research paper entitled "Predicting Sales Revenue by using ANN in Grocery Retailing Industry—A case study in Turkey" which was published in the journal of Trade, Economics and Finance in the year 2014 provided us with the knowledge that in AI the best way for prediction model is ANN.

Saleh [8]—Commerce Product Return rate Statistics and trends—https://www.invespcro.com/blog/ecommerce-product-return-rate-statistics provided us with various important data regarding how return has become an important daily task for the retailers in e-commerce sector.

Schatteman [9], in their research paper entitled "Reverse Logistics"—*Gower handbook of supply chain management*—highlighted and provided various reasons for returns in the reverse logistical environment.

Temur et al. [10] in their research paper entitled "A fuzzy expert system design for forecasting return quantity in reverse logistics network", *Journal of Enterprise Information Management*, provided with the answer that fuzzy expert system is the best solution available for forecasting return quantity in a reverse logistical environment.

Zhang et al. [11] Michael in their research paper entitled "Forecasting with Artificial Networks—The State of the Art" which was published in the International Journal of Forecasting in the year 1998 provided us with the basic knowledge. This particular paper can be called as a visionary paper because in the year 1998 they were able to see the future of forecasting and artificial networks.

4 Objectives

1. To examine the necessity for the prediction of right ordering quantity
2. To find out a technique of AI using which we can predict right ordering quantity.

5 Scope

The study is critically focused on the companies and their problems and how AI concepts helped them to solve their problems. The scope is limited to the storage function of the organization only.

6 Limitations

One of the vital limitations of the study is that the study has a proposed model only using the literatures available, and ANN programming and other aspects which are necessary are yet to be explored.

7 Research Methodology and Data

The current study is of the type of exploratory, where the related literatures have been explored. The study is based on secondary information collected from various sources such as journals, company's Web site, newspaper, articles.

8 Key Findings

8.1 Findings for Objective 1

Using various literature, it has been found that although EOQ technique can help us out with how much quantity to order, but in this competitive era, only EOQ will not serve the purpose, and for this, we need the essentiality of a robust system for the prediction of right order quantity which becomes essentially important in this regard.

8.2 Findings for Objective 2

From the various literatures, it has been found out that artificial neural network as the only solution under the domain of AI which can solve the issue of predicting right order quantity.

9 Conclusion

From the paper, it can be concluded that if right quantity is ordered, the profitability aspect will be maintained and artificial neural network (ANN).

References

1. Ali, M.S., Paul, K.S., Ahsan, K., Azeem, A.: Forecastiong of optimum raw materials using artificial neural network. Int. J. Oper. Quant. Manage. **17**(4), 333–348 (2011)
2. Bottani, E., Centobelli, P., Gallo, M., Kavani, M.A., Jain, V., Murino, T.: Modelling Wholesale distribution operations: an artificial intelligence framework. J. Ind. Manage. Data Syst. **119**(4), 698–718 (2019)
3. Willppu, E.: Neural Network in Logistics
4. Kochak, A., Sharma, S.: Demand forecasting using neural network for supply chain management. Int. J. Mech. Eng. Robot. Res. **4**(11) (2015)
5. Lo, H., Ghasemi, A., Diallo, C., Newhook, J.: Comparative Study on logical analysis of data (LAD), Artificial neural networks(ANN), and proportional hazards model (PHM) for maintenance prognostics. J. Qual. Maintenance Eng. **25**(1), 2–24 (2019)
6. Paul, S., Azeem, A.: An artificial neural network model for optimization of finished goods inventory. Int. J. Ind. Eng. Comput. **2**(2), 431–438 (2011)
7. Penpece, D., Elma, E.O.: Predicting sales revenue by using artificial neural network in grocery retailing industry—a case study in Turkey. Int. J. Trade Econ. Finance **5**(5) (2014)
8. Saleh, K.E.: Commerce Product Return rate Statistics and trends. https://www.invespcro.com/blog/ecommerce-product-return-rate-statistics. Retrieved on 22 July 2019 at 4.30 PM
9. Schatteman, O.: Reverse logistics. In: Gower handbook of supply chain management, pp. 267–279. Aldershot [u.a.]: Gower (2003). ISBN 0-566-08511-9
10. Temur, G.T., Balcilar, M., Bolat, B.: A fuzzy expert system design for forecasting return quantity in reverse logistics network. J. Enterp. Inf. Manage. **27**(3), 316–328 (2014)
11. Zhang, G., Patuwo, E., Michael, HuY: Forecasting with artificial neural network—the state of the art. Int. J. Forecast. **14**(5), 35–62 (1998)
12. Dey, S., Ghose, D.: Predictive fuzzy system an useful tool for handling returns in a reverse logistical environment for the logistics sector. Infokara Res. **9**(1), 887–893 (2020)
13. Dey, S., Ghose, D.: Artificial neural network a proper prediction to answer right order quantity. Infokara Res **9**(1), 879–886 (2020)
14. Dey, S., Ghose, D.: E-retailing & its associated paradigms: a study in India. Int. J. Anal. Exp. Modal Anal. **12**(1), 2511–2528 (2020)
15. Kudo, F., Akitomi, T., Moriwaki, N.: An artificial intelligence computer system for analysis of social infrastructure data. In: IEEE Conference on Business Informatics (CBI) (June 2015)
16. Kimura, J., et al.: Framework for collaborative creation with customers to improve warehouse logistics. Hitachi Rev. **65**, 873–877 (2016)
17. Kumar, P., Herbert, M., Rao, S.: Demand forecasting using artificial neural network on different learning methods—a comparative analysis. Int. J. Res. Appl. Sci. Eng. Technol. **2**(4) (2014)
18. Obe, O.O., Shangodoyin, D.K.: Artificial neural network based model for forecasting of sugar cane production. J. Comput. Sci. **6**(4), 439–445 (2010)
19. https://www.ssischaefer.com/resource/blob/504606/06d87a3eff1abfbdd7af3875404b724a/white-paper-artificial-intelligence-in-logistics–dam-download-en-16558-data.pdf

Multi-Arc Processor—Harnessing Pseudo-concurrent Multiple Instruction Set Architecture (ISA) Over a Single Hardware Platform

Vishal Narnolia, Uddipto Jana and Tufan Saha

Abstract Developers, in order to quench the needs of powerful yet efficient processors, are developing processors with multiple cores where each core may or may not serve the same purpose. To achieve that different instruction set architectures (ISAs) are incorporated in some specific cores. Such an example is the modern-day artificial intelligence-central processing unit (AI-CPU), which not only have multiple cores but also dedicated neural processing units (NPUs) and graphics processing units (GPUs). These special purpose processing units (PU) are designed with different ISAs to handle a complex job like matrix multiplication as a dedicated instruction. Such augmentation comes at a cost of extra hardware, multiple types of interfacings and resource requirements. To reduce this, research is being done to try to develop customizable ISAs which can be customized in various dimensions and can be thrown into different cores. The aim of this work is to introduce Multi-Arc, a new approach to design processors with heterogeneous ISAs by enabling the same hardware design to support multiple and totally different ISAs which can be switched both implicitly and explicitly as per need.

Keywords Processor design · Instruction set architectures · Heterogeneous ISAs · Pseudo-concurrent architecture implementation · Generic ASIP · Switching architectures

1 Introduction

By the fall of the twentieth century, computational requirements were already fueling the race for the development of powerful processors which can meet the demands and should cost less in resources. The start was with the idea of incorporating more complex instructions [1] and processing elements in the processors. But that comes at a huge cost that is more hardware and even higher latency for each instruction. An improvisation was made using pipelined architectures [2] which enabled concurrent processing of the instructions.

V. Narnolia (✉) · U. Jana · T. Saha
Department of CSE, Institute of Engineering & Management, Kolkata, India

© Springer Nature Singapore Pte Ltd. 2020
J. K. Mandal and S. Mukhopadhyay (eds.), *Proceedings of the Global AI Congress 2019*, Advances in Intelligent Systems and Computing 1112,
https://doi.org/10.1007/978-981-15-2188-1_42

However, pipelined architectures were also unable to fulfill the demands of a faster way to compute; hence, the idea of shared architectures (architectures with multiple copies of the same computational resource) was introduced which enabled faster computation but came at a cost of extra devices and complex networking. To reduce this problem, multi-core processors [3] were developed where similar cores were embedded in a single chip, hence creating an illusion of multiple devices in one. But there is no end to demands, and sooner graphics became a major part of processed data which required higher computation and even slower than the other basic works. This paved the way for graphics processing units (GPUs) [4].

Nowadays, the design of hardware depends more on the instruction set architecture (ISA) [5] configured (Refer to Fig. 1). Thus, different ISAs are incorporated in some specific cores within the main CPU chip. Such an example is modern-day AI CPUs [6] which have not only multiple cores but also dedicated NPUs and GPUs, which are nothing but a CPU core with a different ISA to handle a complex job like matrix multiplication as a dedicated instruction. However, these processors might serve the purpose but it comes at a cost of extra hardware, multiple types of interfacings and resource requirements.

To reduce this, researchers nowadays are trying to develop customizable ISAs [7] which can be customized in various dimensions and can be thrown into different cores. Thus, heterogenous ISAs can be created and incorporated without altering the hardware itself, and hence, also obliterating the need of a dedicated NPUs and GPUs as it can be designed as a subset of the main ISA. This ingenious idea enables the working of CPUs efficient in various aspects as the resources needed for work can be optimized by customizing the ISA optimally. But practically, if a microprocessor needs to be able to cater a variety of different programs, its ISA needs to include a more generalized version of the instructions. These instructions cannot be optimized for some particular type of programs while maintaining generalized nature of the processor, which could lead to keeping multiple versions of the instructions which is redundant and often not a good approach. Moreover, an extra core will always be

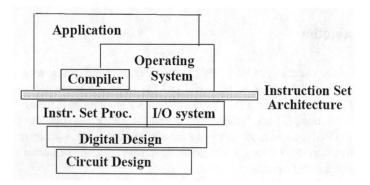

Fig. 1 Logical positioning of instruction set architecture (ISA) in various parts of system design (Image Courtesy [21])

dedicated to house the super ISA and perform customizability. It even means that some very small and less resource heavy ISA will have an abundance of hardware elements and will lead to inefficiency in hardware allocation.

The aim of this work is to introduce a new approach to design processors through Multi-Arc which is a hardware design to support multiple and totally different ISAs that can be switched both implicitly and explicitly as per need. The hardware needs to be designed in such a way that it is a union of all the processing elements required by the participating ISAs. At the time of allocation of a particular ISA, it will be allotted only the needful elements while others can have their independent work going on like direct memory access (DMA) and heavy processing elements not needed by the current ISA. It will even be able to switch between ISAs without effecting the work of another ISA. Thus, a single hardware which is able to switch between different ISAs exclusively will be able to provide an illusion of different CPUs in a single CPU and provide the user to select and work with multiple types of CPUs concurrently without altering the hardware or requiring extra hardware. This type of architecture will need maximum resource as same as the requirement of the most resource hungry ISA present and will not need anything extra to provide the customizability.

2 Related Works

R. Kumar, D. M. Tullsen, P. Ranganathan, N. P. Jouppi and K. I. Farkas, in their work [8] introduced single ISA heterogeneous multi-core architectures, which employed cores of different sizes and workload capabilities. The architecture allowed the application to dynamically determine the most efficient core and thereby direct the execution to that core, thus maximizing performance and energy utilization.

A. Venkat and D. M. Tullsen in their work [9] explored another dimension of heterogeneity, in terms of heterogeneous ISAs. Implemented on top of already micro-architecturally heterogeneous cores, this architecture allows the cores to implement different instructions for achieving the same result. It can be perceived that certain regions of the code prefer certain ISAs more than others. Thus, cores that can leverage on this insight show a significant increase in performance, both in terms of execution time and power efficiency.

A. Venkat, H. Basavaraj and D. M. Tullsen in their work [10] went one step further, exploring the possibilities of implementing multi-ISA heterogeneity using a single ISA. In this work, they explore the customizability of ISAs, so that the benefits of multiple existing ISAs can be achieved within one superset ISA, which consists of multiple versions of the same instruction differing in register width and depth, opcode size and instruction complexity, as well as instruction decoding format. This approach seeks to improve the performance through subtle variations in micro-operation [11] executions that favors the execution of the specific code.

All these works progressively improve on the shortcomings of their predecessors. There are a few points to be noted about these works. Firstly, all of them are primarily focused on multi-core architectures. Secondly, all of them are all about designing better ISAs and optimizing migration of executions to better cores. All of these are definitely an improvement.

But, let us take one step backward. What if we do not go into all these complexities and are able to achieve a processor design where the processor is capable of switching between multiple, often very different, ISAs, and that too on the instruction of the programmer? The main selling point of this idea is that we can integrate the ISAs of multiple application-specific instruction set processors (ASIPs) [12] into a single processor hardware and switch between them as and when necessary, thus allowing the processor to rival the capabilities of a general purpose processor (GPP) [13], if not surpass it, in a more specifically optimized way than a GPP can.

3 Persistent Problems

To better understand the idea which has been brought forward via this work, one needs to understand why the idea is needed. The best way to make things understandable is to put forth the basic problems by far:

- In the work highlighted in [8], specific cores of the processor are designated to process specific instructions so that the execution is faster. The use of different capacity cores to process instruction definitely optimizes the use of power and hardware. But this optimization comes at a cost of extra processing element (PE). Now, the question is, can we do it in one core?
- The second work [9] highlights the use of hetero-ISAs on top of heterogeneous cores. The hetero-ISA contains multiple instructions for the same achieving the same result, each optimized for a different purpose, in different cores. While it provides diversity to the ISA, different instructions for doing the same thing clearly consumes a lot of opcode space, making the opcode size larger. Can it be taken care of somehow?
- Similarly, going through [10] one can see the benefits of ISA customizability. Definitely the best of the lot, this work allows the same instruction to have multiple modes of execution, the interesting thing being that the programmer has the choice to select the optimal instruction mode to use, giving him/her a lot more freedom. However, this still cannot work in a uni-core environment, and the mode selected is a part of the opcode so the opcode space could not be reduced either. There lies another question hidden: What if we always use the master-ISA on its full capabilities? It will simply mean the customizable ISA is doing nothing but blocking extra capabilities of the PE when customized to a lower requirement version of master-ISA, thus creating inefficient utilization of hardware. So, the question arises: 'Can it be done better?'

4 Proposed Idea

Now, the answer to these above questions is what is given by our idea of Multi-Arc. The key features of our idea can be summarized as follows:

- The idea is to use multiple different ISAs which are exclusive of each other instead of combining them together. This allows for the reuse of the opcodes used by one instruction in one ISA as another instruction in different ISA thus making way for reuse and reducing the opcode space.
- The switching between ISAs is to be brought about by a dedicated instruction, giving user the freedom to change the way the processor interprets the instruction.
- The ISAs used will not be existing simultaneously from the processor's perspective. Thus, the different ISAs can reuse the same processor hardware and other units. Thus, this idea can work in a uni-core environment.
- The idea does have the constraint that the ISAs should have the same hardware requirements. Rather there will be a set union of all the PEs required in the processor so as to take the most advantage out of this scheme, but this idea can theoretically allow infinite number of ISAs on the same chip. This gives us the possibility of integrating multiple ASIPs into the same hardware and maybe with development this scheme can compete with a full-fledged GPP in the future.

5 Methodology of Design

Thinking from the perspective of design, there are not many things to be said about the design of the individual ISAs and their implementation. So, the discussion will be restricted to only those sections of the design that are of utmost relevance to the topic.

Here are some of the questions that can be raised about the design:

- How is the machine going to distinguish between the multiple ISAs?
- How is opcode space reusable?
- Can ISAs requiring different instruction register (IR) [14] sizes be incorporated?
- How are the existing instructions being processed in the instruction pipeline be dealt with?
- How is the register depth and width being resized with the change in ISA?
- What makes it efficient and faster?

Let us answer them one by one while we refer you a generalized block diagram of the system as in Fig. 2.

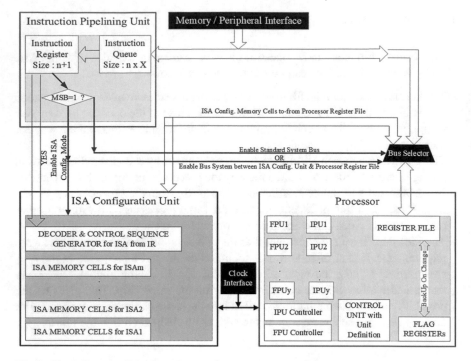

Fig. 2 Block diagram of Multi-Arc System ('n' is maximum word size of the supported architecture; 'X' is the maximum number of instructions to be held in queue; MSB: most significant bit of IR; 'm' is maximum number of ISA to be supported and *remembered*; FPU: floating point unit; IPU: integer processing unit)

5.1 ISA Switching Process and ISA Control

Each ISA will have an ISA code just like every instruction in an ISA has opcodes. This ISA code will be held in a special purpose ISA register. Now, the programmer will be provided with an instruction that will set the value of the ISA register to the remainder left when whatever value provided by the programmer as the operand is divided by the number of ISAs used, which is design specific. The remainder concept is there so that the programmer might not assign the value of the ISA register to some invalid value.

Whenever an ISA switching will be done by the programmer, the current status information of every instruction under process together with the contents of the registers will be saved in an ISA control block of the memory used, and the instruction pipeline will be flushed. This is done because whenever there will be a change in the ISA, the control signals, the register sizes and numbers will be changed, which will result in malhandling of the instructions already in the pipeline. When the ISA will be switched back again, the instruction pipeline will be restored to its previous

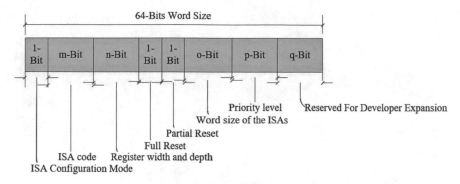

Fig. 3 Multi-Arc ISA configuration control word format

state. This idea works in the same way as an OS performs a context switch [15] when giving each process a share of CPU time.

Basic steps to implement the idea might be

1. There will always be a dedicated bit to state whether an ISA switch is to be done or not.
2. On encountering an ISA switch request, the next groups of bits will denote (as depicted in Fig. 3)

   a. ISA code
   b. Register width and depth (encoded as layout numbers)
   c. A dedicated bit to demand a full reset
   d. A dedicated bit for partial reset
   e. Word size of the ISAs (also encoded)
   f. Priority level.

3. The processor will read the priority bit and then will complete the remaining instructions in the instruction pipeline abiding the priority stated.
4. Then, the full reset [16] bit will be read and the following cases are:

   a. Case 1: When the bit is 0 that is no reset.
      (1) Save the contents of special purpose registers (SPRs) [17] in the dedicated memory for the ISA in the ISA configuration unit, including the instruction pipeline.
      (2) The contents of the general purpose registers (GPRs) [18] are flushed.
      (3) Instruction pipeline is flushed.
   b. Case 2: When reset bit is 1 that is full reset
      (1) Every memory element is flushed.
      (2) The dedicated memory blocks of ISAs in ISA control block are set to default.

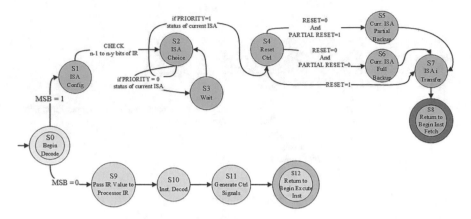

Fig. 4 Opcode decode cycle

5. The partial reset bit is read only if full reset is 0. In case of a successful read of value 1, only the instruction pipeline, the GPR bank and the memory dedicated to the immediate ISA in action will be set to default.
6. After successful reset or save, the ISA code will be decoded, and all the contents of the SPRs of the decoded ISA will be loaded from the ISA configuration unit.
7. A high impedance will be sent to only and all the PEs required by the new ISA.
8. The processor resumes its normal action.

A glimpse of opcode decode cycle is provided in Fig. 4.

5.2 Opcode Size Management

Opcodes are decoded by a normal finite state machine (FSM) [19] or decoder which defines the immediate next few steps. Despite having same primary function, every slight change in functionality means a new opcode. For example, in Intel 8085: ADI B will have different opcode then ADI C or ADC B or ADD B C.

Similarly, in case of customizable ISAs, each customization or variation will eat a certain amount of bit space in instruction word hence leaving lesser opcode space. In Multi-Arc, a dedicated switching instruction not only provides a full instruction word for complete variations but also does not eat any bits in the instruction word during the processing of other instructions hence reducing opcode size or allowing more space in the opcode for data or other variations.

A simple mathematical example will be enough to prove the point (as depicted in Fig. 5):

Let us assume,

Word size = 64 bits where,
No. of bits for data handling = 32 bits.

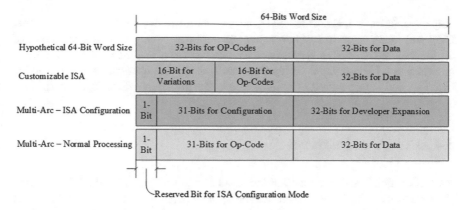

Fig. 5 Hypothetical 64-bit word size system as represented through customizable ISA versus Multi-Arc

In case of customizable ISA suppose,

No. of bits used for variations = 16 bits
Then no. of bits left for opcodes = 16 bits
Hence, 2^{16} possible opcodes.

In Multi-Arc, 1 bit is reserved for determining configuration state,

Hence, bits available for configuration = 31–63 bits.
While in case of normal processing,
No. of bits for opcodes = 31 bits

Thus, catering 2^{31} possible opcodes which is 2^{15} times the normal effective opcode size.

5.3 Generic Abilities

A processor is defined by the type of instructions it is able to process which is decoded from the opcodes of the incorporated ISA. Now, in Multi-Arc for each ISA there may or may not be same decoding pattern from other ISAs which is done by the decoding unit which is activated when respective ISA is switched to. These decoder unit can be both grouped or individual or both for every ISA, for example normal arithmetical ISAs may have a similar decoding pattern hence, same decoder unit, while GPUs needs more powerful computational instructions thus may have different decoder. Meanwhile, the decoding pattern will always be based on the hierarchy of the bits from MSB to LSB as done in normal CPUs which will enable grouping of opcodes as per functionality

This feature provides even more and specific opcodes for each type of ISA. For example, in an 8-bit opcode space A2 for one ISA mean ADD R1, R2 while in other may mean MUL R3, R4. This will harness capabilities of a GPU, NPU and normal CPU and many more in a single hardware platform having PEs as a set union of all the PEs required, in a way creating a generic form of processor which can be switched as a GPU or CPU during runtime. Moreover, a user can even develop his own ISA and incorporate it by abiding one of the predefined decoding patterns.

5.4   Pseudo-concurrency

As earlier mentioned, the total hardware set is a set union of all the PEs required by the participating ISAs so there maybe cases where the more complex PE is not required or busy for a long time at that time the present ISA has no more instructions to process then the OS can instruct work of a different ISA which does not require the PE in use and the registers or memory units involved. This is not only concurrency [20] on the OS level, but it can be assumed that two different processors are working simultaneous while one of the processors has lost all controls over the hardware.

For example, suppose there is an ISA of GPU which is refreshing the screen at a rate of 20 fps that means each frame needs 0.05 s while the hardware has a clock of 5 GHz. So, each refresh needs 10^7 clock cycles. There is also another ISA of an NPU which needs a matrix multiplication requiring 5000 s or 10^{12} cycles. There is one more ISA of a simple calculator which needs ten cycles per instruction. All these can be done together in a single hardware platform simultaneously, while in other times the hardware platform is either only an GPU or only an NPU or only a simple calculator harnessing full power of the hardware.

5.5   Other Customizations

The opcode which tells the processor how many of the next few fetches are its operands, which do not require decoding. This opcode-specific part is not much of our concern. But the different IR size requirements can be solved by having the IR size equal to the requirements of the ISA with the greatest requirements. The decoding of the opcodes of different sizes will be optimized using a multilevel decoding, so that some of the decoders which will not be required for lower opcode sizes.

The register depth and widths can be altered by using the same register bank by keeping the individual register's size equal to the size requirement of the ISA with the lowest size requirement. When considering ISAs with larger requirements, doublets or triplets of registers will be considered as a single register. The interpretation will change with the contents of the ISA register.

6 Comparison

On successful implementation of Multi-Arc idea, it can be compared to any conventional CPU or GPU or AI-CPU on a diverse set of fields.

6.1 Based on Opcode Space and Size

For any ISA, two operations with the slightest difference call for a new opcode. In any conventional multi-core CPU there is a huge number of opcodes required, where it is seen all the opcodes are never used at once for a type of work. But this means a bigger opcode space. In Multi-Arc, all the ISAs effectively reside in the same space because all ISAs are not active at the same time. Besides, in the other types of CPUs mentioned, there are lesser opcodes per bit of opcode size to account for the variations. Multi-Arc resolves this problem as it provides a separate instruction for setting the variations. Thus, it provides a greater number of opcodes in the same opcode size.

6.2 Based on ISA Optimization

Multi-Arc allows us to optimize the different ISAs for specific purposes. In GPP, there is no scope for optimizing the ISA to suit some specific purpose. And when using one single ISA keeping multiple instructions for the same purpose increases redundancy and increases opcode size unnecessarily. Since our design uses multiple ISAs, we can easily optimize each ISA to suit different needs. We can keep different instructions for the same purpose without adding redundancy. The only constraint here is that the chosen ISAs should re-utilize the same hardware resources, and as with the addition of more ISA-specific hardware the purpose of our design gets defeated little by little.

6.3 Register File Customizability

Our design can view the same register file differently for different ISAs, or to be more specific, for different operand sizes of different ISAs. For example, a pair of 8-bit registers as seen by one ISA can be treated as a single 16-bit register by a different ISA. That is, we can trade off register width for register numbers and vice versa.

6.4   Generic Ability

Apart from just a normal customization, Multi-Arc will cost some extra overhead for configuration and switching. But this gives a complete control over the ISA and gives the user to choose whether to use as a CPU or GPU or anything else from a single hardware platform, or all of them pseudo-concurrently.

Thus, overall Multi-Arc on full ability can cater to a great deal of optimization, customizability, generalization and reusability of hardware for multiple purposes. It is also able to perform concurrent processing of different instructions at the execution stage of the pipeline itself. Every new development comes with a cost, and there will be an overhead for configuration although it will not affect normal working as the whole configuration unit is exclusive of the main data processing unit. There will be a hardware cost for memory in ISA control block and multiple decoders, but it will save the overlays for a bigger word size for same amount of functions.

7   Conclusion

To conclude Multi-Arc, it can be one of the pillars to support a new breed of processors where being powerful is not the only aspect but the quantum of efficient usage is also taken into consideration. By switching between architectures, we can definitely provide programmer a way to exploit multiple optimized instructions without actually making any of the instruction sets larger. If development proceeds along this line someday in the future, we might see extremely powerful pseudo-GPPs which have the potential to surpass the present-day GPPs. They might also find use in the scientific and research community to develop processors which can have an instruction set designed to be optimized for performing specific complex calculations and can also have the option to switch over to a more lax instruction set and perform simpler calculations that do not require much precision and resources. Meanwhile, while performing simple calculation, it can also perform complex calculations for different ISAs at the same time as each and every processing element is mutually exclusive and can be used simultaneously. This technology if implemented properly will also help general consumers as they will be able to use a single device for all the various paradigm of works, where each work can be totally different. In the era of high waste production, this idea will also contribute to lesser production waste as each piece of the tech can be reused by the user. This provides us with better and efficient use of hardware as well as a power-efficient processing device as at any instant the work is optimized at the hardware level.

References

1. Bhandarkar, D., Clark, D.W.: Performance from architecture: comparing a RISC and a CISC with similar hardware organization. In: ACM SIGARCH Computer Architecture News, vol. 19, no. 2, pp. 310–319. ACM (1991, April)
2. Dennis, J.B., Gao, G.R.: An efficient pipelined dataflow processor architecture. In: Proceedings of the 1988 ACM/IEEE Conference on Supercomputing, pp. 368–373. IEEE Computer Society Press (1988, November)
3. Kumar, R., Zyuban, V., Tullsen, D.M.: Interconnections in multi-core architectures: understanding mechanisms, overheads and scaling. In: 32nd International Symposium on Computer Architecture (ISCA'05), pp. 408–419. IEEE (2005, June)
4. Owens, J.D., Houston, M., Luebke, D., Green, S., Stone, J.E. and Phillips, J.C.: GPU computing (2008)
5. Anagnostopoulos, P.C., Michel, M.J., Sockut, G.H., Stabler, G.M., van Dam, A.: Computer architecture and instruction set design. In: Proceedings of the June 4–8, 1973, National Computer Conference and Exposition, pp. 519–527. ACM (1973, June)
6. Hanyu, T., Higuchi, T.: Design of a highly parallel AI processor using new multiple-valued MOS devices. In: [1988] Proceedings. The Eighteenth International Symposium on Multiple-Valued Logic, pp. 300–306. IEEE (1988)
7. Saghir, M.A., El-Majzoub, M., Akl, P.: Customizing the datapath and ISA of soft VLIW processors. In: International Conference on High-Performance Embedded Architectures and Compilers, pp. 276–290. Springer, Berlin (2007, January)
8. Kumar, R., Farkas, K.I., Jouppi, N.P., Ranganathan, P., Tullsen, D.M.: December. Single-ISA heterogeneous multi-core architectures: the potential for processor power reduction. In: Proceedings of the 36th Annual IEEE/ACM International Symposium on Microarchitecture, p. 81. IEEE Computer Society (2003)
9. Venkat, A., Tullsen, D.M.: Harnessing ISA diversity: design of a heterogeneous-ISA chip multiprocessor. In 2014 ACM/IEEE 41st International Symposium on Computer Architecture (ISCA) (2014)
10. Venkat, A., Basavaraj, H., Tullsen, D.: Composite-ISA cores: enabling multi-ISA heterogeneity using a single ISA. In: 25th IEEE International Symposium on High Performance Computer Architecture. IEEE (2019)
11. Wood, G.: On the packing of micro-operations into micro-instruction words. In: ACM SIGMICRO Newsletter, vol. 9, no. 4, pp. 51–55. IEEE Press (1978, November)
12. Hoffmann, A., Kogel, T., Nohl, A., Braun, G., Schliebusch, O., Wahlen, O., Wieferink, A., Meyr, H.: A novel methodology for the design of application-specific instruction-set processors (ASIPs) using a machine description language. IEEE Trans. Comput. Aided Des. Integr. Circuits Syst. **20**(11), 1338–1354 (2001)
13. Grimes, J.D., Kohn, L., Bharadhwaj, R.: The Intel i860 64-bit processor: a general-purpose CPU with 3D graphics capabilities. IEEE Comput. Graphics Appl. **9**(4), 85–94 (1989)
14. Tredennick, H.L., Gunter, T.G., Motorola Solutions Inc: Instruction register sequence decoder for microprogrammed data processor and method. U.S. Patent 4,342,078, 1982
15. Nuth, P.R., Dally, W.J.: A mechanism for efficient context switching. In: [1991 Proceedings] IEEE International Conference on Computer Design: VLSI in Computers and Processors, pp. 301–304. IEEE (1991, October)
16. Reset (computing)—Wikipedia. https://en.wikipedia.org/wiki/Reset_(computing). Last accessed 10 July 2019
17. Special function register—Wikipedia. https://en.wikipedia.org/wiki/Special_function_register. Last accessed 9 July 2019
18. Pomerene, J.H., Puzak, T.R., Rechtschaffen, R.N., Sparacio, F.J., International Business Machines Corp: Method and apparatus for guaranteeing the logical integrity of data in the general-purpose registers of a complex multi-execution unit uniprocessor. U.S. Patent 4,903,196, 1990

19. De Micheli, G., Brayton, R.K., Sangiovanni-Vincentelli, A.: Optimal state assignment for finite state machines. IEEE Trans. Comput. Aided Des. Integr. Circuits Syst. **4**(3), 269–285 (1985)
20. Almasi, G.S., Gottlieb, A.: Highly parallel computing (1988)
21. Difference between RISC and CISC Architecture|Firmcodes. http://firmcode.blogspot.com/2015/01/difference-between-risc-and-cisc.html. Last accessed 3 July 2019

Uncertainty Estimation Using Probabilistic Dependency: An Extended Rough Set-Based Approach

Indrajit Ghosh

Abstract Estimation of uncertainty in knowledge extracted from a repository of data is a challenging issue in artificial intelligence and cognitive sciences. The rough set theory is a prominent technique for data mining from a repository of domain data. A rough set is a formal approximation of a crisp set in terms of a pair of sets with respect to equivalence relations. The concept of lower and upper approximations is based on the assumption that information is associated with every object described by a set of attributes. The classical rough set theory with the equivalence class concept and static dependency cannot handle any uncertainty in attribute level which seems to be a major limitation. Several extensions on classical rough set theory have remarkable contributions in different applications. But excavation of the nature of dependency of the attributes and the estimation of uncertainty is still an unsolved issue. This paper proposes an extension in the classical rough set theory by introducing probabilistic relevancy of the attributes. Mining the correlation between attributes in terms of the probabilistic dependency coefficient is an important way out in discovering the uncertainty. The roughness in contributing knowledge by an attribute can be represented using approximations in the probabilistic dependency space. Using the approximations in the values of the probabilistic dependency coefficient, the inherent uncertainty can be estimated and handled more precisely.

Keywords Uncertainty estimation · Probabilistic dependency · Extended rough set theory

1 Introduction

Estimation of uncertainty is a challenging issue in artificial intelligence and cognitive sciences. Several machine learning (ML) techniques have been suggested for different domains like agriculture, health care, engineering, bioinformatics, etc. In most of the cases, the domain knowledge is represented as the repository of the set of

---

I. Ghosh (✉)
Department of Computer Science, Ananda Chandra College, Jalpaiguri 735101, India
e-mail: ighosh2002@gmail.com

© Springer Nature Singapore Pte Ltd. 2020
J. K. Mandal and S. Mukhopadhyay (eds.), *Proceedings of the Global AI Congress 2019*, Advances in Intelligent Systems and Computing 1112,
https://doi.org/10.1007/978-981-15-2188-1_43

objects having a set of attributes. But each attribute may not have equal contribution and certainty in representing a real object. Estimation of relevancy and uncertainty in attribute level is a prime area of research for object-based intelligent systems.

The rough set theory proposed by Pawlak [1, 2] is a prominent technique of knowledge extraction from the repository of domain data. A rough set is a formal approximation of a crisp set in terms of a pair of sets with respect to equivalence relations. The concept of lower and upper approximations is based on the assumption that information is associated with every object described by a set of attributes. But the classical rough set theory with the equivalence class concept cannot handle any uncertainty in the attribute level [3, 4].

As a result, several important techniques in different dimensions have been proposed as the extensions of the basic rough set theory to extract knowledge from experimental or real domain data [4]. Several extensions have been attempted for better handling of the uncertainty in practical applications. Prof. Wong and Ziarco proposed a probabilistic model by introducing the probability approximation space into the classical rough set theory [5]. Three variants of this model were suggested for further improvement. These were the decision-theoretic rough set (DTRS) model [6–8], the variable precision rough set (VPRS) model [9, 10] and the Bayesian rough set model [11, 12]. The main problem associated with the DTRS model was that the loss functions used for boundary region are arbitrarily provided by the user [13, 14]. A remarkable extension was proposed as game-theoretic rough sets (GTRS) [15]. GTRS model incorporated the uncertainty levels of positive, negative, and boundary regions with pairs of threshold values. But the determination of the optimal thresholds is a critical problem to be solved. The variable precision rough set (VPRS) model proposed by Prof. Ziarko was a generalized model where a subset of operators was relaxed to minimize the degree of failure. It was an effective approach to deal with noisy data and uncertain information based on the rough membership of objects. But one of the major demerits of this model was that the rough membership was based only on the static set of objects [9, 10]. As a generalization to VPRS model, a frequency distribution-based parameterized rough set (PRS) model was proposed [16]. It considered the absolute and relative rough membership in object level but not in the attribute level. Bayesian rough set model and the Stochastic dominance-based rough set [17] models were designed for improving classification performance based on data itself. An important extension was proposed by Greco et al. [18] where the equivalence relation of classical rough set theory was replaced by the dominance relation. To deal with the incomplete system, the tolerance-based rough set model was proposed by Kryszkiewicz where a similarity relation to minimize data replaced the indiscernibility relation [19, 20]. Undoubtedly, the several extensions on classical rough set theory have remarkable contributions in different applications. But no effective extension has been suggested for the estimation of dependency of attributes and the uncertainty in a dynamic decision system. It is still an unsolved problem.

This paper suggests a way out for uncertainty estimation in a decision system by applying the probabilistic relevance of attributes expressed in terms of probabilistic dependency coefficient. Replacement of static dependency coefficient with the probabilistic relevance of attributes, proposed by Ghosh [21], can be an effective extension in classical rough set theory. The roughness in contributing knowledge by an attribute can be represented using approximations in probabilistic dependency space. The proposed approximations in probabilistic dependency space capture the uncertainty in deeper details so that the uncertainty of a decision system can be estimated more precisely.

The next section describes the proposed extension over classical rough set theory. Section 3 describes the architecture and mathematical foundations of the extension. Several experiments have been conducted to validate the theoretical model. Due to the limitation of space, only one of them has been chosen arbitrarily and presented in Sect. 4, as a case study. Finally, the conclusions are summarized in Sect. 5.

2 Proposed Extension

2.1 Dynamic Decision System

The main goal of the classical rough set theory is to synthesize approximations of concepts from acquired data structured in a static table, called an information system. In rough set literature, an information system is a pair $I = (U, A)$ where U is a non-empty finite set of objects called the universe of discourse and A is a non-empty finite set of attributes. With every attribute $a \in A$, a set Va is associated which is known as the domain of a. A decision system is basically an information system where the set of attributes are distinguished in two disjoint classes called the set of condition attributes C and the set of decision attributes D, respectively. Thus, the tuple $S = (U, C, D)$ is called a decision system.

In the classical rough set theory, the ability of classification is based on the completeness of a static set of objects known as the universe of discourse of a given decision system where the cardinality possess a static value. The universe of discourse is not able to incorporate any further objects to capture the inherent dynamicity present in several application domains.

This paper proposes a dynamic decision system as an extension of the static universe of discourse as stated in the classical rough set theory. The system is designed for the addition or deletion of objects and the cardinality changes accordingly. The dynamic decision system consists of a set of objects representing each type of target classes. Examples may be collected from the experimental domain or from some benchmark datasets. Each object representing an example is represented with a set of condition attributes and the corresponding decision attribute. To avoid biasness, objects are sequentially added in the corresponding decision system as first come first serve basis.

2.2 *Probabilistic Dependency*

Estimation of dependencies between attributes of a decision system is one of the most significant issues in data analysis. In classical rough set literature, the set of decision attributes D depends totally on a set of attributes C, denoted as $C \Rightarrow D$, if all attribute values from D are uniquely determined by values of attributes from C. In general, the dependency of D on C is expressed in terms of degree of dependency $k (0 \leqslant k \leqslant 1)$ [4, 20].

For (C, D), it is said that D depends on C in a degree k, denoted $C \Rightarrow_k D$, if

$$k = \frac{|\text{POSc(D)}|}{|U|}$$

where

$$\text{POSc}(D) = \bigcup_{X \in U/D} C.(X)$$

and called the positive region of the partition U/D with respect to C.

If $k = 1$, then D depends totally on C, and if $k < 1$, D depends partially in a degree k on C. The coefficient k expresses the ratio of all elements of the universe, which can be properly classified to blocks of the partition U/D [4, 20].

It is to be noted that, the value of the degree of dependency k is derived from the static universe of discourse. For real field applications, the dependency between attributes is application specific and dynamic. Thus, the static dependency coefficient k is not appropriate to reflect the dynamicity of the dependency among the attributes of a dynamic decision system. The static dependency coefficient should be redefined and reformulated to capture the dynamicity inherent in application domains.

As an alternative to the static classical degree of dependency, the concept of probabilistic dependency has been incorporated as an extension to the classical rough set theory. The basic conceptual extension is that the probabilistic dependency is dynamic by nature as the inherent knowledge of a decision system may change with new observation. The probabilistic relevance function (Eq. 1), proposed by Ghosh [21], is applied to estimate the probabilistic dependency of attributes. The probabilistic relevance of each attribute is expressed in terms of a parameter, called *Probabilistic Dependency Coefficient (p)*. It is a function of both, the subset of conditional attributes for each observation as well as the cardinality of the decision system.

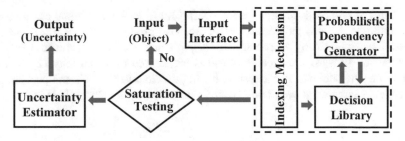

Fig. 1 System architecture

3 Methodology

3.1 Architecture

The architecture of the system is presented in Fig. 1. The central part of the system is designed with the three components; the decision library, an indexing mechanism, and a probabilistic dependency coefficient generator. The decision library consists of a set of decision systems. Each decision system consists of the target decision attribute and a set of condition attributes along with the corresponding values of probabilistic dependency coefficients for each target class. An input interface is designed to append the objects to the system. When an object is inserted through the input interface, the indexing mechanism selects the respective decision system in the decision library. For each addition, the probabilistic dependency coefficient generator measures the values of the probabilistic dependency coefficients of each attribute of the respective decision system. The saturation testing component tests whether the saturation is obtained or more objects are required for saturation. When the system becomes saturated, the uncertainty estimator evaluates both the uncertainty associated with each attribute and the uncertainty of the decision system as a whole. The estimated uncertainties are the outputs.

3.2 Mathematical Foundation

The dynamic decision system consists of m number of possible conditional attributes where the cardinality increases with the addition of new objects. When a new object appears, the initial values of probabilistic dependency coefficient of attributes of the object are not known. So, an unbiased value of probabilistic dependency coefficient β is assigned to each observed attribute. This unbiased probabilistic dependency coefficient can be estimated as $\beta = (1/b)$, where $b \leq m$, is the number of attributes present in the new object. When the object is added in the corresponding decision system, the evaluation function for probabilistic dependency coefficient (p) measures the values of probabilistic dependency coefficient of each conditional attribute corresponding to the targeted decision attribute. This process continues till saturation. Saturation is defined by a probabilistic dependency coefficient threshold $\Delta(p)$ in the order of 10^{-3}, as the accuracy level is considered to be sufficient up to three digits after the decimal. For a system having cardinality $n = |U|$, the probabilistic dependency coefficient evaluation function, as proposed by Ghosh [21] is:

$$p_{n,i} = \frac{\gamma_{n,i}}{n.m} \times \frac{\sum_{j=1}^{n} \beta_{j,i}}{\sum_{i=1}^{m} \beta_{n,i}}$$

where $\gamma_{(n,i)}$ is the frequency of the ith attribute of the system with cardinality n.

For ith attribute, the updating factor $\Delta(p_i)$ for each iteration is given as:

$$\Delta(p_i) = p_{i(new)} - p_{i(old)}$$
$$\text{or} \quad \Delta(p_i) = p_{n,i} - p_{(n-1),i}$$

For experimental validation, this model has been applied to multiple standards and benchmarked decision systems. One of the outcomes of the experiments is presented in the next section. A curve $(n\text{-}p)$ is obtained when values of the probabilistic dependency coefficient (p) of an attribute are plotted in vertical axis against the cardinality (n) of the decision system in horizontal axis. The $(n\text{-}p)$ curves of five attributes of an example are shown in Figs. 2, 3, 4, 5, and 6.

A large fluctuation (roughness) is observed in the lower cardinality region, indicating the instability of the system. But as the cardinality increases, the system gradually becomes stable and gives an almost linear region parallel to the cardinality axis. This linear region is the saturation region (Fig. 2).

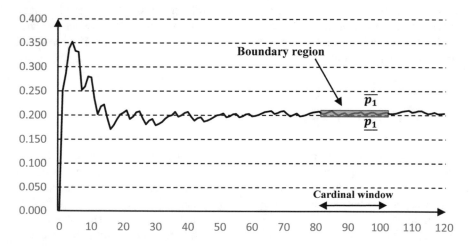

Fig. 2 $n\text{-}p$ curve of the first attribute a_1

Fig. 3 *n-p* curve of attribute a_2

Fig. 4 *n-p* curve of attribute a_3

Fig. 5 *n-p* curve of attribute a_4

Fig. 6 *n-p* curve of attribute a_5

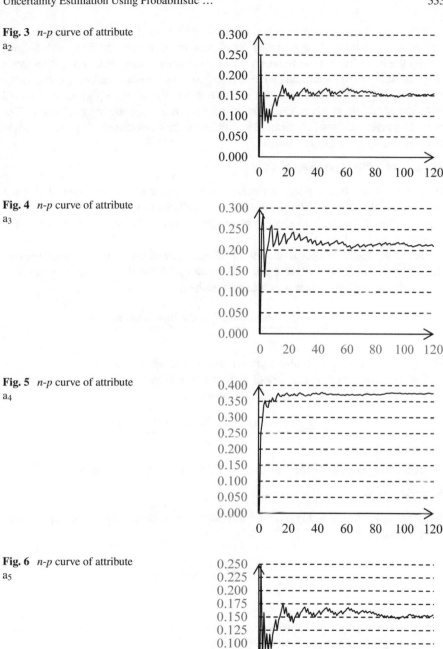

When the decision system becomes saturated with sufficient examples, the values of the probabilistic dependency coefficient of each attribute become confined within the boundary region between lower and upper approximations, respectively (Fig. 2). Several small minima and maxima appear in the saturation region due to the roughness in probabilistic dependency. This boundary region (roughness) is obtained with lower and upper approximations of probabilistic dependency coefficient. The boundary region is used for estimation of uncertainties associated with each attribute and the decision system as a whole.

Lower and upper approximations

A concept of cardinal window has been incorporated to define the lower and upper approximations of probabilistic dependency coefficient of each attribute on its cardinality axis. A cardinal window is basically a span on the cardinality axis in the saturation region (Fig. 2).

Let $\underline{p_i}$ be the lower approximation of dependency of ith attribute within the cardinal window in the saturation area, then $\underline{p_i}$ can be defined as the arithmetic mean of all minima present within the cardinal window:

$$\underline{p_i} = \frac{\sum (\text{minima within the cardinal window})}{x}$$

where x is the number of minima present within the cardinal window.

Similarly, the upper approximation $\bar{p_i}$ can be defined as the arithmetic mean of all maxima obtained within the cardinal window:

$$\bar{p_i} = \frac{\sum (\text{maxima within the cardinal window})}{y}$$

where y is the number of maxima present within the cardinal window.

Boundary region

As per rough set terminology, for ith attribute, the boundary region of dependency is defined as:

$$\boldsymbol{B}_i = \left(\bar{p_i} - \underline{p_i} \right)$$

Uncertainty

Thus, the percentage of uncertainty associated with the ith conditional attribute leading to the decision attribute is estimated as:

$$u_i = \frac{\left(\bar{p_i} - \underline{p_i} \right)}{\underline{p_i}} \times 100$$

4   Validation with Experimental Data

For validation, the proposed model has been applied on different benchmarked and consistent decision systems obtained from field experiments. An application of the model on an experimental and published data set of tea cultivation [22] is presented as an example.

Tea is a major crop in Darjeeling and Jalpaiguri district of India. As tea is a perennial plantation crop, tea plants are subject to the infestation of various insect pests. When a tea plant is infested by an insect pest, some definite sign and symptoms are observed which lead to the identification of that particular insect pest. A decision system can be designed with the observed set of attributes (sign and symptoms) as the conditional attributes and the insect pest as the decision attribute. As the observed attributes are inserted throughout the year, it is a good example of a dynamic decision system. As an example, *Red spider* is a major insect pest of tea which makes considerable damage to the crop. The specific sign and symptoms considered as the conditional attributes for the identification of *Red spider* infestation are presented in Table 1.

Multiple sets of attributes leading to the identification of *Red spider* infestation have been inserted in the *Red spider* decision system sequentially as first come first serve bases to avoid any biasness. The probabilistic dependency coefficient function (Eq. 1) measures the values of probabilistic dependency coefficient of all five conditional attributes (a_1–a_5). As an example, the values of probabilistic dependency coefficient (p_1) obtained for the first attribute a_1 is presented in Table 2.

The values of probabilistic dependency coefficient p_1 are plotted against the cardinality n of the decision system. This (n-p) curve obtained is shown in Fig. 2.

Now, let us calculate the approximations, boundary, and the uncertainty associated with the first attribute a_1.

Let us consider that the left boundary of the cardinal window is at 81 and the right boundary is at 100, as it is within the saturation region (see Fig. 2). Thus, the width of the cardinal window is 20.

Table 1 *Red spider* decision system	Conditional attributes	Technical term and value domain	Decision attribute
	a_1	Fingertip test: (*Red smears*)	*Red spider*
	a_2	Bush appearance: (*New flush stunted*)	
	a_3	Site of damage: (*Matured leaf*)	
	a_4	Spot on leaf: (*Radish brown*)	
	a_5	Leaf color: (*Brown*)	

Table 2 Values of the probabilistic dependency coefficient p_1 of attribute a_1 with increasing cardinality n

n	p_1	n	p_1	n	p_1	n	p_1	n	p_1	n	p_1	n	p_1	n	p_1
1	0.250	16	0.171	31	0.182	46	0.189	61	0.199	76	0.199	91	0.203	106	0.206
2	0.286	17	0.179	32	0.186	47	0.192	62	0.201	77	0.201	92	0.205	107	0.208
3	0.338	18	0.191	33	0.193	48	0.196	63	0.204	78	0.204	93	0.205	108	0.209
4	0.352	19	0.201	34	0.198	49	0.200	64	0.207	79	0.206	94	0.207	109	0.210
5	0.333	20	0.205	35	0.200	50	0.201	65	0.208	80	0.207	95	0.202	110	0.206
6	0.331	21	0.210	36	0.207	51	0.204	66	0.209	81	0.208	96	0.203	111	0.207
7	0.252	22	0.192	37	0.197	52	0.196	67	0.203	82	0.203	97	0.206	112	0.209
8	0.259	23	0.197	38	0.200	53	0.198	68	0.205	83	0.204	98	0.206	113	0.209
9	0.280	24	0.206	39	0.205	54	0.202	69	0.208	84	0.207	99	0.202	114	0.206
10	0.278	25	0.208	40	0.207	55	0.203	70	0.209	85	0.209	100	0.203	115	0.203
11	0.234	26	0.193	41	0.197	56	0.196	71	0.203	86	0.204	101	0.205	116	0.204
12	0.202	27	0.181	42	0.189	57	0.198	72	0.198	87	0.200	102	0.206	117	0.206
13	0.218	28	0.189	43	0.194	58	0.202	73	0.201	88	0.203	103	0.203	118	0.203
14	0.222	29	0.192	44	0.196	59	0.203	74	0.204	89	0.205	104	0.203	119	0.203
15	0.194	30	0.179	45	0.187	60	0.198	75	0.198	90	0.198	105	0.200	120	0.204

The lower approximation $\underline{p_1}$ is (from Table 2):

$$\underline{p_1} = \frac{\sum(0.203 + 0.200 + 0.200 + 0.202 + 0.202)}{5} = 0.201$$

The upper approximation $\overline{p_1}$ is:

$$\overline{p_1} = \frac{\sum(0.208 + 0.209 + 0.205 + 0.207 + 0.206)}{5} = 0.207$$

The boundary region (roughness) of dependency coefficient for attribute a_1 is:

$$B_1 = \left(\overline{p_1} - \underline{p_1}\right) = (0.207 - 0.201) = 0.006$$

Thus, the percentage of uncertainty associated with the attribute a_1, leading to the decision attribute *Red spider* is:

$$u_1 = \frac{\left(\overline{p_1} - \underline{p_1}\right)}{\underline{p_1}} \times 100 = \frac{0.006}{0.201} \times 100 = 2.985$$

Similarly, the values of probabilistic dependency coefficient for other four attributes (a_2 to a_5) are calculated and plotted against the cardinality. The *n-p* curves for other four attributes are shown in Figs. 3, 4, 5, and 6.

The approximations, boundary regions, and the uncertainties of the other four attributes are obtained and presented altogether with first attribute in Table 3.

The total uncertainty of the *Red spider* decision system is $u = \Sigma u_i = 9.062$.

Table 3 Approximations, boundary regions, and uncertainties of the five attributes of the *Red spider* decision system

Conditional attributes	Lower approximation	Upper approximation	Boundary region	Uncertainty (%)
a_1	0.201	0.207	0.006	2.985
a_2	0.214	0.218	0.004	1.869
a_3	0.214	0.217	0.003	1.402
a_4	0.372	0.375	0.003	0.806
a_5	0.150	0.153	0.003	2.000
Total uncertainty of the decision system = 9.062				

5   Conclusion

The proposed work is a thriving extension in classical rough set theory for estimation of uncertainty in attribute level. It makes the rough set theory more pertinent and adequate for a dynamic domain where the decision systems gradually become enriched with real-world examples.

The proposed probabilistic dependency coefficient evaluation function p is derived from the probabilistic point of view. Both the unbiased probability and the conditional probability of each and every attribute are considered for derivation. Probabilistic dynamic dependency coefficient can be treated as a promising parameter for estimation of uncertainty inherent in a stored data. It suggests a way to understand the significance and characteristics of the conditional attributes in defining the decision attribute of a decision system. This model is also capable to estimate the uncertainty of real-valued attributes expressed in the linguistic term. For example, the fourth attribute (a_4) is the most significant attribute with the lowest uncertainty.

This model proposes a way to measure the quality of a decision system based on the total percentage of inherent uncertainty. The proposed system has been applied to various standard data sets obtained from the UCI data repository and definite real field domains which are inherently dynamic. As the system gets trained with standard and real field examples, no heuristics is concerned with uncertainty estimation. The only uncertainty present may be associated with the finding of correct examples. For better performance, more intensive care should be taken to collect field examples.

Another useful feature of this model is that the saturation region suggests whether the decision system is stable or not based on the collected objects. The probabilistic dependency coefficient of each attribute may also help to design the dynamic decision algorithm for better classification.

Due to space limitations, the final form of the dynamic relevancy coefficient function and only one case study has been presented. The future efforts will be to conduct a comparative study for performance evaluation with other existing popular techniques.

References

1. Pawlak, Z.: Rough sets. Int. J. Comput. Inf. Sci. **11**(5), 341–356 (1982)
2. Pawlak, Z.: Rough Sets: Theoretical Aspects of Reasoning about Data. Kluwer Academic Publishing (1991)
3. Zhang, Q., Xie, Q., Wang, G.: A survey on rough set theory and its application. CAAI Trans. Intell. Technol. **1**, 323–333 (2016)
4. Shen, Q., Jensen, R.: Rough sets, their extensions and applications. Int. J. Autom. Comput. **04**(1), 100–106 (2007)
5. Wong, S.K.M., Ziarko, W.: Comparison of the probabilistic approximate classification and the fuzzy set model. Int. J. Fuzzy Sets Syst. **21**(3), 357–362 (1987)
6. Yao, Y.Y., Wong, S.K.M., et al.: A decision theoretic framework for approximating concepts. Int. J. Man-Machine Stud. **37**(6), 793–809 (1992)

7. Yao, Y.Y.: Probabilistic approaches to rough sets. Expert Syst. **20**(5), 287–297 (2003)
8. Yao, Y.Y.: Information granulation and approximation in a decision theoretic model of rough sets. In: Rough-Neural Computing, pp. 491–516. Springer, Berlin, Heidelberg (2004)
9. Katzberg, J.D., Ziarko, W.: Variable precision rough sets with asymmetric bounds. In: International Workshop on Rough Sets and Knowledge Discovery. Rough Sets, Fuzzy Sets and Knowledge Discovery, pp. 167–177, Springer (1993)
10. Ziarko, W.: Variable precision rough sets model. J. Comput. Syst. Sci. **46**(1), 39–59 (1993)
11. Slezak, D., Ziarko, W.: The investigation of the Bayesian rough set model. Int. J. Approx. Reason **40**(40), 81–91 (2005)
12. Greco, S., Matarazzo, B., Slowinski, R.: Rough membership and Bayesian confirmation measures for parameterized rough sets. In: Slezak, D. et al. (eds.) RSFDGrC, LNAI, vol. 3641, pp. 314–324 (2005)
13. Herbert, J.P., Yao, J.T.: Game-theoretic risk analysis in decision-theoretic rough sets. In: Wang, G., Li, T., Grzymala-Busse, J.W., Miao, D., Skowron, A., Yao, Y. (eds.) RSKT 2008, LNCS (LNAI), vol. 5009, pp. 132–139. Springer, Heidelberg (2008)
14. Zhou, X., Li, H.: A multi-view decision model based on decision-theoretic rough set. In: Wen, P., et. al. (eds.) RSKT 2009, LNCS (LNAI), vol. 5589, pp. 650–657. Springer, Heidelberg (2009)
15. Herbert, Joseph P., Yao, Jing Tao: Game-theoretic rough sets. Fundamenta Informaticae **108**, 267–286 (2011)
16. Greco, S., Matarazzo, B., Slowinski, R.: Rough membership and Baysian confirmation measures for parameterized rough sets. In: Rough Sets, Fuzzy Sets, Data Mining and Granular Computing, pp. 285–300. Springer, Berlin, Heidelberg (2005)
17. Kotlowski, W., Dembezynski, K., Greco, S., Slowinski, R.: Stochastic dominance-based rough set model for ordinal classification. Inf. Sci. **178**(21), 4019–4037 (2008)
18. Greco, S., Matarazzo, B., Slowinski, R.: Rough sets theory for multicriteria decision analysis. Eur. J. Oper. Res. **129**(1), 1–47 (2001)
19. Kryszkiewiez, M.: Rough set approach to incomplete information Systems. Inf. Sci. **112**(1–4), 39–49 (1998)
20. Swiniarski, R.W., Skowrm, A.: Rough set methods in feature selection and recognition. Pattern Recogn. Lett. **24**, 833–849 (2003)
21. Ghosh, I.: Probabilistic feature selection in machine learning. In: ICAISC 2018, LNCS, vol. 10841, pp. 623–632. Springer (2018)
22. Ghosh, I., Samanta, R.K.: Teapest: an expert system for insect pest management in tea. Appl. Eng. Agric. **19**(5), 619–625 (2003)

Extension of TOPSIS and VIKOR Method for Decision-Making Problems with Picture Fuzzy Number

Amalendu Si, Sujit Das and Samarjit Kar

Abstract In multiple criteria decision-making (MCDM) methods, the criteria are often conflicting, uncertain, imprecise and non-commensurable. Due to the presence of neutral membership degree, picture fuzzy numbers (PFN) have been successfully used to express uncertain and imprecise information. To deal with conflicting and non-commensurable criteria in MCDM problems, both VIKOR and TOPSIS methods are proven to be efficient. VIKOR and TOPSIS methods are applied in compromise ranking method which is based on aggregation functions representing closeness to the reference points. VIKOR method finds compromise solution whereas TOPSIS method finds a solution which has the shortest distance to the ideal solution and greatest distance from the negative ideal solution. This paper extends VIKOR and TOPSIS methods to solve MCDM problems using picture fuzzy information. Finally, a comparative analysis of these two methods is provided using an illustrative example.

Keywords Picture fuzzy set · VIKOR · TOPSIS · MCDM

1 Introduction

Fuzzy set (FS) was introduced by Zadeh [1] in 1965, where the degree of belongingness of an element in the set is represented by a membership function which in turn defined by a single membership value within the interval zero and one. To consider non-membership degree along with the membership degree of an element in the set, Atanassov [2] extended FS into the intuitionistic fuzzy set (IFS). Although fuzzy

A. Si
Department of CSE, Mallabhum Institute of Technology, Bishnupur 722122, India
e-mail: amalendu.si@gmail.com

S. Das (✉)
Department of Computer Science and Engineering, NIT Warangal, Warangal 506004, India
e-mail: sujit.das@nitw.ac.in

S. Kar
Department of Mathematics, NIT Durgapur, Durgapur 713209, India
e-mail: kar_s_k@yahoo.com

© Springer Nature Singapore Pte Ltd. 2020
J. K. Mandal and S. Mukhopadhyay (eds.), *Proceedings of the Global AI Congress 2019*, Advances in Intelligent Systems and Computing 1112,
https://doi.org/10.1007/978-981-15-2188-1_44

set was introduced as a pioneering idea to deal with the imprecise and incomplete information, still it is difficult to solve many real-life decision problems [3–5] with the help of simply fuzzy sets due to the continuous increment of complexity in the available data set which are obtained from ambiguous and incomplete source of information, lack of statistical analysis of the data set, decision makers experience and knowledge and impact of surroundings, etc. On the other hand, IFS has been established as a strong mathematical tool to evaluate the imprecise or uncertain kind of decision-making information and handle the vagueness in the static or dynamic decision-making problem [6, 7]. As a result, IFS has been widely used in different emerging and active field like logic programming [8], decision making [3–5, 9], pattern reorganization [10], medical diagnosis [9] and cluster analysis [11, 12]. The main importance of the IFS is that it assigns the membership and non-membership value of each element within set. Irrespective of this strength, still it is difficult to represent a situation with IFS when the neutral and refusal degrees are considered with the elements of a set. Recently, a new generalization of FS and IFS, named as picture fuzzy set (PFSs), was proposed in [13]. PFS-based approaches are more suitable in those cases when our opinions involve more options type like yes, abstain, no and refusal. One good example to describe PFS is general election, where a voter can cast his vote for the candidate (yes), against the candidate (no), may not cast his vote (abstain), or may refuse to cast his vote for the candidates and cast in nota (refusal) [14].

Since its introduction in 2014, many researchers have been contributing to the development of decision-making problem using PFS. Some significant research works on PFSs are narrated below. By extending the intuitionistic fuzzy aggregation operator, Wang et al. [15] defined the picture fuzzy aggregation operator. Si et al. [16] proposed a ranking approach for PFNs. In [17], the authors extended the VIKOR method where the criteria information is represented in stochastic format. The extended version manages a number of stochastic criteria and their weight is calculated with the help of fuzzy analytic hierarchic process (AHP). This technique tactfully handles the ambiguity in measurement over the conventional VIKOR procedure. In this article [18], the authors presented MCDM technique in the fuzzy environment. In this approach, the decision makers can get more flexibility to provide the opinion in linguistic terms. Another MCDM method was developed in [19] for risk evaluation in fuzzy mode to reduce the margins of PFS in different way. The risk assessment information was embodied using the PFSs to mitigate information loss during decision making. This article also presented the PFSs projection model into picture fuzzy normalized projection (PFNP) model to manage the limitation of PFSs projection. The weight vectors of criteria are estimated by recommended entropy weight technique. The linguistic picture fuzzy TOPSIS method was extended by Zeng et al. [20] who introduced the Hamming distance between two linguistic PFNs and defined different properties.

Decision making [21–23] is the process to determine the best feasible solution based on the set of criteria and alternatives regarding the problem. Earlier this process was simple and easy but day by day the real-life problems are becoming more complex as multiple numbers of conflicting criteria and constraints are involving with the problem. Sometimes, the criteria of the problems are non-commensurable and there may be no feasible solution that satisfying all the criteria simultaneously. Hence, the solution becomes like a set of nonstandard solutions or a compromise solution according to adjustment or compromise of the user requirement. To manage this kind of situation, VlseKriterijumska Optimizacija I Kompromisno Resenje (VIKOR) method was developed by Opricovic and Tzeng in 2004 [24–26] as a MCDM method. VIKOR [27–30] method produces a compromise solution, which is achieved by the mutual understanding which is closest to the ideal solution. Some significant studies on VIKOR method are found in [29, 31, 32]. Technique for Order of Preference by Similarity to Ideal Solution (TOPSIS) was introduced by Hwang and Yoon in 1981 [33] as another MCDM method which is based on finding the smallest distance from the positive ideal solution (PIS) and the largest distance from the negative ideal solution (NIS) [34]. However, a few studies are found which extend the TOPSIS and VIKOR method using PFNs. Compromise ranking methods like VIKOR and TOPSIS are considered to have more importance when the decision makers prescribe their preferences using more options like "yes", "abstain", "no" and "refusal", and PFS provides the platform to include such situations in decision-making paradigm. Therefore, an extension of TOPSIS and VIKOR using PFN becomes obvious.

In this paper, we extend the VIKOR and TOPSIS method in the framework of PFS to develop a new methodology called PFS-VIKOR and PFS-TOPSIS method to solve MCDM problems in which the assessment values of alternatives are represented in picture fuzzy number. Both the extended method is applied to rank the tiger reserves, which are considered by the expert groups based on five predefined criteria. Remaining of the paper is structured as follows. Section 2 recalls some basic concepts and operators related to picture fuzzy sets. Section 3 presents the traditional method of VIKOR and TOPSIS. In Sect. 4, we extend VIKOR and TOPSIS method using PFNs. Section 5 presents an illustrative example followed by the discussion of results in Sect. 6. We give the key conclusions in Sect. 7.

2 Preliminaries

In this section, we briefly discuss the relevant ideas. A picture fuzzy set (PFS) C on the universe Z is an object in the form of

$$C = \{ (z, \mu_C(z), \eta_C(z), \nu_C(z)) | z \in Z \} \tag{1}$$

where $\mu_C(z) \in [0, 1]$ is called the degree of positive membership of z in C, similarly $\eta_C(z) \in [0, 1]$ and $\nu_C(z) \in [0, 1]$ are called the degrees of neutral and negative membership of z in C respectively. These three parameters $(\mu_C(z), \eta_C(z)$ and $\nu_C(z))$ of the picture fuzzy set C satisfy the following condition $\forall z \in Z, 0 \le \mu_C(z) + \eta_C(z) + \nu_C(z) \le 1$.

Then, the refusal membership degree $\rho_C(z)$ of z in C can be estimated accordingly,

$$\forall z \in Z, \rho_C(z) = 1 - (\mu_C(z) + \eta_C(z) + \nu_C(z)) \tag{2}$$

The neutral membership ($\eta_C(z)$) of z in C can be considered as the degree of positive membership as well as the degree of negative membership whereas refusal membership ($\rho_C(z)$) can be explained as not to take care of the system. When $\forall z \in Z, \eta_C(z) = 0$, then the PFS reduces into IFS.

For a fixed $z \in C$, $(\mu_C(z), \eta_C(z), \nu_C(z), \rho_C(z))$ is called picture fuzzy number (PFN), where $\mu_C(z) \in [0, 1]$, $\eta_C(z) \in [0, 1]$, $\nu_C(z) \in [0, 1]$, $\rho_C(z) \in [0, 1]$ and

$$\mu_C(z) + \eta_C(z) + \nu_C(z) + \rho_C(z) = 1 \tag{3}$$

For the sake of simplicity, we will present PFN as $(\mu_C(z), \eta_C(z), \nu_C(z))$.
For two PFS C and D, then some operations are defined as follows [17, 35]

1. $C \subseteq D$ If $(\forall z \in Z, \mu_C(z) \leq \mu_D(z), \eta_C(z) \leq \eta_D(z), \nu_C(z) \geq \nu_D(z))$ \hfill (4)

2. $C = D$ If $(C \subseteq D$ and $C \subseteq D)$ \hfill (5)
3. $C \cup D = \{(z, \max(\mu_C(z), \mu_D(z)), \min(\eta_C(z), \eta_D(z)), \min(\nu_C(z), \nu_D(z)))|z \in Z\}$ \hfill (6)
4. $C \cap D = \{(z, \min(\mu_C(z), \mu_D(z)), \min(\eta_C(z), \eta_D(z)), \max(\nu_C(z), \nu_D(z)))|z \in Z\}$ \hfill (7)

5. $comC = \overline{C} = \{(z, \nu_C(z), \eta_C(z), \mu_C(z))|z \in Z\}$ \hfill (8)

According to [36] some properties of PFS of these operations are follows:

1. If $C \subseteq D$ and $D \subseteq F$ then $C \subseteq F$;
2. $\overline{\overline{C}} = C$;
3. Operations $\cup$ and $\cap$ are commutative, associative and distributive;
4. Operations $\cup$, com and $\cap$ satisfy the De Morgan's law;

Distance between picture fuzzy sets
Distances between the two PFSs were defined by Cuong [13, 14]. Let C and D in $Z = \{z_1, z_2, \ldots, z_n\}$ the distance between C and D is calculated following ways:

1. Normalized Hamming distance

$$d_H(C, D) = \frac{1}{n} \sum_{i=1}^{n} (|\mu_C(z_i) - \mu_D(z_i)| + |\eta_C(z_i) - \eta_D(z_i)| + |\nu_C(z_i) - \nu_D(z_i)|) \tag{9}$$

2. Normalized Euclidean distance

$$d_E(C, D) = \sqrt{\frac{1}{n}\sum_{i=1}^{n}\left((\mu_C(z_i) - \mu_D(z_i))^2 + (\eta_C(z_i) - \eta_D(z_i))^2 + (\nu_C(z_i) - \nu_D(z_i))^2\right)}$$

(10)

Operation on PFSs

Let $\alpha = (\mu_\alpha, \eta_\alpha, \nu_\alpha)$ and $\beta = (\mu_\beta, \eta_\beta, \nu_\beta)$ be two picture fuzzy numbers, then

1. $\alpha.\beta = \left((\mu_\alpha + \eta_\alpha)(\mu_\beta + \eta_\beta) - \eta_\alpha\eta_\beta, \eta_\alpha\eta_\beta, 1 - (1 - \nu_\alpha)(1 - \nu_\beta)\right)$ (11)

2. $\alpha^\lambda = \left((\mu_\alpha + \eta_\alpha)^\lambda - \eta_\alpha^\lambda, \eta_\alpha, 1 - (1 - \nu_\alpha)^\lambda\right), \quad \lambda > 0$ (12)

Picture fuzzy weighted geometric operators

Let $p_j = (\mu_j, \eta_j, \nu_j, \rho_j)$ $(j = 1, 2, \dots, n)$ be a collection of PFNs, then the picture fuzzy weighted geometric (PFWG) operator [14] is defined as

$$\text{PFWG}_w(p_1, p_2, \dots, p_n) = \prod_{j=1}^{n} p_j^{w_j}$$

$$= \left(\prod_{j=1}^{n}(\mu_j + \eta_j)^{w_j} - \prod_{j=1}^{n}\eta_j^{w_j}, \prod_{j=1}^{n}\eta_j^{w_j}, 1 - \prod_{j=1}^{n}(1 - \nu_j)^{w_j}\right)$$ (13)

where $w = (w_1, w_2, \dots, w_n)$ be the weight vector of p_j $(j = 1, 2, \dots, n)$ and $w_j > 0$ and $\sum_{j=1}^{n} w_j = 1$.

Picture fuzzy weighted averaging operators

Let $\alpha_j (j = 1, 2, \dots, n)$ be set of PFNs, then the picture fuzzy weighted averaging operator (PFWA) [17] is defined as

$$\text{PFWA}(\alpha_1, \alpha_2, \dots, \alpha_n) = \omega_1\alpha_1 \oplus \omega_2\alpha_2 \oplus \cdots \omega_n\alpha_n$$

$$= \left\{\left(1 - \prod_{j=i}^{n}(1 - \mu)^{\omega_j}, \prod_{j=i}^{n}\eta^{\omega_j}, \prod_{j=i}^{n}\nu^{\omega_j}\right)\right\}$$ (14)

Where $\omega = (\omega_1, \omega_2, \dots, \omega_n)^{\text{T}}$ be the weighted vector of α_j such that $\omega_1 > 0$ and $\sum_{j=1}^{n} \omega_j = 1$.

3   VIKOR and TOPSIS Method

VIKOR method implies more effort on ranking and selecting from a set of alternatives with the existence of conflicting and non-commensurable criteria, which are common in real-life problems. For these kinds of criteria, no perfect solution is possible that

satisfy all the requirements, rather a compromise solution is feasible, which is close
to the ideal solution. VIKOR method was developed based on the following Lp metric

$$L_{p,i} = \left\{ \sum_{j=1}^{n} \left[w_j \left(f_j^* - f_{ij} \right) / \left(f_j^* - f_j^- \right) \right]^p \right\}^{1/p} \quad 1 \le p \le \alpha \qquad (15)$$

Here m be the number of alternatives, n be the number of criteria functions, w_j
be the weight of jth criteria function which signifies the relative importance of the
criteria and $f_{ij} (i = 1, 2, \ldots, m, j = 1, 2, \ldots, n)$ is the value of ith alternative for
jth criteria functions. In VIKOR, $L_{1,i}$ and $L_{\infty,i}$ are respectively denoted by S_i and
R_i which have an important role in the ranking process. Maximum group utility and
minimum individual regret of the opponent are supported by S_i and R_i, respectively.
In TOPSIS method, alternative(s) are selected in such a way that it should have
less distance from the positive ideal solution more distance from the negative ideal
solution. Here initially normalized decision matrix and then normalized weighted
decision matrices are found. Then the positive ideal and negative ideal solutions
are determined and distances of each of the alternatives from those solutions are
computed. Finally, the alternatives are ordered based on the relative closeness.

4   Extended VIKOR and TOPSIS Method Using PFS

PFS is important in decision-making situations when human thoughts involve more
options like "yes", "abstain", "no" and "refusal". The idea of PFS is outside the
scope of IFS and FS, since FS and IFS do not support neutrality. In this section,
we extend VIKOR and TOPSIS method to solve MCDM problems with picture
fuzzy information. Let $A = \{ A_i | i = 1, 2, \ldots, m \}$ be a finite set of alternatives and
$C = \{ C_j | j = 1, 2, \ldots, n \}$ be a finite set of criteria. The information about the
alternatives is represented in terms of criteria functions, which are often uncertain
and contains multiple options like "yes", "abstain", "no" and "refusal". This situation
can be managed by the picture fuzzy number using positive, neutral, negative and
refusal membership degrees. Let f_{ij}^k be the estimate of alternative A_i with respect to
criteria C_j for the kth decision maker, where $f_{ij}^k = \left\{ \left(\mu_{ij}^k, \eta_{ij}^k, v_{ij}^k \right) \right\}$. Here μ_{ij}^k, η_{ij}^k,
and v_{ij}^k respectively denote the degree of positive, neutral and negative membership
degrees of the alternative $A_i \in A$ with respect to the criteria C_j by the kth expert. Let
$F_k = \left[f_{ij} \right]_{mn}$ be an assessment matrix in the form of picture fuzzy number which is
expressed as

$$F_k = \begin{bmatrix} \left(\mu_{11}^k, \eta_{11}^k, v_{11}^k \right) & \cdots\cdots\cdots & \left(\mu_{mn}^k, \eta_{mn}^k, v_{mn}^k \right) \\ \vdots & & \vdots \\ \left(\mu_{m1}^k, \eta_{m1}^k, v_{m1}^k \right) & \cdots\cdots\cdots & \left(\mu_{mn}^k, \eta_{mn}^k, v_{mn}^k \right) \end{bmatrix}_{m \times n}$$

4.1 Extended VIKOR Method

Step 1. A set of assessment matrices $F_k(k = 1, 2, ..., p)$ are considered to accumulate the opinions of total p number of experts/decision makers.

Step 2. An aggregated picture fuzzy assessment matrix is constructed based on the opinions of the expert group. Let $F_k = \left[f_{ij}^k \right]_{mn}$ be the picture fuzzy matrix of the kth expert. In group decision making, all individual opinions are combined into a collective opinion to construct an aggregated picture fuzzy assessment matrix F with the help of PFWA operator mentioned in Sect. 2. The aggregated picture fuzzy matrix is given below.

$$F' = \begin{bmatrix} (\mu_{11}, \eta_{11}, \nu_{11}) \dots \dots (\mu_{1n}, \eta_{1n}, \nu_{1n}) \\ \vdots \\ (\mu_{m1}, \eta_{m1}, \nu_{m1}) \dots \dots (\mu_{mn}, \eta_{mn}, \nu_{mn}) \end{bmatrix}_{m \times n} \tag{16}$$

Step 3. The respective best f_j^+ and worst f_j^- for $j = 1, 2, \ldots, n$ are computed. When the criteria j characterizes a benefit then $f_j^* = \max_i f_{ij}$, $f_j^- = \min_i f_{ij}$. When the criteria j characterizes a cost, then $f_j^* = \min_i x_{ij}$, $f_j^- = \max_i f_{ij}$.

$$f_j^+ = \{(\mu^+, \eta^+, \nu^+)\} = \left\{ \left(\max_i \mu_{ij}, \min_i \eta_{ij}, \min_i \nu_{ij} \right) \right\} \tag{17}$$

$$f_j^- = \{(\mu^-, \eta^-, \nu^-)\} = \left\{ \left(\min_i \mu_{ij}, \min_i \eta_{ij}, \max_i \nu_{ij} \right) \right\} \tag{18}$$

Step 4. S_i and R_i for $i = 1, 2, \ldots, m$ are calculated as given below.

$$S_i = \sum_{j=1}^n w_j * \frac{f_j^+ - f_{ij}}{f_j^+ - f_j^-} \tag{19}$$

$$R_i = \max \left[w_j * \frac{f_j^+ - f_{ij}}{f_j^+ - f_j^-} \right] \tag{20}$$

Step 5. Compute the Q_i value by the following relation.

$$Q_i = v \left(\frac{S_i - S^+}{S^- - S^+} \right) + (1 - v) \left(\frac{R_i - R^+}{R^- - R^+} \right) \tag{21}$$

here $S^+ = \min_i S_i$, $S^- = \max_i S_i$, $R^+ = \min_i R_i$, $R^- = \max_i R_i$, and v is considered as weight of the strategy of the majority criteria (or the maximum group utility) which is assumed to be 0.5.

Step 6. The values of S, R and Q are sorted in decreasing order and their ranking list is obtained.

Step 7. Alternative $\overline{A}$, which is ranked best by Q measure (minimum) is selected as a compromise solution based on the following two conditions.

C1: "Acceptable advantage": $Q(\overline{\overline{A}}) - Q(\overline{A}) \geq DQ$, $DQ = 1/(1-m)$, where m be the number of alternatives and $\overline{\overline{A}}$ be second position alternatives in the rank list provides by Q.

C2: "Acceptable stability in decision making": Alternative $\overline{A}$ must also be the best ranked by S or/and R.

If one of the conditions C1 and C2 is not fulfilled, then a set of compromise solutions is proposed as mentioned below.

- Alternatives $\overline{A}$ and $\overline{\overline{A}}$ are chosen as a compromise solution if only condition C2 is not fulfilled, or
- Alternatives $\overline{A}, \overline{\overline{A}}, \ldots, A^m$ if condition C1 is not fulfilled; and A^m is determined by the relation $Q(\overline{\overline{A}}) - Q(\overline{A}) < DQ$, $DQ = 1/(1-m)$ for maximum m.

4.2 Extended TOPSIS Method

This section extends TOPSIS method in the context of PFN which is shown below stepwise.

Step 1. Consider that $F_k = \left[f_{ij}^k\right]_{mn}$, $k = 1, 2, \ldots, p$ be the set of picture fuzzy decision matrices which are formed based on the opinions of the experts.

Step 2. Calculate the respective positive and negative ideal solution M^{k+} and M^{k-} for the kth decision maker as follows.

$$M^{k-} = \{f_1^{k-}, f_2^{k-}, \ldots, f_n^{k-}\} \text{ where } f_j^{k-} = \left(\min_i(\mu_{ij}^k), \max_i(\eta_{ij}^k), \max_i(v_{ij}^k)\right)$$

$$M^{k+} = \{f_1^{k+}, f_2^{k+}, \ldots, f_n^{k+}\} \text{ where } f_j^{k+} = \left(\max_i(\mu_{ij}^k), \min_i(\eta_{ij}^k) \min_i(v_{ij}^k)\right)$$

Step 3. Estimate the distances for each of the alternatives from the respective positive and negative ideal solution using the expressions given below.

$$d_i^{k+} = \frac{1}{n} \sum_{j=1}^{m} \left(f_j^{k+} - f_{ij}^k\right), \quad i = 1, 2, \ldots, m \tag{22}$$

$$d_i^{k-} = \frac{1}{n} \sum_{j=1}^{m} \left(f_{ij}^k - f_j^{k-}\right), \quad i = 1, 2, \ldots, m \tag{23}$$

Step 4. Aggregate the corresponding positive and negative distances of the entire alternatives to form the group positive ideal solution (PIS) d_i^{*+} and group negative ideal solution (NIS) d_i^{*-} respectively as shown below.

$$d_i^{*+} = \frac{\sum_{k=1}^{k} d_i^{k+}}{K}, \, d_i^{*-} = \frac{\sum_{k=1}^{k} d_i^{k-}}{K}$$

Step 5. Using d_i^{*+} and d_i^{*-}, relative closeness for the alternative A_i is computed as given below.

$$R_i^* = \frac{d_i^{*-}}{d_i^{*-} + d_i^{*+}} \text{ for } i = 1, 2, \ldots, m \tag{24}$$

Step 6. The more the value of R_i^*, the better will be the rank of the alternative.

5 Practical Example

In this section, we present a practical example related to the ranking of tiger reserve national parks. In order to do so, a group of five experts are selected from five regions such as North, South, Central, Eastern and Western of India to visit seven national parks for surveying the field study based on varieties criteria which are play an importance role in tiger reserve like density of tiger, natural hazards tendency, wildlife population, eco-development activity, both side security, tourism rush, social activities and water resources. The experts' group visited the national parks for all-round investigation and collecting the information. But it might not be possible for the experts groups to visit the whole area of the park, point to point counting and analysis of all the requirement and natural source of the reserve. Also, it might not be possible for the experts group to completely collect the entire available sample due to lack of resource, shortage of time and some physical difficulties. Among those parameters, some are favorable for the reserve as well as some are having negative impact of the tiger reserve. Such as huge tiger density increases the rate of reproduction but this creates the problem like food criticize. At present, there are twenty-eight tiger reserve national parks around the seventeen states in India. Among those tiger reserve national parks, we consider seven for our investigation purpose which are "Bandhabgarh", "Bandipur", "Sunderban", "Palamau", "Indra-vati", "Nagarjunasagar-Srisailam" and "Manas" and these national parks are respectively denoted by NP$_1$, NP$_2$, NP$_3$,..., NP$_7$ in our example. The group of experts consider five parameters of the national park as per their own assessments which are denoted as "density of tiger" (CR$_1$), "natural hazards tendency" (CR$_2$), "wildlife population" (CR$_3$), "both side security" (CR$_4$) and "tourism rush" (CR$_5$) and give their opinions in the form of picture fuzzy numbers $f^k = \left\{ \left(\mu_{ij}^k, \eta_{ij}^k, \nu_{ij}^k \right) \right\}$, where

μ_{ij}^k be the quantity present the favorable event or positive effect, η_{ij}^k present the not visited or dark portion of the investigation and v_{ij}^k represent the negative side or harmfulness of the ith criteria for the jth park, where $\mu_{ij}^k + \eta_{ij}^k + v_{ij}^k \leq 1$. Hence the refusal membership value is $\pi_{ij}^k = 1 - \mu_{ij}^k + \eta_{ij}^k + v_{ij}^k$. The criteria of national park are not important in the method is considered as refusal membership. There are five picture fuzzy decision matrix provided by five experts denoted by F_k ($k = 1,\ldots, 5$). $F_k = \left[f_{ij}^k \right]_{7 \times 5}$ is represented in Tables 1, 2, 3, 4 and 5 showing the seven tiger reserve and their present information for the predefined five criteria in the form of picture fuzzy information f_{ij} ($i = 1, 2, \ldots, 5$ and $j = 1, 2, \ldots, 7$). Initially, we illustrate the example using extended VIKOR method and then the extended TOPSIS method is described.

To determine the aggregated picture fuzzy decision matrix, the PFWA operator is applied over the picture fuzzy decision matrices obtained from the group of five experts. Table 6 shows the aggregated picture fuzzy decision matrix in tabular form. Note that all the decision matrices in the paper are presented in tabular form.

Next, we compute best f_j^+ and worst f_j^- for each criterion $j = 1, 2, \ldots, 5$ using (17) and (18) based on the aggregated decision matrix (Table 6) and shown in Table 7.

Using (19) and (20), we compute the S_i and R_i. The resultant values are presented in Table 8.

Table 1 Picture fuzzy decision matrix for expert 1

	CR$_1$	CR$_2$	CR$_3$	CR$_4$	CR$_5$
NP$_1$	(0.62,0.2,0.14)	(0.6,0.2,0.15)	(0.75,0.15,0.05)	(0.7,0.15,0.1)	(0.7,0.1,0.15)
NP$_2$	(0.7,0.15,0.1)	(0.75,0.1,0.1)	(0.85,0.1,0.05)	(0.7,0.2,0.06)	(0.65,0.2,0.1)
NP$_3$	(0.6,0.2,0.15)	(0.82,0.1,0.08)	(0.72,0.14,0.14)	(0.81,0.11,0.08)	(0.8,0.11,0.09)
NP$_4$	(0.7,0.1,0.1)	(0.73,0.15,0.12)	(0.75,0.13,0.12)	(0.85,0.12,0.03)	(0.72,0.13,0.15)
NP$_5$	(0.81,0.1,0.07)	(0.76,0.17,0.07)	(0.83,0.15,0.02)	(0.9,0.05,0.05)	(0.76,0.2,0.04)
NP$_6$	(0.65,0.15,0.1)	(0.68,0.24,0.08)	(0.65,0.18,0.17)	(0.7,0.17,0.13)	(0.66,0.25,0.09)
NP$_7$	(0.55,0.17,0.18)	(0.7,0.2,0.08)	(0.7,0.16,0.07)	(0.65,0.13,0.12)	(0.6,0.0,0.3)

Table 2 Picture fuzzy decision matrix for expert 2

	CR$_1$	CR$_2$	CR$_3$	CR$_4$	CR$_5$
NP$_1$	(0.64,0.22,0.14)	(0.65,0.14,0.2)	(0.8,0.11,0.09)	(0.75,0.15,0.1)	(0.7,0.15,0.15)
NP$_2$	(0.74,0.15,0.11)	(0.75,0.12,0.13)	(0.85,0.1,0.05)	(0.74,0.2,0.06)	(0.75,0.12,0.13)
NP$_3$	(0.72,0.2,0.08)	(0.82,0.1,0.08)	(0.72,0.14,0.14)	(0.81,0.11,0.08)	(0.8,0.11,0.09)
NP$_4$	(0.7,0.2,0.1)	(0.73,0.15,0.12)	(0.75,0.13,0.12)	(0.85,0.12,0.03)	(0.72,0.13,0.15)
NP$_5$	(0.81,0.12,0.07)	(0.76,0.17,0.07)	(0.83,0.15,0.02)	(0.9,0.05,0.05)	(0.76,0.2,0.04)
NP$_6$	(0.75,0.15,0.1)	(0.68,0.24,0.08)	(0.65,0.18,0.17)	(0.7,0.17,0.13)	(0.66,0.25,0.09)
NP$_7$	(0.65,0.17,0.18)	(0.7,0.22,0.08)	(0.77,0.16,0.07)	(0.75,0.13,0.12)	(0.65,0.0,0.35)

Table 3 Picture fuzzy decision matrix for expert 3

	CR$_1$	CR$_2$	CR$_3$	CR$_4$	CR$_5$
NP$_1$	(0.64,0.15,0.14)	(0.65,0.2,0.15)	(0.7,0.2,0.1)	(0.55,0.25,0.2)	(0.56,0.14,0.25)
NP$_2$	(0.62,0.15,0.11)	(0.7,0.2,0.1)	(0.62,0.25,0.1)	(0.65,0.2,0.1)	(0.65,0.2,0.1)
NP$_3$	(0.65,0.2,0.08)	(0.55,0.3,0.1)	(0.65,0.17,0.15)	(0.7,0.1,0.1)	(0.72,0.1,0.1)
NP$_4$	(0.7,0.15,0.1)	(0.6,0.2,0.12)	(0.85,0.1,0.05)	(0.6,0.2,0.2)	(0.8,0.1,0.1)
NP$_5$	(0.8,0.1,0.07)	(0.7,0.2,0.07)	(0.8,0.1,0.1)	(0.5,0.3,0.2)	(0.75,0.2,0.05)
NP$_6$	(0.75,0.15,0.1)	(0.75,0.2,0.05)	(0.75,0.1,0.1)	(0.55,0.3,0.1)	(0.5,0.2,0.2)
NP$_7$	(0.8,0.0,0.18)	(0.6,0.2,0.2)	(0.65,0.2,0.1)	(0.75,0.12,0.1	(0.56,0.24,0.2)

Table 4 Picture fuzzy decision matrix for expert 4

	CR$_1$	CR$_2$	CR$_3$	CR$_4$	CR$_5$
NP$_1$	(0.65,0.25,0.1)	(0.65,0.2,0.2)	(0.7,0.15,0.15)	(0.5,0.3,0.2)	(0.9,0.0.05,0.05)
NP$_2$	(0.75,0.15,0.1)	(0.7,0.2,0.1)	(0.6,0.2,0.2)	(0.6,0.15,0.25)	(0.6,0.1,0.3)
NP$_3$	(0.5,0.2,0.3)	(0.65,0.25,0.1)	(0.6,0.15,0.2)	(0.7,0.1,0.1)	(0.7,0.15,0.1)
NP$_4$	(0.55,0.12,0.23)	(0.68,0.15,0.15)	(0.55,0.2,0.2)	(0.75,0.1,0.13)	(0.8,0.1,0.1)
NP$_5$	(0.85,0.1,0.05)	(0.8,0.1,0.1)	(0.8,0.1,0.05)	(0.6,0.15,0.1)	(0.65,0.1,0.2)
NP$_6$	(0.7,0.2,0.1)	(0.56,0.25,0.15)	(0.9,0.05,0.05)	(0.55,0.15,0.2)	(0.75,0.1,0.1)
NP$_7$	(0.6,0.2,0.15)	(0.9,0.05,0.05)	(0.75,0.1,0.15)	(0.6,0.3,0.1)	(0.5,0.2,0.3)

Table 5 Picture fuzzy decision matrix for expert 5

	CR$_1$	CR$_2$	CR$_3$	CR$_4$	CR$_5$
NP$_1$	(0.64,0.22,0.14)	(0.8,0.1,0.1)	(0.6,0.15,0.2)	(0.5,0.3,0.2)	(0.8,0.1,0.1)
NP$_2$	(0.74,0.15,0.11)	(0.75,0.15,0.15)	(0.7,0.1,0.15)	(0.65,0.2,0.1)	(0.6,0.15,0.1)
NP$_3$	(0.72,0.2,0.08)	(0.6,0.2,0.2)	(0.75,0.1,0.1)	(0.9,0.05,0.05)	(0.75,0.1,0.1)
NP$_4$	(0.7,0.2,0.1)	(0.75,0.1,0.1)	(0.85,0.1,0.05)	(0.55,0.25,0.1)	(0.6,0.25,0.1)
NP$_5$	(0.81,0.12,0.07)	(0.45,0.3,0.2)	(0.5,0.2,0.3)	(0.45,0.35,0.1)	(0.9,0.05,0.05)
NP$_6$	(0.75,0.15,0.1)	(0.5,0.2,0.2)	(0.7,0.1,0.2)	(0.6,0.3,0.1)	(0.5,0.25,0.25)
NP$_7$	(0.65,0.17,0.18)	(0.65,0.15,0.1)	(0.75,0.05,0.2)	(0.75,0.15,0.1)	(0.65,0.2,0.1)

Table 6 Picture fuzzy aggregated decision matrix

	CR$_1$	CR$_2$	CR$_3$	CR$_4$	CR$_5$
NP$_1$	0.64,0.21,0.13	0.68,0.16,0.16	0.71,0.15,0.11	0.62,0.22,0.15	0.76,0.1,0.12
NP$_2$	0.71,0.14,0.1	0.73,0.15,0.11	0.75,0.14,0.09	0.67,0.19,0.1	0.65,0.15,0.13
NP$_3$	0.65,0.2,0.12	0.71,0.17,0.11	0.69,0.14,0.14	0.8,0.09,0.08	0.76,0.11,0.1
NP$_4$	0.67,0.15,0.12	0.7,0.15,0.12	0.77,0.13,0.09	0.75,0.15,0.07	0.74,0.13,0.06
NP$_5$	0.82,0.11,0.07	0.71,0.18,0.09	0.77,0.14,0.06	0.74,0.13,0.09	0.78,0.13,0.06
NP$_6$	0.72,0.16,0.1	0.64,0.21,0.12	0.75,0.14,0.12	0.63,0.21,0.13	0.63,0.2,0.13
NP$_7$	0.66,0.14,0.17	0.74,0.15,0.09	0.73,0.12,0.1	0.71,0.16,0.11	0.6,0.16,0.17

Table 7 Best f_j^+ and worst f_j^- for each criterion $j = 1, 2, \ldots, 5$

	CR$_1$	CR$_2$	CR$_3$	CR$_4$	CR$_5$
f_j^+	0.82,0.11,0.07	0.74,0.15,0.09	0.77,0.12,0.06	0.8,0.09,0.08	0.78,0.1,0.06
f_j^-	0.64,0.21,0.17	0.64,0.21,0.16	0.69,0.15,0.14	0.62,0.22,0.09	0.6,0.2,0.17

Table 8 Calculated values of S_i and R_i of the national parks

	NP$_1$	NP$_2$	NP$_3$	NP$_4$	NP$_5$	NP$_6$	NP$_7$
S_i	0.692	0.458	0.459	0.362	0.142	0.726	0.536
R_i	0.2	0.13	0.19	0.13	0.06	0.18	0.18

Thereafter, Q_i is computed by using (21) and finally, the estimated Q_i values are displayed in Table 9. Finally, the ranking of the national parks is given according to computed Q_i value. Lower values indicate a higher rank.

The ranking list is obtained as NP$_5$ > NP$_3$ > NP$_4$ > NP$_7$ > NP$_2$ > NP$_1$ > NP$_6$.

To solve the example using TOPSIS method, firstly, we determine the PIS and NIS of each national parks for each of the five experts as shown in Table 10.

Next we estimate the distance of the national parks from the respective PIS and NIS and aggregate those distances to generate the group positive ideal solution (d_i^{*+}) and the group negative ideal solution (d_i^{*-}) respectively as shown below in Table 11. Then the relative closeness (R_i^*) of the national parks is computed using (24) and the computed value is shown in Table 11.

Table 9 Evaluated values of Q_i

	NP$_1$	NP$_2$	NP$_3$	NP$_4$	NP$_5$	NP$_6$	NP$_7$
Q_i	0.357	0.228	0.092	0.121	0.0005	0.448	0.21

Table 10 PIS and NIS of the national parks

		CR$_1$	CR$_2$	CR$_3$	CR$_4$	CR$_5$
Expert 1	PIS	(0.81, 0.1,0.07)	(0.82,0.1,0.07)	(0.85, 0.1,0.05)	(0.9,0.05,0.03)	(0.8,0.1,0.04)
	NIS	(0.55,0.2,018)	(0.6,0.24,0.15)	(0.65,0.18,0.2)	(0.65,0.2,0.13)	(0.6, 0.25,0.3)
Expert 2	PIS	(0.81,0.12,0.07)	(0.82,0.1,0.07)	(0.85, 0.1,0.02)	(0.9,0.05,0.03)	(0.8,0.05,0.04)
	NIS	(0.64,0.22,0.18)	(0.65,0.24,0.2)	(0.65,0.18,0.17)	(0.7,0.2,0.13)	(0.65,0.25,0.3)
Expert 3	PIS	(0.8, 0.05,0.07)	(0.75,0.2,0.05)	(0.85, 0.1,0.05)	(0.75,0.1, 0.1)	(0.8, 0.1,0.05)
	NIS	(0.62,0.2,014)	(0.55, 0.3, 0.2)	(0.62,0.25,0.15)	(0.5,0.3,0.2)	(0.5,0.24,0.25)
Expert 4	PIS	(0.85, 0.1,0.05)	(0.9,0.05,0.05)	(0.9, 0.05,0.05)	(0.75,0.1, 0.1)	(0.9,0.05,0.05)
	NIS	(0.5,0.25,0.3)	(0.56,0.25,0.2)	(0.55,0.2,0.2)	(0.5,0.3,0.25)	(0.5, 0.2,0.3)
Expert 5	PIS	(0.81,0.12,0.07)	(0.8, 0.1,0.1)	(0.85,0.05,0.05)	(0.9,0.05, 0.03)	(0.9, 0.05,0.05)
	NIS	(0.64,0.22,0.18)	(0.45, 0.3, 0.2)	(0.5,0.2,0.3)	(0.45,0.35,0.2)	(0.5, 0.25,0.25)

Table 11 Group PIS and NIS values of the national park

	NP$_1$	NP$_2$	NP$_3$	NP$_4$	NP$_5$	NP$_6$	NP$_7$
d_i^{*+}	0.338	0.271	0.244	0.241	0.194	0.342	0.311
d_i^{*-}	0.065	0.079	0.098	0.0804	0.0512	0.0444	0.0784
R_i^{*}	0.162	0.227	0.286	0.25	0.2081	0.114	0.201

Hence the preferable ranking sequence is NP$_3$ > NP$_4$ > NP$_2$ > NP$_5$ > NP$_7$ > NP$_1$ > NP$_6$.

6 Result and Discussion

In this study, we compare the two MCDM methods, VIKOR and TOPSIS in the domain of PFS. VIKOR method follows the linear normalization, whereas TOPSIS method follows the vector normalization. The ranking sequences are generated by the two MCDM methods which show comprehensive comparison given in Table 12 which contains the maximum and minimum positive membership values and average neutral membership value of each national park. We can understand from this table that the effect of the maximum and minimum positive membership value is less in this study. The maximum and minimum value of national park NP$_7$ is 0.74 and 0.6, respectively. According to VIKOR method, the rank of NP$_7$ is 4, whereas maximum and minimum value of NP$_1$ is 0.76 and 0.62 respectively, but the rank of NP$_1$ is 6. The ranking sequence of VIKOR method is directly dependent on the average neutral membership value. The rank of the national park decreases due to the increase of the neutral value.

Table 12 Comparative analysis of extended VIKOR and TOPSIS

	NP$_1$	NP$_2$	NP$_3$	NP$_4$	NP$_5$	NP$_6$	NP$_7$
Rank of TOPSIS	6	3	1	2	4	7	5
Rank of VIKOR	6	5	2	3	1	7	4
Max of positive membership value	0.76	0.75	0.8	0.77	0.82	0.75	0.74
Min of positive membership value	0.62	0.65	0.65	0.67	0.71	0.63	0.6
Avg of neutral membership value	0.168	0.154	0.142	0.142	0.138	0.184	0.146

7 Conclusion

In this study, we have extended the VIKOR and TOPSIS method in the context of the picture fuzzy set and then applied it in a practical example related to the ranking of tiger reserve national parks. The main benefit of the proposed approach is that a compromise solution can be evaluated even in the presence of neutral and refusal information. Both of the TOPSIS and VIKOR method have a huge contribution in uncertain decision making as both provide solutions in the presence of conflicting, uncertain, imprecise and non-commensurable criteria. In the future, one can use these extended methods in real-life applications such as health care, sustainable supplier selection, disaster management and can investigate the impact of neutrality in critical decision making.

References

1. Zadeh, L.A.: Fuzzy sets. Information. Control **8**, 338–356 (1965)
2. Atanassov, K.: More on intuitionistic fuzzy sets. Fuzzy Sets Syst. **33**, 37–46 (1989)
3. Bhatia, N., Kumar, A.: A new method for sensitivity analysis of fuzzy transportation problems. J. Intell. Fuzzy Syst. **25**, 167–175 (2013)
4. Kumar, A., Kaur, A.: Optimization for different types of transportation problems with fuzzy coefficients in the objective function. J. Intell. Fuzzy Syst. **23**, 237–248 (2012)
5. Li, D.F.: Extension of the LINMAP for multi-attribute decision making under Atanassov's intuitionistic fuzzy environment. Fuzzy Optim. Decis. Making **7**, 17–34 (2008)
6. Xu, Z.S., Yager, R.R.: Some geometric aggregation operators based on intuitionistic fuzzy sets. Int. J. Gen. Syst. **35**, 417–433 (2006)
7. Xu, Z.S., Yager, R.R.: Dynamic intuitionistic fuzzy multi-attribute decision making. Int. J. Approximate Reasoning **48**, 246–262 (2008)
8. Atanassov, K.T., Georgiev, C.: Intuitionistic fuzzy prolog. Fuzzy Sets Syst. **53**, 121–128 (1993)
9. Liu, H.W.: Multi-criteria decision-making methods based on intuitionistic fuzzy sets. Eur. J. Oper. Res. **179**, 220–233 (2007)
10. Hung, W.L., Yang, M.S.: Similarity measures of intuitionistic fuzzy sets based on Hausdorff distance. Pattern Recogn. Lett. **25**, 1603–1611 (2004)
11. Wang, Z., Xu, Z.S., Liu, S.S.: A netting clustering analysis method under intuitionistic fuzzy environment. Appl. Soft Comput. **11**, 5558–5564 (2011)
12. Xu, Z.S., Chen, J., Wu, J.J.: Clustering algorithm for intuitionistic fuzzy sets. Inf. Sci. **178**, 3775–3790 (2008)
13. Cuong, B.C., Kreinovich, V.: Picture fuzzy sets. J. Comput. Sci. Cybern. **30**(4), 409–416 (2014)
14. Cong, C.B., Son, L.H.: Some Selected Problems of Modern Soft Computing
15. Wang, C., Zhou, X., Tu, H., Tao, S.: Some geometric aggregation operators based on picture fuzzy sets and their application in multiple attribute decision making. Italian J. Pure Appl. Math. **37**, 477–492 (2017)
16. Si, A., Das, S., Kar, S.: An approach to rank picture fuzzy numbers for decision making problems. In: Decision Making: Applications in Management and Engineering (2019). https://doi.org/10.31181/dmame1902049s
17. Tavana, M., Mavi, R.K., Santos-Arteaga, F.J., Doust, E.R.: An extended VIKOR method using stochastic data and subjective judgments. Comput. Ind. Eng. **97**, 240–247 (2016)
18. Wu, Z., Ahmad, J., Xu, J.: A group decision making framework based on fuzzy VIKOR approach for machine tool selection with linguistic information. Appl. Soft Comput. **42**, 314–324 (2016)

19. Wanga, L., Zhanga, H., Wanga, J., Li, L.: Picture fuzzy normalized projection-based VIKOR method for the risk evaluation of construction project. Appl. Soft Comput. **64**, 216–226 (2018)

20. Zeng, S., Qiyas, M., Arif, M., Mahmood, T.: Extended version of linguistic picture fuzzy TOPSIS method and its applications in enterprise resource planning systems. Math. Prob. Eng. Article ID 8594938 (2019)

21. Si, A., Das, S.: Intuitionistic Multi-fuzzy convolution operator and its application in decision making. In: Proceedings of the International Conference on Computational Intelligence, Communications, and Business Analytics (CICBA) (2017); Mandal, J., Dutta, P., Mukhopadhyay, S. (eds.): Communications in Computer and Information Science, vol. 776, pp. 540–551. Springer

22. Das, S., Malakar, D., Kar, S., Pal, T.: A brief review and future outline on decision making using fuzzy soft set. Int. J. Fuzzy Syst. Appl. **7**(2), 1–43 (2018)

23. Das, S., Kumar, S., Kar, S., Pal, T.: Group decision making using neutrosophic soft matrix: An algorithmic approach. J. King Saud Univ. Comput. Inf. Sci. (2017). http://dx.doi.org/10.1016/j.jksuci.2017.05.001

24. Opricovic, S., Tzeng, G.H.: Extended VIKOR method in comparison with outranking methods. Eur. J. Oper. Res. **178**, 514–529 (2007)

25. Opricovic, S., Tzeng, G.H.: The Compromise solution by MCDM methods: A comparative analysis of VIKOR and TOPSIS. Eur. J. Oper. Res. **156**(2), 445–455 (2004)

26. Opricovic, S., Tzeng, G.H.: Multicriteria planning of post-earthquake sustainable reconstruction. Comput. Aided Civ. Infrastruct. Eng. **17**(3), 211–220 (2002)

27. Yu, P.L.: A class of solutions for group decision problems. Manage. Sci. **19**(8), 936–946 (1973)

28. Zeleny, M.: Multiple Criteria Decision Making. McGraw-Hill, New York (1982)

29. Vahdani, B., Hadipour, H., Sadaghiani, J.S., Amiri, M.: Extension of VIKOR method based on interval-valued fuzzy sets. Int. J. Adv. Manuf. Technol. **47**, 1231–1239 (2010)

30. Cuong, B.C., Kreinovich, V.: Picture Fuzzy Sets-a New Concept for Computational Intelligence Problems. Departmental Technical Reports (CS). Paper 809. http://digitalcommons.utep.edu/cs-techrep/809 (2013)

31. Sayadi, M.K., Heydari, M., Shahanaghi, K.: Extension of VIKOR method for decision making problem with interval numbers. Appl. Math. Model. **33**, 2257–2262 (2009)

32. Sanayei, A.S., Mousavi, A., Yazdankhah, F.: Group decision making process for supplier selection with VIKOR under fuzzy environment. Expert Sys. Appl. **37**, 24–30 (2010)

33. Hwang, C.L., Yoon, K.: Multiple Attribute Decision Making: Methods and Applications. Springer, New York (1981)

34. Yoon, K.: A Reconciliation among discrete compromise situations. J. Oper. Res. Soc. **38**, 277–286 (1987)

35. Shih, H.S., Shyur, H.J., Lee, E.S.: An extension of TOPSIS for group decision making. Math. Comput. Model. **45**, 801–813 (2007)

36. Cuong, B.C.: Picture fuzzy sets rest results. Part 1: Seminar "Neuro-Fuzzy Systems with Applications", Preprint 03/2013, Institute of Mathematics, Hanoi (2013)

37. Mohamed, A.B., Saleh, M., Gamal, A., Smarandache, F.: An approach of TOPSIS technique for developing supplier selection with group decision making under type-2 neutrosophic number. Appl. Soft Comput. J. **77**, 438–452 (2019)

A Study of Gathering of Location-Aware Mobile Robots

Soumik Banerjee and Sruti Gan Chaudhuri

Abstract Gathering of *Autonomous Mobile Robots* is very popular topic on the field of Robotics. Many researchers have addressed the problem through various frameworks of the robots. Distributed model of robots is one such framework which considers the robots to run a distributed algorithm independently by their own at each robot site and finally approach to gather at a single point. This paper has presented the most recent works on this topic with a plenty number of open problems.

Keywords Gathering · Autonomous · Location-aware mobile robots

1 Introduction

Robotics is inspired by observing characteristics of human or natural entities. Cooperative behavior is one of the most important parameters of a well-built society. Similarly, a multi-robot system can perform many complex tasks efficiently by executing them cooperatively. In hostile situations, deploying a big robot might be difficult or impossible and also not cost-effective. Hence, the concept of a group of small robots working together has become a trend in recent robotic research. Small robots are less expensive and easy to design, configure and deploy. Their cooperative abilities enable them to perform many complicated tasks more efficiently and effectively than an expensive machine working alone. Swarm Robotics is the new approach to the coordination of multi-robot system which consists of large numbers of mostly simple physical robots.

The task of two or more mobile entities, starting at distinct initial positions, meeting at some other point, is called rendezvous. This has many applications in multiple domains ranging from human interaction and animal behavior to programming of

S. Banerjee · S. G. Chaudhuri (✉)
Jadavpur University, Kolkata, India
e-mail: sruti.ganchaudhuri@jadavpuruniversity.in

S. Banerjee
e-mail: soumikbanerjee34@gmail.com

© Springer Nature Singapore Pte Ltd. 2020
J. K. Mandal and S. Mukhopadhyay (eds.), *Proceedings of the Global AI Congress 2019*, Advances in Intelligent Systems and Computing 1112,
https://doi.org/10.1007/978-981-15-2188-1_45

autonomous mobile robots and software agents. There are mainly two situations for algorithmic problems requiring accomplishing a rendezvous task in a man-made environment;

- (i) Autonomous mobile robots that start in different locations of a plane and have to meet.
- (ii) Software agents, that travel in a communication network in order to perform maintenance of its components or to collect data distributed within the nodes of the network.

To exchange collected data and plan further moves, these agents also need to meet periodically. Rendezvous algorithms do not depend on the physical nature of the mobile entities, but only on their characteristics, such as perceiving capabilities, memory, mobility and also the nature of the environment. In the case of more than two agents, the rendezvous problem is also called *Gathering*. In this paper, we will present a survey on the reported work on gathering of autonomous, homogeneous, location-aware mobile robots.

- Autonomous: Robots can take moving decisions by themselves.
- Homogeneous: Within the system, all robots have the same characteristics.
- Location Aware: Each robot knows its own initial position in the system given by its coordinates or nodes.

2 Framework

The main purpose of developing a computational model for swarm robots is to model simple robots that can perform a given task correctly within a finite time. The traditional model is referred to as the weak model, as the robots in this model have minimal memory, communication power, etc. The traditional model was originally used for gathering in plane. The traditional distributed computational model for swarm robots is described next. A robot can sense its surrounding by observing 360° around itself up to a certain range. This range is called the Visibility range and this is denoted as V_r, when the robot is r. Working principle of swarm robots in traditional model is as follows. At any point of time, a robot is either active or inactive. Each robot executes a cycle consisting of four states as follows.

- Wait: This is the state when swarm robots are inactive. That is why this state is called Wait or Sleep state.
- Look: In this state, swarm robots will observe other robots present in the visibility range V_r at the time t. In this state, it also collects the coordinate positions of other robots present in V_r at time t.
- Compute: In this state, a swarm robot computes its destination point based on the location of other visible robots present in V_r at time t.
- Move: In this state, the swarm robots move toward the points computed in Compute phase. If the destination computed by a robot is its current position, then the robot

executes a null movement, i.e., they stay at the same location. Usually, the robots move in straight lines. But the robots may have guided trajectory as well [19, 23].

If a robot movement stops before reaching its destination, it is called non-rigid motion. In rigid motion, a robot is guaranteed to reach its destination. A robot movement is neither infinite nor infinitesimally small. Suppose t is the computed destination by a robot r. Let D be the distance between t and the current position of r. There exists a small constant d such that, if $D < d$, r is guaranteed to reach t; otherwise r will move at least a distance d toward t. Without these assumptions, it is impossible to guarantee that a robot will reach its destination in finite time.

All robots may not be active at the same time. Depending on their activation schedule, robots will be active and complete their cycle. The duration of the cycles of the robots also varies from each other. There are two types of Schedulers:

- **Asynchronous**: In this kind of scheduling, the cycle of individual robot is independent to each other. Each robot executes Wait-Look-Compute-Move states cycle wise. Moreover, the duration of Compute, Move and Wait states is finite but they are unpredictable. Duration of each state might be different for different robots. This model is called cooperative distributed asynchronous (CORDA) model and was introduced by Flochini et al. [18].
- **Semi-Synchronous**: In this case, timeline is logically divided into some global rounds. In each round, one or more robots are activated and they obtain snapshot in their Look phase. Based upon these snapshots, the robots compute their destination points in their Compute phase and move to their destinations in their Move phase. As all active robots take snapshots at the same time and all operations are instantaneous, the movement of the robots is not observable by other robots. This model is called Sym or ATOM and was introduced by Suzuki and Yamashita [24]. If in each round all robots are activated, it is called **Synchronous** or **Fully Synchronous** model.

No Memory: These robots are usually oblivious robots. It means each robot has a local memory to store information regarding the location of other robots within its visibility range during the cycle, but after completing one cycle, all the information is deleted.

No Message Passing: A robot cannot communicate with other robots by passing messages through wired or wireless medium. A robot gathers only positional information about its surrounding robots by observing or taking snapshot in its Look state.

3 Objective of the Paper

In the traditional model, most of the cases consider no communication or message passing between the robots. However, they gave more importance on taking the snapshots of the surroundings by which the robots compute the locations of the other robots with respect to their own coordinate systems. This model is purely theoretical and the main interest of studying this model is to understand the minimal capabilities required to solve a problem. The researchers have also reported many not-solvable problems under the traditional model [1, 5–7]. One of the recent trends of research is to modify the existing model in an optimized way to solve the unsolvable problems. Gathering is one of such unsolvable problems. Incorporating memory solved this problem [10, 13, 17, 20]. We started finding the way of solving gathering without increasing persistent memory but using some existing real-time technological facilities in theoretical point of view. We also realize that finding the exact location of a robot by collecting snapshots by camera is not actually feasible or realistic in many cases and it actually gives some errors and retractions in harsh environment. Nowadays implementing GPS is not an big issue. By making the robots aware of the location using this GPS facility is our one of the interests. Hence, our objective is to survey on this topic and find open scope of research in this domain.

4 Reported Research Works

Andrew Collins et al. [8] have studied efficient rendezvous of two mobile agents moving asynchronously in the Euclidean 2d-space. Each agent has limited visibility and knows only their own initial position (coordinates) in the plane (i.e., location aware). The agents possess coherent compasses and the same unit of length. The sum of lengths of the trajectories of both agents is the cost of the rendezvous algorithm. This cost is calculated as the maximum over all possible asynchronous movements of the adversary controlled agents. The proposed algorithm by the authors allows the agents to meet in a local neighborhood of diameter $O(d)$, where d is the original distance between the agents. The cost of their algorithm is $O(d^2 + \epsilon)$, for any constant $\epsilon > 0$, which is almost optimal, since a lower bound of $\Omega(d^2)$ is straightforward. In contrast to [12] on asynchronous rendezvous of bounded-visibility agents in the plane which provides the feasibility proof for rendezvous and proposes a solution exponential in the distance d and in the labels of the agents. Authors of this paper have shown that an almost optimal solution is possible when the identity of the agent is based solely on its original location. A space-filling curve preserves the locality visits specific grid points in the plane to provide a route. The mobile agents adopt this to search other robots. This concept was also somewhat counterintuitive in view of the result from [21] stating that for any simple space-filling curve, there always exists a pair of close points in the plane, such that their distance along the space-filling curve is arbitrarily large. Each agents' route is a sequence of segments which are subsequently traversed during its movement. The entire route depends uniquely on its initial position. The actual walk of each agent along every segment is asynchronous and the agents meet if they eventually get within the visibility range of each other. The adversary initially places both agents at any two points in the plane at the same moment and it is assumed that since that time any moving agent may find the other agent, even if the other agent did not start its walk yet. Given its initial location a_0, the route chosen by the agent is a sequence of segments $(e1, e2)$, such that in stage i the agent traverses segment $ei = [a_{i-1}, a_i]$, starting at a_{i1} and ending at ai and these are repeated indefinitely until rendezvous. The walk f is interpreted as follows: at time t the agent is at the point $f(t)$ of its route. The adversary may arbitrarily vary the speed of the agent, as long as the walk of the agent in each segment is continuous. Agents with routes $R1$ and $R2$ and with walks $f(1)$ and $f(2)$ meet at time t, if points $f(1)(t)$ and $f(2)(t)$ are identical. Since the actual portions of trajectories may vary depending on the adversary, the authors considered the cost of the rendezvous algorithm as the maximum of the sum of the lengths of the trajectories of both agents from their starting locations until the rendezvous, i.e., the worst-case over all possible walks chosen for both agents by the adversary. In this paper, they were looking for the rendezvous algorithm of the smallest possible cost with respect to the original distance d between the agents. Their algorithm may be extended in a standard way in order to solve gathering of a set of agents. This left the following open problems.

- Instead of instantly stopping while a meeting occurs, a group of $n \geq 2$ agents involved in the meeting chooses a leader (e.g., the agent with the lexicographically smallest initial position) and all agents follow its route.
- If the total number of agents is known in advance, they gather after traversing a route of length $O(d^2 + \epsilon)$, provided they were initially within a d-neighborhood of some point in the $2d$-space.

Czyzowicz et al. [11] have proposed algorithm for gathering in the plane of location-aware robots in the presence of spies. They considered a set of mobile robots(represented as points) including unknown number of byzantine robots which are indistinguishable from the reliable ones, is distributed in the Cartesian plane. A byzantine robot may report a wrong position, fail to report any, or fail to follow its assigned route. The reliable robots need to gather. The robots are equipped with GPS devices and at the beginning they communicate their initial positions (Cartesian coordinates) to the central authority. With this information and without knowing which robots are faulty, the central authority provides a trajectory for every robot which aims to result in the shortest possible gathering time of the healthy robots. The efficiency of a gathering strategy is measured by its competitive ratio and the role of the adversary controlled byzantine robots, is to act to delay the gathering and maximizing competitive ratio. The goal of this paper is to design a strategy resulting in gathering of all reliable robots within the smallest possible time by minimizing the competitive ratio. They proposed efficient algorithms for the following cases when the central authority is aware of an upper bound on the number of byzantine robots: (i) Optimal algorithms for collections of robots known to contain at most one faulty robot. (ii) Algorithms with small constant competitive ratios when the proportion of byzantine robots is known to be less than one-half or one-third. (iii) Algorithms with bounded competitive ratio in the case where the proportion of faulty robots is arbitrary. Their main results are: (i) Algorithms providing constant competitive ratio for all but a small bounded region in the space of possible n (total no. of robots) and F (faulty/byzantine robots) pairs. (ii) Having knowledge of the upper bound of the number of byzantine robots in the subset (represented by the parameter F) permits fine-tuning of the gathering algorithm, resulting in better competitive ratios. (iii) Considering the gathering problem for collections involving only a single byzantine robot, their algorithm gives optimal competitive ratio for any number of robots, at most one of which is byzantine. (iv) Next, two algorithms are presented. Here, the number of byzantine robots is bounded by a small fraction of n, specifically, competitive ratios of 2 and $2\sqrt{2}$ when $F < \frac{n}{3}$ and $F < \frac{n}{2}$ respectively. (v) Finally, two gathering algorithms solve the problem for any n and any F. The competitive ratio of one of these algorithms is constant, while the other is bounded by $F + 2$. They left the following scope open for the future research work,

- Improvement of competitive ratio and/or complexity of these algorithms.
- Allowing the robots to communicate/exchange their positions at any time during the gathering process.
- Consider robot gathering (in the presence of byzantine robots) under local (limited) communication range.

In a very similar manner to [11], Chuangpishit et al. [4] have proposed algorithm for Rendezvous on a line by location-aware robots despite the presence of Byzantine Faults where a set of mobile robots is placed at points of an infinite line. This system also contains a collection of robots equipped with GPS devices (to communicate their initial positions on the line to a central authority)including an unknown subset of indistinguishable byzantine robots. The set of the non-faulty robots need to rendezvous in the shortest possible time in order to perform some task, while the byzantine robots may try to delay their rendezvous for as long as possible. The goal of the central authority is to determine trajectories for all robots so as to minimize the time until the non-faulty robots have rendezvoused without knowledge of which robots are faulty and to minimize the competitive ratio. The authors provided (i) an algorithm with bounded competitive ratio, where the central authority is informed only of the set of initial robot positions, without any knowledge of which ones or how many of the robots are byzantine and (ii) algorithms with improved competitive ratios when an upper bound on the number of byzantine robots is known to the central authority. In some instances, these algorithms are optimal. Their main results are: (i) two general rendezvous algorithms for $n > 2$ robots with $f \leq (n - 2)$ faulty ones. Both algorithms assume no knowledge of the actual value of f and the second algorithm stops as soon as sufficiently many robots are available to perform the task. They also proved a lower bound of 2 on the competitive ratio for arbitrary $n > 2$ and $1 \leq f \leq (n - 2)$. (ii) Algorithms for the cases where the central authority possesses some knowledge concerning the number of faulty robots. For the case where the ratio of the number of faulty robots to the total number of robots is strictly less than $\frac{1}{2}$, they provided an optimal algorithm and when this number is strictly less than $\frac{2}{3}$ they gave an algorithm that beats the general case algorithms above unless f is known to be less than 5. (iii) Optimal algorithms for the particular cases where $f \in 1, 2$. The main case is when $n = 4$ and $f = 2$ where they showed the exact value of the competitive ratio is $1 + \phi$, where ϕ is the golden ratio. They left the following scope open for the future research work,

– Improvement of competitive ratio and/or complexity of these algorithms.
– The model they presented ignores any communication beyond the broadcasting of the initial positions of the robots. It might be of interest to consider algorithms in a richer communication model where the robots may broadcast information as they follow their trajectories.

Das et al. [14] considered the mobile agents gathering problem in a grid in the presence of an adversarial malicious agent which tries to prevent honest agents from entering an empty node by occupying it. The honest agents move in synchronous rounds and at each round an agent can move to an adjacent node only if this node is not occupied by the malicious agent. The model presents the honest agents as identical finite state automata that move in an anonymous oriented grid topology and have no information about the size of the graph. On the other hand, the malicious agent is presented as arbitrarily fast and to have full knowledge of the locations and the strategy of the honest agents at all times. The agents can communicate with each other only when they meet at a node. Assuming that the agents have constant memory, the goal was to find the minimum number k of synchronous mobile agents that are able to gather in a grid despite the presence of the malicious agent and to design an algorithm for gathering of k or more honest agents in any grid network. The authors proved that (i) rendezvous is impossible for $k = 2$ agents, which holds even if the agents have unbounded memory and global visibility, (ii) gathering is possible for $k = 3$ agents with global visibility and finite memory. They gave a universal deterministic algorithm for gathering problem of any $k \geq 4$ agents when the agents have only local visibility and constant memory and proved that (i) $k > 3$ synchronous agents with only local visibility capability can gather starting from any configuration without multiplicities, while for asynchronous agents it has been previously proved that the agents can gather only if they start from a connected configuration and have visibility at distance two. This left the following open problems and future scopes:

– Whether $k = 3$ agents can still gather in a grid starting from any configuration when they have only local visibility.
– Study of this problem in other topologies such as oriented multidimensional grids and other well-structured graphs that are easy to explore by constant memory agents.
– This problem can also be studied considering a less powerful malicious agent which has limited speed capabilities, or with multiple such mobile adversaries. For instance since the problem cannot be solved in a k-connected graph with k mobile adversaries, an interesting question is to determine the maximum number of mobile adversaries for which the problem is solvable in a k connected graph.

Das et al. [15] also considered the problem of gathering mobile agents in a graph in the presence of mobile faults that can appear anywhere in the graph. Here, they considered synchronous agents and presented new algorithms for the un-oriented ring graphs that solve strictly more cases than the ones solvable with asynchronous agents. Also, they showed that previously known solutions for asynchronous agents in the oriented ring can be improved when agents are synchronous. They (i) proved that in an un-oriented ring network of n nodes the problem is not solvable when n is odd and k (synchronous agents) is even, and presented an algorithm that solves the problem in all the other cases for $k > 2$, also in the case when n and k both are odd which was unsolvable with asynchronous agents and also (ii) discussed how to solve the problem when $k = 2$ agents and n is even, which was not solvable in the asynchronous setting, and how to solve the problem in the oriented ring by improving the already known algorithm for asynchronous agents. They left the following scope open for the future research work

– Problem of gathering in an un-oriented ring with $k = 2$ agents and constant memory, and the gathering of synchronous agents in other topologies.

Bampas et al. [2] have considered the problem of two anonymous mobile agents (robots) moving in an asynchronous manner have to meet in an infinite grid of dimension $\delta > 0$, starting from two arbitrary positions at distance at most d. Since the problem is clearly infeasible in such general setting, the authors have assumed that the grid is embedded in a δ-dimensional Euclidean space and that each agent knows the Cartesian coordinates only of its own initial position (and not the one of the other agents). The authors designed an algorithm which allows the location-aware agents to meet after traversing a trajectory of length $O(d^\delta \text{ polylog } d)$. This bound for the case of 2D-grids subsumes the main result of [8]. This algorithm is almost optimal, since the (d^δ) lower bound is straightforward. They also applied their rendezvous method to the following network design problem where the ports of the δ-dimensional grid have to be set such that two anonymous agents starting at distance at most d from each other will always meet, after traversing a $O(d^\delta \text{ polylog } d)$ length trajectory if they move in an asynchronous manner. They further applied their method to the geometric rendezvous problem of two anonymous location-aware agents moving asynchronously in the δ-dimensional Euclidean space (The agents have the radii of visibility of $r1$ and $r2$, respectively. Each agent knows only its own initial position and its own radius of visibility. The agents meet when one agent is visible to the other one) and proposed an algorithm designing the trajectory of each agent, so that they always meet after traveling a total distance of $O(\frac{d}{r})^\delta \text{polylog } (\frac{d}{r})$, where $r = \min(r1, r2)$ and for $r \geq 1$. This left the following open problems:

– Is it possible to design an optimal $O(d^\delta)$ rendezvous algorithm in δ-dimensional grids?
– Is it possible to extend the location-aware approach to some classes of graphs other than grids?
– Is it possible to use a more sophisticated port arrangements to enable the mobile agent using $O(\log d)$ or even a constant number of its memory bits?

5 Conclusion

In this paper, we have presented an extensive survey on the reported works on gathering of autonomous, homogeneous, location-aware mobile robots. We have studied gathering in different systems, like, a 2D Euclidean plane, a Line, a Ring, a Grid and systems with or without mobile faults and also both synchronous and asynchronous systems. Some of the models have provided algorithm for improved competitive ratio with the help of communication, whereas others have provided optimal or near-optimal algorithms to solve various gathering problems with the help of location-aware agents. From this survey, we can see that use of location-aware robots or providing robots with communication capabilities helps to solve some of the gathering problems which were previously unsolvable by the traditional model. Still there are many unsolved problems and a lot of scope of improvements in terms of communication and other abilities of the robots leaving plenty of scope for further research in this area.

References

1. Agmon, N., Peleg, D.: Fault-tolerant gathering algorithms for autonomous mobile robots. SIAM J. Comput. **36**(1), 5682 (2006)
2. Bampas, E., Czyzowicz, J., Gsieniec, L., Ilcinkas, D., Labourel, A.: Almost optimal asynchronous rendezvous in infinite multidimensional grids. In: Lynch, N.A., Shvartsman, A.A. (eds.) Distributed Computing (DISC). LNCS, vol. 6343 (2010)
3. Bampas, E., Blin, L., Czyzowicz, J., Ilcinkas, D., Labourel, A., Potop-Butucaru, M., Tixeuil, S.: On asynchronous rendezvous in general graphs. Theor. Comput. Sci. (2018). https://doi.org/10.1016/j.tcs.2018.06.045
4. Chuangpishit, H., Czyzowicz, J., Killick, R., Kranakis, E., Krizanc, D.: Rendezvous on a line by location-aware robots despite the presence of Byzantine Faults. In: Fernndez, A.A., Jurdzinski, T., Mosteiro, M., Zhang, Y. (eds.) Algorithms for Sensor Systems (ALGOSENSORS). LNCS, vol. 10718 (2017)
5. Cieliebak, M., Flocchini, P., Prencipe, P., Santoro, N.: Solving the robots gathering problem. In: Proceedings of 30th International Colloquium on Automata, Languages and Programming (ICLAP), pp. 1181–1196 (2003)
6. Cohen, R., Peleg, D.: Convergence properties of the gravitational algorithm in asynchronous robot systems. In: Albers, S., Radzik, T. (eds.) Algorithms. LNCS, vol. 3221 (2004)
7. Cohen, R., Peleg, D.: Convergence of autonomous mobile robots with inaccurate sensors and movements. In: Durand, B., Thomas, W. (eds) STACS 2006 LNCS, vol. 3884 (2006)
8. Collins, A., Czyzowicz, J., Gasieniec, L., Labourel, A.: Tell me where I am so I can meet you sooner. In: International Colloquium on Automata, Languages, and Programming. LNCS, pp. 502–514 (2010)
9. Collins, A., Czyzowicz, J., Gasieniec, L., Kosowski, A., Martin, R.: Synchronous rendezvous for location-aware agents. In: Peleg, D. (ed.) Distributed Computing (DISC). LNCS , vol. 6950 (2011)
10. Cord-Landwehr, A., Fischer, M., Jung, D., Meyer auf der Heide, F.: Asymptotically optimal gathering on a grid. In: 28th ACM Symposium on Parallelism in Algorithms and Architectures (SPAA), pp. 301–312 (2016)
11. Czyzowicz, J., Killick, R., Kranakis, E., Krizanc, D., Morales Ponce, O.: Gathering in the plane of location-aware robots in the presence of spies. In: Lotker, Z., Patt-Shamir, B. (eds.) Structural Information and Communication Complexity (SIROCCO). LNCS, vol. 11085 (2018)
12. Czyzowicz, J., Labourel, A., Pelc, A.: How to meet asynchronously (almost) everywhere. In: Proceedings of SODA 2010, pp. 22–30 (2010)
13. Czyzowicz, J., Labourel, A., Pelc, A.: How to meet asynchronously (almost) everywhere. ACM Trans. Algorithms **8**(4), 37 (2010). https://doi.org/10.1145/2344422.2344427

14. Das, S., Giachoudis, N., Luccio,F.L., Markou, E.: Gathering of robots in a grid with mobile faults. In: Catania, B., Krlovi, R., Nawrocki, J., Pighizzini, G. (eds.) Theory and Practice of Computer Science (SOFSEM). LNCS, vol. 11376 (2019)
15. Das, S., Focardi, R., Luccio, F.L., Markou, E., Moro, D., Squarcina, M.: Gathering of robots in a ring with mobile faults. Theor. Comput. Sci. **764**, 42–60 (2018). https://doi.org/10.1016/j.tcs.2018.05.002
16. Das, S., Liu, H., Kamath, A., Nayak, A., Stojmenovic, I.: Localized movement control for fault tolerance of mobile robot networks. In: Orozco-Barbosa, L., Olivares, T., Casado, R., Bermdez, A. (eds.) Wireless Sensor and Actor Networks (WSAN). IFIP International Federation for Information Processing, vol. 248 (2007)
17. Flocchini, P., Santoroy, N., Vigliettay, G., Yamashitaz, M.: Rendezvous of two robots with constant memory. In: Moscibroda, T., Rescigno, A.A. (eds.) Structural Information and Communication Complexity (SIROCCO). LNCS, vol. 8179 (2013)
18. Flocchini, P., Prencipe, G., Santoro, N., Widmayer, P.: Hard tasks for weak robots: the role of common knowledge in pattern formation by autonomous mobile robots. In: Algorithms and Computation, vol. 1741 of LNCS, pp. 93–102 (1999)
19. Flocchini, P., Prencipe, G., Santoro, N., Widmayer, P.: Arbitrary pattern formation by asynchronous, anonymous, oblivious robots. Theor. Comput. Sci. **407**, 412–447 (2008). https://doi.org/10.1016/j.tcs.2008.07.026
20. Flocchini, P., Santoro, N., Viglietta, G., Yamashita, M.: Rendezvous with constant memory. In: Structural Information and Communication Complexity*–20th International Colloquium, SIROCCO. 8179 (2013). https://doi.org/10.1007/978-3-319-03578-9-16
21. Gotsman, C., Lindenbaum, M.: On the metric properties of discrete space-filling curves. IEEE Trans. Image Process. **5**(5), 794–797 (1996)
22. Kranakis, E., Krizanc, D., Rajsbaum, S.: Mobile agent rendezvous: a survey. In: Flocchini, P., Gsieniec, L. (eds.) Structural Information and Communication Complexity (SIROCCO). LNCS, vol. 4056 (2006)
23. Peleg, D.: Distributed coordination algorithms for mobile robot swarms: new directions and challenges. In: Distributed Computing (IWDC). LNCS, vol. 3741, pp. 1–12 (2005)
24. Suzuki, I., Yamashita, M.: Distributed anonymous mobile robots—formation and agreement problems. In Problems. 3rd International Colloquium on Structural Information and Communication Complexity (SIROCCO), pp. 1347–1363 (1996)

An Energy Efficient Autonomous Street Lighting System

Pragna Labani Sikdar and Parag Kumar Guha Thakurta

Abstract An energy efficient street lighting system is proposed in this paper to reduce the energy consumption obtained by the streetlights. An autonomous adjustment of the brightness for the street lights is adopted by the sensing of pedestrians and vehicles. The street lights are lit with different levels of brightness in accordance with the circumstances. The parameters for estimating the energy consumption are discussed and subsequently, it is reduced. Various simulation results of our proposed work indicate an improvement over the existing approaches.

Keywords Streetlight · Energy · Brightness · Sensor · Traffic

1 Introduction

The street lights in a city are considered as an important facility to provide safety for the pedestrian and vehicles [1]. An efficient street lighting system plays a major role to reduce the percentage of road accidents. It has a potential impact to draw attention on the crime happening in the street. In order to operate street lights manually, most of the countries use ON-OFF strategies. As a result, there can be various circumstances where road lights are ON unnecessarily [2]. In conventional street lighting system, lights are continuously on with 100% brightness for total operational hours, and therefore, consuming a huge amount of electricity [3]. Especially, it causes wastage of power consumption during less traffic volume during midnight. It is known that the road lights use 19% of global electric energy [5]. Another study in [4] reports that the continuous emitting of road light consumes 114 TeraWatt hour (TWh) annually. It leads to the emission of 69 million tons of CO_2. According to statistics in 2006 [6], around 90 million streetlights are installed in all over the world and this number is growing day-by-day. In around 2050, the amount of streetlight is assumed to

P. L. Sikdar (✉) · P. K. G. Thakurta
Department of CSE, NIT Durgapur, Durgapur, India
e-mail: pragnait2013@gmail.com

P. K. G. Thakurta
e-mail: parag.nitdgp@gmail.com

© Springer Nature Singapore Pte Ltd. 2020
J. K. Mandal and S. Mukhopadhyay (eds.), *Proceedings of the Global
AI Congress 2019*, Advances in Intelligent Systems and Computing 1112,
https://doi.org/10.1007/978-981-15-2188-1_46

be increased by 300% [7]. In effect, it demands more power supply, and therefore increasing more CO_2 emission.

An energy efficient street lighting system is proposed in this paper to reduce the energy consumption obtained by the streetlights which, in turn, facilitate proper utilization of energy savings. In this work, it is assumed that the street lights are controlled by sensors where autonomous adjustment of brightness is adopted by the sensing of pedestrians and vehicles. Here, a specific light with its two neighboring lights, in left and right, lit with full brightness when the sensors can sense a pedestrian or vehicles in its sensing area. The brightness of other lights except those is adapted accordingly. In order to facilitate such scenario, a timely dimmed concept is used. As a result, the brightness of each light at specific time is considered as a parameter in order to calculate the energy consumption for the proposed system. Another parameter for estimating energy consumption is considered as the active time which signifies the duration of the light to be lit. The light posts in the proposed work are installed in a distance apart such that it can reduce the active time and average brightness of the lights. The simulation result of our proposed work shows an improvement over the existing approaches. The terms 'brightness' and 'illumination' are used interchangeably in this work.

The rest of this paper is organized as follows: System model is presented in Sect. 2. The problem is formulated in Sect. 3. Next, Sect. 4 describes the proposed approach. Various simulation results are shown in Sect. 5. The proposed work is concluded with a scope of future enhancement in Sect. 6.

2   System Model

Each streetlight (ST_i), as shown in Fig. 1, is attached with a sensor (S_i) and the

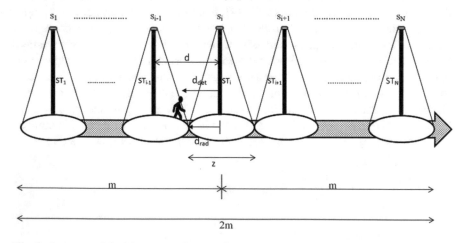

Fig. 1 System model of the proposed approach

brightness of the lights is adjusted by these sensors according to the Euclidian distance (d_{det}) from the pedestrian. Here, d_{rad} denotes the detection range of a sensor. Hence, the approximate distance d_{approx} from the pedestrian is determined in terms of d_{det} and d_{rad} as follows:

$$d_{approx} = \begin{cases} 0 & d_{det} \le d_{rad} \\ d_{det} - d_{rad} & \text{otherwise} \end{cases} \quad (1)$$

The total distance $2m$ meter (m meter for both ahead and back) is assumed to be lit with different illumination level for one pedestrian. This level is dependent on the number of illumination zones (Z_i) present on each side. If the inter-distance between two streetlights is d meter, then each 'm' meter is divided into different sub-regions $\text{SR} = \frac{m}{z} - 0.5$. It denotes the number of lights required to cover each 'm' meter. Here, one light is at the location of the pedestrian and the SR numbers of lights are assumed to be installed at the rest $\left(m - \frac{z}{2}\right)$ meter. Accordingly, the total number of lights N covering such 2m meter is required as follows:

$$N = 2 \times \text{SR} + 1 \quad (2)$$

Here, one light would be lit at the location of the pedestrian and SR number of lights would be lit at 'm' meter ahead as well as 'm' meter back of the pedestrian. The brightness of the streetlights on each side of a pedestrian is adjusted according to the illumination zone $Z(d_{approx})$ of the streetlight with respect to d_{approx}. Here, $Z(d_{approx})$ can be determined as follows:

$$Z(d_{approx}) = \begin{cases} 0 & \frac{d_{approx}}{d} = 0 \\ \frac{d_{approx}}{d} - 1 & 0 < \frac{d_{approx}}{d} \le \text{SR} \\ \text{SR} & \frac{d_{approx}}{d} > \text{SR} \end{cases} \quad (3)$$

3 Problem Formulation

In order to provide safety to the pedestrians and vehicles during night, a huge amount of energy by the streetlights is wasted for conceiving "always_ON" strategy. To reduce this energy consumption, it is better to turn off the lights occasionally. However, it does not provide safety to the passersby particularly at that time. Hence, an auto adjustment of the brightness level for the streetlights can be applied for further enhancement. Using this strategy, the illumination level of the nearest lights would lit with full brightness and the rest of the lights would gradually decrease by a factor of 1/SR. Hence, for any inter-distance d, between the street lights, the maximum number of sub-regions that are going to be lit on each side is SR. So, the brightness (φ) of each streetlight can be presented by the following.

$$\varphi = \left\{1 - \frac{1}{\text{SR}} Z(d_{\text{approx}})\right\} \times 100\% \tag{4}$$

In (4), the value 'φ' of the streetlight is adjusted in accordance with reducing by a factor of 1/SR from the full brightness, depending on the illumination zone $Z(d_{\text{approx}})$ they belong to.

The energy consumption $E(N)$ of the streetlights is dependent on the product of maximum power (P_{max}) of the light, the illumination level (φ) and the active time (T), i.e., the duration of a single time step n. Therefore, $E(N)$ can be estimated as follows:

$$E(N) = \min \sum_{n=0}^{N} P_{\text{max}} \varphi T \tag{5}$$

In (5), keeping P_{max} constant, the values of the other parameters such as φ and T can be reduced if the value of d is changed. Therefore, it can be presented by the following.

$$\text{minimize } E(N) = \min \sum_{n=0}^{N} P_{\text{max}} \varphi T \tag{6}$$

s.t:

$P_{\text{max}} = \text{constant}$

φ and T may vary for varying the values of 'd'.

Another important parameter for estimating $E(N)$ is the traffic volume (TV) per day. Hence, the problem addressed in this work on minimization of the value of $E(N)$ can be formulated by next:

$$\text{minimize } E(N) = \min \sum_{n=0}^{N} P_{\text{max}} \varphi T \times (\text{TV}) \tag{7}$$

4   Proposed Approach

The inter-distance (d) of the streetlights, distance (m) to be lit before and after a pedestrian, the speed (S) of a pedestrian and the detection range (d_{rad}) of the sensors are initialized by the user. The system would calculate the number of lights (N) to be lit with the illumination level (φ) for each pedestrian/vehicle according to the illumination zone $Z(d_{\text{approx}})$ on which a pedestrian belong to. Accordingly, the value of $E(N)$ is minimized. This procedure is described by the following algorithm.

initialize $m, d, S, d_{rad}, d_{det}, d_{approx} = 0$
if $d_{det} \leq d_{rad}$
    set $d_{approx} = 0$
else
    $d_{approx} = d_{det} - d_{rad}$
set $z = d$
calculate $SR = \left| \frac{m}{z} - 0.5 \right|$ and $N = 2 \times SR + 1$

if $\left\lceil \frac{d_{approx}}{d} \right\rceil = 0$
    then $Z(d_{approx}) = 0$
else if $\left\lceil \frac{d_{approx}}{d} \right\rceil > 0$ and $\left\lceil \frac{d_{approx}}{d} \right\rceil \leq SR$
    then $Z(d_{approx}) = \left\lceil \frac{d_{approx}}{d} \right\rceil - 1$
else
    $Z(d_{approx}) = SR$
calculate $\varphi = \left\{ 1 - \frac{1}{SR} Z(d_{approx}) \right\} \times 100\%$ and $T = \frac{d}{S}$

$$minimize\ E(N) = min \sum_{n=0}^{N} P_{max} \varphi T \times (TV)$$

4.1 Example

Two arbitrary values of 'd' are taken as 20 m (>30) and 40 m (>30) for illustrating the proposed work. Now for both cases, the values of φ and T are calculated accordingly which in turn effects the value of $E(N)$.

Case 1 (<30 m)

    When d is set to 20 m, then the value of N to cover $2m$ meter distance is 15 accordingly. Even if here N is increased, but the average value of both φ and T are reduced in this case. Here, the parameter φ for the nearest three lights from the pedestrian would be lit with 100% and for the other lights would gradually decrease by a factor of $1/SR$, i.e., 14.28%. It results into the average brightness of all 15 lights to be 60%.

    The active time in this case would be $T = \frac{20}{S}$, which is lower than the active time of the existing algorithm [4]. The following Fig. 2 shows the changing of the illumination level φ of different streetlights (Si) at different time. At time $t = 1$, the pedestrian is at S_8. Accordingly, φ for S_7, S_8, S_9 is 100% and it is gradually decreasing for other lights. At time $t = 2$, the pedestrian is at S_9 and the value of φ for the other lights changes accordingly and it is continued for the rest of the time.

Case 2 (>30 m)

    If d is set to 40 m, then accordingly the value of N to cover $2m$ meter distance is 9. For these 9 lights, three nearest lights from the pedestrian are lit with 100%

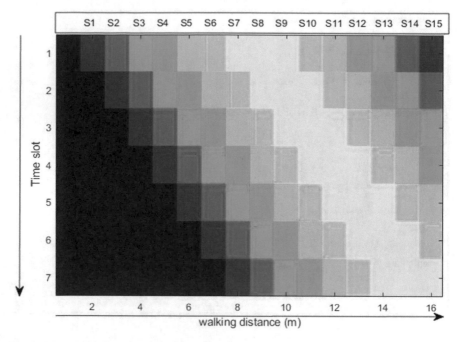

Fig. 2 Illumination level according to the position of pedestrian for inter-distance 20 m

brightness and the value of φ for other lights is gradually decreased by a factor of 1/SR or 25%. Thus, the resulting average brightness of all these 9 lights is 66.6667%.

Here, the value of T is equal to $\left(\frac{40}{S}\right)$ which is greater than the active time of existing algorithm [4]. Although it is observed that both φ and T are greater than the existing algorithm; however, the value of N is less than the existing one that in turn balancing the overall energy consumption in this case. In Fig. 3, the illumination level for different streetlights (Si) is shown at different time. At time $t = 1$, the pedestrian is at S_5 and according φ for S_4, S_5 and S_6 is at full brightness that of the other is gradually decreased by 25%. At time $t = 2$, the position of the pedestrian is shifted to S_6 and the value of φ is changed accordingly.

5 Simulation Studies

This section is divided into two sub-sections as discussed next.

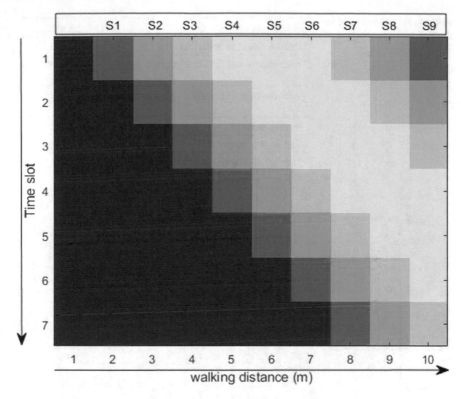

Fig. 3 Illumination level according to the position of pedestrian for inter-distance 40 m

5.1 Simulation Setup

In the proposed work, a road segment of 150 m is considered for both before and after a pedestrian to be lit when passing on the road. Generally, the value of d is considered as 30 m. The number of streetlights required to cover 300 m according to (2) is 11. In this case, the value of φ is 63.63%. The value of T becomes as $\left(\frac{30}{S}\right)$ unit. The following Table 1 summarizes the parameters with their values used for simulations.

5.2 Simulation Result

In order to estimate the value of $E(N)$, one of the essential factor is the traffic ratio per day. For the simulation purpose, a normalized traffic ratio from 12:00 midnight to the next day for both weekdays and weekends is used which is shown in Fig. 4. It is used for different traffic levels for a total of 7 days in a week. In Fig. 4, it is

Table 1 Parameters used in simulations

Parameters	Value
Road segment to be lit (m)	150 m
Inter-distance of the streetlight (for less than 30)	20 m
Inter-distance of the streetlight (for greater than 30)	40 m
Region covered	3.5 km
Traffic volume per day	438 vehicles
Number of streetlights (for less than 30)	152
Number of streetlights (for greater than 30)	92

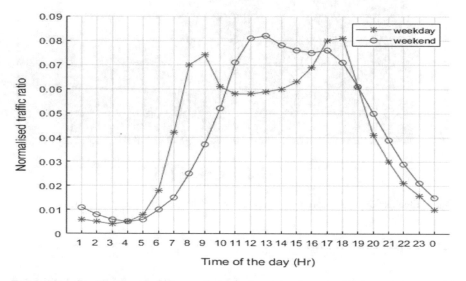

Fig. 4 Normalized traffic ratio during weekends and weekdays

observed that high volume of traffic uses the roads during office hours, i.e., from 06:00 to 08:00 and from 16:00 to 20:00 in weekdays. The traffic volume other than these time periods is comparatively less. On the other hand, high traffic volume is usually observed to use the road from 09:00 to 20:00 in weekend.

For the completeness of the work, Table 2 shows the existing manual lighting schemes with their operational hours.

The duration from 16:00 of a day till 8:00 in the next morning is considered for estimating the energy consumption by the streetlights in the proposed work. Here, the value of φ is considered as 100% in the rush hours, i.e., from 16:00 to 20:00 and from 6:00 to 8:00. Further, this value of φ becomes as average while the traffic volume is considerably less. The energy consumption for various number of traffic volume per day is determined as well as compared with different existing approaches which is shown in Fig. 5. It is observed here that the energy consumed by the streetlights in

Table 2 Existing manual lighting schemes with operational hours

Manual lighting schemes	Operational hours
Conventional	All the streetlights are on with 100% brightness in total operational hours
Chronosense	The brightness of the streetlights is reduced to 65% from 22:00 to 05:00
Part-night	All the streetlights are switched off from 12:00 to 5:30 and lit with 100% brightness for the rest of the operational hours
Dynadimmer	Streetlights kept on with 65% from 20:00 to 23:00, 40% from 23:00 to 05:00, 55% from 05:00 to 06:00 and 100% for the rest of the operational hours

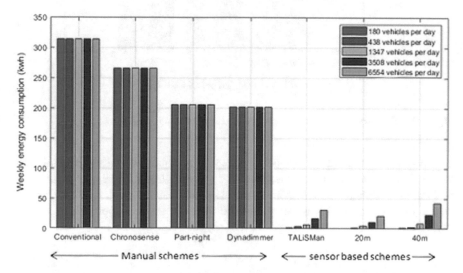

Fig. 5 Comparison between proposed work and existing approaches in terms of weekly energy consumption

manual lighting scheme is not dependent on the traffic volume. Here, the conventional scheme consumes highest energy as compared to the other schemes. Although the existing schemes such as chronosense, part-night and dynadimmer consume 15.33%, 34.50% and 35.78% less energy than the conventional one, respectively. However, the proposed work in sensor-based approach consumes less amount of energy for case 1 and case 2 discussed earlier over the existing work TALiSMaN [4].

For better understanding, we have compared the weekly energy consumption with the existing approach by creating a normalization factor for a traffic volume of 438 vehicles per day. For the sensor-based proposed work, the number of streetlights for different brightness may vary to obtain the simulation result. To cover 3.5 km regional area, the number of streetlights required for TALiSMaN, case 1 and case 2 discussed earlier are 112, 152 and 92, respectively. So, at any point of time, even in

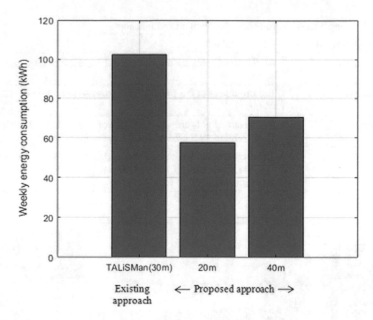

Fig. 6 Weekly energy consumption for normalized number of streetlights

rush hours, the value of the number of streetlights would not exceed beyond these values. Under such scenario, it is impossible to calculate the energy consumption without normalization as it provides an ambiguous result. In Fig. 6, the case 1 and case 2 for the proposed work are compared over the existing approach by using a normalization factor.

6 Conclusion

In this paper, an autonomous lighting system is proposed for minimizing the energy consumption by the streetlights. Here, the proposed approach is compared over the existing manual approaches and also sensor-based approaches in terms of energy consumption. Different illumination level and active time as well as the traffic volume are taken into consideration for minimizing the energy consumption. The installation cost of the streetlights for different inter-distance would be the future focus of this work.

References

1. Mohandas, P., Dhanaraj, J.S.A., Gao, X.Z.: Artificial neural network based smart and energy efficient street lighting system: a case study for residential area in Hosur. Sustain. Cities Soc. **48**, 101499 (2019)
2. Marino, F., Leccese, F., Pizzuti, S.: Adaptive street lighting predictive control. Energy Procedia **111**, 790–799 (2017)
3. Lau, S.P., Merrett, G.V., White, N.M.: Energy-efficient streetlighting through embedded adaptive intelligence. In: 2013 International Conference on Advanced Logistics and Transport, pp. 53–58. IEEE (2013)
4. Lau, S.P., Merrett, G.V., Weddell, A.S., White, N.M.: A traffic-aware street lighting scheme for Smart Cities using autonomous networked sensors. Comput. Electr. Eng. **45**, 192–207 (2015)
5. Castro, M., Jara, A.J., Skarmeta, A.F.: Smart lighting solutions for smart cities. In: 2013 27th International Conference on Advanced Information Networking and Applications Workshops, pp. 1374–1379. IEEE (2013)
6. IEA, P.W.: Light's Labour's Lost. Policies for Energy-Efficient Lighting (2006)
7. Northeast Group, LLC, Global LED and Smart Street Lighting: Market Forecast (2014–2025), Washington: Northeast Group, LLC (2014)

A Survey on Gathering of Distributed Mobile Agent in Discrete Domain

Himadri Sekhar Mondal and Sruti Gan Chaudhuri

Abstract This paper presents a survey on gathering problem of a set of autonomous programmable mobile agents popularly known as robot swarm, moving on discrete plane, i.e. along the ages of a graph and gather at a particular node. The mobile agents are distributed in nature, i.e. they are not controlled by any centralized authority. The main objective of these robots is to execute a job in a collaborative manner. Gathering is one form of such coordination problems. Gathering of swarm robots is a very popular field of research in swarm robotics where the researchers have addressed the problem through artificial intelligence aspects. Many popular simulations and heuristics have been developed for swarm robots considering the learning capabilities of the robots. There also exists a parallel theoretical field of swarm robotics which studies deterministic algorithms for the swarm robots. This field is more interested about the feasibility and solvability of a problem along with algorithm analysis. In this paper, we are going to explore these theoretical studies on gathering problem.

Keywords Mobile robots · Gathering · Discrete domain

1 Introduction

The mobile entities in distributed system are often known as robots or mobile agents or swarm robots. A distributed system is a collection of multiple computing devices that work in collaboration to achieve a common goal. If a big expensive centralized robot is considered, the system will depend on that robot. Now, if this robot fails, then the system becomes inactive. The replacement or repairing of this robot is expensive and time consuming. The system remains inactive till this robot is repaired. On the other hand, if the system of a single robot can be replaced by a system of

H. S. Mondal · S. G. Chaudhuri (✉)
Jadavpur University, Kolkata, India
e-mail: sruti.ganchaudhuri@jadavpuruniversity.in

H. S. Mondal
e-mail: himadri.mondal93@hotmail.com

© Springer Nature Singapore Pte Ltd. 2020
J. K. Mandal and S. Mukhopadhyay (eds.), *Proceedings of the Global
AI Congress 2019*, Advances in Intelligent Systems and Computing 1112,
https://doi.org/10.1007/978-981-15-2188-1_47

multiple robots working together, then the failure of a single robot will not make the system stop. In a system of multiple robots, the robots are less expensive, and their maintenance is also easier. The multi-robot system can work collectively to execute some job involving motion [19, 20], e.g. moving a big object, cleaning a big area or covering a big area, etc. The robots can also make a 3D structure by themselves [15]. U.S. Navy has created a swarm of boats which can track an enemy boat and surround it and destroy it [14, 18].

The distributed model of mobile robots consists of a set of small identical robots having similar capabilities. They cannot communicate with each other by passing explicit messages through wired or wireless medium. Robots move on a 2D or 3D plane or sometimes they have been considered to be placed on the nodes of a graph and move along the edges. They get the locations of the other robots by sensing/observing (taking snapshot) their surroundings. Depending upon the collected positions of the other robots, they compute a destination to move to and move there. After reaching the destination, they flash out the computational data and again start a fresh observation. Their sensing/observing range may vary depending upon the quality of their sensing devices. All robots may not observe, compute or move simultaneously. A robot can observe when other robots are moving. Hence, they may get inconsistent positions of the moving robots. The main challenge to design movement strategies or algorithms (which are feed to the robots to be executed during computation) will be free from the effect of this inconsistency of fetching positional data of the robots. The jobs performed by these robots need to form several geometric pattern by changing their positions or after completing the job, meeting to a location. Since, the robots cannot remember any computational data, unless this meeting point is not given in advance, there is no straightforward way to fix this point and keep this intact. The distributed algorithms for mobile robots address these types of problems which is also the scope of this paper.

2 Computational Model

The robots under the discussed model are having motorial capabilities and sensorial capabilities by which they can freely move along the ages of a graph and sense the position of the other robots in the graph. Additionally, they are

– Homogeneous (all executing the same deterministic algorithm)
– Autonomous (no centralized control)
– Oblivious (no memory of the past events)
– Anonymous (no unique identifiers).

The following are the description of the special features of these robots.
Physical characteristics: The robots are autonomous and identical in appearance and operations. They do not have any identification marks. The robots cannot differentiate between other robots by their appearance.

Communication: The robots cannot send or receive explicit messages to or from the other robots. However, they are enabled by sensing capability by which they can collect the positional information of the robots.

Operational cycle: The robots neither need any human intervention, nor are they controlled by any central system. The robots may be either active or inactive. When active, they are operated by executing a cycle of their own repeatedly. The cycle consists of the following phases (Fig. 1):

- **Wait**: The robots are inactive
- **Look**: The robot observes its surroundings and takes snapshots of the position of the other robots. They mark these positions in their local coordinate system.
- **Compute**: According to the positions of the robots found in LOOK phase, the robot computes a destination to move to.
- **Move**: The robot moves to its computed destination.

Oblivious: The robots remove all their computed data of the current cycle after reaching their destinations and start a fresh cycle.

Scheduling: There are three schedulings of the activation cycle.

- Fully synchronous model [FSYNC]: There is a global clock. At each cycle, all robots are activated and execute the cycle simultaneously.
- Semi-synchronous model [SSYNC]: There is a global clock but, at each cycle, a robot may be activated or not. When activated, the robots are FSYNC.
- Asynchronous model [ASYNC]: There is no global clock. Each phase of a cycle may have any finite duration, and different robots' executions are completely independent.

Environment: The robots arc deployed on a graph in which they move along the edges and reside on nodes.

Agreement: Each robot has its own compass in which the direction and orientation of the axes (South–North–East–West) may or may not be the same for all the robots.

Fig. 1 Operation cycle of robot

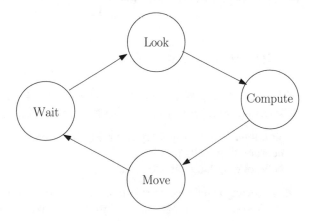

Note that the concept of agreement is getting its meaning when we refer the *graph drawing* otherwise the graph in the environment has no such agreement.

Visibility: Each robot senses or observes the presence of other robots around itself up to a certain range which is the visibility of the robot.

- **Global**: When the robots can observe or sense all the robots present in the system.
- **Local**: When the robots can observe or sense only up to their k hop neighbours, where $k = 1, 2, \ldots$.

Multiplicity detection: A node of the graph may consist of more than one robot. This is called multiplicity of a node [8]. When a robot is observing, it may or may not differentiate between a node having single robot and a node having multiple robots. When a robot can only identify a node with multiple robots but cannot count the exact number of the robots in that node, it is called *weak multiplicity*. On the hand, when a robot can also compute the exact number of robots on a node of multiplicity, it is called *strong multiplicity*. A robot may or may not be able to identify the multiplicity of the node it resides. When a robot can detect the multiplicity of the node it resides, it is known as **local multiplicity**. When a robot can detect the multiplicity of other nodes, it is known as **global multiplicity**. There are four possible variations.

- **Global-Weak**
- **Global- Strong**
- **Local- Strong**
- **Local-Weak.**

The model just described may be seems to be unrealistic and having many limitations in terms of resource usability such as communication power or memory use. However, the main objective of developing this model is to use the minimal resource as much as possible. Sometimes adding a resource is not a big issue but maintaining this resource efficiently requires another attention. Thus, one of the interests of making this model is to study the possibility of solving a problem with very minimal available resources. A new resource is only taken when it is proved that without that resource the problem is unsolvable.

3 Reported Research Works

In this section, we present a brief survey of the existing recent research works. There exists a huge number of works in this field in various aspects such as

- variations in graph topology like grid, ring, tree.
- incorporating faults in robots.
- incorporating faults in networks.

The following results are chosen to cover all these domains.

Baldoni et al. [3] proposed an algorithm to explore an anonymous, partial grid by a set of mobile anonymous robots. The grid is partial means the grid is finite, but some nodes of the grid may be missing. Robots cannot communicate with each other. Robots cannot see the total grid; a part of grid is visible to them. The aim of this paper is to design an algorithm executed by each robot to visit all nodes of the grid. No node can hold more than one robot at a time. This paper focuses to solve constrained perpetual graph exploration problem, where perpetual exploration means robots have to revisit the nodes of the graph for exploration.

Bonnet et al. [5] proposed an algorithm to investigate the exclusive perpetual exploration of grid-shaped network. They considered a set of anonymous, oblivious and fully asynchronous robots. No axis agreement is considered for this problem. The main focus of this paper is to get the minimum number of robots that are necessary and sufficient to solve the problem of exploration in general grids. It is proved that three deterministic robots are necessary and sufficient, provided that the size of the grid is $n \times m$ with $3 \leq n \leq m$ or $n = 2$ and $m \geq 4$. The authors have also got a result that exclusive perpetual exploration requires as many robots in the ring as in the grid. Their proposal on defining and classifying configurations considerably simplifies the design and verification of the proposed algorithms. It is believed that it can be extended to address an arbitrary number of robots and be a first step in providing a complete framework to study coordination problems in mobile robots networks.

D'Angelo et al. [8] studied the gathering problem on grid networks. A team of asynchronous robots placed at different nodes of a grid has to meet at some node and remains there. Robots operate in LOOK-COMPUTE-MOVE cycles. In this paper, a full characterization about gatherable configuration for grid is shown without multiplicity detection. In this paper, they gave a brief about gatherable configuration of grid. According to this paper, multiplicity detection is not required for this process. If there is a 2×2 grid or it is periodic on a grid with at least an even side, or it is symmetric with the axis passing through edges, then this process fails to gather. It also raises a question that whether the grid topology is the least structured class of graphs that permits to avoid the multiplicity detection assumption.

Stefano and Navarra [9] have set a goal to provide algorithms with respect to different assumptions about the capabilities of the robots as well as the topology of the underlying graphs like trees and ring. In this paper, they were interested in minimizing the number of movements of the robots. However, it opens a vast interest in segment of twofold objective function with the help of previously briefed technology and algorithms as they not only require the gathering objective but also to reduce the moves.

Stefano and Navarra [10] have considered a group of robots randomly placed on various nodes of the grid. The proposed method is about gathering robots at a point within minimum numbers to gather robot in a point. They have considered infinite grids as input graph to perform this task.

Abshoff et al. [1] considered a two-dimensional gathering problem for n number of robots. Given a swarm of n indistinguishable, point-shaped robots on a two-dimensional grid. Initially, the robots form a closed chain on the grid and must keep this connectivity during the whole process of their gathering. Connectivity means that neighbouring robots of the chain need to be gathered at the same or neighbouring points of the grid. In this model, gathering means to keep shortening the chain until the robots are located inside a 2×2 sub-grid. The model is completely local (no global control, no global coordinates, no compass, no global communication or vision). Each robot can only see its next constant number of left and right neighbours on the chain. It needs $O(n^2)$ run time.

Cord-Landwehr et al. [7] have solved the gathering problem for n indistinguishable, point-shaped robots on a 2D grid in $O(n)$ time. Robots are arbitrarily distributed in the grid. Visibility is limited (one hop), i.e. it can see its all neighbours in the one hop visibility range. Robots are merged without breaking the connectivity of the robots.

Dutta et al. [11] have presented gathering algorithm for ring and grid network. Here, for grid network, they have considered that the robot has two hop visibility range and complete axes agreement of the robots. The gathering takes place by maintaining the visibility connection between the robots. For gathering in ring, the authors considered that the robots agree on orientation of the ring. Both the cases, the robots do not need to detect multiplicity of a node.

Bhoumik and Gan Chaudhuri [4] have proposed a gathering algorithm in tree considering that the gathering node is given in advanced. However, this paper considers that the robots can see only its neighbours.

Chen et al. [21] have analyzed mobile agent routing plan on the basis of two main key factors: node energy consumption and time consumption in the process of mobile agent migration. An improved Dijkstra algorithm is designed to solve this model. They have simulated the algorithm where it is shown that the proposed algorithm can get the optimal route between any two nodes in wireless sensor network. In this paper, characterization of a single mobile agent in WSN is calculated. However, there is a future scope for studying and developing about multiple mobile agents routing in the wireless sensor network. There is also scope for balancing the energy consumption and the time consumption for mobile agent migration.

Chalopin et al. [6] have described an process to gather group of asynchronously moving identical mobile agents to gather in a single node of graph. The graph has faulty or malicious edges which destroy the robots when the robots cross those edges. The objective of this paper is to minimize the number of agents to be destroyed and gather the survived ones in a single node of the graph. Their algorithm does not need to know about the full network topology. A tight bound on the network size is enough for solving the problem. The paper considered two optimization criteria minimizing the number of agents that are destroyed and the number of moves taken by the surviving agents to gather at the same point and same time. As an extension, it would be interesting to consider optimizing the whiteboard memory or agent memory when solving the same problem in faulty networks.

Dovrev et al. [12] have considered the presence of malicious nodes also called black holes which destroy the robots and have shown that two mobile agents can search the black holes with O(n) number of moves in hypercube and related network. Black hole is a biggest threat for any network environment. It destroys the agents and does not leave any traces. This paper raises an open question to find out networks for which the cost (moves performed by the agents) will be $\theta(n)$.

Kawai et al. [13] have considered the gathering of multiple mobile agents in anonymous unidirectional ring networks under the assumption that for each robot, the number of nodes and the number of agents will be unknown. Firstly, they have proved that for any small constant $p(0 < p \leq 1)$ that there exists no randomized algorithm that solves, with probability p, the rendezvous problem with terminal detection. For this type of instances, they have developed the relaxed rendezvous problem called the rendezvous problem without detection that does not require termination detection. They have proved that there exists no randomized algorithm that solves, with probability 1, the rendezvous problem without detection. For the remaining cases, they showed the possibility, that is, a randomized algorithm that solves, with any given constant probability $p(0 < p < 1)$, the rendezvous problem without detection.

4 Future Scope

The previous section gives a snap of the various types of problems on theoretical studies on swarm robots. All the theories have one common objective that is to use the resources efficiently. There exists a vast domain of open problems, such as

- To characterize various kind of graphs(tree, grid, ring) for gathering problems in the presence of faulty nodes or faulty links (for dangerous graph [6]) or in the presence of malicious robots.
- Gathering in graph in the presence of faulty robots is another emerging field.
- Gathering in graph considering the robots have limited visibility, e.g. k hop visibility is a direction of the future scope which is addressed by very few researchers.

5 Conclusion

This paper presents a distributed model of swarm robots and related research works on that. This field is an emerging field of robotics on which less researchers work comparative to traditional popular filed of robotics. However, there exists a huge amount of open problems to be solved and implemented in real time to minimize the resource usages. This paper addresses only gathering problem, but arbitrary pattern formation algorithms or scattering algorithms are other open scopes of investigations.

References

1. Abshoff, S., Cord-Landwehr, A., Fischer, M., Jung, D., Meyer auf der Heide, F.: Gathering a closed chain of robots on a grid. In: 2016 IEEE International Parallel and Distributed Processing Symposium(IPDPS), Chicago, IL, USA, pp. 23–27 (2016)

2. Arora, P., Gupta, R.: Application of mobile agents: a canvas. Int. J. Adv. Res. Comput. Sci. (IJARCS) **1**(3) (2010)
3. Baldoni, R., Bonnet, F., Milani, A., Raynal, M.: On the solvability of anonymous partial grids exploration by mobile robots. In: 12th international conference on principals of distributed system (OPODIS), lecture notes in computer science, vol. 5401, pp. 428–445 (2008)
4. Bhoumik, S., Gan Chaudhuri, S.: Gathering of asynchronous mobile robots in a tree. In: IEEE Conference, Applications and Innovations in Mobile Computing (AIMoC), pp. 97–102 (2015)
5. Bonnet, F., Milani, A., Potop-Butucaru, M., Tixeui, S.: Asynchronous exclusive perpetual grid exploration without sense of direction. In: 15th International Conference on Principles of Distributed Systems (OPODIS), Lecture Notes in Computer Science, vol. 7109, pp. 251–265. Springer (2011)
6. Chalopin, J., Das, S., Santoro, N.: Rendezvous of mobile agents in unknown graphs with faulty links. In: Pelc, A. (eds.) Distributed Computing (DISC). LNCS, vol. 4731 (2007)
7. Cord-Landwehr, A., Fischer, M., Jung, D., Meyer auf der Heide, F.: Asymptotically optimal gathering on a grid. In: SPAA 2016, pp. 301–312 (2016)
8. D'Angelo, G., Di. Stefano, G., Navarra, A.: Gathering of robots on anonymous grids without multiplicity detection. In: 19th International Colloquium on Structural Information and Communication Complexity (SIROCCO), LNCS, vol. 7355, pp. 327–338 (2012)
9. Di.Stefano, G., Navarra, A.: Optimal gathering of oblivious robots in anonymous graphs. In: Moscibroda, T., Rescigno, A. (eds.) Structural Information and Communication Complexity (SIROCCO). LNCS, vol. 8179 (2013)
10. Di.Stefano, G., Navarra, A.: Optimal gathering on infinite grids. In: Felber, P., Garg, V. (eds.) Stabilization, Safety, and Security of Distributed Systems. SSS 2014, LNCS, vol. 8756 (2014)
11. Dutta, D., Dey, T., Gan Chaudhuri,S.: Gathering multiple robots in a ring and an in nite grid. In: Distributed Computing and Internet Technology, ICDCIT 2017. LNCS, vol. 10109, pp. 15–26. Springer, Cham (2017)
12. Dovrev, S., Floccini, P., Kralovic, R., Principe, G., Ruzicka, R., Santoro, N.: Black hole search by mobile agents in hypercubes and related networks. In: On Principles of Distributed Systems (OPODIS), pp. 171–182 (2002)
13. Kawai, S., Ooshita, F., Kakugawa, H., Masuzawa, T.: Randomized rendezvous of mobile agents in anonymous unidirectional ring networks. In: Even, G., Halldórsson, M.M. (eds.) Structural Information and Communication Complexity (SIROCCO). LNCS, vol. 7355 (2012)
14. Lendon, B.: U.S. Navy could 'swarm' foes with robot boats. https://edition.cnn.com/2014/10/06/tech/innovation/navy-swarm-boats/index.html (2014)
15. Michael, N., Fink, J., Kumar, V.: Cooperative manipulation and transportation with aerial robots. Auton. Robots **30**(1), 73–86 (2011)
16. Prakash, A.: Swarm Robotics: New Horizon of Military Research. https://www.roboticsbusinessreview.com, May 23, 2018
17. Rahmani, M.: Search and Rescue by Swarm Robots. Semantic Scholar (Submitted for publication)
18. Steinberg, M.: Intelligent autonomy for unmanned naval systems. In: Proceedings of Defense and Security Symposium, p. 623013. International Society for Optics and Photonics (2006)
19. Sze Kong, C., Ai Peng, N., Rekleitis, I.: Distributed coverage with multi-robot system. In: 2006 IEEE International Conference on Robotics and Automation, pp. 2423–2429 (2006)
20. TresanchezRibes, M.: Optimization of floor cleaning coverage performance of a random path-planning mobile robot. http://hdl.handle.net/2072/4296
21. Zhu, D., Yuan, H., Yan, J., Qing, Y., Yang, W.: Data fusion analysis with optimal weight in smart grid. In: 2019 International Conferenceon Artificial Intelligence in Information and Communication (ICAIIC), pp. 326–331 (2019)

Restaurant Recommendation System Based on Novel Approach Using K-Means and Naïve Bayes Classifiers

Shreya Joshi and Jigyasu Dubey

Abstract Human opinion has become the most useful metric for recommendation systems in recent years. Before buying any product, we are looking for ratings. Growth in social media has provided an opportunity to know what interests like-minded people. In this paper, we present a k-means nearest neighbor and Naïve Bayes classifier-based systems for recommendation of restaurants. To make precise predictions and provide proficient recommendations, the data and method are most important factors. By thoroughly analyzing the literature, we came across the fact that the Facebook and Yelp are most successfully used datasets. While working, we realized the issues related to Yelp in searching from outcome of first search and time consumption. We chose Zomato that is most used restaurant-finding application which provides many attributes and users ratings. We have used Zomato data along with k-means and Naïve Bayes and achieved overall accuracy of 93%.

Keywords Recommendation system · Classification · Content-based method · Collaborative approach · Accuracy · Naïve Bayes

1 Introduction

This cultural impact nowadays performs a significant part in choice as we depend on people's views. Traditional systems were not the situation. The recommendation scheme offers a variety of alternatives for choosing the finest alternatives accessible. We discover many rating-based advice schemes. These schemes are focused on mammoth data filtering. One field, which receives comparatively little more attention in terms of recreation and enjoyment, is the advice of food products [1].

It is hard to choose which place to go to when you want to eat out. It even gets hard when you find something appropriate for a bigger community because the selection

S. Joshi (✉) · J. Dubey
Shri Vaishnav Vidhyapeeth Vishwavidhyalay, Indore, India
e-mail: shreya.270892@gmail.com

J. Dubey
e-mail: jigyasudube@yahoo.co.in

© Springer Nature Singapore Pte Ltd. 2020
J. K. Mandal and S. Mukhopadhyay (eds.), *Proceedings of the Global AI Congress 2019*, Advances in Intelligent Systems and Computing 1112,
https://doi.org/10.1007/978-981-15-2188-1_48

time increases proportionally with enormous options available [2]. We are providing a study of accessible methods and assessment there. This motivates us to generate a suggestion request centered on personal information or dataset to suggest the finest restaurant depending on the characteristics needed.

2 Literature Review

Contemporary work done by authors in different sub-domains related to our approach is being presented in Table 1.

2.1 Naïve Bayes

Naïve Bayes is a basic, yet powerful and regularly utilized, artificial intelligence (AI) classifier. It is a probabilistic classifier that makes classifications utilizing the maximum, which uses a posteriori choice standard in a Bayesian setting. It can also be represented to utilizing a basic Bayesian system. Naïve Bayes classifiers have been particularly mainstream for text classification, and are traditional solutions for issues, for example, spam detection. The Naïve Bayes classification is an integrated approach using a supervised-learning method and statistical method for categorization [8]. Naïve Bayes classifiers are renowned algorithms that figure out how to classify images.

The Naïve Bayes classifier being a generative classifier technique that works on Bayes' Theorem, it calculates the posterior probability $P(c|x)$ from prior probability of class $P(c)$ and predictor $P(x)$, and the probability of predictor given class $P(x|c)$. The result of the prediction is the class containing the highest posterior probability [9].

Naïve Bayes is examined comprehensively ever since the 1950s and has continued to be a standard strategy for text classification. The issues of making a decision about reports like being appropriate to one gathering or the other, along with word frequencies have been the features of the algorithm. With appropriate preprocessing, it is self sufficient in this research area with additional creative techniques that comprise support vector machines [10].

2.2 K-Means Algorithm

As indicated by numerous researchers, k-means is amongst one of the top ten algorithms used in data mining [11]. The algorithm uses an iterative clustering technique that partitions the available dataset into K clusters where the cluster size K is chosen by the user. Throughout iterative regression, it locates an ideal situation for K centers

Table 1 Work of contemporary authors

S. No.	Title	Year	Methods	Key points
1	Recommendation System based on Item and User Similarity on Restaurants Directory Online [3]	2018	Recommendation system based on item and user similarity	Models were built over a longer time span which produced better F1-measure values
2	An Improved Restaurant Recommendation Algorithm Based on User's Multiple Features [4]	2018	Improved collaborative filtering method (ICFM)	Similarity of user preferences plays the most important role in influence between users
3	Social Graph based Location Recommendation using User's behavior by locating the best route and dining in best restaurant [4]	2016	Sentiment Mining	Accuracy is 70% matched with the Zomato rating
4	Location, Time, and Preference Aware Restaurant Recommendation Method [5]	2016	User preference and restaurant popularity	Customer reviews can be used in metric for better recommendation
5	Restaurants Review Star Prediction for Yelp Dataset [6]	2015	Linear regression, random forest tree and latent factor model	Random forest regression has a better performance, since it uses some reasonable features extracted
6	A Chinese Restaurant Recommendation System Based on Mobile Context-Aware Services [7]	2013	Combination of personal preferences and locations to rank restaurants	Improved capacity to accomplish the user requirements and preferences for restaurant suggestions and recommendations

and computes the summation of all distances and finds the least in the category. It is a supervised-learning algorithm with the goal that the underlying classes may be not identified.

3 Data Source Comparison

Among the many websites which host reviews for local businesses, Yelp is the most popular choice and the most influential local review site. This was based upon a survey conducted by Yelp and Nielson in 2014 [12]. Now, since websites like Yelp have been built, people can check information before directly visiting a restaurant. The available information can be reviews, ratings, price ranges, and hours of operation. After evaluating prices, times, and reviews the decisions about physically going for visit or not can be made.

Although reviews provide valuable recommendations about restaurants, scrolling multiple pages to decide on which restaurant is best, user's potential to satisfy their personal appetite or their own food preference is much more important. Above that, when people visit a new place, move or travel, users have to go through the same search procedure again. This can be tedious and sometimes annoying [12]. A problem we found is that Yelp does not provide proper search results when people conduct their sub search.

Social network data is another major source of information. People mostly rely on for recommendation from the social media platform. Zomato is an application which covers almost every small or big restaurant. Its interface as API analysis is one of the most useful analyses for foodies who want to taste the best cuisines of every part of the world which lies in their budget. This analysis is also for those who want to find the value for of restaurants in various parts of the country for the cuisines. Additionally, this analysis caters the needs of people who are striving to get the best cuisine of the country and which locality of that country serves those cuisines with maximum number of restaurants.

4 Process Flow for Food Recommendation System

The prime goal of our proposed work is to analyze, understand, and recommend the restaurant on the basis of restaurant ratings and services. Figure 1 represents the process flow of our restaurant recommendation system.

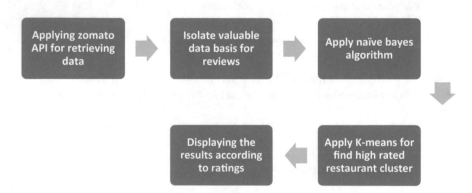

Fig. 1 Process flow for recommendation system

Table 2 Selected variables

Average cost for two	Has table booking	Has online delivery	Is delivering now	Switch to order menu	Price range	Aggregate rating	Votes	Rating category (RC)

4.1 Applying Zomato API for Retrieving Data

We have used Zomato API for extracting data from big and small restaurant of every part of the world.

4.2 Isolate Valuable Data Basis for Reviews

After gathering of information dependent on the kind of information created regarding number of times the keyword referenced or state the appraisals on a specific scale, we will choose the parameter for scaling the data. This data had unstable rating variables. There were two variables, "rating_cat" and "rating_num" for rating but sometimes, they did not indicate similar score. It was causing a severe harm to proposed model. The reason, creating the difference between ratings, looked like a small votes. Dropping off categorical variables and where the difference is niche and make new categorical variables. The selected variables are shown in Table 2.

The rating category was defined as:

'Not rated': -1, 'Poor': 0, 'Average': 2, 'Good': 3, 'Very Good': 4, 'Excellent': 5

4.3 Apply Naïve Bayes Algorithm

We have applied Gaussian Naïve Bayes classifier on data for it was split in the ratio of 70 and 30%, respectively, for training and testing (Fig. 2).

For example, the conditional probability in our case is given as

$$P(X) = (P(RC \geq 3 | \text{cost for two} = \text{Average}) / P(RC \geq 3)$$

4.4 Apply K-Means for Find High-Rated Restaurant Cluster

We have used k-means nearest neighbor algorithm to discover high-evaluated restaurant clusters in city. Rating-based different clusters were made and information was

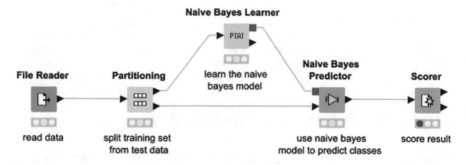

Fig. 2 Process flow of Naïve Bayes learner [13]

gathered. We had $K = 7$ and were having seven different cluster named as 0–6. With cluster-based analysis, we were able to find out

- The areas/localities having restaurants with high rating, It will help in avoiding areas having below average rating. We can divide the locality on basis of rating such as area where restaurants with average rating of two are situated.
- Restaurant with more number of unique cuisines have higher acceptability.
- Restaurants providing more services are having higher ratings.
- Appropriate average cost for two better rating distribution.

4.5 *Displaying the Results According to Ratings*

This is the last stage where the required information from the Zomato data will be shown with the suggestions and finally, the rating is displayed on the basis of services, ambience, and food ratings.

5 Implementation and Result Analysis

The evaluation of the highest-rated and lowest-rated restaurants of the city in all the countries was done but only for countries with maximum restaurants (India) were used. This was done using the grouping functions like pandas which helped in grouping 'Country' and 'City' to calculate the aggregate rating of selected country or city. Later, the top- and least-rated restaurants were found for every city in that country. The figure below describes the maximum ratings and the voters count for the selected geographic domain (Fig. 3).

According to the chosen dataset, Zomato, most rated data was found in the Delhi NCR region. Thus, the top most ratings and number of restaurants in this region were

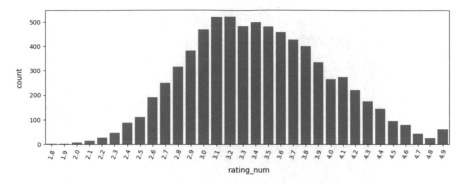

Fig. 3 Normal distribution of restaurant ratings

very high compared to other areas. According to the analysis, the top six cities are mentioned in Fig. 4.

The present dataset of Zomato was modified on basis of ratings. On the application of Gaussian Naïve Bayes algorithm, it was classified as 70% of the total as training data and 30% as the testing data. Total number of samples was listed as 17,185. Amid of all, only 53 were misclassified (Fig. 5).

In order to find high-rated restaurant clusters in city, we used k-means nearest neighbor. As a result, seven clusters were created on the basis of the ratings and the items were grouped. It was observed that the average rating of each cluster varies in accordance with the designated clusters.

From Fig. 6, it can be interpreted as the Local area 2, 4 have slightly higher scores than the suburbs area in terms of ratings. And through the vote counts, we can state

Fig. 4 Top six city based on number of restaurant in India

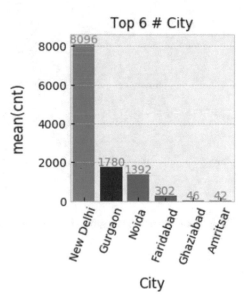

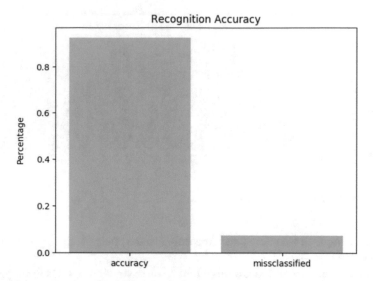

Fig. 5 Classification accuracy of Naïve Bayes classifier

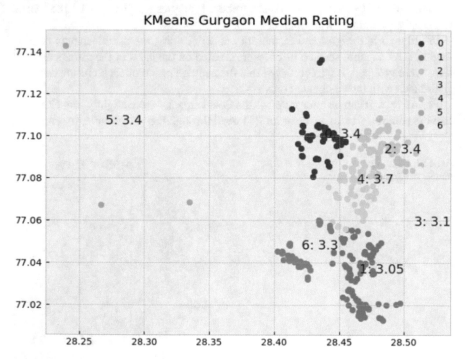

Fig. 6 Localities where rating-based clusters of restaurants were gathered

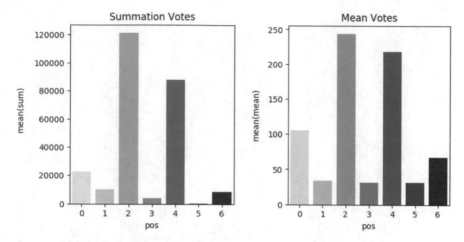

Fig. 7 Cluster of city has higher votes and higher values

that the center Area, 2, 4, and 6 are more attractive than the North and South of the city (Fig. 7).

Thus, the low-rated areas within the clusters having ratings score 2–3 such as the subpart of city can be avoided, while the areas with rating 4–5 must be selected as preference for the visit to the restaurants (Fig. 8).

The more is the merrier, if the restaurant gives more additional services its impact on rating is higher. The preference could be given to the restaurants that provide higher quality of service, table bookings, and other privileges to provide a better dining experience to the customer. The top reviewed and rated restaurants become a choice for the dining very often (Fig. 9).

Cuisine is complex variable for restaurant selection. The selection is dependent upon Country and kind of food people want to eat. It can be further divided into various options that define the taste of an individual. People love to have ample of choices and thus, restaurants having more cuisine has higher rating (Fig. 10).

6 Conclusions and Future Work

Our recommendation system works to distinguish in Zomato users' different tastes in choosing a restaurant. Using specific information about each area-based user, we are able to predict with 93% accuracy their highest rankings of restaurants. In this work, we can find which part of city has good restaurant and which is not that good on basis of cluster of high and low rating. We also come across the fact that number of cuisine and combination of cuisine such as fusion one has positive impact on rating. If a restaurant has additional services than also its rating is higher. All these analyses

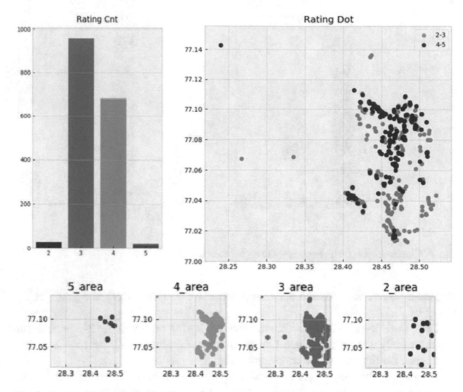

Fig. 8 Rating-wise area selection for opting restaurant

were drawn on the basis of experiments performed using Naïve Bayes and k-means algorithm.

With this model, we can further investigate how other effective features can be used to predict users' choice of restaurants. This allows more analysis to be done on which factors are more relevant to users in making their decisions which in turn reflects the kind of information Zomato should include in restaurant descriptions to enable users better utilize this application to search a restaurant that will delight them.

More algorithms and its performance can be analyzed; further, a combination of more approaches can be taken into consideration. Such an approach to group recommendation systems provides an alternative to other group recommendation systems users; and the top restaurant can be selected.

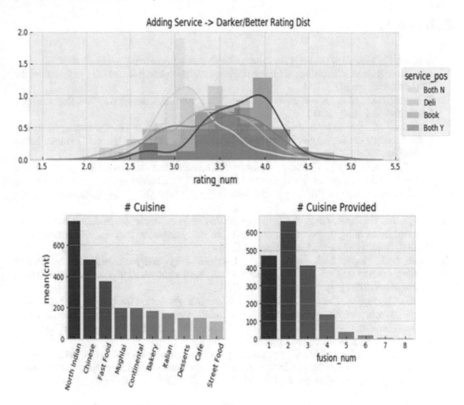

Fig. 9 Additional services help in increasing overall rating

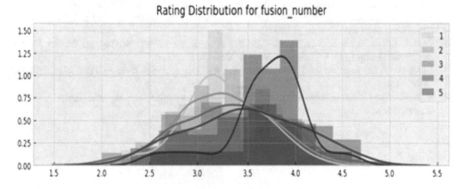

Fig. 10 More fusion cuisine better ratings

References

1. Trattner, C., Elsweiler, D.: Food Recommender Systems Important Contributions, Challenges and Future Research Directions of Literature Exists. Table 1 provides a list of important rearXiv: 1711.02760v2 [cs.IR] (2017)
2. Shashank, S.U., et. al.: A survey on food recommendation systems. Int. J. Adv. Res. Comput. Sci. Softw. Eng. **6**(4) (2016)
3. Mustafa, A.A., Budi, I.: Recommendation system based on item and user similarity on restaurants directory online. In: 6th International Conference on Information and Communication Technology (ICoICT), Bandung, pp. 70–74. 10.1109/ICoICT.2018.8528775 (2018)
4. Shi, Y., Zhao, Q., Wang, Y., Cao, J.: An improved restaurant recommendation algorithm based on user's multiple features. In: IEEE 9th International Conference on Software Engineering and Service Science (ICSESS), pp. 191–194, Beijing, China (2018). https://doi.org/10.1109/icsess.2018.8663850
5. Taneja, A., Gupta, P., Garg, A.: Social graph based location recommendation using user's behavior by locating the best route and dining in best restaurant. In: Fourth International on Parallel, Distributed and Grid Computing (2016)
6. Habib, M.A., Rakib, M.A., Hasan, M.A.: Location, time, and preference aware restaurant recommendation method. In: 19th International Conference on Computer and Information Technology (ICCIT), pp. 315–320, Dhaka, (2016)
7. Yu, M., Ouyang, O., Xue, M.: Restaurants Review Star Prediction for Yelp Dataset (2015)
8. Chu, C., Wu, S.: A Chinese restaurant recommendation system based on mobile context-aware services. In: IEEE 14th International Conference on Mobile Data Management, pp. 116–118, Milan (2013). https://doi.org/10.1109/mdm.2013.78
9. Karthick, G., Harikumar, R.: Comparative performance analysis of Naive Bayes And SVM classifier for oral X-Ray images. In: 4th International Conference on Electronics and Communication Systems (ICECS), pp. 88–92, Coimbatore, (2017). https://doi.org/10.1109/ecs.2017.8067843
10. Moe, Z.H., San, T., Khin, M.M., Tin, H.M.: Comparison of Naive Bayes and support vector machine classifiers on document classification. In: IEEE 7th Global Conference on Consumer Electronics (GCCE), Nara, pp. 466–467. https://doi.org/10.1109/gcce.2018.8574785 (2018)
11. Turney, P.: Thumbs Up or Thumbs Down? Semantic Orientation Applied to Unsupervised Classification of Reviews. ACL (2002)
12. Ghosh, J., Liu, A.: K Means. In: The Top Ten Algorithms in Data Mining, pp. 21–36. Taylor & Francis Group LLC. Florida, Chapman & Hall/CRC (2009)
13. Jiang, R.: A Customized Real Time Restaurant Recommendation System. Pennsylvania State University
14. https://stackoverflow.com/questions/10059594/a-simple-explanation-of-naive-bayes-classification

Low-Cost Optical Illusion Fluid Display Device

Arjun Dutta and Debika Bhattacharyya

Abstract A human being depends on taking decisions after sensing situations of the external environment upon five senses. Among that, visual sense plays a key role. If we look at a stationary object after looking at the moving object, it continues to move for some time. The study of illusion is also related to pathology as it is used for the interpretation of the body's abnormal functioning to understand proper functioning. We perform magic tricks in front of our family and friends, a trick which we will be able to explain and enjoy a certain amount of success. This paper deals with antigravity water illusion and its practical applications. The authors have shared physical demonstration which will make people understand about the proposed work. Researchers from various backgrounds like physics, physiology, and data science and neurobiological doctors have a continuous quest upon the similar frame "illusion". It is also observed that in cinematic shoots such as backward spin of car wheel even if the car is moving in forward direction. This can be apparently clarified by the uphill movement of water droplets which is counterintuitive with the etiquette of water droplets. The fluid display device consists of a reservoir for storage of water, outward nozzle for dispensing of fluid, a pump pulsating the fluid, an inlet for receiving the fluid from outward nozzle and a light source.

Keywords Antigravity · Doctors · Data science · Illusion · Physics · Physiology

A. Dutta (✉)
Department of Computer Science Engineering, Institute of Engineering and Management,
Kolkata, India
e-mail: arjuncode47@gmail.com

D. Bhattacharyya
Dean Academics, Institute of Engineering and Management, Kolkata, India
e-mail: bdebika@iemcal.com

© Springer Nature Singapore Pte Ltd. 2020
J. K. Mandal and S. Mukhopadhyay (eds.), *Proceedings of the Global
AI Congress 2019*, Advances in Intelligent Systems and Computing 1112,
https://doi.org/10.1007/978-981-15-2188-1_49

1   Introduction

In the ancient times, people have feared what they do not understand. A logical explanation was necessary. They considered them to be the work of god as in miracles; then, they thought of magic. In the 1980s, the secrets were passed one person to another and now are universally available through the use of Internet. Magic has become science. This illusion has physiological justification which implicates neurons becoming less sensitive at heterogeneous sites throughout the brain. This sometimes occurs because neurons become fatigued. It also tells us about the nature of perception and experience as referred in Fig. 1.

There are three types of illusions:

Optical Illusion: In this type of illusion, the images comprehend that the sense of sight tends to mislead and causes error in perception, e.g., Blivet and Bezold effect.
Auditory Illusion: It occurs due to sounds which actually do not exist. But it is heard by the ears and presumed as a sound related to the stimulus in the environment.
Tactile Illusion: It is experienced by the patients who have undergone amputation.

Our eye is considered as an important part of the brain, and almost all the visual information is controlled by the brain. In the past, the study of brain interaction was carried out by anatomists, nerve physiologist and psychologist. But nowadays with the advancement of technology, research and new discoveries, the conventional methods are replaced by informational science and others. Illusion is therefore used as a powerful tool to explore the mechanism of visual senses guided by the brain [20].

In this paper, we have proposed a working prototype of a low-cost optical illusion device. We proposed a microcontroller-based electronic device programmed in a descent manner, producing illusion of fluid droplets uphill.

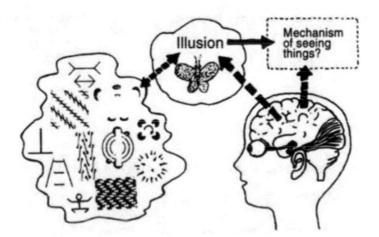

Fig. 1 Mechanism of perception of illusion [20]

The invention is akin towards development of an optical illusion fluid display device. An improved magical illusion fluid setup device produces an ideal, pleasing and natural view of rising and falling, levitating and dancing droplets. Such fluid can be displayed vertically. The illusion is observable in ordinary room lighting conditions. The entire illusion phenomenon may be examined and even touched by the audience.

This paper is organized as follows: the next section tells us about the related works. Section 3 describes the block diagram and proposed architecture in detail. Section 4 shows the experimental results and observations on a benchmark real-time dataset. The last section consists of conclusion and references.

2 Related Works

A short review of different facet of several domains of research relevant to this study is highlighted. Researchers levitate water droplets to carry out thermostat experiment. They also used sound waves to levitate water droplets in midair which promotes the detection of toxic heavy metal contaminants such as cadmium and lead in water. In this detailed study, the author has proposed the use of variation in the frequency of light and brightness to apparently levitate water droplets uphill.

Illusion is something which plays trick with our vision. If we look at a scene for a period of time which is moving in one direction, then quickly turning to the scene which is at stationary, the object appears to move in opposite direction to the original object. Same observation would occur if we see a stationary object after staring towards a waterfall. The Waterfall Illusion accomplish discrepancy as the object viewed seem to both in stationary as well as in motion, as the viewed object moved seems to be both moving and not moving at the same time, unlike the Muller-Lyer illusion [1].

The Muller-Lyer stimulus excites perceptual effects along with its variants which are accurately predicted by the probability distribution of the physical sources. The statistical relationships of the stimulus elements and there possible physical sources overviewed the atypical which associates with the identical lines in geometrical variations examined [2]. Pierre Schott highlighted that magic tricks were first practiced in middle ages for entertaining passersby. He also illustrated about the scope of Project of Engineering Science (P.E.S) in professional, educational and research domains. His objective was not only related to knowledge acquired during course work but also spreading an awareness about the necessity of such a subject. His intention was also to inspire and gain interest towards the subject "Physics" for the young learners [3]. Illusions serve as an important tool into the neurobiology of vision and also have pointed towards new experimental techniques. He overviewed how every vision appears as an illusion is one of the cons as explained by David. The resolution of peripheral vision is approximate counterpart for looking through a frosted shower door, and the viewers appreciate the illusion of watching the periphery clearly. His primary objective was to highlight how illusions guided the path for neuroscience research [4]. A phenomenon containing a flash and a moving object will become visible and is observed to be displaced from one another [5]. Bright and dark illusory bands were discovered by Ernst Mach, at the edges of a luminance ramp separating various luminance regions [6].

The scientists of the team of Bath found a way to change the directions of the water on a ratcheted heated surface by changing heat and texture of the surface. Their process allows the water droplets to levitate and to climb uphill [7]. Many researches showed that people is aware of different types of biases of their own social judgement by experiencing visual illusion which emphasized them regarding sensory perceptions. Ayumi Kambara concluded that visionless to biases results interpersonal conflicts when different people have dissimilar views [8]. The splitting of the huge water droplets was explored using high-speed video and vertical wind tunnel. The study concluded to probe upon breakup fragment distribution of bigger drops. Upon the observation of splitting of 25 droplets of water, 13 were bags and 8 were filaments [9].

Giving light to the term "Hologram" which tells about something unreal which appears in front of our eyes but actually does not. This illusion technique was invented many years ago by a philosopher and alchemist Giovanni Battista della Porta which was later modified by Prof. J. H Pepper. Holograms in museums used to create a new form of scenography and footlight art all over the exhibited objects. Holographic showcases offer an augmented experience of mixed reality which offers an intellectual engagement of the visitors [10]. After many researches, scientists used acoustic levitators for liquid levitation. It makes easy for scientists to manufacture medicines with lesser side effects. It also has an important impact in medical science as they do not crystallize upon levitating the liquid medicine solution; the water evaporates leaving behind the amorphous medicine [11]. The planet Mars is believed to be populated with intelligent life due to discovery of lines on the surface of the planet. These lines were first spotted by an Italian Astronomer, Giovanni Schiaparelli named as canali. In the beginning of the twenty-first century due to the advancement of space exploration and research, it was proved to be an optical illusion [12].

To get human behaviour accustomed with the surrounding objects, human cerebral cortex predicts visual motions. Many researchers accepted the assumption that the predictive coding theory is one of the bases of illusion generation. The DNN predicted the motion vectors of the untrained rotating propeller. It also predicted at par human perception about the rotational motion in the rotating snake illusion [13, 14]. Machine vision systems can recognize face and even can create real synthetic faces alike humans. After many researches, it was found that the same systems cannot predict optical illusion. It was concluded that the dataset of illusion images might be incomplete so till date optical illusions acts as a safeguard of human experience which machines cannot defend [15]. Many scientists look for evidences for different studies related to brain. Brain researchers call illusion perceptual, not optical as the visual system is more dominant. Researchers at the California Institute of Technology and University of Sussex have a conflict about the brain's adaptive ability [16]. Researchers at MIT's Media Lab and Harvard University delve into a new structure for developing diffusivity of the optical device or the extent till which the light scatters. Researchers also believed that their modifications would finally lead holographic video screens [17]. Michael Jackson's move was an illusionary trick, which broke the relationship between physiology and physics [18, 19].

Stroboscopic fluid display devices as mentioned in U.S Pat. No. 3387782 and U.S. Pat. No. 5,820,022 have the following disadvantages. They produce unstable and small water droplets, and hence, it does not fit for displaying in household lighting conditions. The white lighting is poorly used; as a result, the visualization becomes difficult. These devices fail to grab the true beauty of the phenomenon because of the improper background settings as mentioned in U.S Pat. 9472128 B2 [21].

The proposed circuit is designed to research about the illusion phenomenon of antigravity water levitation. The author[1] experimentally demonstrates the illusion phenomenon which is seen and accepted by many people.

3 Device Methodology

3.1 Block Diagram

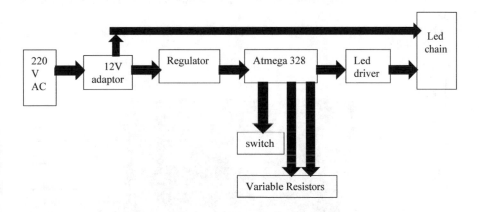

3.2 A Low-Cost Optical Illusion Device Comprises of the Following Three Stages

Upper level: As referred in Fig. 2, the upper layer consists of the brain of the proposed prototype, i.e., the circuit design of the system. The upper level consists of following parts: L7805CV MOSFET, I.C-ATmega328P, variable resistor. On the top of the model, author[1] has mounted and configured the circuit. To ensure minimum flaws with the system, author[1] has implemented an LED which will glow whenever current passes through the system. The proposed model relates for displaying an optical illusion with fluid under certain ambient light conditions.

Middle Level: This consists of the base of the model such that there is a gap of 1.25 foot between the top and the bottom for the illustration view. It consists of the display system, where the observers will look into and enjoy the vision of antigravity water illusion. It contains of the following parts: LED strip, funnel, and nozzle which facilitates the inlet and outlet of fluid.

Lower level: It consists of a tank made to store the water so that it can flow through the pipe attached with the pump. The lower level is primarily made for the storage of water. It also allows for the passage of fluid from the middle to the lower layer through a hole. Then, through a pump, water is carried from the tank in the lower level to the upper level. The bottom of the model is made such that it can resist the entire load of the setup.

The major components used are:

L7805CV MOSFET: It is a 3-terminal positive voltage regulator, with fixed output voltages. It functions alike a semiconductor switch which operates to vary the resistance between two pins in response to a voltage on the third pin.

I.C-ATmega328P: It is a low-power CMOS, 8-bit microcontroller based on AVR enhanced RISC architecture. It executes instructions in single clock cycle. The ATmega328P achieves amount of items passing through the system at 1 MIPS per MHz. This permits the user to develop power consumption adjacent to the processing speed.

REGULATOR: Regulator plays an important role; it is used to lower the voltage from 12 to 5 V.

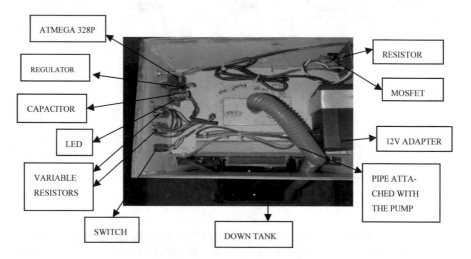

Fig. 2 Circuit diagram of the developed prototype

3.3 Top View of the Developed Setup

4 Control Algorithm

Variable List:

mode: It is used to toggle between three frequency-brightness modes 1–3. (as shown in Fig. 3).
mode_changed: It represents the stage formed after transition from one mode to another.

Constant List:

BASE_FREQ: It is the initial frequency range with a value of 80 Hz.

Function List:

led_on(): It determines the onset condition of the light intensity.
led_off(): It determines the offset condition of the light intensity.
digit Read: It reads the value from a specified digital pin.

If((mode = 1)and(mode_changed = 1))
Then

  Frequency <- BASE_FREQ;
  led_on();
  Mode-changed = 0;

Endif
If((mode = 2)and(mode_changed = 1))
Then

  Mode_changed <- 0;

Endif
If((mode = 3)and(mode_changed = 1))
Then

  Mode_changed <- 0;

endif
If((mode = 4)and(mode_changed = 1))
Then

  led_off();
  Mode_changed <- 0;

Endif

If(digitRead(SW)=HIGH)
Then

   If(mode >=5)
   Then
      mode <- 1;
   endif
   Mode_changed <-1;

endif

5   Results and Analysis

The simulated data is obtained real time and verified from the experimental results as shown in Fig. 4.

Figure 3a–c shows that the experiment is carried out in dark room in order to obtain better result. The figure demonstrates about three conditions upon which the illusion phenomenon occurs. The above figures are different from each other with respect to the frequency of light, brightness, and formation of dark bands.

(a)   **(b)**   **(c)**

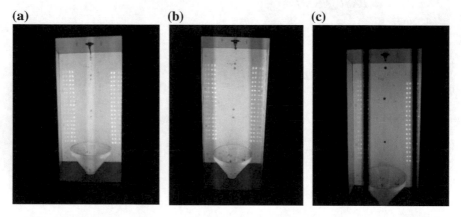

Fig. 3   **a** Frequency of light is high as well as brightness which corresponds to mode 1; **b** Frequency of light is high, and brightness is low vice versa which corresponds to mode 2; **c** Frequency of light and brightness are low which correspond to mode 3

5.1 Visual Illustration of the Three Modes

See Fig. 3.

5.2 Graphical Representation as per the Visual Illustration of Fig. 3a–c

Figure 4 clearly shows a linear characteristics, and the graph was plotted phase shift against brightness with the variation of load, where phase shift is basically mentioned in terms of π, whereas the phase difference is related to the path difference as,

$$(\Delta x/\lambda) = (\Delta \Phi/2\pi)$$

where

- Δx is the path difference between the two waves.
- $\Delta \Phi$ is the phase difference between two waves.

Also, phase difference $(\Delta \Phi) = (2\pi * \Delta x)/\lambda$; unit is in radian or degree

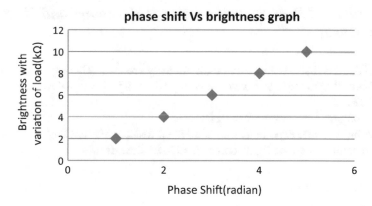

Fig. 4 Graphical representation between phase shift and brightness

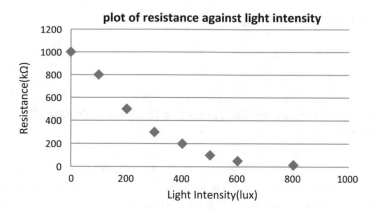

Fig. 5 Graphical representation between resistance and light intensity

where λ is the wavelength path difference $(\Delta x) = (\lambda*\Delta\Phi)/2\pi$; unit is metre.

And the brightness is varied with the variation of load resistors connected with the system in terms of ohm. In normal cases, the intensity and brightness of light is measured in candela and lux.

5.3 Graphical Representation Between Resistance and Light Intensity

Figure 5 depicts that resistance and light intensity are inversely proportional to each other. According to the Ohm's law, the voltage drop is proportional to the resistance, i.e. $V = IR$.

V = voltage; I = current; R = resistance

If the light source experiences an equivalent resistance more than the amount of current reaching, then the light source appears to glow dimmer.

6 Observation

As per the name suggests, we will be observing that the water droplets will be moving upwards. The master device for the above effect illustration is the variable resistors. There are two variable resistors used simultaneously. First variable resistor is used to set the brightness according to the need of the user which varies from one environment to another. Second variable resistor is used to change the frequency of light.

As shown in Fig. 3a–c, dark bands are observed and the water droplets are visible properly when we decrease the frequency and brightness of the working model.

7 Conclusion

In the past, illusions were considered as an indecorous domain for research. But nowadays after many researches, it is found having a greater impact in various domains like in embedded systems, predictive coding theory, and neuroscience research. The objective of this study is to levitate water antigravity which appears to make water defy gravity.

In the beginning, it was hard to meet the desired results, but after few changes in circuit and other considerations as estimated by author[1], the experiment worked out successfully. Our another objective is to put a light to the reader that there is a vast area for research related to "illusion". The proposed design was built cost effective at 1500 INR which adds a positive fact towards our experiment. This beautiful work of art will enhance the home or office for decoration purposes. It also can be used as a low-cost unique display device in amusement parks and museums.

References

1. Crane, T.: The waterfall illusion. Analysis **48**(3), 144–147 (1988). Peterhouse, Cambridge
2. Howe, C.Q., Purvas, D.: The Muller-Lyer illusion explained by the statistics of image–source relationship. PNAS **102**(4), 1234–1239 (1988). Centre for Cognitive Neuroscience, Duke University, 25 Jan 2005, @Adam Mortan
3. Schott, P.: The use of magic in optics in higher education. Creative Education, **1**, 11–17 (2010). https://doi.org/10.4236/ce.2010.11003. Published Online June 2010 (http://www.SciRP.org/journal/ce) Copyright © 2010
4. Eagleman, D.M.: Visual illusions and neurobiology. Timeline, Perspectives, 2001. Macmillan Magazines Ltd. Accessed on 30/02/2019
5. Nijhawan, R.: Motion extrapolation in catching. Nature **370**, 256–257 (1994)
6. Ratliff, F.: In: Cohen, R.S., Seeger, R.J. (eds.) Ernst Mach Physicist and Philosopher, pp. 165–184. Reidel, Dordrecht (1970)
7. Leidenfrost thermostat uses levitating water droplets to keep cool. www.bath.ac.uk/research/news/2015/05/19/leidenfrost-thermostat/. Accessed on 15/02/2019
8. Kambara, A.: Effects of experiencing visual illusions and susceptibility to biases in one's social judgments. SAGE Open. October–December 2017: 1–6 © The Author(s) 2017 https://doi.org/10.1177/2158244017745937 https://journals.sagepub.com/home/sgo
9. Emersic, C., Connolly, P.J.: The breakup of levitating water drops observed with a high speed camera. Atmos. Chem. Phys. **11**, 10205–10218 (2011). www.atmos-chem-phys.net/11/10205/2011/. https://doi.org/10.5194/acp-11-10205-2011
10. Pietroni, E.: The illusion of reality inside museum—a methodological proposal for an advance Museology using holographic showcases. Informatics **6**, 2 (2019). https://doi.org/10.3390/informatics6010002. www.mdpi.com/journal/informatics. Published: 4 Jan 2019
11. Scientists levitate water using sound. http://www.planet-science.com/categories/over-11s/technology/2012/09/scientists-levitate-water-using-sound.aspx. Accessed on 20/05/2019
12. The Canals of Mars|Science Blogs. https://scienceblogs.com/universe/2012/09/28/the-canals-of-mars. Accessed on 15/07/2019
13. Watanabe, E., Kitaoka, A., Sakamoto, K.: Illusory motion reproduced by deep neural networks trained for prediction. Front. Psychol. https://doi.org/10.3389/fpsyg.2018.00345, 15 Mar 2018
14. Artificial intelligence tricked by optical illusion, just like humans. https://blog.frontiersin.org/2018/04/26/artificial-intelligence-tricked-by-optical-illusion-just-like-humans/. Accessed on 15/01/2019

15. Neural networks don't understand what optical illusions are—MIT Technology Review. https://www.technologyreview.com/s/612261/neural-networks-dont-understand-what-optical-illusions-are/. Accessed on 15/01/2019
16. Optical Illusion Show How Brain Anticipates the Future To 'See' The Present—'The New York Times'. https://www.nytimes.com/2008/06/10/health/research/10mind.html. Accessed on 25/12/2018
17. Mixing solids and liquids enhances optical properties of both. MIT News. http://news.mit.edu/2016/mixing-solids-liquids-enhances-optical-properties-0609. Accessed on 25/12/2018
18. Yagnick, NS.: How did Michael Jackson challenge our understanding of spine biomechanics? Neurosurg. Spine **29**, 344–345 (2018). Published online 22 May 2018. https://doi.org/10.3171/2018.2.spine171443
19. The physical feat behind Michael Jacksons anti-gravity illusion. https://www.mnn.com/lifestyle/arts-culture/blogs/physical-feat-behind-michael-jacksons-anti-gravity-illusion. Accessed on 30/12/2018
20. Idesawa, M.: A study on visual mechanism with optical illusion. J. Robot. Mechatron. (1997)
21. Rosenthal, L.K.: Enhanced optical illusion fluid display device. United States Patent, US 9,472,128, 18–20 Oct, last Accessed 2016/11/21

Smart Grid Demand-Side Management by Machine Learning

Rebeka Bhattacharyya and Avijit Bhattacharyya

Abstract The smart grid is a complex electrical network system comprising of different subsystems at different levels of aggregation. It facilitates a bidirectional information flow among all the actors such as producers of electricity, end users of electrical energy, transmission and distribution system operators (TSO/DSO), and demand response (DR) aggregators. Smart grid contains smart meters that send user statistics to the server. Accurate forecasting of the electricity usage is required in order to take controlled actions to balance the supply and demand of electricity. This forecasting can be achieved using machine learning-based predictive models. This paper deals with the forecasting of short-term and mid-term load for the grid entity using machine learning. A predictive system is designed using machine learning techniques in order to process the smart meter data which in turn is used as the training data for the model. The outcomes are then shown with data and results to make it more understandable by the reader.

Keywords Smart grid · Electrical network · Bidirectional energy flow · Forecasting · Machine learning · Predictive model · Smart meter

1 Introduction

Electricity is one of the driving forces of the modern world. But electricity is something that many of us take for granted. Around 1.3 billion people still do not have access to electricity. With the increase in the number of customers, the demand for electricity is steadily increasing. Figure 1 shows a graph of rise of customers from the year 2005–2015.

R. Bhattacharyya (✉)
Electronics and Communication Engineering, Institute of Engineering and Management, Kolkata, India
e-mail: rebeka.bhattacharyya@gmail.com

A. Bhattacharyya
Tata Consultany Services, Kolkata, India
e-mail: avijit.bhattacharyya@gmail.com

© Springer Nature Singapore Pte Ltd. 2020
J. K. Mandal and S. Mukhopadhyay (eds.), *Proceedings of the Global AI Congress 2019*, Advances in Intelligent Systems and Computing 1112, https://doi.org/10.1007/978-981-15-2188-1_50

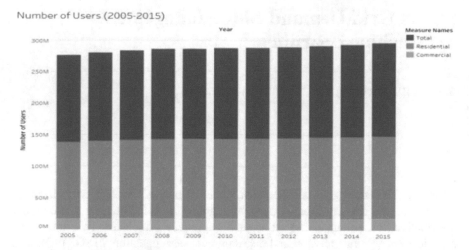

Fig. 1 Number of electric grid users from 2005 to 2015

Since electricity plays a vital role in human society, appropriate management strategies for the grids are important. The need for the management strategies has led to the development of smart grids (SG) and demand-side management (DSM). Smart grids are generally used for generation–transmission distribution pattern of the power system. For this, the distributed energy resources (DER) can be thought of as a potential solution. The two most important components of smart grids are constant connectivity and consumption. These are achieved by a network of devices which are called Internet of things (IoTs). These devices provide the detailed data which are required to provide accurate information which gives the smart grid the unique capabilities over other systems. All these data are handled carefully which are then used to make decisions based on certain cases.

Smart grids are bidirectional and have no strict structure. Energy flow can take place from consumer side and also such as wind and solar farms. Power flow is also bidirectional as demonstrated by energy storages and houses in Fig. 2.

Dynamic energy management (DEM) is an approach to manage the load at the demand side of the grid. Dynamic energy management has two parts—demand-side management (DSM) and demand response (DR). Demand-side management is the planning, implementation, and monitoring of the customers' activities in order to manage the consumption of electricity which in turn controls the utility load shape. An important component of demand-side management is the demand response which deals with the change in electricity consumption by the customers with the change in electricity prices.

The entire project consists of four steps. The first step or the data loading loads data for the project. The second step or data preprocessing processes the data and transforms it into usable format. The data preprocessing forms the raw data from the data mining techniques. The third step or the load profile generation takes into account the heating/cooling energy consumption, hot water usage, and set-point temperature of the property. A good profile is required for training the predictive model accurately. The fourth step involves the formulation of predictive model. The methodology of the project has been shown in Fig. 3.

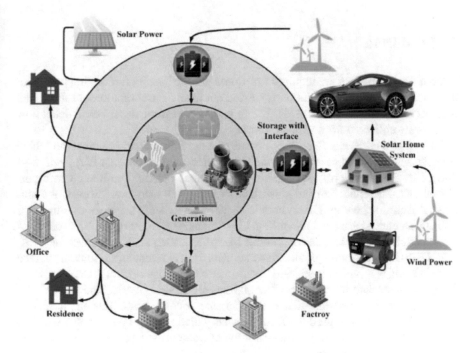

Fig. 2 Demonstration of bidirectional nature of smart grids

Fig. 3 Proposed algorithm model

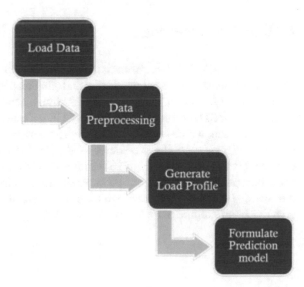

2　Proposed Algorithm Model

See Fig. 3.

3　Load Data

A large amount of data are required for profile generation and forecasting model. In this paper, an empirical validation is carried out from an assimilation of established datasets for the State of Utah (US). The following paragraph describes in detail how the data are collected from various sources:

EIA or United States Energy Information Association is an organization which collects energy related data all over the USA. There are two types of EIA datasets—Form EIA-861 and Form EIA-861S which includes information like electric purchases, sales, revenue, demand-side management, peak load, etc. CBECS or Commercial Building Energy Consumption Survey provides the statistics of consumption of water and space heating/cooling. US National Observatory or USNO provides information about the weather conditions from the weather stations all over the USA. USU Utah Climate Centre provides weather data from the weather stations all over the State of Utah. National Oceanic and Atmospheric Administration or NOAA provides hourly weather data from the weather stations all over the world. The Department Of Energy or DOE provides data related to energy consumption, its end uses, etc. In this project, we have used the DOE datasets which provides data related with hot water usage profile on hourly basis, variation of water usage in different regions, and probability of each end use.

4　Data Preprocessing

Data preprocessing involves forming the usable data from the datasets available. The first task is to make datasets of residential and climatic data. The second task is to form cluster of utility in the same region. Similar climatic conditions will have similar hot degree days (HDD) and cold degree days (CDD). Residential data are collected from the EIA datasets which contain the load characteristics all over the USA. Weather data are collected from the two sources—the USU (Utah State University) Climate Centre which collects data from 16 weather stations and NOAA (National Oceanic and Atmospheric Administration) which collects data from weather stations all over the world. For the utility areas which cover more than one weather station, the effective hourly temperature T_{eff} is calculated as:

$$T_{\text{eff}} = \sum_{i=1}^{W} N_i * t_i \bigg/ \sum_{i=1}^{W} N \tag{1}$$

Here, N_i is the number of users in the ith weather station and t_i is the temperature of the ith weather station. N is the total combined population of the selected weather stations.

Many areas share the same weather stations, and it is assumed that the utility areas have the same weather conditions. If the end customers of the utilities are similar, then the utilities will have the same predictive models. So it is inefficient to formulate similar predictive models. This problem is solved by formulating predictive models for the clusters of similar customers. It should be noticed that Smithfield City has been chosen and assigned the code letter 'a' for further references in the paper.

Domestic hot water (DHW) represents the total share of energy load ranging from 25 to 40% of total energy consumption. Hence, proper evaluation of domestic water usage is very important for developing an accurate load profile. The average daily consumption of water depends upon the number of people and their way of living.

5 Load Profile Generation

Load profile provides the data of electric consumption from time to time. An accurate load profile is required in order to train the data for the predictive model in order to do the predictions. As the predictions depend on the service area of the utility, a unique profile is formed for all the individual clusters. The first stage of profile generation is to form a load data for heating and cooling purposes. Once the load data are generated, it is then possible to form the final load data for the building.

5.1 Heating and Cooling Load

Weather and temperature are the most important components of electricity consumption. The energy consumption depends upon the heating and cooling of residential areas. The prediction models use threshold temperature in order to indicate at what time the heating and cooling of building are required. This threshold temperature is called the set-point temperature which has value above 72 °F and below 65 °F. As the set-point temperature increases, the cooling and heating load also increases. In this project, we have taken the set-point temperature to be 65 °F.

5.2 Cooling Degree Days (CDD) and Heating Degree Days (HDD)

Once the set-point temperature is fixed, we can find the cooling degree days (CDD) and heating degree days (HDD). CDD is a measure of the energy required to cool a building if the outside temperature is higher than the temperature inside the building. HDD is a measure of the energy required to heat a building if the outside temperature is lower than the temperature inside a building. So, daily CDD is calculated as the difference between the outside temperature and set-point temperature summed up over 24 h. Similarly, daily HDD is calculated as the difference between the set-point temperature and outside temperature summed up over 24 h.

Mathematically,

$$HDD = \frac{\sum_{i=1}^{t} (T_i^{out} - T_i^{set})}{24} \qquad (2)$$

$$CDD = \frac{\sum_{i=1}^{t} (T_i^{set} - T_i^{out})}{24} \qquad (3)$$

Here, T_i^{out} and T_i^{set} are the outside and set-point temperature, respectively.

The annual energy consumption for heating purpose (kWh) for a given building is expressed as

$$Q = \frac{M' * AD * 24}{\eta} \qquad (4)$$

Here, AD is the annual HDD of the building, η is the performance efficiency of the heating source, and M' is the heat loss coefficient of the building.

Taking the default values, M' can be calculated as 0.350 kW/K and the efficiency η can be averaged at 0.9, which means 90% efficiency is assumed for the heating device. Thus, the final equation for calculation of the energy consumed for HDD becomes

$$Q = \frac{0.350 * AD * 24}{0.9} \qquad (5)$$

The annual energy consumption for cooling purpose (kWh) for a given building is expressed as

$$Q = \frac{mC_p * ACD * 24}{\Psi} \qquad (6)$$

Here, ACD is the annual CDD, Ψ is the coefficient of performance of the cooling source, Cp is the specific heat capacity of air, and m denotes the mass flow rate of kilogram of air cooled per second. For a 12,000 BTU air-conditioning unit, m and Ψ can be assumed to be 0.109 kg/s 3 and 3.5, respectively. Also, the specific heat of air is 1.005 kJ/kg/K. By using these constants, the CDD and HDD values can be readily calculated (Fig. 4).

CDD increases for hotter months and diminishes for cooler months. A negative value of CDD shows that the set-point temperature is already less than the outside temperature and so no cooling is required (Fig. 5).

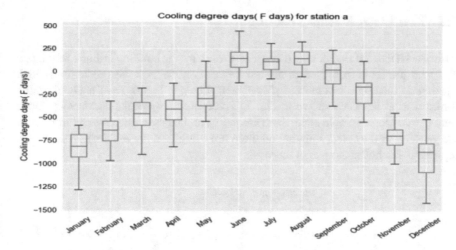

Fig. 4 Cooling degree days (CDD) for cluster a

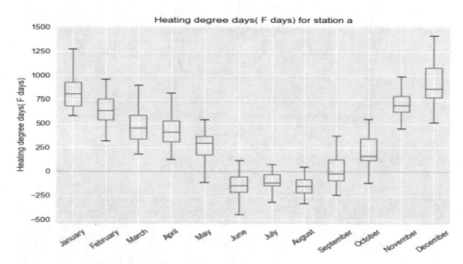

Fig. 5 Heating degree days (HDD) for cluster a

HDD is proportional to the difference between the set-point temperature and outside temperature. For warmer months, there is no need to heat the building and so HDD is zero. On the other hand, during cold months HDD increases extensively during the beginning and ending months of the year. Once the energy consumption is calculated using CDD/HDD, it is then combined with the hourly load data for DHW consumption and the final load profile is calculated for a specific cluster.

5.3   Load Profile Data Training

After the HDD/CDD is calculated, the load profile for the building is then formulated. The load profile is then used as the training dataset to train the prediction model. The hourly load profile of cluster 'a' has been shown in Fig. 6. The final load profile for cluster 'a' has been shown in Fig. 7. There are many peaks and trough. Since the State of Utah is a cold state, there are peaks at the beginning and end months of the year that is the colder months and there are troughs in the middle portion which represents the hotter months of the year.

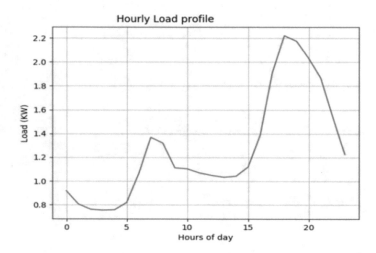

Fig. 6 Hourly load profile of cluster a—January 1

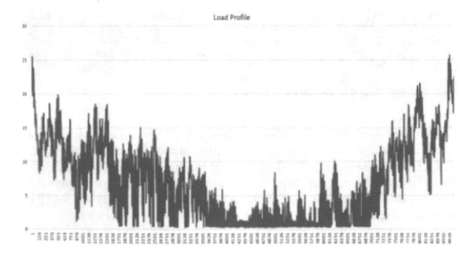

Fig. 7 Final load profile for cluster a

```
print df.head()
```

	Load	Temprature (F)	CDD	HDD	Day of Week
0	23.905050	4.5	0.0	60.5	Thursday
1	24.685159	2.5	0.0	62.5	Thursday
2	24.840359	2.1	0.0	62.9	Thursday
3	24.719850	2.4	0.0	62.6	Thursday
4	25.147472	1.3	0.0	63.7	Thursday

Fig. 8 First five rows of the preliminary training dataset

6 Predictive Modeling

A predictive model for the load forecasting is formulated using feature selection. Feature selection is the process of selecting a subset of relevant features for constructing the model. It helps to form accurate model as fewer attributes reduce the number of errors.

In this project, the attributes are reduced by forming the load profile. Since the training set contains most of the information required, the load profile can be used to form the model. One of the attribute can be taken as the hourly temperature due to its high correlation with the load. The second and third attributes can be taken as HDD/CDD. As it depends largely on the days of the week, day can be taken as the fourth attribute. Thus, regression equation can be expressed as:

$$L_{\text{pred}} = \alpha_0 + \alpha_1 X_{\text{temperature}} + \alpha_2 X_{\text{CDD}} + \alpha_3 X_{\text{HDD}} + \alpha_4 X_{\text{day}} \tag{7}$$

Here, α_i are the prediction coefficients and denote the weight of each independent variable on dependent variable, and X_i denote the independent variables (Fig. 8).

For seven days of the week, seven new columns were added to the dataset. For a sample case, the true case is one and all other cases are zero. The regression equation can be written as

$$L_{\text{pred}} = \alpha_0 + \alpha_1 X_{\text{temperature}} + \alpha_2 X_{\text{CDD}} + \alpha_3 X_{\text{HDD}} + \alpha_4 X_{\text{mon}} + \alpha_5 X_{\text{tue}}$$
$$+ \alpha_6 X_{\text{wed}} + \alpha_7 X_{\text{thurs}} + \alpha_8 X_{\text{fri}} + \alpha_9 X_{\text{sat}} + \alpha_{10} X_{\text{sun}} \tag{8}$$

7 Predictive Outcomes and Performance Evaluation

Equation (8) gives the predictive coefficients and the intercept constants. Figure 9 shows the hourly load profile of cluster a, for the year of 2016. Using the training data of year 2015, Fig. 14 shows the load plot of the predicted load for the year

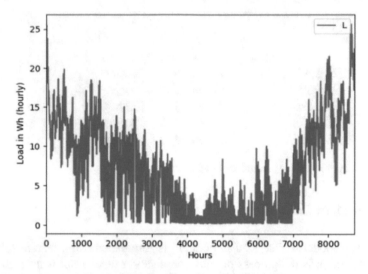

Fig. 9 Predicted load profile for cluster 'a' (City of Smithfield Utility)

of 2016. The predicted load profile closely resembles the true load profile mainly during winter months (Fig. 10). Low energy consumptions and domestic hot water (DHW) consumptions may lead to this anomaly.

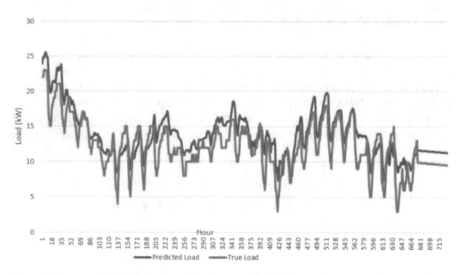

Fig. 10 Predicted load profile versus. true load profile for cluster 'a' (City of Smithfield Utility)

8 Conclusion

This paper develops an energy forecasting model for the State of Utah. This forecasting model helps the companies to operate grid in a more economical way. On the other hand, load forecasting can help the end use customers to estimate their net loads, normalize the load distribution, and manage their utility bills by avoiding the effects of dynamic pricing. Thus, formulating an accurate load forecasting model would improve and assist the existing DSM techniques on both, provider's and consumer's end. As a part of this paper, an hourly load profile is generated, which is the backbone of the predictive model. An accurate and error-free hourly load profile helps to effectively train the model and reduce the probability of error. For a completely digital smart grid network, the need for generating hourly load profile is minimized.

References

1. Dunn, L.N., Berger, M.A., Sohn, M.D.: Demand response forecasting methodology—berkeley lab. Available at https://openei.org/doe-opendata/.../drforecastingmethodology20160624.pptx
2. Global status of modern energy access. In: International Energy Agency—World Energy Outlook. pp. 23–52, Academic Press (2012)
3. der Hoeven, V., Birol, M.: World energy outlook 2012 presentation. In: International Energy Agency—World Energy Outlook (2012)
4. Faisal, M.A., Aung, Z., Williams, J., Sanchez, A.: World energy outlook 2012 presentation. In: International Energy Agency—World Energy Outlook (2012)
5. Lee, J., Jung, D.-K., Kim, Y., Y.-W., Lee, K.
6. Liotta, A., Geelen, D., Kempen, G.V., van Hoogstraten, F.: A survey on networks for smart-metering systems. Int. J. Pervasive Comput. Commun. **8,** Acad. Press. 23–52 (2012)
7. Geelen, D., Kempen, G.V., Hoogstraten, F.V., Liotta, A.: A wireless mesh communication protocol for smart-metering. In: International Conference on Computing Networking and Communications ICNC. pp. 343–349, Academic Press (2012)
8. Lai, C.S., Lai, L.L.: Application of big data in smart grid. In: IEEE International Conference on Systems, Man, and Cybernetics, Kowloon. pp. 665–670 Academic Press (2015)
9. Parmenter, K.E., Hurtado, P., Wikler, G.: Dynamic Energy Management, p. 53 (2008)
10. K.D, G.G, I.R, M.C, M.N. et al.: Assessment of demand response and advanced metering—staff report (2008)
11. Siano, P.: Demand response and smart grids?—A survey. Renew. Sustain 461–478 (2014)
12. Diamantoulakis, P.D., Kapinas, V.M., Karagiannidis, G.K. Big data analytics for dynamic energy management in smart grids. Big Data Res. 94–101 (2015)
13. Khan, A.R.Mahmood, A. Safdar, A. Khan, Z.A. Khan, N.A.: Load forecasting, dynamic pricing and dsm in smart grid: a review. Renew. Sustain. Energy Rev. 1311–1322 (2016)
14. Barakat, E.H., Qayyum, M.A., Hamed, M.N., Rashed, S.A.A.: Short-term peak demand forecasting in fast developing utility with inherit dynamic load characteristics. I. Application of classical time-series methods. II. Improved modelling of system dynamic load characteristics. IEEE Trans. Power Syst. **5**(3), 813–824 (1990)
15. Macedo, M., Galo, J., de Almeida, L., de Lima A.C.: Demand side management using artificial neural networks in a smart grid environment. Renew. Sustain. Energy Rev. **41** (2010)
16. Zang, H.T., Xu, F.Y., Zhou, L.: Artificial neural network for load forecasting in smart grid, pp. 128–133 (2015)

17. Hong, T.: Short term electric load forecasting. Ph.D. Dissertation, North Carolina State University (2010)
18. Cetinkaya, N.: A new mathematical approach and heuristic methods for load forecasting in smart grid. In: 2016 12th International Conference on Natural Computation, Fuzzy Systems and Knowledge Discovery (ICNC-FSKD), vol. 54, pp. 1103–1107, Aug 2016

An Integrated Domestic Sensing and Control System with Supervisory Check and Real-Time Data Acquisition

Tanishq Banerjee, Purbayan Chowdhury, Shuvam Ghosal, Suvrojit Kumar Saha and Ranit Bandyopadhayay

Abstract This project implements IoT, machine learning, and cloud computation to devise an extensive monitoring system for regular domestic appliances. The system is voice-controlled and supports regional languages. This project has five main subparts—secure door locking/unlocking system, intensive fire and gas alarm system for notifying the user in case of any gas leak or fire mishap, auto-adjustable temperature monitoring system for maintaining an optimum room temperature, intelligent water tap control system, and smart room lighting system which automatically switches on the room lights in the presence of any person especially when light entering through the windows is insufficient. The entire system is controlled by a central brain that uses predefined dataset with an alternative option for manual override. It also provides the features of speech recognition through ANN and various language models for multilingual operations. There are options for voice/video calls and separate user-admin and guest session logs.

Keywords Internet of Things (IoT) · Image processing · Automation · Machine learning · Neural network · Microcontroller · Cloud computing · Sensors · Facial recognition · Artificial neural networks (ANN)

T. Banerjee · P. Chowdhury (✉) · S. Ghosal
Department of Computer Science and Engineering, Institute of Engineering and Management, Kolkata, India
e-mail: pur.cho.99@gmail.com

S. Ghosal
e-mail: shuvamghosal98@gmail.com

S. K. Saha · R. Bandyopadhayay
Department of Electronics and Communication Engineering, Institute of Engineering and Management, Kolkata, India
e-mail: sahasuvrojitkumar@gmail.com

R. Bandyopadhayay
e-mail: ranitbanerjee43@gmail.com

© Springer Nature Singapore Pte Ltd. 2020
J. K. Mandal and S. Mukhopadhyay (eds.), *Proceedings of the Global AI Congress 2019*, Advances in Intelligent Systems and Computing 1112,
https://doi.org/10.1007/978-981-15-2188-1_51

1   Introduction and Literary Review

With the emerging ideas of IoT, machine learning, and cloud computation, future presents with outstanding opportunities and ways to enhance our way of living and mechanize the work processes around us such to increase the efficiency of the whole system. The Internet of things is the network of physical devices, vehicles, home appliances, and other items embedded with electronics, software, sensors, actuators, and connectivity which enable these things to connect and exchange data, creating opportunities for more direct integration of the physical world into computer-based systems, resulting in efficiency improvements, economic benefits, and reduced human exertions [1]. Machine learning is a subset of artificial intelligence in the field of computer science that often uses statistical techniques to give computers the ability to "learn" with data, without being explicitly programmed [2]. Cloud computing is an information technology paradigm that enables ubiquitous access to shared pools of configurable system resources and higher-level services that can be rapidly provisioned with minimal management effort, often over the Internet [3].

Automation has been a long driving force in most of these mechanization problems such that minimal human effort is required. It focuses on the use of various control systems for operating equipment such as machinery, and processes in factories. A distributed control system (DCS) is a computerized control system for a process or plant with a large number of control loops, and there is a central operator supervisory control. Supervisory control and data acquisition (SCADA) is a control system architecture that uses computers, networked data, and graphical user interface for communication for high-level supervisory management and uses a programmable logic controller (PLC) and discrete proportional–integral–derivative (PID) controllers [4]. A real-time sensing environment is implemented with a number of sensors acquiring varied data and working as an automatic supervisory control.

A Naive Bayes (NB) language classifier, a recurrent neural network, and a convolutional neural network (CNN) image classifier are used for speech-to-command mapping, audio-to-text processing, and image processing, respectively [5]. Till date, a large number of works have been done on how to improve the ease by which one leads one life at home. Large number of projects include simple home automation systems which involve inter connectivity of various devices used in our houses. Some include controlling different devices in the house through an android application or through predefined values. Some include simple voice-recognized activation systems to control the entire house, or in some cases, the entire interface is a Web application. This is where this project stands out. The idea of this system is to integrate all the above-mentioned procedures and be available to the user as an alternative method to operate the system. The system consists of a predefined dataset possessed by each device which will control the system automatically if not manually overridden. As alternatives the user will have access to monitor real-time situations prevailing in the system and surroundings and accordingly overriding it as and when necessary through the IoT Web server through his smartphone if he is at a distant place from the system or via other methods such as speech, Bluetooth, or Wi-fi module from his cell phone [6]. The details of each nodes and command over them via various technologies mentioned are given in the upcoming sections of the paper.

Usage of photovoltaic cells to generate solar energy during the daytime and using that stored energy at the time of power cuts and also during those hours of the day when the power consumption rate of all the electrical equipments is less as compared to the rest of the day [7]. In addition to this, using energy switch as well as other combinations of sensors and microcontrollers can efficiently assist the entire process of monitoring the energy consumption rate as well as the power generated from the solar energy. The difference between these two values will indicate the amount to be charged from the consumer at the end of each month. A detailed report of the above values is to be shared via SMS with both the consumer and the electricity supplier company. This process is not only energy-efficient but also at the same time eases the system of monitoring the entire process of electricity supply and consumption [8].

Next, there is a requirement of using highly secure communication protocol. This is the area of prime concern as the system is fully automated and is least dependent on manual operation. Therefore, in order to maintain the highest security protocol, along with the camera module the door locking/unlocking process is equipped with a Bluetooth connectivity system. As soon as the Raspberry Pi will detect the presence of the user in front of the door using the camera and the image processing algorithm, it will simultaneously activate the Bluetooth module. If the user's cell phone gets paired with the module without facing any trouble, the door will get unlocked successfully. The password required for the pairing action will be updated regularly at the end of each week. Only the user will get to know about the new updated password. This will provide additional security to the entire system.

In case of any discrepancy or if someone tries to break into the house forcibly, the Raspberry Pi will sense it using a certain combination of sensors and will activate the "Burglar Alarm System." The user will get notified in no time. Also, using the GSM module connected to it, it will make a recorded call to the nearest police station mentioning the address of that particular location.

2 Integrated System Overview

See Fig. 1.

Case 1: Operating Inside the House

When a person tries to enter a room, his/her face is scanned and the door is opened automatically if there is a user account in the local microcontroller; else, one has to register with some specific details with a generated pin from admin user. As soon the person enters the room, a client session is started where the client activities within the room are tracked, and in case of security breach, such as breaking in or any mishap caused by client, the admin client is notified duly and for increased threat, local police station is informed and in case of fire, the local fire station is informed.

Fig. 1 Distributed system layout

The client can operate all electrical appliances both through command and by using any kind of remote or button provided by appliance. In case of automated systems like light, temperature, water and door locking/unlocking functions are performed, as predefined by the admin or the programmer. One can override this automated control by the use of voice commands—"Close the door," "Turn off the light," "Switch off AC," etc., or by remote devices or just manual switch press. Each and every such operation is recorded for every user separately, and for a multiuser system, same operations are recorded for each user. Such a client session is halted, as one exits the room.

Case 2: Operating Outside the House
When the admin user is outside the house, he/she can control the entire system from his android or iOS app in his/her smartphone. If admin user wants to give access to someone, he can do so, by pressing yes after a notification arrives.

When a new/temporary client wants to enter the house. Also, every appliance can be duly controlled remotely through this app. Suppose a scenario, where a user wants to have a cooler room after a long day work, he can do so through the app, or in case he forgot to switch off some appliances, he can do so by using this app. The whole system is simulated over the Web, and it is the only abiding factor in the workings of the system when outside of the house. The command generated by the admin user, through his application in his cellular device, gets sent across via server to the microcontroller in the house, and the command is forward passed to the executing device and the operation is executed.

3 Subsystem Overview

3.1 Door Locking/Unlocking System

When a person rings a bell at the main door, a 10 MP camera captures the image of the person and matches it with the CNN model of image classifier trained, in the local server, and if the person is admin, then the door opens using a servo motor and closes automatically after one enters as detected by IR sensor module. In case of any other user other than admin, the admin is asked whether to grant permission along with details of user that local server has in its local server. For each room door, multiple PIR sensor works to locate the user presence and carry out the multiple client session for each user separately for each room. These details are deleted after 30 days and can be viewed by admin user in a log format, i.e., timestamp and operation (Fig. 2).

3.2 Automatic Temperature Adjusting System

There is a predefined dataset consisting of room ids and optimum temperature, and the system tries to maintain this optimum temperature for the given room. The temperature of the room is detected through a DHT sensor and modified as per command via the potentiometer connected to an IR emitter. An extensive air circulation system, where a central AC/heater with some other heating or cooling units, adjust the temperature accordingly detecting the presence of number of persons in each room, in case of a vacant room, temperature is not maintained unless instructed by the user. Such a system not only effectively maintains the pleasant weather around but also cuts off extra expense of cooling/heating multiple rooms with multiple cooling/heating systems (Fig. 3).

3.3 Smart Lighting System

Two IR sensor modules are installed on the main door frame. For every entry of a person, the microcontroller counts the number of entries made and stores the corresponding value. Depending on the room size, multiple number of passive infrared (PIR) sensors are installed at the corners of the room which get activated only when positive feedback is received from IR sensor module in a certain fixed order. In

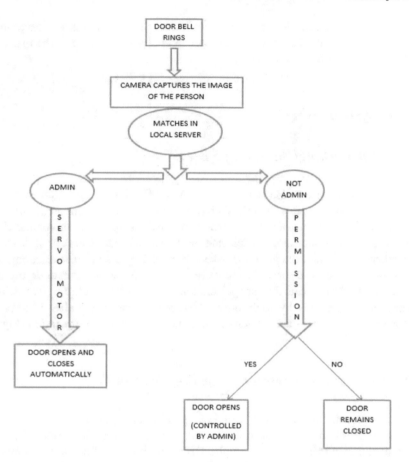

Fig. 2 Door locking/unlocking system layout

addition to this, at each and every window frame, LDR modules are installed at a suitable angle so that they can properly receive the Sun's rays at any point of time. A definite threshold value is set for the entire network of LDR sensor modules. Once all the conditions get fulfilled by the IR sensor modules and the PIR sensors, and the LDR senses a value less than the threshold value (i.e., the light entering the room from outside is not sufficient), the lights get turned on automatically. For the entire duration unless there is no one else left in the room, the room lights continue to glow. At some point of time, when the combination of IR modules gets a signal in opposite order (which happens when a person is leaving the room), the PIR modules continue to search for body heat of a human being. If neither of the PIR sensors can sense anything, it means that there is no one inside the room, and the lights get turned off automatically (Fig. 4).

Fig. 3 Automatic
Temperature Adjusting
system layout

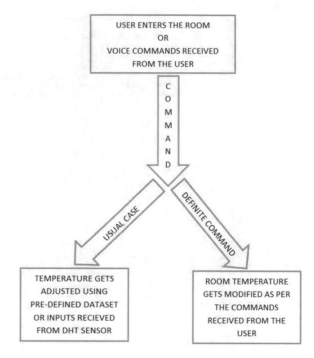

3.4 Water Tap Control System

This particular system consists of a combination of motors (for controlling the flow of water), a sensor (for measuring the level of water), and a local microcontroller (which will control the water flow from the tap as per the commands received from the main Raspberry Pi and the water level sensor). If the user commands to open the tap, the motor will be rotated clockwise and the water will flow. As soon as the bucket gets filled, the sensor will send a signal to the local microcontroller, the motor will be rotated anticlockwise, and the tap will be closed (Fig. 5).

3.5 Gas Leak/Fire Alarm System

Gas and fire sensors are to be installed in every part of the house. In case of any gas leak (LPG, CNG, etc.) or fire resulting into smoke, the alarm system will continue to ring to alert the neighborhood. Using a GSM module, an SMS will be sent to the user's phone instantly so that he/she can call the fire brigade and take all necessary actions (Fig. 6).

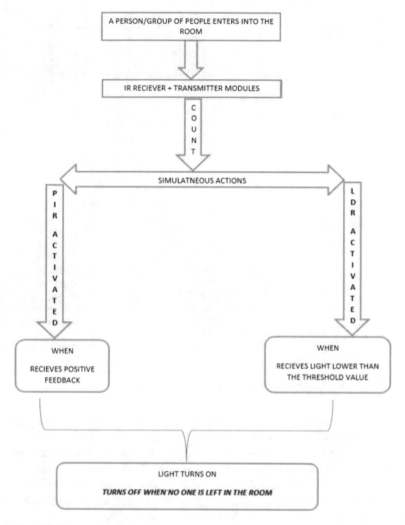

Fig. 4 Smart lighting system layout

4 Algorithms and Software Description

4.1 RNN Speech Recognition

Long short-term memory (LSTM), a variant of RNN, is the state-of-the art as it consists of a cell state which enables it to retain information over long periods of time and is the main component of speech recognition. A LSTM model supplemented with attention layers has been found to be highly efficient in this context [9] (Fig. 7).

Fig. 5 Water tap control
system layout

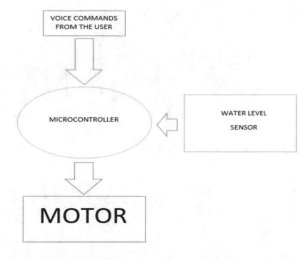

Fig. 6 Gas leak/fire alarm
system layout

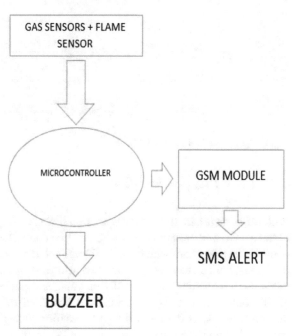

The mel-frequency coefficient (MEC) spectrogram is taken as inputs to a bidirectional RNN or LSTM which can recognize the relationship between past and future portions of the speech spectrogram which aids in the speech-to-text conversion process [10] (Fig. 8).

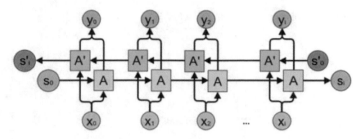

Fig. 7 Bidirectional RNN [13]

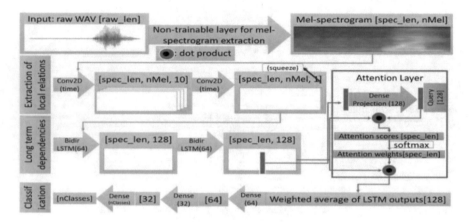

Fig. 8 Speech recognition model [14]

4.2 CNN Image Classifier

A convolution is a mathematical operation that is widely used in signal processing to filter signals, find patterns in signals, etc. In a convolutional layer, all neurons apply convolution operation (which can be thought of as a combination of two functions) to the inputs; hence, they are called convolutional neurons. This special type of neural networks has yielded good results on images especially image classification and localization [11]. It uses a special fixed size filter or kernel which is moved over different regions of the input image to perform convolutions for feature extraction and a fully connected neural network (FCN) for the purpose of classification [12] (Fig. 9).

The image classifier used in this case will be trained on the different images of the end user, and some data augmentation can be done to improve the accuracy of the model. This trained CNN-based model will take input from a camera and perform binary classification based on the presence of the user in the captured image.

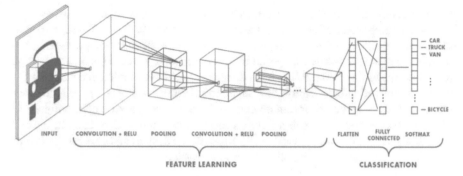

Fig. 9 Basic CNN image classifier architecture [15]

Table 1 System requirements

Language	Python 3.7.2
Modules	TensorFlow, PyTorch
IDE	JupyterLab, Spyder
GPU	NVIDIA GeForce RTX 2070

4.3 *Naïve Bayes Speech Classifier*

The classifier classifies text into nouns, verbs, and adjectives, compares the text with given labeled speech command, and using multinomial Naïve Bayes algorithm recognition, selects the command. The probability of a document d in class c is computed as

$$P(c|d) \propto P(c) \prod_{1 \leq k \leq n_d} P(t_k|c) \tag{1}$$

where $P(t_k|c)$ is the conditional probability of term t_k occurring in a document of class c.

Laplace smoothing is used to prevent zero error (Table 1):

$$\hat{P}(t|c) = \frac{T_{ct} + 1}{\sum_{t' \in V} (T_{ct'} + 1)} = \frac{T_{ct} + 1}{\left(\sum_{t' \in V} T_{ct'}\right) + B'} \tag{2}$$

5 User-End Analysis

The customer primarily needs to set up a user-admin account which will establish a connection to a remote server in an IoT cloud and a local server in a Raspberry

Pi for transmitting the local data retrieved from the various input interfaces. The individual components of the IoT system will be connected through Wi-fi. In case of the image classifier, the user has to give several images of his/her own profile in order to train the classifier, while for speech recognition, some voice clips in.wav or.mp3 formats to the bidirectional RNN. Each door contains a camera that takes images for identification and authentication of user-admin. Five subsystems as a whole work on commands or switches or remote-control inputs. The system works online, and efficient electric system is required for its proper functioning. User can always add commands to its database for various functions of the system and cherish the fact of enjoying a safe, sound, and intelligent home ecosystem.

6 Future Prospects

Similar kinds of autonomous techniques can be utilized in other fields related to industrial, domestic, and professional environments. Usage of Google Assistant Services, microcontrollers like Arduino Uno, Nano, and Mega, and different shields for these microcontrollers will make the system far more cost-efficient without disturbing its effectiveness.

References

1. Kodali, R.K., Jain, V., Bose, S., Boppana, L.: IoT based smart security and home automation system. In: 2016 International Conference on Computing, Communication and Automation (ICCCA), pp. 1286–1289, April 2016. IEEE (2016)
2. Kool, I., Kumar, D., Barma, S.: Visual machine intelligence for home automation. In: 2018 3rd International Conference on Internet of Things: Smart Innovation and Usages (IoT-SIU), pp. 1–6, February 2018. IEEE (2018)
3. Dickey, N., Banks, D., Sukittanon, S.: Home automation using Cloud Network and mobile devices. In: 2012 Proceedings of IEEE Southeastcon, pp. 1–4, March 2012. IEEE (2012)
4. Chauhan, R.K., Dewal, M.L., Chauhan, K.: Intelligent SCADA system. Int. J. Power Syst. Optim. Control. 2(1), 143–149 (2010)
5. Kaviani, P, Dhotre, S.: Short survey on Naive Bayes Algorithm. Int. J. Adv. Res. Comput. Sci. Manag. 4 (2017)
6. Mandula, K., Parupalli, R., Murty, C.A., Magesh, E., Lunagariya, R.: Mobile based home automation using Internet of Things (IoT). In: 2015 International Conference on Control, Instrumentation, Communication and Computational Technologies (ICCICCT), pp. 340–343, December 2015. IEEE (2015)
7. Sora, D.: Energy Switch: A Home-Automation System for Renewable Energy Self-Consumption Optimization (No. 2015-13). Department of Computer, Control and Management Engineering, Universitàdegli Studi di Roma "La Sapienza"
8. Ahmed, T., Miah, M., Islam, M., Uddin, M.: Automatic electric meter reading system: A cost-feasible alternative approach in meter reading for Bangladesh perspective using low-cost digital wattmeter and Wimax technology. arXiv preprint arXiv:1209.5431 (2012)
9. Sherstinsky, A.: Fundamentals of recurrent neural network (RNN) and long short-term memory (LSTM) network. arXiv preprint arXiv:1808.03314 (2018)

10. Tripathi, S., Bhatnagar, S.: Speaker recognition. In: 2012 Third International Conference on Computer and Communication Technology, pp. 283–287, Nov 2012. IEEE
11. Jmour, N., Zayen, S., Abdelkrim, A.: Convolutional neural networks for image classification, pp. 397–402 (2018). https://doi.org/10.1109/aset.2018.8379889
12. Krizhevsky, A., Sutskever, I., Hinton, G.E.: ImageNet classification with deep convolution neural networks. In: Advances in Neural Information Processing Systems, pp. 1097–1105 (2012)
13. https://miro.medium.com/max/1200/1*6QnPUSv_t9BY9Fv8_aLb-Q.png
14. https://miro.medium.com/max/1000/1*YOoWJrXhU3FcFp4EPVi5UA.png
15. https://miro.medium.com/max/1838/1*XbuW8WuRrAY5pC4t-9DZAQ.jpeg
16. Schuster, M., Paliwal, K.K.: Bidirectional recurrent neural networks. IEEE Trans. Signal Process. **45**(11), 2673–2681 (1997)

Development of a BCI-Based Application Using EEG to Assess Attentional Control

Sayan Dutta, Tanishq Banerjee, Nilanjana Dutta Roy
and Bappaditya Chowdhury

Abstract The main objective of this paper is to introduce a new and interactive way of assessing an individual's attentional capabilities with the help of EEG-based signals running a BCI serious game. Brain–computer interface technology is used to realize this objective by studying the EEG readings emitted by the brain and process the data in real time to play games and also study the brain activity for medical research. The goal is to create a seamless interface that can be used both by the medical and entertainment industry. In this paper, we have devised a convenient and effective process to assess attentional value and developed a game that takes advantage of the focus parameter provided by the BCI headset.

Keywords Brain–computer interface (BCI) · Electroencephalogram (EEG) · BCI serious game · Real-time data processing · Attentional control

1 Introduction

With the dawn of emerging technology in the medical and unbelievable advancement in gaming technologies in particular, the idea of incorporating a completely new way of playing games is an extremely difficult task. Advancements such as facial recognition, voice recognition, gesture control, virtual reality, augmented reality, and cloud gaming are various technologies which have been incorporated into the gaming industry to make gaming much more dynamic and interactive. However, with technology such as brain–computer interface (BCI), fiction has become a step closer to reality. Brain computing is a part of human–computer interactions (HCI) where the signals generated by the human brain are modelled into data which in turn is streamed into the computer for controlling either a hardware device, for instance moving a wheelchair, or, as in our case, controlling a software application. There

S. Dutta (✉) · T. Banerjee · N. D. Roy
Institute of Engineering & Management, Salt Lake Electronics Complex,
Gurukul, Y-12, Sector V, Kolkata 700091, India

B. Chowdhury
AMRI Hospital, Kolkata, India

© Springer Nature Singapore Pte Ltd. 2020
J. K. Mandal and S. Mukhopadhyay (eds.), *Proceedings of the Global
AI Congress 2019*, Advances in Intelligent Systems and Computing 1112,
https://doi.org/10.1007/978-981-15-2188-1_52

are many non-invasive techniques available to study the activities of the brain, out of which the EEG method is the most common due to its cost-effectiveness and portability [1].

This paper emphasizes on the attention parameter of an individual when the brain is functionally active. Here, the data is collected from the temporal region of the brain via two non-invasive dry non-spikey electrodes which mainly deal with persons' focus levels. This collected data is then directed to the computer via Bluetooth module from which the corresponding alpha and beta wave strength is obtained. This stream of data is used as a parameter to move the object in our proposed game. The game is a two-dimensional game, developed with the ability to operate in real time. It is aimed to not only find an alternate and exciting way to play games but also to serve the medical needs of a player as seamlessly as possible.

Till date, a large number of BCI devices have been developed which are beneficial to medical industry [2]. Games which deal with entertainment as well as detection of some medical issues are popularly known as serious games. Very few number of serious games actually deal with BCI controlled gaming [3]. A literature survey conducted shows the availability of limited number of methods for implementing BCI technology in serious games. One such approach features brain cognitive control over the game by processing various neural signals along with a neurofeedback system to get feedback on the users' experience and difficulty level [4]. Another approach describes the users adapting to frequently changing background images. Initially, the users' scores were weak due to their frequent defocusing because of the random change in background colour. But with more practice and gameplay experience by the user, they noticed prominent increase in scores and focus level of the user [5]. The proposed system mentioned in this paper is inspired by both of the aforementioned papers. Real-time data streaming of electric pulse signals generated by the brain [6] into the game for entertainment purpose and assessment of attention/concentration level of the person playing the game are the two important aspects that were narrowed upon during the literature survey. This paper thus aims to benefit both the gaming and medical industry with the help of its proposed game and gaming system.

2 Methodology

The entire project has been subcategorized into three subsections for relative ease of processing and troubleshooting.

The subsections can be described as:

(1) Data Acquisition.
(2) Data Processing and Error Management.
(3) BCI-based Game Script.

Figure 1 describes the flow of the project.

The various subsections mentioned in the figure above have been elucidated hereafter.

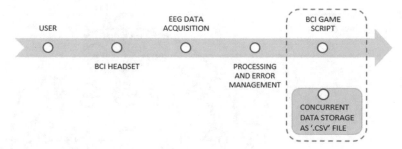

Fig. 1 Flow of the proposed system

(1) Data Acquisition:

The hardware that has been used for gathering and computing the EEG signals by the system is the open-source OPENBCI ULTRACORTEX MARK IV headset [7] that has been 3D printed for our project. The device is able to record and process electroencephalogram (EEG) signals and many other signals. Only EEG signals have been discussed since it is the only signal that is being utilized to capture brain activity and concentration strength of any individual in our project. The device is able to target 35 electrode locations of the 10–20 system.

The version of the aforementioned OPENBCI ULTRACORTEX MARK IV headset that has been used in this project is an 8-channel headset. It accommodates a total of 8 electrodes (6 spikey dry electrodes and 2 non-spikey or flat electrodes). The node locations in this device are set up based on the internationally accepted 10–20 electrode placement system in the context of EEG.

The electroencephalogram or EEG signals are waveforms that showcase the cortical electrical activity. The human brain is composed of millions and millions of neurons. And each of these neurons interacts with each other using considerable magnitude of electrical activity and is responsible for carrying information which is later perceived as actions by the human body. The EEG signals help in studying this miniscule activity occurring in our brain and are categorized as delta, theta, alpha, beta, and gamma waves based on frequencies of signals ranging from 0.1 Hz to over 100 Hz [8]. The OPENBCI ULTRACORTEX MARK IV helps us to easily acquire these EEG signals by the corresponding electrodes in the headset placed in correspondence with the 10–20 electrode placement scheme.

The 10–20 system (Fig. 2) describes the relation between electrode placement and the portion of cerebral cortex [9]. Each point in Fig. 2 relates to one acceptable position of electrode placement. The letters F, T, C, P, and O stand for frontal, temporal, central, parietal, and occipital. Even numbers (2, 4, 6, 8) refer to the right hemisphere, and odd numbers (1, 3, 5, 7) refer to the left hemisphere. The Z refers to an electrode placed on the midline [9].

The version of the device utilized for this project has 8 channels. The electrode placement is done accordingly and is described in Fig. 3.

Fig. 2 10–20 system of electrode placement

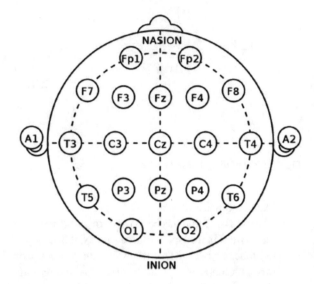

Fig. 3 Electrode placement in this project

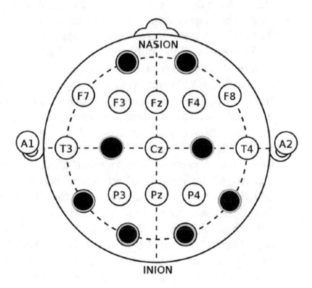

In Fig. 3, the nodes coloured in solid black represent the corresponding electrode placement in the headset. The primary objective of this project is to utilize the focus or concentration parameter. Since it is considered that the beta and alpha waves are mostly concerned with the conscious mind [10–12], only the electrodes associated with capturing significant amount of these waves are considered (Fig. 4).

The nodes highlighted in black are utilized for studying the alpha and beta waves emitted by the brain at any given instant. OPENBCI provides an open-source GUI that is utilized to acquire and study the signals captured by the

Fig. 4 Electrode placement
in Fp1 and Fp2 node

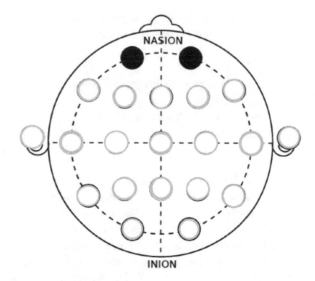

device. An inbuilt focus detection widget is also provided that assesses the
alpha and beta waves captured by the device. A simple fast Fourier transform
provides us with an FFT plot. The FFT plot decomposes any periodic data
in different frequency components. In this FFT plot, the alpha band can be
defined as frequency components between 7.5 and 12 Hz, and beta band is
defined as frequency components between 12.5 and 30 Hz [13]. The final data
is the readings of alpha and beta frequency range provided by channel 1 and
channel 2 electrodes associated with the Fp1 and Fp2 region (Fig. 4) of the brain.
Certain modifications were made in the GUI algorithm originally provided by
the OPENBCI organization, in order to acquire the appropriate data in CSV
format.

(2) Data Processing and Error Management:

The entirety of data processing and error management is performed in Python
environment due to relative ease of processing. Since achieving results in real
time is an important aspect of this project, the concept of socket is put into
use. Socketing is an effective way to achieve interprocess communication. The
acquired data is simultaneously stored in a CSV file for future model training
purpose.

An established and defined algorithm is then used to assess the data and
classify as focused and non-focused state. The alpha and beta values in this
algorithm utilize their corresponding amplitude levels also.

The algorithm can be simplified as:
Repeat for all acquired values:

{
Alpha_average = {average(FFT_value_in - microvolts) | FFT_frequency_Hz in
[7.5, 12.5]}
Beta_average = {average(FFT_value_in − microvolts) | FT_frequency_Hz in
[12.5, 30.0]}
            },

where FFT stands for fast Fourier transform.

The above-mentioned algorithm, which is established in OPENBCI documentation [14], gives us the average alpha and beta values. However, there always remains certain amount of instrumentation error and other extrinsic interferences because of which some amount of error margin is always introduced. To tackle this issue, an error factor of ± 0.1 is introduced.

A range of [0.7–4.0] for Alpha_average and a range of [≤ 0.7] for Beta_average is considered the optimum range for classifying an individual as truly focused, as found by experimental study [13].

We further modified the lower bound of Alpha_average and upper bound of Beta_average with an error factor of ± 0.1 that provided us with a 9.94% increase in the concluding result. This change was made to cope up with the challenge of providing a streaming influx of data for the game which is discussed in the next section. Once the error factor is introduced in the established algorithm and the results are obtained, these data(s) further needed to be communicated to another Python game script running consecutively with these scripts. The concept of socket programming is again utilized here to facilitate interprocess communication.

(3) BCI-based Game Script:

This section deals with the final subsection of this project. An interactive game is designed that can actively access the final processed readings from other script and use it as an input to run the game engine.

A simple concept is followed while designing the game. The objective of the game is to provide a visual medium for the user to interact and engage in cortical activity. The game acts as a stimulus for the user to engage in an activity that targets the brain activity.

The background is kept as minimal as possible with a simple scenic image. The foreground uses a figure of an UFO (Fig. 5).

The basic concept that encapsulates the game is that, if and when an individual will focus or try to concentrate, the UFO will move along the vertical axis by some amount. The amount of UFO movement is not dependant on the amplitude of the concentration or focus levels. Whenever a 'true focus' is detected by the algorithm, the UFO moves up along the vertical axis by a fixed amount and whenever the values are not in the range described by the algorithm, it is

Fig. 5 Background and foreground of the BCI game

considered as a 'non-focused' state and the UFO drops down the vertical axis by a fixed amount. A score is also calculated and displayed as an incentive for the user to try to maximize their focusing ability and achieve highest score. The objective of the game is to assess the attentional capability of any individual and use an aesthetic medium to visually represent their attentional capabilities.

3 Results

The OPENBCI headset is calibrated, and the wirings are adjusted initially. The electrodes are also checked to prevent any derailment. Figure 6 is a snapshot that depicts a test user wearing the headset, getting ready for the neural data to be streamed and processed.

The first observation that is made is the FFT or fast Fourier transform plot that is obtained from the OPENBCI widget.

Figure 7a shows the FFT plot during a typically unfocused state, and Fig. 7b shows the relative FFT plot during a typical focused state. The two curves only consider the input from electrodes in the Fp1 and Fp2 region of the brain since only alpha and

Fig. 6 User wearing the OPENBCI headset

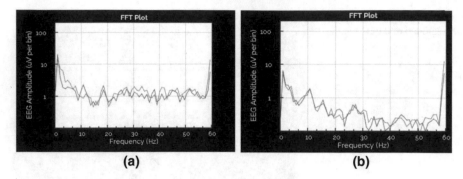

Fig. 7 FFT plot observed during a typical unfocused state (**a**) and focused state (**b**)

Fig. 8 Main menu of the BCI game

beta waves are considered for this system. The following data was retrieved from the FFT plot, and the Alpha_average and Beta_average values were ready for further processing (Fig. 8).

Table 1 depicts the subset of the dataset that is used to manoeuvre the object in the game (UFO). The columns in Table 1 depict the corresponding Alpha_average and Beta_average values computed from the FFT plots, obtained by the electrodes placed in the frontal cortex region of the BCI headset. If the user is in a focused state, the resultant Alpha_average value (the first column) should be within [0.6–4] and the corresponding Beta_average value (the second column) should be less than 0.8. If both these conditions satisfy, the individual is validated to be in a focused state and the UFO in the game will ascend along the vertical axis (Fig. 9a); otherwise, it will start its descent gradually (Fig. 9b) and crash (Fig. 11a). This data is constantly being processed by the script where the game is concurrently running.

Figure 8 depicts the main menu of our designed game. The main menu includes two buttons: 'START' and 'EXIT' buttons. The 'START' button fires the game script, the data that is being streamed and processed by our proposed system is analysed by the game script, and the necessary actions are performed. The 'EXIT' button

Table 1 Subset of dataset obtained from BCI headset (Column 1 denotes Alpha_average and Column 2 denotes Beta_average)	Alpha_average	Beta_average
	1.9190886	0.96383286
	1.9487019	0.9791486
	1.9785846	0.7969459
	1.9815729	0.99744034
	2.0065045	0.3142573
	2.0272954	0.3306373
	2.0377433	1.0462238
	2.0366743	1.057771
	2.027125	1.0648961
	2.0261703	0.605094
	2.0047317	0.7679742
	1.9671408	0.8653191
	1.9175301	0.7564997
	1.8834324	0.4462542
	1.8579422	1.0364553

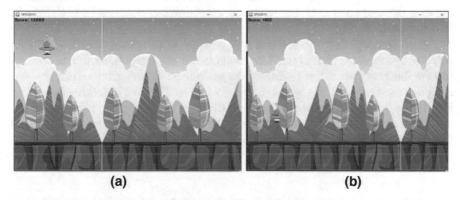

(a)   (b)

Fig. 9 Movement of UFO as a function of alpha and beta values

immediately shuts down the entire game process, and the actions to be followed are terminated.

Figure 9a depicts a situation in which the UFO ascends along the vertical axis. This happens when the algorithm computes a data within the specified focus range. Figure 9b on the other hand depicts a situation in which the UFO descends along the vertical axis. This happens when the algorithm receives data that is an outlier with respect to the specified range. An important fact here is that the movement of the UFO is independent of the magnitude of the values computed and solely depends on the range of value gathered.

Fig. 10 Ground level
considered for UFO crash

The red line highlighted in Fig. 10 denotes the lowest possible range of the UFO. If the UFO probes further below the demarcated line, it will be considered as a crash situation. This is an analogy for the decreasing concentration. If the concentration of the individual falls below a particular range, it will be visually notified by the crashing of the UFO. This mechanism is provided to ensure an interactive environment between the user and the system where the attentional control can be visualized in real time aesthetically.

Figure 11a depicts the dialog box that pops up as soon as a crash condition is detected by the game algorithm. This results in the game algorithm to reset and re-run the function responsible to display the main menu of the game so that the user can play another round if they feel the need to do so. Figure 11b follows as soon as Fig. 11a is displayed in-game. Figure 11b displays the final score achieved by the user to provide a competitive edge in the game. This will ensure that the user is able to analyse their attentional control performance after every round.

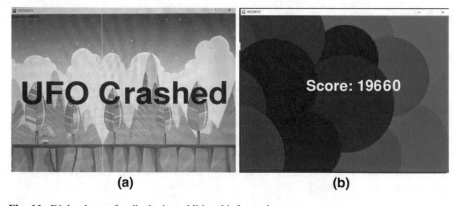

Fig. 11 Dialog boxes for displaying additional information

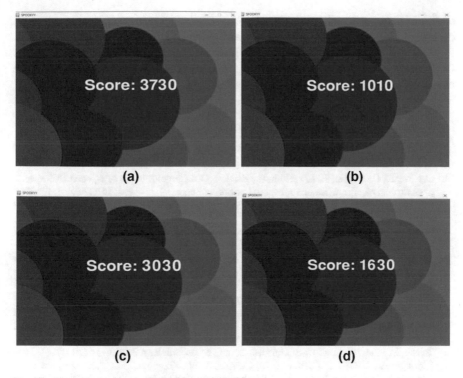

Fig. 12 Score(s) obtained by four different individuals

Figure 12a–d displays the score that was obtained by four different test subjects. The individual(s) were made to sit in isolation and no external interference or distraction was ensured. The individuals were then asked to play the game that was designed. Their corresponding attentional parameter was obtained by the headset, and the data was processed by the game script and the UFO was manoeuvred accordingly. The game script computed the total score accordingly which was then displayed for further assessment.

4 Conclusion

We have thus successfully devised a system that uses EEG-based BCI technology to introduce a completely new way to play games for the gaming industry while also making use of such games to study the attentional pattern of any individual. This is a culmination of medical and entertainment industry and has enormous potential for growth and can help people to enjoy an immersive gaming experience, while also helping them to monitor and assess their concentration level by playing a game. Alongside gathering data and processing them, the dataset obtained, consisting of alpha and beta wave readings, is also stored as CSV file. Upon considerable gathering of dataset, a classification algorithm can be implemented to train and create our own focus detection classifier. Integrating machine learning algorithm with this project to further extend the possibilities of attentional control is within the domain of future works for this project.

References

1. Carrino, F., Dumoulin, J., Mugellini, E., Khaled, O.A., Ingold, R.: A self-paced BCI system to control an electric wheelchair: evaluation of a commercial, low-cost EEG device. In: 2012 ISSNIP Biosignals and Biorobotics Conference: Biosignals and Robotics for Better and Safer Living (BRC), pp. 1–6 (2012)
2. Laamarti, F., Eid, M., Saddik, AE.: An overview of serious games. Int. J. Comput. Games Technol. **2014**, 358152:1–358152:15 (2014)
3. Moreno-Ger, P., Martínez-Ortiz, I., Freire, M., Manero, B., Fernández-Manjón, B.: Serious games: a journey from research to application. In: 2014 IEEE Frontiers in Education Conference (FIE) Proceedings, pp. 1–4 (2014)
4. Nijholt, A.: BCI for games: a 'state of the art' survey. In: ICEC, pp. 225–228 (2008)
5. Diaz, B.A., Sloot, L.H., Mansvelder, H.D., Linkenkaer-Hansen, K.: EEG-biofeedback as a tool to modulate arousal: trends and perspectives for treatment of ADHD and insomnia. Neuroimaging Cogn. Clin. Neurosci., 431–454 (2012)
6. Teplan, M.: Fundamentals of EEG measurement. Meas. Sci. Rev. **2**(2), 1–11 (2002)
7. Aldridge, A., Barnes, E., Bethel, C.L., Carruth, D.W., Kocturova, M., Pleva, M., Juhár, J.: Accessible electroencephalograms (EEGs): a comparative review with OpenBCI's ultracortex mark IV headset. In: 2019 29th International Conference Radioelektronika (RADIOELEK-TRONIKA), pp. 1–6 (2019)
8. Kumar, J.S., Bhuvaneswari, P.: Analysis of electroencephalography (EEG) signals and its categorization—a study. Procedia Eng. **38**, 2525–2536 (2012)
9. Klem, G.H., Lüders, H.O., Jasper, H.H., Elger, C.: The ten-twenty electrode system of the international federation. The international federation of clinical neurophysiology. Electroen-cephalogr. Clin. Neurophys. **52**, 3–6 (1958)
10. Cvijetic, S.: What are beta brainwaves? improve focus and motivation with beta brainwave entrainment. (2019, March 23). Retrieved from https://owlcation.com/stem/What-are-Beta-Brain-Waves-Focus-and-Motivation-with-Beta-brainwave-entrainment
11. Craig, A., Tran, Y., Wijesuriya, N., Nguyen, H.T.: Regional brain wave activity changes associated with fatigue. Psychophysiology **49**(4), 574–582 (2012)
12. Dustman, R.E., Boswell, R., Porter, P.B.: Beta brain waves as an index of alertness. Science **137**(3529), 533–534 (1962)
13. Sun, W.: OpenBCI Focus detection algorithm and Focus Visualization Widget. (2019, Mar 19). Retrieved from https://openbci.com/community/focus-visualization-widget/
14. OpenBCI Software Documentation. (2019, Mar 19). Retrieved from https://docs.openbci.com/OpenBCI%20Software/00-OpenBCISoftware

Framework for Appraisal
of Twenty-Twenty League Players

Sannoy Mitra, Tiyash Patra, Raima Ghosh, Souham Ghosh and Avijit Bose

Abstract In the following paper, we propose to design a framework, which uses concrete data source to analyze the player statistics. Data collected from various websites is filtered according to each player and their corresponding T-20 statistics. A rank system on the basis of playing statistics of each player is devised through our proposed algorithm. The proposed algorithm then categorizes the players according to their roles in the team. This categorization leads to better implementation of strategies according to pitch type. We broadly classify the players with respect to the predominant pitch type and create an ideal model for each pitch type. Hence, the best playing eleven can be shortlisted on the basis of a proposed framework consisting of the roles assigned by the model. It illustrates the probabilistic best playing eleven.

Keywords Player statistics · T-20 · Rank system · Playing eleven

1 Introduction

Cricket has its roots in our country since the period of colonization. This game has manifested itself in different formats ranging from 5-day test match to T-20 internationals. Though the sport previously catered only to international matches, in

S. Mitra · T. Patra · R. Ghosh · S. Ghosh (✉) · A. Bose
Institute of Engineering and Management, Kolkata, India
e-mail: joe.sg123@gmail.com

S. Mitra
e-mail: officialsannoy@gmail.com

T. Patra
e-mail: patra.tiyash6@gmail.com

R. Ghosh
e-mail: raima.ghosh01@gmail.com

A. Bose
e-mail: avijit.bose@iemcal.com

© Springer Nature Singapore Pte Ltd. 2020
J. K. Mandal and S. Mukhopadhyay (eds.), *Proceedings of the Global
AI Congress 2019*, Advances in Intelligent Systems and Computing 1112,
https://doi.org/10.1007/978-981-15-2188-1_53

recent years, we have seen the growth of cricket premier leagues where franchisees are owned by the bureaucrats of the society. The most popular T-20 League in the world is currently the Indian Premier League (IPL) [2, 4].

The league featured ten teams at maximum till date, each representing a certain state. The players are drafted into the team via auction. Hence, auction plays a very important role in these leagues all over the world. There are multiple franchises bidding for a marquee player so that the team can get the best of the world in their squad. Cricket is a game where both statistics and talent interplay in a strong manner. On the basis of all of the previous IPL matches, we have analyzed individual player statistics, with the help of devised algorithms. This analysis led to ranking of the players; calculated combinations of best 11 players can further be selected for D-day [3]. Any work in the given field is done following the basic parameters. Many researchers have taken the traditional route of using MVP to formulate the players' credentials from the statistics [1]. But, we have presented a more concise and logical approach to this paper. Aim of our paper trickles down to sorting players into their specializations on the basis of their game statistics, analyzing respective performances in particular genre and listing out the best playing 11 out of 20 [5].

Main analysis and scoring systems usually involve data on batsman's run-scoring capabilities and bowler's wicket-taking abilities. It was investigated that winning chances of teams by utilizing career bowling figures to determine each team's probability of dismissing the opposition, combined with each batsman's career average run scoring to predict the likely team score [6]. A model was developed for 50 over international cricket that utilizes overs and wickets remaining combined with runs scored so far to forecast the total score of a team that bats first during game. Their model can also be applied during the second innings to determine the probability of the team batting second winning the match [7].

2   Proposed Framework

The elaboration of the corresponding flowchart is our suggested framework for selecting the best possible playing eleven from a squad of 20 players.

2.1   Detailed Procedure

Data Source: The data is sourced and collected from various sports websites which have personal data of each player for every season.
Cleaning: The data is then cleaned and processed as per program's analysis. For a batsman, we stress on their average runs scored per match, runs in boundaries and strike rate. Similarly, for a bowler, we stress on their wickets taken, economy rate per over and the runs given away in their entire career.

Fig. 1 Flowchart

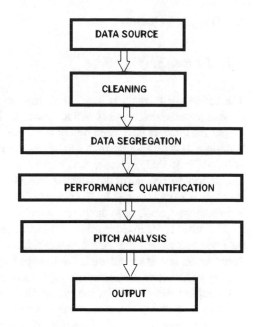

Data Segregation: T-20 cricket is heavily dependent on powerplay overs. During these six overs, maximum runs are seen to be scored, and wickets are taken. Hence, the team combination is heavily dependent on pinch hitters and death over bowlers. Similarly, the middle overs of a T-20 match require statistically and technically sound batsmen and wicket taking bowlers. Hence, getting the perfect combination right requires a specific algorithm. The following algorithm proposed helps in getting the best possible team combination.

Pitches also play a vital part in selecting the perfect team combination. For example, a batting pitch has to be approached with more pinch-hitting batsmen and less statistically sound batsmen. On the other hand, for a bowling pitch, an equal ratio of wicket-taking bowlers and economic bowlers has to be considered. Similarly, more number of technically sound batsmen and lesser pinch hitters have to be selected. The following algorithm assigns a particular bowling and batting score to each player to help the program choose the perfect team combination (Fig. 1).

3   Data Analytics

3.1   Player Segregation

The following formula is used to normalize and the players based on which the players as batsmen, bowlers and all-rounders have been segregated on plotting the graph; the clusters have been obtained which defined the groups.

Batsman score: The batsman's score evaluation involves two major components: AVERAGE RUNS, STRIKE RATE. A ratio of player's average run throughout the tournament to average of the net runs scored in the tournament is multiplied with the sum total of runs bagged by the player. Thus, a quantized value is obtained, which is directly proportional to the player's technical performance.

The other counterpart involves, a ratio of player's strike rate in the tournament to the average strike rate in the tournament, times the sum of the runs hit as boundaries. Tournament average is calculated as total of the average of each player by the number of players involved. Similarly, average strike rate is total strike rate of all players in the specific case; divided by the total number of players taken.

Runs as boundaries are calculated as number of 4s or 6s hit by the player; times their respective weight. For example: if a player hits 50 (4s) and 30 (6s), runs as boundaries = 4 * 50 + 6 * 30 = 380.

$$BattingScore = (PAR/TAR) * TRS + (PSR/TSR) * RIB \qquad (1)$$

where, PAR = Player Average Runs; TAR = Tournament Average Runs; TRS = Total Runs Scored; PSR = Player Strike Rate; RIB = Runs In Boundaries; TSR = Total Strike Rate

Bowler score: In order to evaluate the bowler's score, three parameters are considered: WICKETS TAKEN, ECONOMY RATE, RUNS GIVEN AWAY. The first component involves the ratio of total wickets taken by the bowler in the tournament to the average wickets taken in the tournament . The other ratio is between the player economy rate in the whole tournament and the average economy rate of the tournament. These two parameters are given priorities of 40 and 60

$$BowlingScore = (WTP/TTW) * 0.4 + (PER/TER) * 0.6 \qquad (2)$$

where WTP = Wickets Taken by Player; TTW = Total Tournament Wickets; PER = Player Economy Rate; TER = Tournament Economy Rate

Explanation In Fig. 2, the batting score has been plotted against the bowling score. Again, k-means clustering has been applied on the dataset by which the specialist batsmen, specialist bowlers, batting all-rounders and bowling all-rounders have been sorted. The algorithm helps the machine learning model to differentiate between the various categories of players and their specific roles. Figure 2 shows the original plot while Fig. 3 depicts the results after the algorithm was applied.

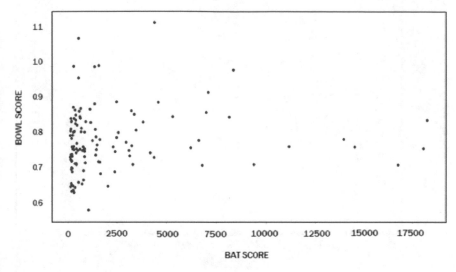

Fig. 2 Player segregation

Fig. 3 K-Means cluster for player segregation

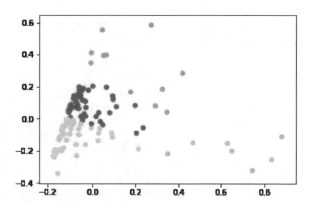

3.2 Classification of Batsmen

Explanation In Fig. 4, the strike rate has been plotted against the total runs. K-means clustering has then been applied on the dataset by which the pinch hitters, technically good batsmen and complete batsmen (pinch hitter + technically good) have been segregated to help the model differentiate between the various categories of batsmen and their specific roles in a T-20 match. Figure 4 shows the original plot while Fig. 5 depicts the results after the algorithm was applied.

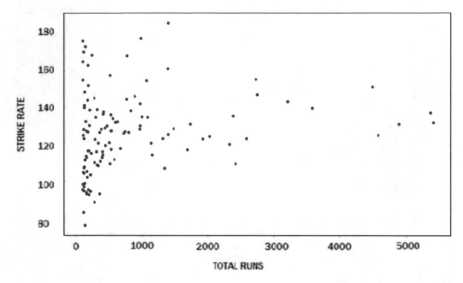

Fig. 4 Classification of batsman

Fig. 5 K-means cluster for
the classification of batsman

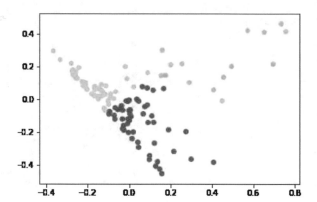

3.3 Classification of Bowlers

Explanation In the above graph, the wickets taken/ balls bowled against the econ-
omy/balls bowled have been plotted. K-means clustering has been applied on the
dataset by which the economical bowlers and the wicket-taking bowlers have been
separated. The algorithm helps the machine learning model to differentiate between
the various categories of bowlers and their roles in a T-20 match. Figure 6 shows the
original plot while Fig. 7 depicts the results after the algorithm was applied.

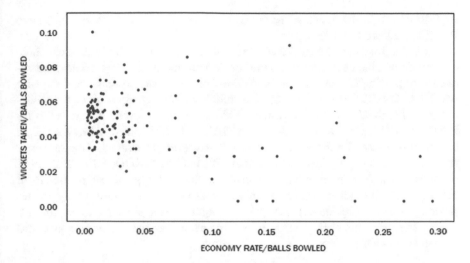

Fig. 6 Classification of bowlers

Fig. 7 K-means cluster for the classification of bowlers

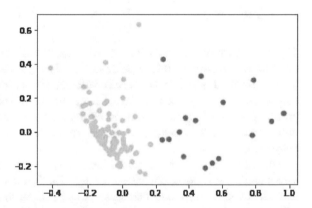

4 Pitch Analysis

The selection of best playing 11 for different pitch types have been designed, on the basis of impacts of the field on the team. The dataset has a separate column allocated for the roles of each player. It contains specifications of a pacer, spinner or a wicket-keeper. The first category is FLAT pitch. There has been a common trend of flat pitches: they provide suitable batting conditions. Two of the opening batsmen must selected as: A STRATEGIC BATSMAN and another STRATEGIC + PINCH-HITTING BATSMAN. The top and middle order contain a total of four batsmen. An ALL-ROUNDER is strictly recommended in the sequence of four. An extended research from papers and player statistics, a ratio of 2:1, has been devised, for PINCH HITTERS to BOTH (pinch hitting+ strategic batsman). The sixth and seventh positions would consist of a BATTING ALL-ROUNDER and BOWLING

ALL-ROUNDER, respectively. The tail end of the sequence might be formed by two PACERS and two SPINNERS.

The second and third categories would be for GREEN and DUSTY pitches. There are certain notable characteristics common in both the pitch types. Hence, the sequence of playing eleven is quite same. The opening is suggested to be done with two STRATEGIC batsmen. The top and middle orders contain a set of four players. An ALL-ROUNDER is suggested, in the set of four players. A 1:1:1 ratio for PINCH HITTERS to STRATEGIC BATSMAN to BOTH (pinch hitting+ strategic batsman) is advised for the rest set. The sixth and seventh positions of the sequence might be taken up by the BATTING and BOWLING ALL-ROUNDER. The tail of the sequence makes the difference according to the pitch type. Green pitches favor better swing of the ball; hence, the tail might consist of PACE bowlers. Dusty pitches favor better spin of the ball; hence, the tail might consist of SPINNERS. The following categorization of players has to be organized from the clustering algorithms implemented on the dataset.

5 Conclusion

The paper thereby proposes framework which when executed shall help any IPL team strategize every match. Once this model is implemented using a machine learning algorithm, this will make work for the support staff easier as we shall take into consideration the statistics of players from all the previous IPL seasons. The paper proposed a framework for the best possible playing eleven in a particular pitch condition based on the roles assigned and found out by the unsupervised machine learning algorithm.

Further, fielding can be introduced as a new parameter which might just give this idea a cutting edge toward wicket-keeping. The fact that young players do not have enough statistics to support their case is to be taken to consideration. This system will give them a chance to show their skill sets. Similarly, the system might just be automated for a next-level purpose. This shall enable us to ensure equal rotation of players; hence, the players would not tire out during hectic sessions.

References

1. Jayanth, S.B., Anthony, A., Abhilasha, G., Shaik, N., Srinivasa, G.: A team recommendation system and outcome prediction for the game of cricket. J. Sports Anal. **4**(4), 263–273 (2018)
2. Perera, H., Davis, J., Swartz, T.B.: Optimal lineups in Twenty20 cricket.
3. Khan, J.R., Biswas, R.K., Kabir, E.: A quantitative approach to influential factors in One Day International Cricket: Analysis based on Bangladesh. J. Sports Anal. **5**, 57–63 (2019)
4. Davis, J., Perera, H., Swartz, T.B.: Player evaluation in Twenty20 cricket. J. Sports Anal. **1**(1), 19–31 (2015)

5. Deep Pradesh, C., Patvardhan, C., Singh, S.: A new machine learning based deep performance index for ranking IPL T20 Cricketers. Int. J. Comput. Appl. **137**(10), 42–49 (2016)
6. Cohen, G.L.: Cricketing chances. Abstracts from the 6th Australian Conference on Mathematics and Computers in Sport, 1–3 July 2002
7. Gray, S., Le, T.: How to fix a one-day international cricket match. In: Abstracts from the 6th Australian Conference on Mathematics and Computers in Sport, Bond University Queensland, Australia. Sports Engineering, 5, 239, 1–3 July 2002

Analysing Pearl's Do-Calculus

Shreya Guha

Abstract Here, we analyse Pearl's do-calculus. The paper illustrates the proof of the three rules of the do-calculus. The concepts and notations of Kronecker delta function and d-separation theorem have been discussed. The concepts of uprooting and mowing a node have been explored in this paper. We present graphs to explain the concepts of subgraphs and augmented graphs. The paper also highlights the differences between Bayes' conditional probability and do-operator. Shafer's probability trees are also illustrated in this paper. The nodes of these trees denote events and the edges represent the probabilities of various outcomes. The complicated concepts as described by David Hume and Abraham De Moivre have also been covered in this paper. An example showing a chemical causation of nature has been cited to illustrate these concepts. The paper also explores some of the applications of do-calculus in the real world. The do-calculus can be used in IoT systems. It can be used for fault diagnosis in automobiles and vehicular systems. It also finds applications in medical science by helping patients in choosing treatment options. It also helps to explore whether genes play an important part in inheritance of diseases. It also finds applications in domains of machine learning such as computational advertising.

Keywords Do-operator · Belief propagator models · Dempster–Shafer theory · Shafer's probability tree

1 Introduction

The do-calculus, developed by Pearl [1], facilitates the identification of causal effects in those models, where the model structure is not specified initially but is determined from the data. Such types of models are known as nonparametric models. Judea Pearl's do-calculus is a part of his theory of probabilistic causality, which is a study of Bayesian networks. The do-operator, proposed by Pearl, calculates the probability

S. Guha (✉)
Institute of Engineering and Management, Kolkata, India
e-mail: shreyaguha24@gmail.com

© Springer Nature Singapore Pte Ltd. 2020
J. K. Mandal and S. Mukhopadhyay (eds.), *Proceedings of the Global AI Congress 2019*, Advances in Intelligent Systems and Computing 1112,
https://doi.org/10.1007/978-981-15-2188-1_54

distribution, where we artificially set a value to a variable. The three rules of do-calculus along with the proofs are illustrated in this paper. This paper also highlights some of the applications of do-calculus in the domain of IoT, machine learning and medical science.

2   Some Basic Notations [2]

Kronecker delta function:
   Kronecker delta function can be represented as δ_n^m or $\delta(m, n)$ or $\delta_{m,n}$. It is piecewise function of variables m and n.

$$\delta(m, n) = \begin{cases} 1 \text{ if } m \neq n \\ 0 \text{ otherwise} \end{cases}$$

We can indicate random variables either as $\underline{m}$ or M (underlined or capital letters) and its corresponding values as m (lower case letters). M and $\underline{m}$ are used interchangeably in the text.
   For a probability distribution $P_{M,N}(m, n)$:

$$P(m : n) = P(m, n)/P(m)P(n) = P(m \mid n)/P(m) \tag{1}$$

$$P(m : n \mid q) = P(m, n \mid q)/\{P(m \mid q)P(n \mid q)\} = P(m \mid n, q)P(m \mid n) \tag{2}$$

We can denote n-tuples by a letter followed by a dot, as in $a. = (a_1, a_2, \ldots, a_n)$, or as a boldface letter like $\mathbf{a}$.
   We will consider two n-tuples of random variables as ordinary normal sets. Thus, we can use the standard set system and notation with these two tuples. Sometimes A_i and $\{A_i\}$ may be used to denote the same thing as we may not distinguish between these two notations.
   A classical Bayesian network is represented by a directed acyclic graph (DAG) where vertex represents random variables A_j. Let $\underline{a}. = (a_1, a_2, \ldots, a_n)$. Arrow from $\underline{m}$ to $\underline{n}$ indicates that $\underline{m}$ is the parent node and $\underline{n}$ is the child node. Set of all parent nodes can be denoted as $\underline{\text{pa}}(a_j)$ and set of all children nodes are denoted as $\underline{\text{ch}}(a_j)$.
   Each node is assigned transition probability matrix $P(a_j \mid \text{pa}(\underline{a}_j))$. This transition probability matrix depends on two factors—values of a_j of node $\underline{a}_j$ and values of $\text{pa}(a_j)$ of nodes $\underline{\text{pa}}(a_j)$.
   Therefore, the total probability of entire Bayesian network is

$$P(a) = \prod_{j=1}^{N} P(a_j \mid \text{pa}(\underline{a}_j)) \tag{3}$$

3 Subgraphs and Augmented Graphs

G is a graph with nodes $\underline{m}$. such that $\underline{n}. \subset \underline{m}.$, as shown in Fig. 1.

G_n (Fig. 2) denotes "restriction" of graph G where the $\underline{m}. - \underline{n}.$ vertices and the edges associated with these vertices have been removed as shown by the red cross mark.

$G_{n'}$ (Fig. 3) denotes graph G after removing all arrows entering node set $\underline{n}$. Arrows and nodes with red cross are removed.

$G_{n''}$ (Fig. 4) denotes graph G with edges exiting node set $\underline{n}$. erased.

The notation $G \leftarrow \underline{rt}(\underline{t}.)$ (Fig. 5) denotes the augmented graph obtained by adding node set $\underline{rt}(\underline{t}.)$ to graph G. The arrows are made to point from $\underline{rt}(\underline{t}.)$ to node set $\underline{t}$. For each $\underline{t}_j \in \underline{t}.$, only one twin node $\underline{rt}(\underline{t}_j) \in \underline{rt}(\underline{t}.)$, is added using only one arrow from node $\underline{rt}(t_j)$ to t_j.

Fig. 1 Graph G. The dotted frame signifies a corral where $\underline{n}. \subset \underline{m}.$ of graph G. The frame encloses $\underline{n}.$ nodes and leaves $\underline{m}. - \underline{n}.$ nodes outside

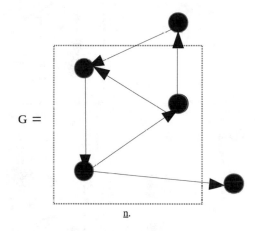

Fig. 2 Graph G_n

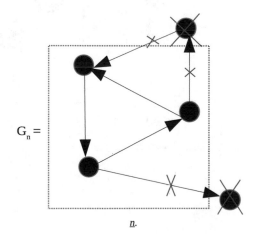

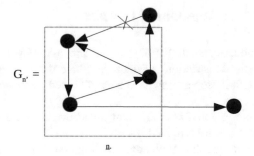

Fig. 3 Graph $G_{n'}$

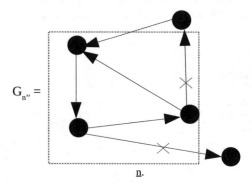

Fig. 4 Graph $G_{n''}$

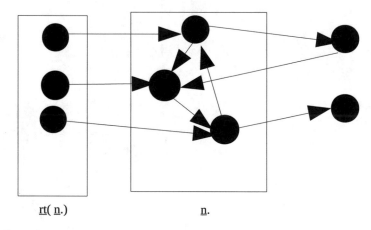

Fig. 5 Graph $G \leftarrow \underline{rt}(\underline{n}.)$

Fig. 6 Figure depicting a graph template TG. Arrows with red cross denote ban on arrows entering $m.$, a ban on arrows exiting $t.$, and a ban on arrows going from $m.$ to $q.$

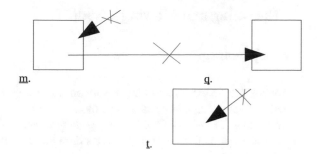

Graph $G \leftarrow \underline{rt}(\underline{n}.)$ (Fig. 5) is modified version of graph G of Fig. 1 after node set $\underline{rt}(\underline{n}.)$ is added to root nodes for $\underline{n}$.

A graph template can be described as a set or family of graphs. In a graph template (Fig. 6), some corrals have bans (denoted by arrows with red cross implying that it is forbidden). Arrows entering, leaving or going from one corral to another are banned.

4   d-Separation Theorem

The d-separation theorem helps us to infer from a graph G, whether $\underline{m}.$ and $\underline{t}.$ are probabilistically conditionally independent given a third set $\underline{q}.$, where $\underline{m}., \underline{t}.$ and $\underline{q}.$ are corrals.

It is denoted as $\underline{m}. \perp \underline{t}. \mid \underline{q}.$

$\underline{m}.$ and $\underline{t}.$ are said to be d-separated for all paths from $m.$ to $t.$, if at least one of the following holds:

Path includes a "chain" with an observed middle node (Fig. 7).

Path includes a "fork" with an observed parent node (Fig. 8).

Path includes a "collider" (Fig. 9).

D-Sep Theorem:

$(\underline{m}. \perp \underline{t}. \mid \underline{q}.)_G$ iff, for all possible values of $m., t., q.,$ $P_G(m.: t. \mid q.) = 1$, or, equivalently, $P_G(m. \mid t, q.) = P_G(m. \mid q.)$.

Fig. 7 Observed middle node

Fig. 8 Observed parent node

Fig. 9 Observed collider node

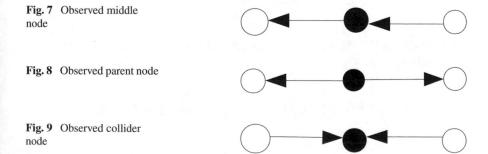

5 Uprooting and Mowing a Node

5.1 Uprooting

Uprooting a node means removing the incoming arrows, i.e. erasing the nodes. Pearl called this uprooting a node as intervention.

If $\underline{m1}. \cap \underline{m2}. = \phi$, $\underline{m1}. \cap \underline{m3}. = \phi$, $\underline{m2}. \cap \underline{m3}. = \Phi$ and $(\underline{m1}. \cup \underline{m2}. \cup \underline{m3}. \cup \underline{m4}.) = \underline{m}$. where $\underline{m4}.$ is the set of other nodes then probability that $\underline{m2}. = m2$ when $\underline{m1}. = m1$ is uprooted:

$$P(m2. \mid m1.') = \frac{(P_{G_{n'}}(m1., m2.))}{P_{G_{n'}}(m1.)} \neq P_G(m2. \mid m1.) \tag{4}$$

where $P_{G_{n'}}(m.)$ denotes probability distribution for subgraph $G_{n'}$.

$P_G(m2. \mid m1.')$ and $P_G(m2. \mid \mathrm{do}(\underline{m1}.) = m1.)$ denotes the same thing. The notation do(.) is known as the do-operator.

Also,

$$P(m. - m1. \mid m1.') = \frac{P(m.)}{\prod_{j:\underline{m}_j \in m1.} P(m_j. \mid pa(m_j))} = \prod_{j:\underline{m}_j \in \underline{m}.-\underline{m}_1.} P(m_j. \mid pa(\underline{m}_j)) \tag{5}$$

If $\underline{m1}. = (\underline{m1}_1, \underline{m1}_2, \underline{m1}_3, \ldots, \underline{m1}_n)$, then $P(m. - m1. \mid m1.')$

$$= P(m. - m1. \mid m1', m2', \ldots, mn') \tag{6}$$

Also,

$$P\big(m2. \mid m1.'\big) = \sum_{m.-(m2. \cup m1.)} P\big(m. - m1. \mid m1.'\big) \tag{7}$$

and

$$P\big(m2. \mid m1.', m3.\big) = \frac{P\big(m2., m3. \mid m1.'\big)}{\prod_{j:\underline{m}_j \in m1.} P(m3. \mid m1.')} \tag{8}$$

where $P(m2. \mid m1.', m3.)$ signifies the probability of $\underline{m2}.$ when $\underline{m1}.$ is uprooted and it is conditioned on $\underline{m3}.$

We also note that

$$P\big(m2. \mid m1., \mathrm{rt}(\underline{m1}.) = 0\big) = P\big(m2. \mid m1.\big) \tag{9}$$

$$P\big(m2. \mid m1., \mathrm{rt}(\underline{m1}.) = 1\big) = P\big(m2. \mid m1.'\big) \tag{10}$$

where $\mathrm{rt}(\underline{m1}.) = n$, where $n \in \{0, 1\}$.

From the above two equations, i.e. (8) and (9) we observe that when $\mathrm{rt}(\underline{m1}.) = 1$, all the roots of $\underline{m1}.$ are sliced off.

5.2 Mowing

Mowing a root means removing or erasing all the outgoing arrows of the node.

If $\underline{m1}. \cap \underline{m2}. = \phi, \underline{m1}. \cap \underline{m3}. = \phi, \underline{m2}. \cap \underline{m3}. = \Phi$ and $(\underline{m1}. \cup \underline{m2}. \cup \underline{m3}. \cup \underline{m4}.) = \underline{m}.$ where $\underline{m}.$ are nodes of graph G, and $\underline{m4}.$ is the set of other nodes then, probability of $\underline{m}. = m$ when $\underline{m1}. = \underline{m1}.$ is mowed:

$$P_{\underline{m1}."(m1.''')}(m.) = \left[P\left(m. - m1. \mid [m1.]'\right)\right]_{m1. \to m1.} P\left(m1.. \mid [m. - m1.]'\right)$$
$$= P\left(m. - m1. \mid [m1.''']'\right) P\left(m1. \mid [m. - m1.]'\right) \qquad (11)$$

Also,

$$\sum_m P_{\underline{m1}."(m1.')}(m.) = 1 \quad \text{and} \quad P_{\underline{m1}."(m1.''')}(m.) = P_{\prod_j \underline{m}_{j}"\left(m_j'''\right)}(m.) \qquad (12)$$

$$P_{\underline{m1}."(m1.''')}(m1., m2.) = \sum_{m. - (m1. \cup m2.)} P_{\underline{m1}."(m1.''')}(m.) \qquad (13)$$

$$P_{\underline{m1}."(m1.''')}(m2. \mid m1., m3.) = \frac{\left(P_{\underline{m1}."(m1.')}(m1., m2., m3.)\right)}{\left(P_{\underline{m1}."(m1.')}(m1., m3.)\right)} \qquad (14)$$

Adding both sides of (10) with $m1$ gives

$$P_{\underline{m1}."(m1.''')}(m. - m1.) = \left[P\left(m. - m1. \mid [m1''']'\right)\right] \qquad (15)$$

Adding both sides of (14) with $m. - (m1. \cup m2.)$ gives

$$P_{\underline{m1}."(m1.''')}(m2.) = \left[P\left(m2. \mid [m1.''']'\right)\right] \qquad (16)$$

6 Pearl's Do-Calculus

Given an outcome y, we know its value is dependent on some feature(s). The do(.) operator calculates the probability of $Y = y$ conditioned that the feature X's values remains constant at $X = x$. It is generally represented as $P(y \mid do(X = x))$. The causal conditioning, $P(y| \operatorname{do}(X = x))$ is one of the two types of conditioning operators frequently used with the other one being Bayes' conditioning, $P(y \mid X = x)$.

7 Difference Between Bayes' Conditional Probability and Do() Calculus

Conditional probability defined by Bayes' relates to observational conditions. It cannot compute a distribution when intervention occurs. Causal effects help us to predict how systems would respond to hypothetical interventions. To mitigate this, Pearl proposed a new operator called do-calculus which seeks to provide probability distribution when we artificially impose a value to variable.

Theorem *Suppose $\underline{m}.$ is the set of all nodes of graph G and $\underline{m}$ is the union of the disjoint subsets $\underline{m1}$, $\underline{m2}$, $\underline{m3}$, $\underline{m4}$. and $\underline{m5}$.*

Rule 1: $(m1. <\text{-}> 1)$:

$$\left(\underline{m2}. \perp \underline{m1}. \mid \underline{m3}., \underline{m4}.\right)_{(G_1)} \text{ where } G_1 = G_{\underline{m3'}}$$

iff for all $m1., m2., m3., m4.$,

$$P(m2. : m1. \mid m3.', m4.) = 1$$

Rule 2: $(m1. <\text{-}> m1.')$:

$$\left(\underline{m2}. \perp \underline{m1}. \mid \underline{m3}., \underline{m4}.\right)_{(G_2)} \text{ where } G_2 = G_{\underline{m3'};\,\underline{m1'}}$$

iff for all $m1., m2., m3., m4.$,

$$P\left(m2. : m1.' \mid m3.', m4\right) = P(m2. : m1. \mid m3.', m4.)$$

Rule 3: $(m1.' <\text{-}> 1)$:
    If

$$\left(\underline{m2}. \perp \underline{m1}. \mid \underline{m3}., \underline{m4}.\right)_{(G_3)} \text{ where } G_3 = G_{\left((m3')\left[\underline{m1}.-\operatorname{an}\left(\underline{m4}., G_{(m3')}\right)\right]'\right)}$$

then for all $m1., m2., m3., m4.$,

$$P\left(m2. : m1.'\right)(m3 : m4.) = 1$$

where an in the expression denotes the set of ancestor nodes.
Pearl's three rules can also be represented as follows:

1. While ignoring the observations:

$$P(b \mid do(a), c, d) = P(b \mid do(a), d) \quad \text{if } (B \perp C \mid A, D)_{\left(G_{(\bar{A})}\right)}$$

2. The Backdoor criterion when action and observation exchange occurs:

$$P(b \mid do(a), do(c), d) = P(b \mid do(a), d) \text{ if } (B \perp C \mid A, D)_{\left(G_{(\bar{A}, \underline{C})}\right)}$$

3. Ignoring actions or interventions:

$$P(b \mid do(a), do(c), d) = P(b \mid do(a), d) \text{ if } (B \perp C \mid A, D)_{\left(G_{((\bar{A}), (\bar{C}(\bar{D})))}\right)}$$

where for graph G,

$$\{A\} \cap \{B\} = \phi, \{C\} \cap \{B\} = \phi, \{C\} \cap \{D\} = \phi, \{A\} \cap \{D\} = \phi, \{D\} \cap \{B\}$$
$$= \phi \text{ and } \{A\} \cap \{C\} = \phi.$$

Here, $G_{(\bar{A})}$ represents the graph in which edges directed towards A are deleted and $G_{(A)}$ represents the graph where edges pointing away from A are sliced off. $C(D)$ denotes the set of nodes belonging to C which are not parents of D.

8 Proofs of the Three Rules

Rule 1:
  From d-sep theorem, $\left(m2. \perp \underline{m1}. \mid \underline{m3}., \underline{m4}.\right)_{(G_1)}$ iff

$$\delta_{\underline{m3'}} P(m2. \mid m1., m3., m4.) = \delta_{\underline{m3'}} P(m2. \mid m3., m4.)$$
$$\text{LHS} = P\big(m2. \mid m1., m3.', m4.\big)$$
$$\text{RHS} = P\big(m2. \mid m3.', m4.\big)$$

Rule 2:
  By the d-sep theorem, $\left(m2. \perp \underline{m1}. \mid \underline{m3}., \underline{m4}.\right)_{(G_2)}$ iff

$$\lim_{m1.'''\to m1} \delta_{(m3.''(m3.'''))} \delta_{(m4.')} P(m2. \mid m1., m3., m4.)$$
$$= \lim_{m1.'''\to m1} \delta_{(m1.''(m1.'''))} \delta_{(m3.')} P(m2. \mid m1., m3., m4.)$$
$$\text{LHS} = P(m2. \mid m1., m3.', m4.)$$
$$\text{RHS} = P(m2. \mid m1.', m3.', m4.)$$

Rule 3:

Let $\widetilde{m1}.$ and denote the following:

$$\widetilde{m1}. = \underline{m1}. - \text{an}(\underline{m4}., G_{m3'})$$

and

$$\breve{m1}. = \underline{m1}. \cap \text{an}(\underline{m4}., G_{m3'})$$

Let S and S' denote the following statements:

$$S = \left(\underline{m2}. \perp \underline{m1}. \mid \underline{m3}., \underline{m4}.\right)_{(G_3)} \quad \text{where } G_3 = G_{(m3'.,[\tilde{s1}].')}$$

$$S' = \left(\underline{m2}. \perp \underline{m1}., \text{rt}(\underline{m1}). \mid \underline{m3}., \underline{m4}.\right)_{(G_3''')} \quad \text{where } G_3''' = [G < -\text{rt}(\underline{m1}).]_{\underline{m3}.''}$$

By d-sep theorem, S' implies for all $m2., m1., \text{rt}(\underline{m1}.), \underline{m3}., \underline{m4}.$,

$$P(m2, m1., \text{rt}(\underline{m1}.), m3.', m4) = P(m2. \mid m1.', m3.'m4.)$$

Now, we need to prove that $S \to S'$. We prove this by doing $\neg S' \to S$. For $\neg S'$ there exists α, unblocked at definite $\underline{m3}., \underline{m4}.$, which satisfies

$$\alpha \in \text{Path}_{TG'''}\left(\underline{M2}, \underline{M1}_1, \underline{M1}_2, \ldots, \underline{M1}_n, \text{rt}(\underline{Mn}_n)\right)$$

where $\underline{M2} \in \underline{m2}., \underline{M1} \in \underline{m1}.$

$\underline{M1}_1$ is the only node in α that is nearest to $\underline{M2}$. Also the path in TG_3''' at a definite $\underline{m3}.$ and $\underline{m4}.$ is also unblocked. It is a shorter path than α. Let us name it as α_0.

Therefore,

$$\alpha_0 \in \text{Path}_T\left(\underline{M2} < \underline{M1}_1, \text{rt}(\underline{M1}_1)\right)$$

Here, we take T as TG_3'. To bring our proof to an end we need to show that the above equation holds true for $T = TG_3'''$ instead of a greater set TG_3'''.

As it can be observed from diagram below, there is an extra ban on arrows entering $\widetilde{m1}.$ implying that these arrows enter node $\underline{M1}$.

No arrow enters $\widetilde{m1}.$ if $\underline{M1}_1 \in \breve{m1}.$. Therefore, we have proven our point. If $\underline{M1}_1 \in \underline{m1}.$ there would have been two options, either $\underline{M1}_1 \in \text{col}(\alpha_0)$ or $(\underline{M1})_1 \notin \text{col}(\alpha_0)$. (Set of collider nodes of path α_0 is col().)

Case 1:

$$(M1)_1 \notin \mathrm{col}(\alpha_0)$$

No arrow points from $\underline{\mathrm{rt}(M1_1)}$ to $\underline{M1}_1$.
Therefore, there is an arrow which points from $\underline{M1}_1$ to a node outside $\widetilde{m1}$.
Therefore, no arrows from α_0 enter $\widetilde{m1}$.
Hence proved.
Case 2:

$$\underline{M1}_1 \in \mathrm{col}(\alpha_0)$$

α_0 is unblocked at definite $\underline{m3}.$ and $\underline{m4}.$
Therefore, $\mathrm{de}(\underline{M1}_1) \cap (\underline{m3}. \cup \underline{m4}.) \neq \Phi.$
Here, $\mathrm{de}(M1_1)$ denotes descendants or set of descendant nodes.
But this is not possible as concluded from Fig. 10. Since in TG_3 as arrows from $\widetilde{m1}.$ can not enter $\underline{m4}.$ and arrows will not be allowed to enter $\underline{m3}.$

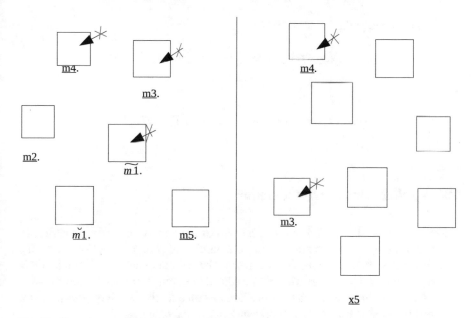

Fig. 10 Here, the set $\underline{m5}.$ denotes the set of other nodes

9   A Short Note on "Path"

While proving rule three in Sect. 7 the term "Path" has often been used. Though this should have been included previously in the text, still this is being illustrated in this section.

Path$_G(\underline{X} < \underline{Y})$ denotes the set of all undirected paths of graph G that starts at node $\underline{X}$ and ends at $\underline{Y}$. < implies that number of arrows are $> = 1$(direction does not matter) and the number between $\underline{X}$ and $\underline{Y}$ is greater than equal to 0.

$\underline{X} = < \underline{Y}$ can also imply that $\underline{X}$ and $\underline{Y}$ are the same node.

When $x_1, x_2, \ldots, x_n$ are disjoint subsets of $\underline{x}$., then Path$_G(\underline{x_1} < \underline{x_2} < \underline{x_3} < \cdots < \underline{x_n})$.

Say G has $\underline{x}$. Nodes and subsets $\underline{x1}.$, $\underline{x2}.$, $\underline{x3}.$, $\underline{x4}.$, where $\underline{x4}.$ is the set of other nodes such that $\underline{x_i}. \cap \underline{x_j}. = \phi$ for $i \neq j$ and $i, j \in [1\text{--}4]$ and $\underline{x1}. \cup \underline{x2}. \cup \underline{x3}. \cup \underline{x4}. = \underline{x}$. Then, $\alpha \in$ Path$_G(\underline{x1}. < \underline{x2}.)$ is blocked at definite $\underline{x3}$ if

$$\text{Either } (\exists \underline{m} \in \alpha)\big[\underline{m} \notin \text{col}(\alpha) \wedge \underline{m} \in \underline{x3}.\big]$$

$$\text{Or,} \quad (\exists \underline{m1} \in \alpha)\big[\underline{m1} \in \text{col}(\alpha) \wedge \overline{\text{de}}(\underline{m1}) \cap \underline{x3}. = \emptyset\big]$$

For unblocked path,

$$(\forall \underline{m} \in \alpha)\big[\underline{m} \notin \text{col}(\alpha) \implies \underline{m} \notin \underline{x3}.\big]$$

and

$$(\forall m1 \in \alpha)\big[\underline{m1} \in \text{col}(\alpha) \implies \overline{\text{de}}(\underline{m1}) \cap \underline{x3}. \neq \emptyset\big]$$

10   Applications of Do-Operator

- Real-time fault diagnosis is an integral part of modern automobile industries. In recent times, the vehicles are built up of a number of programmable components which in turn control the mechanical parts. These components are driven by electronic control units (ECUs). This implies that if one component fails it could lead to mechanical failure. Research has thus focussed on effectively solving this problem in real time using Bayesian networks and intervenal calculus.
- The do-calculus finds many applications in medical sciences from helping patients to choose options for treatment of brain tumour, kidney stones, gall-bladder stones and many others. It also helps to explore whether gene plays an important part in inheritance of diseases.
- The do-calculus also finds major application in machine learning (e.g., computational advertising).

11 Shafer's Probability Trees

In Shafer's probability trees, the nodes indicate events with outcome(s). The branches connecting these nodes indicate respective probabilities.

Suppose, the information required to arrive at a particular decision is available. This information may be obtained from varying sources with non-uniform authenticity levels. The Dempster–Shafer theory helps in the reduction of uncertainty in decisions. Pearl's belief propagation model takes the causal and evidential information for an event as input. This model is applied on a causal graph where nodes indicate the events and the directed arrows represent the causal relationship between two events.

In classical probability theory, an experiment or trial outcome is an event. A sample space is the totality of event outcomes. In Shafer's causal probability trees, a probability tree is a series of events and their potential probabilistic relations. Shafer follows classical probability tree presentations. A probability tree's root is at the top of the trees, and the leaves are below. Shafer adds a twist on how to interpret probability trees to understand dynamic causality. These trees are suitable for describing logical causal derivations starting from their roots and going down [3].

A level of a probability tree comprises of branches coming out from a node. Each level of a probability tree denotes a sample space. Therefore, the total probability of all possible events which are directly below a node is 1. A probability space is formed by each node and its corresponding children nodes. Figure 11 shows node A with children A1, A2 and A3. The values 0.2, 0.3 and 0.5 are probabilities of moving from the event A to each of its children A1, A2 and A3. Let us assume the tree in Fig. 12 is a subtree of a more massive tree. As per Shafer [3], individual events do not represent an entire causal system. Humean events (after David Hume) are local events transitioning from one level to the next in probability trees.

To specify which step is taken in a path down a probability tree, Shafer points to concomitants. A concomitant is something that occurs with an event. It may be a correlation. For instance, if A causes "A3", then the occurrence of A should correlate with the occurrence of "A3". However, if A does not cause "A3", then the occurrence of A may or may not correlate with the occurrence of "A3". In Fig. 12, the value 1 ATM/1000 is one thousandth the atmospheric pressure on the earth at sea level. In

Fig. 11 Three Humean events in a subtree of a more significant probability tree

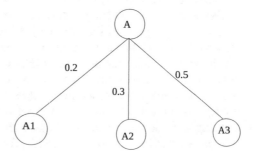

Fig. 12 Probability three of
the phases of water at
different atmospheric
pressures and temperatures

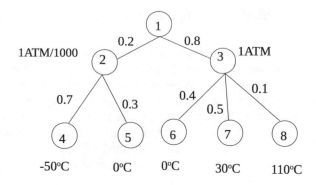

nodes (4) and (6), water is in a solid state. In nodes (5) and (8), water is in gaseous
state, whereas in node (7), water is in liquid state. Given these different atmospheric
pressures and temperatures, this tree represents chemical causation from nature. In
particular, water's phases can be detailed scientifically. This scientific documentation
is established on related information. The temperatures and atmospheric pressures
cause the states of water illustrated. Figure 12 describes a probability tree representing
some phase transitions of water.

At this juncture, 1 ATM describes the atmospheric pressure at sea level on earth. It
also illustrates Moivrean paths (after Abraham De Moivre) from the root node down
to the leaves. Moivrean events are event sequences from the root of a probability tree
to one of its leaves. With concomitant information Moivrean paths describe causality.

Recall, the sum of the probabilities coming out of each set of Humean events at
any node is 1. This is illustrated by the probabilities in Fig. 11 where $1 = 0.2 +
0.3 + 0.5$. This is also true for the subtree of nodes (1, 2 and 3) in Fig. 12 having
probability $1 = 0.2 + 0.8$. The same applies for the two subtrees rooted at nodes (2
and 3). A consequence of these Humean events is that the sum of probabilities at
each level is 1. Also, the sum of all the products of the probabilities down all paths
is 1. That is, computing the probability of a single Moivrean event or path is done by
multiplying the probabilities down the entire path. The sum of the probability of all
Moivrean events is 1. Moivrean paths may split into several separate subtrees. Two
separate subtrees are independent if neither depends on the other. Some probability
trees also have decision points. However, decision points are not causing. They are
external decisions that may drive a system [3].

The next example illustrates Moivrean paths—it further describes some sensors
are monitoring experiments in extremely different environments. These experiments
are sensitive to the state of water in these environments. These environments differ
with temperature and pressure. The different temperature and pressures cause water
to be solid, liquid or a gas. These states of water, in turn, affect the probability
of success of the experiments. In the case of slight pressure, such as 1/1000th of
atmospheric pressure, water jumps from a solid state directly to a gas state. When
warming up the ice at such low pressure, water skips the solid state and goes directly
to its gas state as its temperature increases from ($-50\,°C$) up to the low temperature

Table 1 Experiment outcomes displaying states of water under varying conditions as detected by the sensors

Experiment outcome	H20 state	Atmospheres	Temperature (°C)
Success: 0.9, failure: 0.1	Solid	1/1000	−50
Success: 0.8, failure: 0.2	Gas	1/1000	0
Success: 0.9, failure: 0.1	Solid	1	0
Success: 0.05, failure: 0.95	Liquid	1	35
Success: 0.8, failure: 0.2	Gas	1	110

of (0 °C). Table 1 shows how the liquid state of water dramatically reduces sensor experiments success rates, where solid and gas states have better success rates.

The Moivrean paths to sensors detecting water in liquid states depict the below probability equation:

$$0.32 = 0.8 \times 0.5 = 0.8 \times 0.5(0.05 + 0.95)$$

Summing the first column gives a probability of 0.546 for the success of the experiment for all the sensors. We conclude that the probability of failure for an arbitrary experiment is 0.454. Table 2 explains the logic behind such an experiment. The whole Moivrean paths to success can be described by each of the below probability atmospheric water levels.

Level 1 down to level 2 is node 1 and its edges go down to nodes 2 and 3 in Fig. 13. Subsequently, the levels increase as we go down. Each level is a Humean event. All success states are leaves on the bottom with "**S**" below them.

As depicted above, 1 ATM is the atmospheric pressure at a sea level. The nodes 9, 10, 13 and 14 describe the solid state. The nodes 15 and 16 are displayed as a liquid, whereas the nodes 11, 12, 17 and 18 denote gaseous state.

Table 2 Probability atmospheric water levels

Probability	Level 1 down to level 2	Level 2 down to level 3	Level 3 down to level 4
0.126	0.2	0.7	0.9
0.048	0.2	0.3	0.8
0.288	0.8	0.4	0.9
0.02	0.8	0.5	0.05
0.064	0.8	0.1	0.8

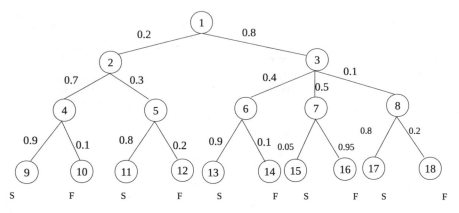

Fig. 13 Exhibit of IoT intelligent devices and sensors in different environments

12 Conclusion

In this paper, the three rules of do-calculus along with their proofs have been high-lighted. The difference between Bayes' probability and do-calculus has been clarified. The theory of causality along with do-calculus and Shafer's probability trees help in detecting anomaly in IoT systems and also in studying machine learning and also implementing medical treatments. We can troubleshoot the problem by changing various parameters and observing the effects they have on the outcome. Thus, the do-calculus provides us with a powerful tool to predict how systems would respond to various interventions in a stronger way than a Bayesian approach.

Bibliography

Cited in the Text

1. Pearl, J.: Causal diagrams for empirical research. Biometrika **82**(4), 669–688 (1995)
2. Tucci, R.R.: Introduction to Judea Pearl's Do-Calculus. arXiv preprint arXiv:1305.5506 (2013)
3. Shafer, G.: The Art of Causal Conjecture. MIT Press, Cambridge, MA (1996)

As Deemed Worthy of Attention

4. Bareinboim, E., Pearl, J.: Causal inference by surrogate experiments: z-identifiability. arXiv preprint arXiv:1210.4842 (2012)
5. Huang, Y., Valtorta, M.: Pearl's calculus of intervention is complete. arXiv preprint arXiv: 1206.6831 (2012)
6. Konar, A.: Belief calculus and probabilistic reasoning. Computational Intelligence: Principles, Techniques and Applications, pp. 353–391 (2005)

7. Koller, D., Friedman, N.: Probabilistic Graphical Models: Principles and Techniques. MIT Press, Cambridge, MA (2009)
8. Peters, J., Janzing, D., Schölkopf, B.: Elements of Causal Inference: Foundations and Learning Algorithms. MIT Press, Cambridge, MA (2017)
9. Pearl, J.: The do-calculus revisited. arXiv preprint arXiv:1210.4852 (2012)
10. Shafer, G.: The Relevance of Trees (1994)
11. Shafer, G.: Mathematical foundations for probability and causality. In: Proceedings of Symposia in Applied Mathematics, vol. 55, pp. 207–270. American Mathematical Society, Boston, MA, USA (1998, July)

A Dynamic Threshold-Based Trust-Oriented Intrusion Detection System in MANET

Khondekar Lutful Hassan, J. K. Mandal and Shukla Mondal

Abstract In order to discard and detect rushing attacker in the network, this article analyses the MANET with dynamic threshold-based trust-oriented AODV (DTT-AODV) model. In the proposed method has been simulated and analyzed with pre-determined threshold-based enhanced trust mechanism under the rushing attack in MANET. So, based on the analysis of recent techniques, a dynamic threshold-based trust-oriented AODV (DTT-AODV) model has been proposed to improve MANET security. The performance of the network is analyzed on different densities of node based on throughput, control overheads, normalized routing overheads, and packet delivery ratio. The main idea in this proposed technique is to detect and discard the malicious node from the network established on the other nodes replied with the route replies and the route requests present in the network. As post-identification measured, evaluation and periodically maintenance of trust calculation are done. Based on the determined threshold dynamically, it is decided whether the node will be excluded or included from the path.

Keywords Rushing attack · Trust · Dynamic · Direct · Indirect · MANET · AODV · Security · Routing protocols

K. L. Hassan (✉) · S. Mondal
Department of Computer Science and Engineering, Aliah University,
Kolkata 700160, West Bengal, India
e-mail: klhassan@yahoo.com

S. Mondal
e-mail: shuklamondal95@gmail.com

J. K. Mandal
Department of Computer Science and Engineering, University of Kalyani, Kalyani,
Nadia 741235, West Bengal, India
e-mail: jkm.cse@gmail.com

© Springer Nature Singapore Pte Ltd. 2020
J. K. Mandal and S. Mukhopadhyay (eds.), *Proceedings of the Global
AI Congress 2019*, Advances in Intelligent Systems and Computing 1112,
https://doi.org/10.1007/978-981-15-2188-1_55

1 Introduction

The key characteristics of MANET like changing topology dynamically, self-maintaining, self-configuring routing protocols, etc., make it very unsafe because of the packet transmission between the nodes following the routing protocols to reach to destination nodes. As MANET is more dynamic than the wired networks so any node can join into the network as well as can leave too. As it is more open, there is lot of security vulnerabilities present in the network. AODV, the reactive routing protocol, is the widely known routing protocols in MANET. In AODV, by flooding the packet on to the network, the route discovery process is done on demand. Several researches have been done to exploits the security and the technique to detect and prevent it. But still many security threats are there in the network.

The malicious or the attacker nodes present in the network can alter the network resources and by carrying its harmful packets into the route makes the network vulnerable. Some of the attacks and the vulnerabilities in the network are as follows.

In *denial-of-service (DoS)* attack, the attacker nodes try to block the access of the legitimate users from accessing the network services. The *repudiation* refers to the malicious node which acts as the denial node in the network. Because of the duplicate suppression technique in the AODV routing protocol, the *rushing attacker* node quickly forwards the RREQ packets to gain control in the network. The other RREQ packets are then discarded, and AODV protocol accepts the first RREQ message as it is legitimate RREQ. By using tunneling, one attacker forwards its control packets to second attacker node so that the RREQ of the attacker can arrive to receiving node before the actual RREQ of other legitimate nodes can reach. This way the attacker can alter or drop a data packet that leads to denial in the network. This attack is known as wormhole attack. In the classic blackhole attack, the attacker node gives a route reply to every route request in the network so that the data packet is dropped. In *gray hole attack*, initially the attacker nodes do not behave like a malicious one so that it can gain trust. *Resource depletion attack* creates congested channels by sending unnecessary routing control packets and uses up the network resources like battery power, space, etc. The attacker nodes could also establish routes and could send unnecessary and fake data repeatedly. This can be very dangerous security threats to various applications.

Those vulnerabilities lead to route disruption, route invasion, node isolation, i.e., trying to isolate the nodes from the rest of the network, resource depletion, etc. The route disruption means the breaking of links and disrupting routing tables.

In the subsequent sections, it is discussed the proposed model to detect such vulnerable nodes and how the route to the destination is secured by avoiding and discarding such nodes in the network. Then, the network is analyzed with various simulation results under various node densities and parameters.

2 Related Work

Many researchers evaluate trust with respect to different levels of contexts. The trust evaluation process proposed by some of the researchers implementing them on MANET has shown improved result by detecting and discarding misbehaving nodes from the network. Many security vulnerabilities of the AODV routing protocol have been surveyed and analyzed [1] some of the attacks on AODV discussed earlier with respect to attacker goals and attack vectors. Possible attacks by unauthenticated and internal vulnerable nodes are identified that can disrupt message flows, attract routes, and prevent route creation. To counter the rushing attack, randomization of RREQ and limit on RREQ has been imposed to the modified AODV to prevent gratuitous flooding.

Another trust-based model with dynamically self-organizing tree in FireCol architecture [2] is designed to enhance the security of the system for better performance. Based on the dominant cluster separation and cluster formation, the trust evaluating model enhances the security and effectiveness of intrusion detection in collaborative wireless Adhoc networks.

In the trust activation process in the secure AODV [3] model, each node collects all activity records of neighbor nodes in the network. Calculation of trust process is done when the node communication is started in the network. The RREQ, RERR, and RREP are calculated by each node by listening in the network whether it is received or forwarded. The trust is calculated by Eqs. (1) and (2) based on local-level (LLT) and global-level (GLT) trust.

$$\text{LLT}_{3-2} = \frac{\sum \text{node 2 receives packet from node 1}}{\sum \text{node 2 forwards the packet to node 3 and received node 1}}; \quad (1)$$

where $\text{LT}_{3-2} = 1$.

$$\text{GLT}_{3-2} = \frac{\sum \text{node 2 receives packet}}{\sum \text{node 2 receives forwarded packet}}; \quad (2)$$

where $\text{GLT}_{3-2} \leq 1$.

TST-AODV [4] explained successful and failure criteria for the packet transmission. The node requests (I_R) rate, data transmission (I_D), and reply (I_P) are normalized ranges between -1 and $+1$; such that I_R, I_D, and I_p and which are also the intermediate parameters, respectively, the trust-level (TRL) calculation is done by Eq. (3).

$$\text{TRL} = \text{TR(RREQ)} * I_R + \text{TR(RREP)} * I_P + \text{TR(DATA)} * I_D \quad (3)$$

EAER–AODV [5] is a trust model which considers opinion of a node represented as a three-dimensional metric. While calculating node's direct trust, it utilizes the concept of average counter rate to create a stable network topology. The three components in the three-dimensional metric, trusts (TR), distrusts (DTR), and uncertain (U), are defined as given in Eq. (4).

$$\varepsilon_{a,b} = \text{TR}_{a,b} + \text{DTR}_{a,b} + U_{a,b} \tag{4}$$

In Eq. (4), trust $\text{TR}_{a,b}$ is the probability that the node a can trust node b. Distrust $\text{DTR}_{a,b}$ refers to the probability that the node a cannot trust the node b. Uncertainty $U_{a,b}$ refers to the interval between trust and distrust components. The sum of these three components is 1.

Hazra and Setua [6] proposed a method where trust is calculated with addition to context and other dependency patterns and then takes decision about the node's vulnerability level and subsequent action with the node. The trust calculating module calculates final value of trust based on the indirect and direct trust provided by the neighbor nodes. After that, decision about the belief and disbelief is taken based on threshold value.

The SEODV model [7] enhances the message authenticity to secure the route of the connection. MBDP-AODV [8] is designed based on dynamic threshold mechanism in MANET. The MANET performance under blackhole attack has been compared with the MBDP-AODV protocol, and it has been found that MBDP-AODV protocol performs well, but its disadvantage is that it has high routing overhead because of the destination nodes transmitting multiple reply packets.

NTPTSAODV [9] is a two-tier trust-based model for intrusion detection technique in MANET to enhance its security mechanism. This approach computes the path trust followed by node trust. The node trust calculation has done from the trust calculation of control packets TR_{CP} and trust calculation from data packets TR_{DP} of each node as given in Eqs. (5) and (6).

$$\text{TR}_{\text{CP}} = w1 * f1 + w2 * f2 + w3 * f3 \tag{5}$$

$$\text{TR}_{\text{DP}} = w4 * * f4 \tag{6}$$

Here, the final values $(f1, f2, f3, f4)$ are used to calculate the nodes RREQ rate, node RREP rate, RERR rate, and data packet transmission rate. The weight factors of control as well as data packets are $w1, w2, w3, w4$. The final node trust value is calculated with $\text{TR}_{\text{CP}} * \text{TR}_{\text{DP}} * \text{NS}$ where NS is the node stability.

Impact of a trust evaluation model [10] for MANET based on node's historical behaviors and fuzzy system rules is analyzed under AOTMDV protocol to detect the malicious nodes and prevent them.

DATEA [10] is a distributed and adaptive trust metric for MANET where it is defined one-hop module and multi-hop module. The calculation of direct trust and recommended trust is contained in one-hop module. The calculation of indirect trust is contained in multi-hop module. The communication trust and energy trust are included in the direct trust.

A friendship-based mechanism [11] is designed to secure AODV routing protocol in MANET to evaluate the nodes using trust concept. The calculation of the friendship value is done by following $F = \sum_{i=1}^{H} \frac{F_i}{H}$ where friendship value F_i is the ith next hop, and H is the intermediate hops between the destination and originator nodes.

3 Dynamic Threshold and Trust Computation

A dynamic threshold-based trust-oriented AODV (DTT-AODV) model has been proposed where degree of trustability, i.e., the trustable and non-trustable value, is defined dynamically based on destination sequence number (D.SecNo) using multiple reply packets. For threshold computation purpose, it calculates the mean and standard deviation from multiple reply packets from the destination sequence number. The degree of trustability (DF) is computed as $DF = \sqrt{\sum_{i=1}^{J} \frac{(D.SeqNo_i)^2}{K}}$, where J is the number of reply packet, and D.SeqNo$_i$ is the ith reply packet.

The nodes which are misbehaving identified based on their routing delay and high transmission range. Figure 1 shows the trust evaluation of the neighbor node (NN) which is in the selected path for communication is done by the trust evaluating node (EN). Based on direct and indirect trust levels, the updated and final trust value of NN is then computed. The computation of final trust may be computed indirectly or directly.

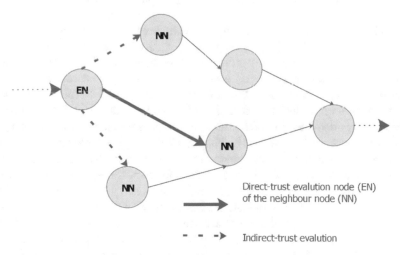

Direct-trust evaluation node (EN) of the neighbour node (NN)

Indirect-trust evaluation

Fig. 1 Trust evaluation components

3.1 Trust Evaluation Process

The RREQ packet from the selected route for communication is received by the NN from the network, and then, it broadcasts the route request response (RResponse) packet to the EN. EN evaluates trusts directly based on the packet receiving time, and then in response of RREQ that is received, the RQresponse is broadcasted. For this RQresponse, the information for route delay time and transmission is collected in trust_store for calculating trust evaluation indirectly and directly. Figure 2 shows how the trust evaluation process of NN is done.

3.2 Discarding Malicious Node

To discard malicious node, a RREQ is received by NN, and it broadcasts RQresponse. EN receives this packet at time $t_1(\text{NN})$. NN broadcasts the received RREQ after some required processing time. The RREQ is received by EN at time of $t_2(\text{NN})$. For the purpose of trust evaluation, the $t_1(\text{NN})$ and $t_2(\text{NN})$ values are then fetched. The initial trust calculation is done by Eq. (5).

$$t_v(\text{NN}) = (t_2(\text{NN}) - t_1(\text{NN})) \tag{4}$$

$$t^{ft}(\text{NN}_{\text{EN}}) = 0.7 \times t^{dt}(\text{NN}_{\text{EN}}) + 0.3 \times t^{it}(\text{NN}_{\text{EN}}) \tag{5}$$

The calculated value $t_v(\text{NN})$ is compared with direct trust evaluation process with a threshold value th. Here, $th = (t_R + t_{\text{const.-k}} + t_{\text{pkt}})$ where t_R is the time taken for route requests, $t_{\text{const.-k}}$ is the time constant in the network at $k = 1, 2, \ldots,$ and t_{pkt} is the time taken for packet transmission, travel, queuing, and routing delay. So, the trust calculation model follows the degree of trustability $t^{ft}(\text{NN}_{\text{EN}}) > th$ and $(t_2 - t_1) \geq th$ implies that the calculation of final trust value for NN by EN has the belief and considered as non-malicious. So, the packet can be forwarded onto the route. $t_v(\text{NN}) < th$ implies that t_R is rushing quickly toward destination avoiding route delays and delays in network constant with comparison to standard time. Hence, it is considered as the vulnerable node as well as the forwarding route is discarded.

The calculation of final trust value from the direct and indirect trust evaluation is calculated by Eq. (6). The computation of direct trusts, $t^{dt}(\text{NN}_{\text{EN}})$, is performed by monitoring NN directly, and the indirect trusts, $t^{it}(\text{NN}_{\text{EN}})$, is calculated indirectly by EN on the basis of trusts calculated by NN. The proposed method for removing malicious node from forwarding path in routing table is described in Algorithm 1.

Algorithm 1 Algorithm for removing malicious node from forwarding path in routing table

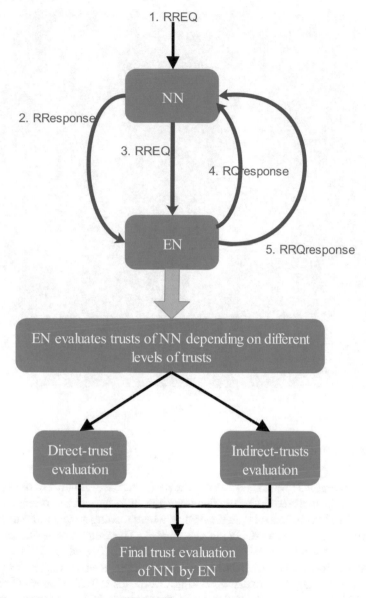

Fig. 2 Trust evaluation process of NN by EN

Algorithm

1. *Initialize*: Upon receiving the route request (RREQ) by NN and RQresponse is broadcasted which EN receives at time $t_i(NN)$. NN broadcasts

the received RREQ after few essential processing times. The RREQ is received by EN at time $t_{i+1}(NN)$

2. Initial trust calculation is done by the following

$$t_v(NN) = (t_{i+1}(NN) - t_i(NN))$$

3. The final trust value, $ft_j = t^{ft}(NN_{EN})$, is calculated based on direct and indirect levels of trust evaluation.

4. Obtain threshold $th = DF$, degree of trustability

$$DT = \sqrt{\sum_{i=1}^{J} \frac{(D.SeqNo_i)^2}{K}}$$

5. Check $if\ ft_j \geq th$

   Then
   Considers it as non-malicious node and forwards the packet on the forwarding route
   Else
   Discard it and the forwarding route as it might be the malicious node.

6. *End.*

4   Results and Analysis

Network Simulator 2 (NS2.35) is used for the performance evaluation of the MANET. The network is analyzed under the dynamic threshold-based trust-oriented AODV (DTT-AODV) in comparison with the AODV where rushing attacker or the misleading nodes are present (RA-AODV) in the network. The simulation results are given with respect to the topological dimension 500×500 m^2. To evaluate DTT-AODV in comparison to RA-AODV in MANET, four different performance metrics like Packet Delivery Ratio (PDR), Normalized Routing Overheads (NRO), Control Overheads (CO) and Throughput. For the evaluation in the network, 20% of the nodes are considered as the malicious nodes which are positioned anywhere in the network. The performance of the network has been analyzed under the four various performance metrics with dimension of 500×500 m^2.

All the performance metrics are analyzed and discussed from low to high density in the scenario 500×500 m^2. Sections 4.1, 4.2, 4.3, and 4.4 describe the analysis of PDR, CO, NRO, and throughput, respectively.

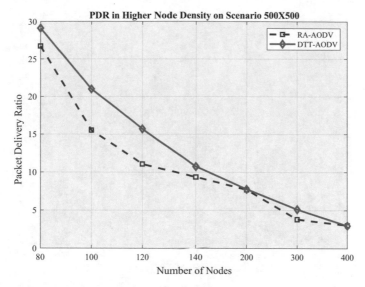

Fig. 3 Comparison of PDR of DTT-AODV and RA-AODV in 500 × 500

4.1 Packet Delivery Ratio (PDR) Analysis

Comparison of PDR of DTT-AODV and RA-AODV is shown in Fig. 3. It is seen from Fig. 3 that the PDR of DTT-AODV is always higher than RA-AODV. It is also clear from the picture that when the node density is increasing, the rate of PDR is decreasing. When node density is increasing in any network at that time, the number of packets in the network is also increasing. It creates traffic congestion in the network. So, with the increasing of node density the rate of PDR is also decreasing.

In DTT-AODV, all the vulnerable nodes are removed from the routing tables. So, the data packets are avoided the malicious nodes successfully to reached the destination. Hence, it may be considered that the PDR of DTT-AODV is always higher than RA-AODV in all cases.

4.2 Analysis of Control Overhead

Comparison of control overhead of DTT-AODV and RA-AODV is given in Fig. 4. It is seen from Fig. 4 that the control overhead of DTT-AODV is less than RA-AODV. It is also seen that control overhead of any network increasing with the increasing of node density in the both routing models.

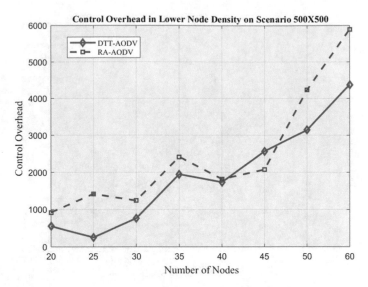

Fig. 4 Comparison of control overhead of DTT-AODV and RA-AODV in 500 × 500

As DTTAODV detects all the malicious nodes as well as removes all the malicious nodes from the routing tables, the rate of drop packets is always lesser than RA-AODV. So, the rate of resend of the packets in RA-AODV is higher than DTT-AODV. That's why the control overheads of DTT-AODV are lower than RA-AODV.

4.3   Analysis of Normalized Routing Overhead (NRO)

Comparison of normalized routing overhead of DTT-AODV and RA-AODV is shown in Fig. 5. From Fig. 5, it is clear that the normalized routing overhead (NRO) of DTT-AODV is less than RA-AODV.

As DTT-AODV detects all the malicious nodes as well as removes all the malicious nodes from the routing tables, the rate of drop packets is always lower than RAAODV. So, the rate of resend of the packets in RA-AODV is always greater than DTT-AODV. So, the normalized routing overheads of DTT-AODV are lesser than RA-AODV.

4.4   Throughput Analysis

Comparison of control overhead of DTT-AODV and RA-AODV is shown in Fig. 6. It is seen from Fig. 6 that the control overhead of DTT-AODV is less than RA-AODV. It is also seen that the rate of throughput and node density is inversely proportional with each other because when the number of nodes increased at that time, the number

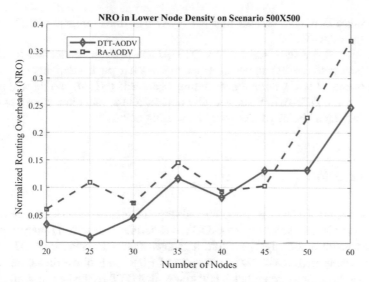

Fig. 5 Comparison of normalized routing overhead of DTT-AODV and RA-AODV in 500 × 500

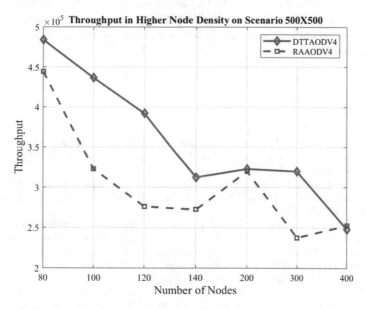

Fig. 6 Comparison of throughput of DTT-AODV and RA-AODV in 500 × 500

of packets increased in the network which create network congestion to decrease the rate of throughput in the both of the cases.

As DTT-AODV detects all the malicious nodes as well as removes all the malicious nodes from the routing tables, the rate of drop packets and the rate of resend of the packets in DTT-AODV are lesser than RA-AODV. So, the throughput of DTT-AODV is more than RA-AODV.

From Figs. 3, 4, 5 and 6, it is seen that DTT-AODV performs better than RA-AODV with respect to four different performance metrics like packet delivery ratio (PDR), control overhead (CO), normalized routing overheads (NRO), and throughput. So, we can say that DTT-AODV successfully detects all the malicious nodes as well as removes all the malicious nodes from the routing tables.

5 Conclusion

In this paper, DTT-AODV is proposed and implemented. The performance of DTT-AODV which is compared with RA-AODV is described in result and analysis section. From the result and analysis section, it is seen that performance of DTT-AODV is always better than RA-AODV in respect of PDR, control overhead, normalized routing overhead, and throughput. It is also seen that DTT-AODV successfully detects the rushing attack and removes the malicious nodes from the routing tables. So, the data packets always avoided the malicious nodes to reach the destination. In the future, dynamic trust mechanism can be implemented in various routing protocols as well as this mechanism can be used to detect the other attacks like blackhole attack, gray hole attack, etc.

References

1. Von Mulert, J., Welch, I., Seah, W.K.G.: Security threats and solutions in MANETs: a case study using AODV and SAODV. J. Netw. Comput. Appl. **35**, 1249–1259 (2012). https://doi.org/10.1016/j.jnca.2012.01.019
2. Poongodi, M., Bose, S.: A novel intrusion detection system based on trust evaluation to defend against DDOS attack in MANET. Arab. J. Sci. Eng. **40**, 3583–3594 (2015). https://doi.org/10.1007/s13369-015-1822-7
3. Simaremare, H., Abouaissa, A., Sari, R.F., Lorenz, P.: Secure AODV routing protocol based on trust mechanism. In: Khan, S., Khan Pathan, A.S. (eds.) Wireless networks and security. Signals and Communication Technology. Springer, Berlin, Heidelberg (2013)
4. Subramanian, S., Ramachandran, B.: QOS assertion in MANET routing based on trusted AODV (ST-AODV). Int. J. Adhoc, Sens. Ubiquitous Comput. **3**(3) (2012)
5. Mukherjee, S., Chattopadhyay, M., Chattopadhyay, S., Kar, P.: EAER-AODV: enhanced trust model based on average encounter rate for secure routing in MANET. In: Chaki, R., Cortesi, A., Saeed, K., Chaki, N. (eds.) Advanced Computing and Systems for Security. Advances in Intelligent Systems and Computing, vol. 667. Springer, Singapore (2018)
6. Hazra, S., Setua, S.K.: Trust oriented secured AODV routing protocol against rushing attack. In: Meghanathan, N., Nagamalai, D., Chaki, N. (eds.) Advances in Computing and Information Technology. Advances in Intelligent Systems and Computing, vol. 176. Springer, Berlin, Heidelberg (2012)
7. Li, C., Wang, Z., Yang, C.: SEAODV: a security enhanced AODV routing protocol for wireless mesh networks. In: Gavrilova, M.L., Tan, C.J.K., Moreno, E.D. (eds.) Transactions on

Computational Science XI. Lecture Notes in Computer Science, vol. 6480. Springer, Berlin (2010)

8. Gurung, S., Chauhan, S.: A dynamic threshold based algorithm for improving security and performance of AODV under black-hole attack in MANET. Wireless Netw. **25**, 1685–1695 (2019). https://doi.org/10.1007/s11276-017-1622-y
9. Patel, S., Sajja, P.S., Khanna, S.: Enhancement of security in AODV routing protocol using node trust path trust secure AODV (NTPTSAODV) for mobile Adhoc network (MANET). In: Satapathy, S.C., Joshi, A. (eds.) Information and Communication Technology for Intelligent Systems (ICTIS 2017), vol. 2, pp. 99–112. Springer International Publishing (2018)
10. Zhang, D., Gao, J., Liu, X., Zhang, T., Zhao, D.: Novel approach of distributed & adaptive trust metrics for MANET. Wireless Netw. (2019). https://doi.org/10.1007/s11276-019-01955-2
11. Eissa, T., Abdul Razak, S., Khokhar, R.H., Samian, N.: Trust-Based routing mechanism in MANET: Design and Implementation. Mobile Netw. Appl. **18**, 666–677 (2013). https://doi.org/10.1007/s11036-011-0328-0

Secured Session Key-Based E-Health: Biometric Blended with Salp Swarm Protocol in Telecare Portals

Arindam Sarkar, Joydeep Dey and Sunil Karforma

Abstract Technological emergence-based user-friendly systems have emerged to cope up the security issues in online medical portals. The Government of India is doing loads of works on online health schemes such as "Ayushman Bharat Yojana." The proposed research technique may be incorporated with existing online portal for secured data transmission. In this proposed methodology, nature-inspired biological algorithm based on salp swarm has been deployed to generate 256-bits session key for secure transmission of medical information. Arruda E.F. et al. have proposed optimal testing policies for diagnosing patients with intermediary probability of disease in the journal of Artificial Intelligence in Medicine (Elsevier: June 2019 (Vol. 97)) with no detailed diagnosis of unknown diseases. Moreover, online expert opinion with secured data communication was not present at their technique. This proposed work resolves the stated issues by encrypting the patients' fingerprints with lower order session key, and higher order is used for *IDEA* encrypting the signals. The resultant of both round of encryption is transmitted to a group of known physicians with threshold limits. Several quality testings such as key sensitivity, signal sensitivity, chi-square, session key space, and statistical tests were carried out, and results are compared with existing benchmark techniques.

Keywords Salp swarm · Crossing number · IDEA encryption · Histogram

A. Sarkar
Department of Computer Science and Electronics, Ramakrishna Mission Vidyamandira, Belur 711202, WB, India

J. Dey (✉)
Department of Computer Science, M.U.C. Women's College, B.C. Road, Burdwan 713104, WB, India

S. Karforma
Department of Computer Science, The University of Burdwan, Burdwan 713104, WB, India

© Springer Nature Singapore Pte Ltd. 2020
J. K. Mandal and S. Mukhopadhyay (eds.), *Proceedings of the Global AI Congress 2019*, Advances in Intelligent Systems and Computing 1112,
https://doi.org/10.1007/978-981-15-2188-1_56

1 Introduction

E-health service portals are used to facilitate and treat the patients remotely from different demographic areas. Assessing the medical diagnostic and treatment facilities by the patients from anywhere and anytime is the blessed part in this fast technological era. Government of India had launched a health scheme, i.e., Ayushman Bharat Yojana [1, 2] in the year of 2018. The objective of this scheme is to promote the rural and urban medical services and proper treatments. Moreover, nascomial infections, body wear and tear, and traveling cost incurred can be appropriately minimized to huge extent. To address any disease/infections, Indian Government is aiming at promoting and supporting such noble activity. The proposed technique may be pushed as an online transmission module under such E-health government portals. Using the interfaces, the patients can send his/her clinical signals to the doctors for better opinion. In such systems, the main key function that needs to be implemented is the security issue on confidential data. On the way of transmission inside the Internet, the intruders can steal the patients' data, and further they damage or distort them badly. In the journal of Artificial Intelligence in Medicine (Elsevier: June 2019(Vol.97), Arruda E.F. et al. have optimal testing policies for diagnosing patients with intermediary probability of disease [3]. The online benefit for getting an expert opinion from physicians is lagging in that paper. The proposed technique has shown a novel approach toward that fault, which may be incorporated on online government health portals like Ayushman Bharat Yojana.

This paper has proposed a session key generation technique based on metaheuristic approach [4] to enhance the security approaches on the E-health portals. Salp swarm protocol [5] is a swarm intelligence-inspired algorithm to find the optimum solution from the salps behavior. Exploration and exploitation are the two key terms used in this salp swarm algorithm, which are fine tuned to obtain the solution.

Encryption technique is the science of transformation of the plain data into non-readable format using a secret key at the senders' terminals. The receivers' job is to decrypt the non-readable data into original data as they have only the access to the secret key. Thus, it resists the Man-In-The-Middle attacks inside the Internet transmission phase. Sarkar et al. [6] have proposed a security enhancement scheme in the electronically controlled health domain.

Mathematical calculations carried on the salp swarms' behavior are approximated to generate the proposed session key. The proposed technique dealt with ECG signal stored at Physio Bank ATM under BIDMC Congestive Heart Failure Database (chfdb) [7]. The analytical emphasis has been given on the quality metrics at this proposed work. Chi-squared test has been deployed to find the difference between the observed frequency character and the actual frequency character in the encrypted signals. Equation 1 shows to find chi-squared values

$$\chi^2 = \sum_{i=1}^{n} \frac{(\text{OCH}_i - \text{ECH}_i)^2}{\text{ECH}_i}, \tag{1}$$

where OCH_i and ECH_i denote actual occurrence and expected occurrence of the ith character, respectively, in the signals. Using the above equation, high chi-squares values were obtained on different session keys. The entropy of an encrypted signal is an index of its character content in terms of binary characters. It is bits per character in a signal. The entropy obtained by using the proposed technique is 7.91, which is as per with benchmark parameter. The proposed technique had also conducted with mean-squared error, peak signal-to-noise ratio (PSNR), and structural similarity index (SSIM) by the following equations, which yielded at par with benchmark values. The mean-squared error (MSE), peak signal-to-noise ratio (PSNR), and structural similarity index (SSIM) were the key metrics considered under test. MSE may be calculated of length N using the following Eq. 2.

$$\text{MSE} = \frac{\sum_{i=1}^{N} (S_i - \text{Enc}_i)^2}{N}, \tag{2}$$

where S_i and Enc_i denote the original clinical signal and the encrypted signal. PSNR is the ratio between the maximum pixel value to the compressed image. It is computed by the following Eq. 3.

$$\text{PSNR} = 10 \log_{10} \left(\frac{\text{MaxPixel}}{\text{MS Error}} \right)^2, \tag{3}$$

where MaxPixel is the maximum number of pixels present in the original signal and MS error as computed from earlier Eq. 2. Structural similarity index (SSIM) is a measuring component to measure the quality of the compressed image by the following Eq. 4

$$\text{SSIM}(X, Y) = \frac{(2\mu_X\mu_Y + P_1) * (2\sigma_{XY} + P_2)}{\left(\mu_X^2 + \mu_Y^2 + P_1\right) * \left(\sigma_X^2 + \sigma_Y^2 + P_2\right)} \tag{4}$$

μ_X and μ_Y denote the mean of the original image and compressed image, respectively. σ_X and σ_Y are the variance of the original image and compressed image, respectively. σ_{xy} is the correlation between the original image and compressed image, respectively,

$$P1 = \{0.01 * (2256-1)\}, \text{ and } P2 = \{0.03 * (2256-1)\}.$$

Loads of extensive research works are carried out on the biometric authentication arena [8, 9]. Use of fingerprints is inbuilt features in smart phones in today's era due to the fact that the patterns of ridges are unique in nature. S. Prabu et al. have proposed a multiple modal authentication upon biometric recognition using artificial intelligence. This work had been published in the journal of Medical Systems in the year of 2019 [10].

2 Problem Statement

Authentications of the transactions are to be verified. Documents such as AADHAR card, birth certificates, and passports can be counterfeiting easily if any fraudulent individual gets those. With his/her sound technical knowledge, any involvement in medical crime may be done silently, while the original person will get imprisonments/penalty.

3 Proposed Technique

The above said problem may be minimized by the use of biometric features in E-health domain. Individual patients will have their unique finger imprints, which may also be encrypted through the generated session key in this proposed technique. The advantage of encrypting such biometric feature are that no threat of lost, stolen, misplaced, guessed randomly, etc. A single patient will have their single ID generated through proposed system.

Mathematically, the salp swarm is classified into two groups, namely leader and followers. Top salp is designated as leader whose job is to direct the entire swarm. Remaining salps are followers, defined in an n-dimensional search space domain where n is the search space variable in count. Each salp is mounted with an additional vector of 256-bits long. This vector for the top most salp present in the salp table will act as session to encrypt the data. SHA-256 algorithm [11] is a random number generating algorithm to have unique and static sized hash of 32 bytes. Such hash is very much suitable in case of password approval, digital signatures, login interfaces, etc. A message digest would be formed using the SHA 256 function. This receives the concatenated coordinates of each salp to determine its weighted vector.

Algorithm 1 Proposed Session Key Generation by Salp Swarm Algorithm

Input(s): Patient's Biometric(PB)&$SIGNAL$
Population of Salp (Sample)
Output(s): **256** bits Session Key

1.   **For** k = 0 to $\left(\text{Sample} - 1\right)$ **do**
2. $SalpM[k][0] \leftarrow FindRandom(lx, ux)$
3. $SalpM[k][1] \leftarrow FindRandom(ly, uy)$
4. $SalpM[k][2] \leftarrow FindRandom(lz, uz)$
5.   SalpM[k][3 ... 130]$\leftarrow$ SHA256 (SalpM [k][0] | ';' | SalpM[k][1] | ';' | SalpM[k][2])
6. $k = k + 1$
7.   **End for**
8.   **For** I = 0 to $\left(\text{FoodCount} - 1\right)$ **do**
9. $Food[I][0] \leftarrow FindRandom(LBX, UBX)$
10. $Food[I][1] \leftarrow FindRandom(LBY, UBY)$
11. $Food[I][2] \leftarrow FindRandom(LBZ, UBZ)$
12. $I = I + 1$
13. **End for**
14. **Assign Counter = 0**
15. **While** (NOT$_{\text{TERMINATING}}$CONDITION)
16.   **For** l = 0 to $\left(\text{Sample} - 1\right)$ **do**
17.   **Set Counter = Counter + 1**
18. $P1 = \lceil \{ (2 * e)^{(-4 * pit)} / (MAX) \}^2$
19. $, where, pit\ is\ the\ current\ iteration\ and\ MAX\ is$
20. $the\ maximum\ iteration$
21.   **If** (P3 >= 0) **Then**
22.   SalpM[0][0...2]= Mean(Food($_n$)+
                [{ $UB_n - LB_n$ }* RAND[0,1]+ LB_n]*RAND[0,1]
23.   **Else**
24.   SalpM [I][0 ... 2] = Mean(Food$_n$) - [{ $UB_n - LB_n$ } * RAND[0,1] + LB_n]* RAND[0,1]
25.   **End if**
26.   SalpM [I][0 ... 2]= 0.5 * (SalpM[I][0 ... 2] − SalpM[I-1][0 ... 2])
27.   **End for**
28. **End while**
29.   KEY1[128]= SalpM[**Counter**][0 ... 127]
30.   KEY2[128]= SalpM[**Counter**] [128 ... 255]
31.   CN ← Call *Cross Number* (Minutiae Matrix, 8)
32.   C1← *XOR*(CN , KEY1)
33.   C2← *IDEA_Encrypt*(SIGNAL, KEY2)
34.   C3← *Secret_Share*(C1 | C2, n, k)

Biometric parameter has been exploited at this technique to ensure the patients' security at the online health services portals. Every individual has unique fingerprints which enable us to encrypt it using the lower order session key. Lee et al. [12] had carried out a template scheme related on fingerprint minutiae bit-string matrix. The idea behind their work was to convert fingerprint minutiae to an original minutiae matrix. Minutiae points are the key features of a fingerprint image scanned through the E-health interface. These are mainly used in the pattern matching, validation, etc. Encrypting through thus generated matrix using the session key prevents the biometric compromisation. Minutiae points are used to determine the uniqueness of any patients' fingerprint image in terms of binary matrix as mentioned here. It may be defined as the points where the ridge lines converge or bifurcate. They are the local ridges discontinuities and may be of different kinds as given below. Finger placing on the sensor and resolution of the sensor determine the good quality of the biometric image. An image having 25 to 80 minutiae points is treated as a good quality image.

- Ridge Endings: It is the point of sudden ridge endings.
- Ridge Bifurcations: It is the point of diversion of a ridge into multiple ridges.
- Ridge Dot: It is a very small ridge.
- Ridge Islands: They are slightly longer than dots and occupy a middle space between two diverging ridges.
- Lakes: They are the empty spaces between any two diverging ridges.
- Spurs: It is a notch extended from the ridges.
- Bridges: It is a small ridge which connects any two adjacent ridges.
- Crossovers: They are created if two ridges cross over each other (Fig. 1).

Due to high computational power, intrinsic, and simple in nature, crossing number-based minutiae extraction [13, 14] has been deployed in this technique. A skeleton image with eight-connected ridge flow as given in the following Fig. 2 has been used here. The local neighbor of each ridge pixel of the patients' image is mapped by a

Fig. 1 Minutiae point representation

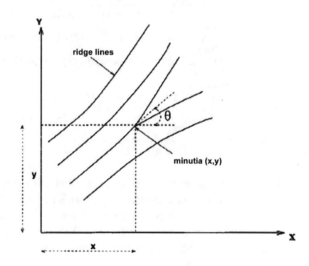

Fig. 2 Anticlockwise eight-connected pixels under evaluation

PIX4	PIX3	PIX2
PIX5	PIXEL	PIX1
PIX6	PIX7	PIX8

[3 × 3] window from which the minutiae points are extracted. The crossing number is estimated as the fifty percent of the sum of the differences in the eight-connected pixels using the following equation in counterclockwise pattern.

$$CN = \frac{\sum_{i=1}^{8} |PIX_{i+1} - PIX_i|}{2} \tag{5}$$

Thus, a binary mapping function is implemented to derive the binary matrix of the patients' fingerprints. Following binary conversion rule illustrates this property.

If [CN=0] then 0000:- Point of Isolation
If [CN=1] then 0001:- Ridge Ending Point
If [CN=2] then 0010:- Ridge Continuation
If [CN=3] then 0011:- Ridge Bifurcations
If [CN=4] then 0100:- Crossing Points

Hence, the binary matrix formed will be successively XORed with the lower order bits [0,127] of the salp swarm guided generated session key.

International Data Encryption Algorithm (IDEA) is a symmetric blocked ciphered-based modified proposed encryption standard (PES). It was published by Lai and Massey in the year of 1990 [15]. IDEA uses a 64-bit block of clinical signals accessed from PhysioBank ATM for BIDMC Congestive Heart Failure Database (chfdb) [7] which produces 64-bit of ciphered text. It includes eight and half rounds of encryption using variation in the rotated key bits by 25. A modification has been done as described in the following algorithm 2. At the end of eighth round, an ascending order sorting will be done on four 16-bits blocks accounting the number of set bits. Furthermore, a swapping would be done between the resultant of first and third blocks, and second and fourth blocks. A higher order 128-bits key generated by the proposed salp swarm algorithm sways this layer of encryption. The same pattern is continuously repeated until the clinical signals get exhausted. Prior to this round of encryption, the clinical signals were decomposed into multiple blocks of 64-bit in order to be compatible with IDEA algorithm. Padding in terms of zeros has been placed on the last block of signals to have the symmetry in size.

Algorithm 2 Proposed Signal Encryption with Modified IDEA

Input(s): Clinical Signals(S), Output of Algorithm 1

Output(s): Ciphered Signal Matrix(CSM)

/* *Partition into H.O.KEY into sub keys* */
For I = 0 *to* 7
For J = 0 *to* 15
$K[I] = SalpM[J++]$
End for
End for
/* *Partition into 64 bits signals* */
While [! *Last_Block(S)*]
$B[64] = S[0 \dots 63], B1[16] = S[0 \dots 15], B2[16] = S[16 \dots 31]$
$B1[16] = S[32 \dots 47], B1[16] = S[48 \dots 63],$
$B = \square, Move\ to\ next\ block$
  For R = 0 *to* 7
a) Multiply(2^{16} + 1): B[1] and sub key K[1].
b) Add(216): B[2] and sub key K[2].
c) Add(216): B[3] and sub key K[3].
d) Multiply(216 + 1): B[4] and sub key K[4].
e) Bitwise XOR(output of Step a and c).
f) Bitwise XOR(output of step b and d).
g) Multiply(216 + 1): output of step e and sub key K[5].
h) Add(216): output of step f and g.
i) Multiply(216 + 1): output of step h and sub key P[6].
j) Add(216 + 1): output of step g and i.
k) Bitwise XOR: output of step a and i.
l) Bitwise XOR: output of step c and i.
m) Bitwise XOR: output of step b and j.
n) Bitwise XOR: output of step d and j.
        o) ARR[4] ← ASC_SORT(output of k, l, m, n, No_of_1s)
        p) Bit_Interchange(ARR[0], ARR[2])
        q) Bit_Interchange(ARR[1], ARR[3])
r) Left_Circular_Shift(K[128], 25)
   End for
   I. Multiply(216 + 1): B[1] and sub key K[1].
   II. Add(216): B[2] and sub key K[2].
   III. Add(216): B[3] and sub key K[3].
   IV. Multiply(216 + 1): B[4] and sub key K[4].
End while
   Concatenation of the blocks

The resultant of the aforementioned encryption would be communicated using the concept of secret sharing [16]. The ciphered signal would be transmitted to a group of n physicians, and out of those *n*, only *k* number of shares (*k<n*) once accumulated would decrypt the original message. All the participating nodes would receive the partial information. Thus, the proposed technique is likely not to be prone with the online Man-In-The-Middle attacks.

Table 1 Quality metrics obtained

Signals	PSNR	MSE	SSIM	Encryption time (s)	Decryption time (s)	Post entropy
ECG [7]	6.4752	10,297.58	0.0361	0.08509	0.1622	7.93

4 Analytical Quality Metrics

The analytical emphasis has been given on the quality metrics at this proposed work. The mean-squared error (MSE), peak signal-to-noise ratio (PSNR), structural similarity index (SSIM), encryption/decryption time, post entropy, etc., were the key metrics considered under test.

In the following Table 1, a comparison between the different clinical signals corresponding to the quality metrics as stated above has been briefly given.

5 Histogram and Autocorrelation Analysis

Histogram is an analysis tool used in this proposed technique. They are basically generated by measuring the variations of the orientations among these sample values in two-dimensional spaces. Autocorrelation gives the information about characteristics like periods, dependence, and multiple pattern appearance. Table 2 contains the histogram of plain signal and proposed encrypted signals in its first row. Second row contains the autocorrelation of plain signal and proposed encrypted signals.

Table 2 Graph of histogram and autocorrelation

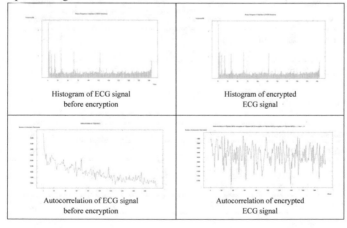

| Histogram of ECG signal before encryption | Histogram of encrypted ECG signal |
| Autocorrelation of ECG signal before encryption | Autocorrelation of encrypted ECG signal |

6 Comparison with Existing IDEA

In the sixth round of existing IDEA algorithm, there is a possibility of detecting the keys by the intruders. This paper modifies the implementation of IDEA algorithm by the nature-inspired metaheuristics salp swarm-based key generation technique. The exchange between the blocks and the keys before the final transformation leads to more diffusion steps to detect the key pattern by the intruders. Thus, it has raised the complexity of the modified algorithm. After such modifications done on the IDEA algorithm, the encryption strength will be better.

7 Statistical Tests Analysis

The NIST Test Suite [17] is a statistical package consisting of 15 tests. It was developed to test the randomness of (arbitrarily long) sequences generated through random or pseudorandom number generators. Their objectives are on different issues of randomness that could persist in a sequence of data. Following tests were carried on our proposed technique with favorable outputs (Table 3).

Table 3 NIST tests on proposed technique

Name statistical test	Sequence having P-value $>=$ 0.05	Proportion	Results
Frequency test	919	0.919	True
Frequency test in block	957	0.957	True
Run test	951	0.951	True
Longest run of ones in block	966	0.966	True
Binary matrix run test	980	0.980	True
Discrete Fourier transformation test	964	0.964	True
Non-overlapping template matching test	959	0.959	True
Overlapping template matching test	981	0.981	True
Maurer's universal statistical test	974	0.974	True
Linear complexity test	974	0.974	True
Serial test	962	0.962	True
Approximate entropy test	958	0.958	True
Cumulative sum test	948	0.948	True
Random excursion test	964	0.964	True
Random excursion variant test	972	0.972	True

8 Analysis on Session Key Space

Salp swarm [5]-based session key has been proposed to encrypt clinical signals [7] by the patients on online portals. Any encryption technique which is agile is needed to be analyzed with respect to the time taken to decrypt a cipher text using the latest supercomputers. For a key size of 128-bits either lower order or higher order, this technique needs $2^{128} = 3.40282 \times 10^{38}$ trials of permutations. IBM Summit at Oak Ridge, USA, had invented the fastest supercomputer in the world with 123 PFLOPS, i.e., means 123×10^{15} floating point computing operations per second in the year of 2018. Perhaps, it may be considered that each trial to decode may take 1000 FLOPS to perform its operations. The number of seconds counted in a non-leap year $= 365 \times 24 \times 60 \times 60 = 31,536,000$. So, the number of trials required per second is 123×10^{12}. Hence, the number of years needed to crack one cipher text is $(3.40282 \times 10^{38})/(123 \times 10^{12} \times 31,536,000) = 8.772 \times 10^{52}$ years. The robustness of this technique is that it needs non-feasible amount of time to decrypt half of the session key.

According to the above-stated calculations, the proposed technique may likely to be joined as a security component on governmental E-health portals.

9 Comparison Study

A comparative study under certain parameters has been done exhaustively at this proposed technique. Table 4 contains the summary of that study. In the journal of Artificial Intelligence in Medicine, June 2019, Arruda E.F. et al. have optimal testing policies for diagnosing patients with intermediary probability of disease [3]. In the year of 2016, Raeiat-ibanadkooki et al. [18] have proposed how to compress and encrypt the ECG with wavelet transformation and chaos-based Huffman code without loss of any data. Lin et al. [19] had proposed a chaotic visual cryptosystem with empirical mode decomposition and logistic maps for the EEG signal. Lin C. et al. [20] had proposed another chaos-based encryption technique on logistic map to encrypt the ECG signals. Moreover, the security analysis was not mentioned clearly there. Ahmad M and et al. had presented a security analysis to justify the quality of the encryption. Security analysis was missing in their paper [21].

10 Conclusion

Online portals must add the patients' data security during transmission of signals or data. Metaheuristic guided salp swarm protocol has been devised to generate the robust session key. Moreover, statistical tests were carried out on those generated keys to ensure its robustness in favor of true randomness and resistant to intruders attacks.

Table 4 Comparative statements with earlier works

Sl. no.	Comparative points	Proposed technique	Ref 3 2019	Ref 19 2016	Ref 20 2016	Ref 21 2014	Ref 22 2011
1	ECG clinical signal	Yes	Yes	Yes	No	No	No
2	Signal database	Physio Bank ATM	Physio Bank ATM	MIT-BIH	UCI KDD	NTOU	Bonn University
3	Data encryption	Yes	Yes	Yes	Yes	Yes	Yes
4	Data compression	No	No	Yes	No	No	No
5	Secret key space analysis	Yes	Yes	No	No	No	No
6	Histogram	Yes	Yes	No	No	No	No
7	Plain signal sensitivity	Yes	Yes	No	No	Yes	No
8	Secret key sensitivity	Yes	Yes	No	No	No	No
9	Entropy analysis	Yes	Yes	No	No	No	No
10	Mean-squares error MSE	Yes	Yes	No	Yes	No	No
11	Peak signal-to-noise ratio PSNR	Yes	Yes	No	No	No	No
12	Structural similarity index SSIM	Yes	Yes	No	No	No	No
13	Pseudorandomness analysis NST 800-22 suites	Yes	Yes	No	No	No	No
14	Comparative study	Yes	Yes	No	No	No	No

Additional round of encryption is done on the minutiae matrix with the lower order session key bits. Hence, it may be said that if the biometrics are leaked, then also this online module designed for various government E-health portals will resist to the differential attacks and chosen plain signal attacks.

Future scope of this proposed technique is to add an artificial diagnosis system through online mode of transmission.

Acknowledgements The authors acknowledge the moral and congenial atmosphere support provided by the Maharajadhiraj Uday Chand Women's College, B.C. Road, Burdwan, India (a constituent college under The University of Burdwan, India).

Conflict of Interest
All the authors declare that they have no conflict of interests.

References

1. Press Information Bureau Ministry of Finance. Ayushman Bharat for a New India-2022, Announced (Press Release): Press Information Bureau, Government of India. 2018. (Accessed on June 2019). Available from: http://www.pib.nic.in/newsite/PrintRelease.aspx?relid=176049
2. Press Information Bureau Ministry of Health and Family Welfare. Cabinet Approves the Largest Government Funded Health Program–Ayushman Bharat National Health Protection Mission (NHPM). (Press Release): Press Information Bureau, Government of India. 2018. (Accessed on June 2019). Available from: http://www.pib.nic.in/newsite/PrintRelease.aspx?relid=177844
3. Arruda, E.F., et al.: Optimal testing policies for diagnosing patients with intermediary probability of disease, In: Artificial Intelligence in Medicine 97, pp. 89–97. Elsevier (2019)
4. Yang, X-S.: Nature-Inspired Optimization Algorithms. Elsevier BV, pp. 23–44 (2014). https://doi.org/10.1016/b978-0-12-416743-8.00002-6
5. Mirjali, S., et al.: Salp Swarm Algorithm: A bio-inspired optimizer for engineering design problems. Adv. Eng. Softw. **114**, 163–191 (2017)
6. Sarkar, Arindam, Dey, Joydeep, Chatterjee, Minakshi, Bhowmik, Anirban, Karforma, Sunil: Neural soft computing based secured transmission of intraoral gingivitis image in e-health care. Indones J Electr Eng Comput Sci **14**(1), 178–184 (2019)
7. Physionet.org (2016). PhysioBank ATM. Available on: https://physionet.org/cgi-bin/atm/ATM. Accessed on June' 2019
8. Lee, H.C., Gaensslen R.E. (eds.): Advances in Fingerprint Technology, CRC Press, Boca Raton (1994)
9. Mehtre, B.M., et al.: Fingerprints image analysis for automatic identification. Match Vision Appl. **6**, 124–139 (1993)
10. Vaijayanthimala, J.: Multi-modal biometric authentication system based on face and signature using legion feature estimation technique. In: Multimedia Tools and Applications, pp. 1–20, Springer, Berlin (2019)
11. Lamberger, A., Mendel, A.: Higher-Order Differential Attack on Reduced SHA-256 (PDF). IACR Cryptology ePrint Archive. **2011**, 37 (2011)
12. Lee, Chulhan, Kim, Jaihie: Cancelable fingerprint templates using minutiae-based bit-strings. J. Netw. Comput. Appl. **33**, 236–246 (2010)
13. Ratha, N.K., et al.: Adaptive flow orientation based feature extraction in fingerprint images. Pattern Reccognition **28**(11), 1657–1672 (1995)
14. Jain, A.K., Hong, L., Pankanti, S., Bolle, R.: An identity authentication system using fingerprints. Proc. IEEE **85**(9), 1365–1388 (1997)
15. Lai, X., Massey, J.L.: A proposal for a new block encryption standard advances in cryptology -EUROCRYPT '90, pp. 389–404. Lecture Notes in Computer Science, Springer, Berlin (1991)
16. Naskar, P.K., Khan, H.N., Chaudhuri, A.: A key based secure threshold cryptography for secret image. Int. J. Netw. Secur. **18**(1), 68–81 (2016)
17. Rukhin, A., Soto, J., Nechvatal, J., Smid, M., Barker, E., Leigh, S., Levenson, M., Vangel, M., Banks, D., Heckert, A., Dray, J., Vo, S.: A statistical test suite for random and pseudorandom number generators for cryptographic applications. NIST special publication 800–22 (2001)
18. Raeiatibanadkooki, M., Quchani, S.R., KhalilZade, M., Bahaadinbeigy, K.: Compression and encryption of ECG signal using wavelet and chaotically huffman code in telemedicine application. J. Med. Syst. **40**(3), 1–8 (2016)
19. Lin, C.-F.: Chaotic visual cryptosystem using empirical mode decomposition algorithm for clinical EEG signals. J. Med. Syst. **40**(3), 1–10 (2016)
20. Lin, C.-F., Shih, S.-H., Zhu, J.-D.: Chaos based encryption system for encrypting electroencephalogram signals. J. Med. Syst. **38**(5), 1–10 (2014)
21. Ahmad, M., Farooq, O., Datta, S., Sohail, S.S., Vyas A.L., Mulvaney, D. In: 4th International Conference on Biomedical Engineering and Informatics, pp. 1471–1475. (2011)
22. Jain, A.K., Hong, L., Bolle, R.: Online fingerprint verification. IEEE Trans Pattern Anal Intell **19**(4), 302–314 (1997)

Author Index

© Springer Nature Singapore Pte Ltd. 2020
J. K. Mandal and S. Mukhopadhyay (eds.), *Proceedings of the Global
AI Congress 2019*, Advances in Intelligent Systems and Computing 1112,
https://doi.org/10.1007/978-981-15-2188-1

Printed in the United States
By Bookmasters